Selected Topics in Organic Chemistry

Selected Topics in Organic Chemistry

Edited by
Francesca Hopkins

MURPHY & MOORE
www.murphy-moorepublishing.com

Published by Murphy & Moore Publishing,
1 Rockefeller Plaza,
New York City, NY 10020, USA

ISBN: 978-1-63987-498-9

Cataloging-in-Publication Data

Selected topics in organic chemistry / edited by Francesca Hopkins.
p. cm.
Includes bibliographical references and index.
ISBN 978-1-63987-498-9
1. Chemistry, Organic. 2. Chemistry. I. Hopkins, Francesca.
QD251.3 .S45 2022
547--dc23

For information on all Murphy & Moore Publications
visit our website at www.murphy-moorepublishing.com

MURPHY & MOORE

Contents

Contents

Preface

This book has been a concerted effort by a group of academicians, researchers and scientists, who have contributed their research works for the realization of the book. This book has materialized in the wake of emerging advancements and innovations in this field. Therefore, the need of the hour was to compile all the required researches and disseminate the knowledge to a broad spectrum of people comprising of students, researchers and specialists of the field.

The branch of chemistry that includes the study of structures, properties, composition, reaction and preparation of organic compounds is known as organic chemistry. Organic compounds are covalently bonded carbon containing compounds including hydrogen, halogen and other elements. They are classified on the basis of functional groups, aliphatic groups, aromaticity and the size of molecule. Organic compounds are structurally diverse and form the functional and structural basis of all living beings and many commercial products. Organic chemistry is an umbrella science which overlaps organometallic chemistry, biochemistry, medicinal chemistry, materials science and polymer chemistry. Organic chemistry is applied in a large number of industries such as pharmaceutical industry, petroleum industry, and chemical industry. This book traces the progress of this field and highlights some of its key concepts and applications. The topics included in this book on organic chemistry are of utmost significance and bound to provide incredible insights to readers. It aims to serve as a resource guide for students and experts alike and contribute to the growth of the discipline.

At the end of the preface, I would like to thank the authors for their brilliant chapters and the publisher for guiding us all-through the making of the book till its final stage. Also, I would like to thank my family for providing the support and encouragement throughout my academic career and research projects.

Editor

1

Di-2-ethylhexylphthalate May Be a Natural Product, Rather than a Pollutant

Aurelio Ortiz and Estibaliz Sansinenea

Facultad de Ciencias Químicas, Benemérita Universidad Autónoma de Puebla, C.P. 72570, Puebla, PUE, Mexico

Correspondence should be addressed to Estibaliz Sansinenea; estisansi@yahoo.com.mx

Academic Editor: Qizhen Du

Di-2-ethylhexylphtalate is an ester of phthalic acid that has been used as plasticizer in many materials. Due to the extended use, it has been persistently found in different environments being classified as a pollutant with some risks for human health. However, in the last years, it has been found that this compound is produced by plants or microorganisms like bacteria or fungi. This finding opened a serious debate about the origin of this compound and questioned if it is a real pollutant or a natural metabolite with some biological activities that could help us in several ways. This review tries to give some data of the different points of view about this question.

1. Introduction

Phthalate compounds are colorless liquid chemicals that have been used as plasticizers to improve the plasticity and the flexibility of materials such as food packages, building, toys, medical devices such as blood storage bags, intravenous fluid bags, and other products [1, 2]. Di-2-ethylhexylphtalate (DEHP) has been persistently found in different terrestrial and aquatic environments resulting in one of the most important pollutants in the world [3]. These compounds are released from industrial products, and their photodegradation are very slow under natural conditions. Therefore, measurable concentrations of phthalate compounds accumulate over time in environments such as air, soil, water, and sediments causing a high human exposure [4]. The human beings can accumulate this compound through inhalation, skin contact, and the food chain. In this sense, many studies have been carried out regarding the toxicity effects of this compound in human health. These compounds are suspected to cause teratogenicity, mutagenicity, and carcinogenicity even at very low concentrations [5]. Increased awareness of phthalates toxicity has led to a dramatic increase in concern about the fate and removal of these pollutants in and from the environment. To reduce the presence of these compounds in the environment, many reactions have been reported to degrade and transform these compounds. One of the most effective methods is employing a great variety of microorganisms to biodegrade pollutants because it is simple, cheaper, and environmentally friendly. Several strains belonging to the genera *Sphingomonas*, *Pseudomonas*, *Rhodococcus*, *Agromyces*, *Acinetobacter*, *Microbacterium*, and *Gordonia* have been reported as DEHP-degrading bacteria [6–10]. These bacteria have been isolated from natural environments such as soil, water, or sediments; therefore, their use as biodegrading bacteria is environmentally friendly since they have natural source, form part of natural chain, and use natural metabolic pathways both under anaerobic and aerobic degradation, converting DEHP in less harmful metabolic secondary compounds [2]. In the same way, a series of phthalate-degrading fungal species has been summarized in previous studies [11] such as, *Aspergillus parasiticus*, *Fusarium subglutinans*, and *Penicillium funiculosum*, which can completely consume intact DEHP, either singly or in groups [12]. However, there is another part in this story that can generate controversy about the origin of this compound. This compound has been isolated from many microorganisms such as bacteria and fungi. The natural biosynthesis of this compound has been a matter of several debates questioning, that is, a pollutant with severe toxicity effects in human health. This review gives a global overview about the occurrence of DEHP naturally or synthetically.

2. Di-2-ethylhexylphtalate (DEHP) Chemical Compound

In general, phthalates are esters of phthalic acid, also known as esters of benzene-1,2-dicarboxylic acid. Phthalates contain a benzene ring with two ester groups. Their solubility in water decreases with an increase in the length of the carbon chain or molecular weight. Phthalates are oily liquids characterized by high boiling temperature, weak solubility in water, and satisfactory solubility in most organic solvents. These compounds are produced due to an esterification reaction to phthalic acid with various alcohols. Particularly, DEHP, also known as dioctyl phthalate (DOP), is obtained by the esterification reaction of 2-ethylhexanol with phthalic anhydride. It is a colorless liquid, and its molecular formula is $C_{24}H_{38}O_4$ with a molecular weight of 390.57 $g{\cdot}mol^{-1}$. The compound is an isomeric substance due to the presence of diastereotopic protons adjacent to the chiral center. It has been widely used as plasticizer in many articles made of polyvinyl chloride (PVC) such as sewage pipe due to its low cost and chemical resistance [3, 13]. It has been also used to improve the plasticity and the flexibility of materials such as food packages, building, toys, and medical devices such as blood storage bags, intravenous fluid bags, and other products. It is present in many plastics, especially vinyl materials, which may contain up to 40% DEHP, although lower levels are common [14]. For example, DEHP has been used in PVC gloves that are used for cooking purposes. Several studies demonstrated the occurrence of phthalate in prepared lunches was mainly caused by DEHP containing PVC gloves used during the preparation of the foods [13–16]. DEHP also finds use as a solvent in glowsticks. As the phthalate plasticizers are not chemically bound to PVC, they can leach, migrate, or evaporate into indoor air and atmosphere, foodstuff, other materials, etc. Therefore, the extended use of DEHP is more than obvious [13, 14].

Industrial production has been carried out following the reaction of phthalic anhydride **1** with 2-ethylhexanol **2**. Esterification of phthalic anhydride by alcohol takes place in two stages. The first stage is so rapid that it can be carried out in the absence of catalyst. However, esterification of the second carboxylic group is very slow and needs to be facilitated by acid catalyst, and the resulting water must be removed from the reaction mixture to achieve di-2-ethylhexylphtalate (DEHP) **3**, as shown in Scheme 1.

The synthesis is very simple; however, there have been some studies about what catalyst is the most convenient and appropriate to give the best yields. Sulfuric acid, methanesulfonic acid, *p*-Toluenesulfonic acid, and sodium hydrogen sulfate are the most conventional homogeneous used catalysts. Among them, sulfuric acid, as a strong acid, has been one of the most employed, both in laboratory and industrial practice. However, the generation of some by-products has leaded to replace it by *p*-toluenesulfonic acid that is less active catalyst but less aggressive. The applications of these catalysts cause some problems such as corrosion, loss of catalyst, and environment problems. Therefore, development of more efficient catalysts will be interesting and useful. A comparison is reported [17] between the homogeneous catalyst, *p*-toluenesulfonic acid, with heterogeneous catalysts such as zeolites and sulfated zirconia, demonstrating that heterogeneous catalysts showed high efficiency for esterification reactions. Another work, to study the kinetic model of this synthesis, uses methanesulfonic acid as a catalyst over the same reaction giving good results [18]. Each catalyst has some advantages over the others. It can be seemed that the heterologous catalysts showed high reaction yield, and they were used solvent free without requiring neutralization and washing steps, being of low cost and environmentally friendly catalysts.

The above-described method is the only way to synthesize DEHP; however, there are some studies on the optimization of synthesis of two starting materials that are used in the DEHP synthesis. There are two main routes to produce phthalic anhydride: the oxidation of *o*-xylene with air or K_2SO_4 as a promoter and oxidation of naphthalene in the same way using in both cases different silica gel and titania-supported catalysts, being V_2O_5/TiO_2 catalyst which gives the best yields [19]. More recently, the oxidation of naphthalene was carried out with ozone to achieve phthalic anhydride; however, the yields were lower than before-used methods [20]. On the other side, 2-ethylhexanol is industrially produced through three consecutive reaction processes such as, propylene hydroformylation to *n*-butyraldehyde, *n*-butyraldehyde self-condensation, and typical aldol condensation reaction, which can be catalyzed by an acid, a base, or an acid-base bifunctional catalyst, to give 2-ethyl-2-hexenal, and a hydrogenation, catalyzed by the supported Cu or Ni catalysts, of this last compound to give 2-ethylhexanol [21]. More recently, the one-step synthesis of 2-ethylhexanol from *n*-butyraldehyde was exploited using $Cu/Mg_{2.5}AlOx$ catalyst. This one-step synthesis of 2EHO could simplify the process, reduce equipment costs and operating costs, and improve the process economy. However, this methodology is under investigation [22].

3. Occurrence of DEHP in the Environment and Its Effects

Phthalate esters are gradually released from industrial products during manufacturing, storage, use, and disposal [23]. The photodegradation and hydrolyzation rates of these pollutants are very slow under natural conditions. For example, DEHP has a half-life of 2000 years [24]. DEHP is used as an excipient in approved pharmaceutical formulations, added into polyvinyl chloride (PVC) to reinforce its flexibility, and to be used as a solvent for dyes. In general, DEHP is considered as high-volume selling plasticizer which could migrate to the environment and penetrate the human body through water, air, food, and medical equipment. It has been detected in many environmental places such as rivers, lakes, groundwater and wastewater, soil, sludge, and sediments. Phthalate esters can be released into the atmosphere as

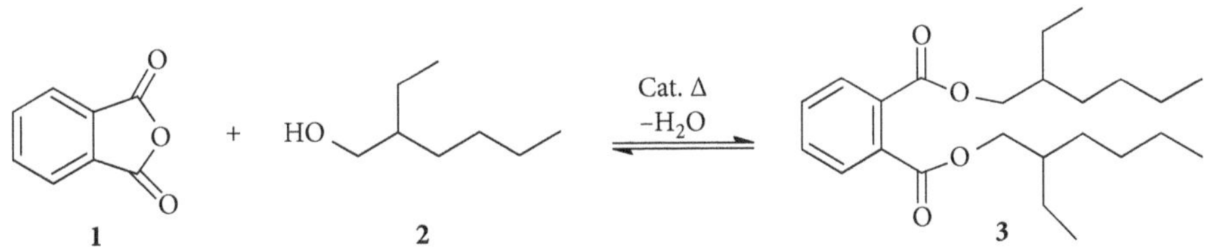

SCHEME 1: Synthesis of DEHP.

gaseous molecules, being temperature an important factor to control phthalate esters behavior. Despite the major phthalate esters are primarily present in vapor phase, DEHP preferred the particle phase. Rainfall transfers phthalate esters from the atmosphere to fresh water on land, causing their wide distribution in rivers, lakes, and sediments due to the natural hydrological cycle. DEHP is the predominant congener in both fresh water and sediments because of urbanization and industrialization. DEHP has low-water solubility; therefore, this compound concentrates from water into soil and sediments. There are many reports indicating the presence of diverse phthalates in agricultural soils being absorbed by crops and vegetables [25, 26]. Due to the plastics used in agriculture, many of phthalates were found in arable soils [25]. The phthalate esters that accumulate in agricultural soils may be absorbed by crops and vegetables, causing direct damage to the human food chain. However, as it can be seen later, DEHP is related with microorganisms since some of them can produce this compound.

Due to the high concentration of this compound in diverse environments, there have been realized some studies about the effect of this compound on the human health. The environmental risk limit (ERL) for DEHP phthalates in soil is 1000 μg/kg fresh weight [27]. Due to their potential health and environmental risks, the World Health Organization (WHO), US, Australia, Japan, and New Zealand have recommended the concentration of DEHP in drinking water to be below 8, 6, 9, 100, and 10 μg/L, respectively [28]. However, its potential toxicity is inconclusive. Some studies highlight its probable carcinogenic effect and its role as emerging endocrine disruptor and epigenetic toxicant. However, its potential toxicity is inconclusive. This compound is not severely toxic over a reasonable period; however, oxidized products of DEHP are probably more harmful than DEHP itself, because for humans the metabolism of DEHP seems to be more complex and involves several oxidative metabolites. Logically, many studies have been done in animals since they are more available for clinical trials; however, only few data are available for human toxicity being sometimes contradictory. There is no clear evidence about the association between exposure to DEHP and liver cancer in humans [29]. There are some studies [30] that have demonstrated that the exposure of a man to DEHP produce sperm abnormalities, such as reduced sperm count and motility, among others. Many other experiments about the effect of DEHP on human reproductive system have failed trying to detect the real adverse impact of DEHP in humans [31, 32]. About the effect of DEHP on several biological functions, there are many experiments that conclude that DEHP induces chromosomal aberrations, such as DNA strand breaks and gaps.

Consequently, there is a need to develop efficient processes to remove DEHP from different environments or degrade this compound. There are some questions about conventional treatment processes to degrade DEHP. One question is about if the physicochemical treatments are better than biological treatments. Normally, advanced oxidation processes have been developed and applied to remove DEHP at laboratory scale; however, for large-scale application, more research is need. This can be overcome by coupling this method with biological processes improving the DEHP biodegradation and reducing the operating cost as well as a shortening retention time [33]. It is important to consider the biodegradation pathway of DEHP that is shown in Figure 1. In Figure 1 are shown the different pathways that can follow DEHP biodegradation. It can be observed that the pathways are different using aerobic bacteria or anaerobic bacteria. Accordingly, bacteria can follow different ways through ortho- or metacleavage pathway [6]. There are many aerobic bacteria that can degrade phthalates primarily *Arthrobacter* sp., *Pseudomonas* sp., *Sphingomonas* sp., *Burkholderia* sp., *Ochrobactrum* sp., *Serratia* sp., *Bacillus* sp., and *Acinetobacter* sp. In addition to bacteria, some fungal species such as *Aspergillus*, sp., *Fusarium* sp., and *Penicillium* sp. can also degrade phthalates. Monoester phthalate and phthalic acid are common central intermediates in the anaerobic mineralization of phthalate esters. Phthalic acid should be converted to benzoate through decarboxylation, but no experimental evidence is available regarding the exact location of decarboxylation in degradation pathway. Benzoate degradation proceeds through ring cleavage to carbon dioxide, hydrogen, and acetate, which is converted into methane in further steps. The bacterial aerobic degradation of phthalate esters involves a sequence of reactions and is common to most microbes. As is shown in Figure 1, the cleavage of esters linkages is the common first step yielding phthalic acid. This acid can follow different routes depending if the bacteria is Gram positive or Gram negative. However, both routes converge in protocatechuate, which can follow different ways through ortho- or metacleavage pathway [34].

4. Natural Product Isolated from Several Organisms and Its Activity

Today, one of the biggest problems in the world related with the human health is that the number of drug-resistant bacteria has alarmingly increased; therefore, the search for new drugs has emerged as an interesting research area. In

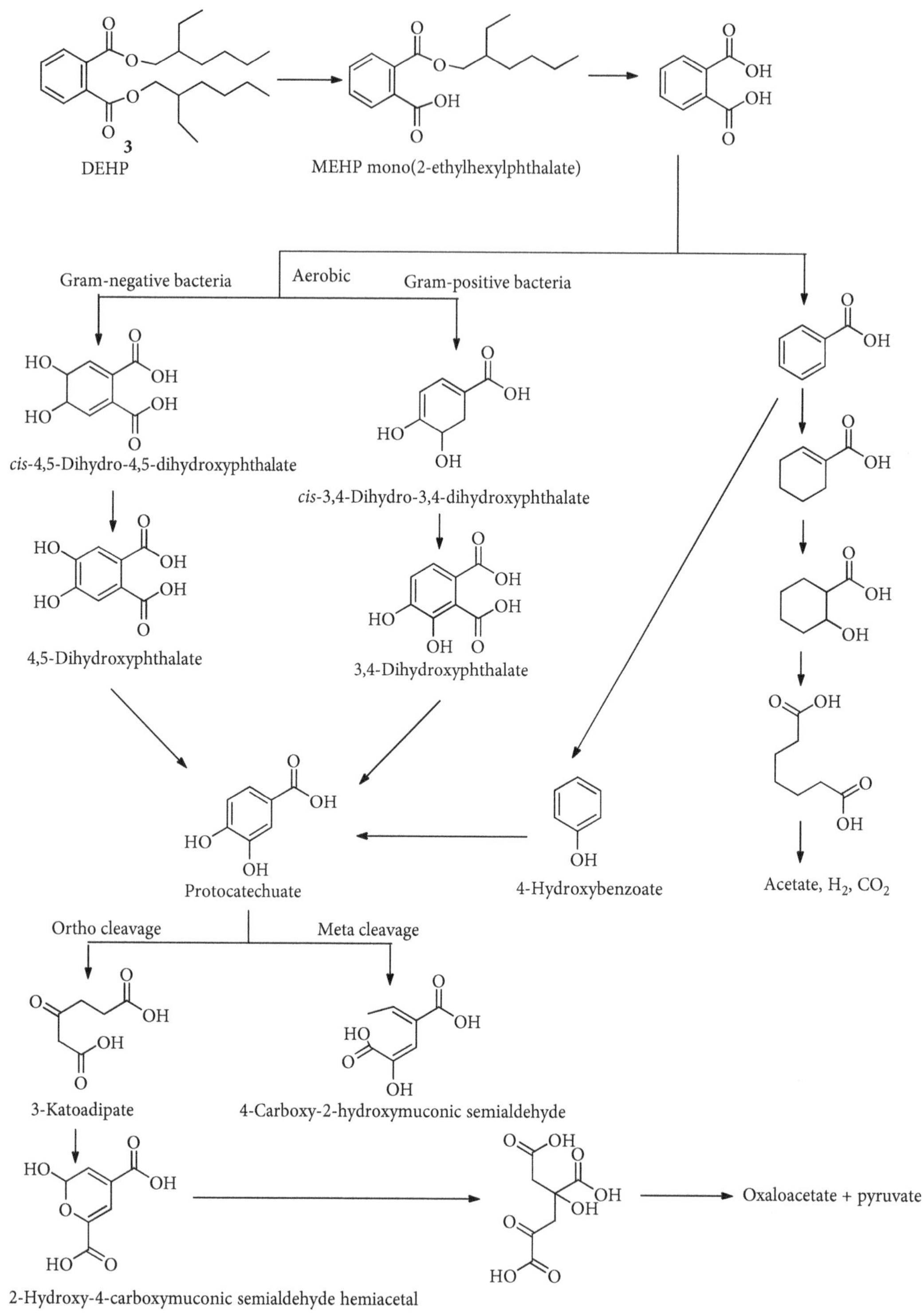

Figure 1: Biodegradation pathway of DEHP.

this sense, microbial secondary metabolites have been the target for this search, since many of the most consumed antibiotics have been isolated from bacteria or fungi. Some of the chemical compounds isolated from microorganisms have other uses such as di-2-ethylhexylphtalate which has been widely used in plastics. As it has been mentioned, this compound has been considered as an environmental polluting due to the industrial origin. However, this idea has changed gradually since it has been found that this compound is synthesized by many organisms such as plants, bacteria, or fungi, and several studies have demonstrated different biological activities for this compound.

A phytochemical study on the flower of *Calotropis gigantea* showed the isolation of di-2-ethylhexyl phthalate [35]. In the same study, this compound showed an antibacterial and antifungal activity against *Staphylococcus aureus*, *Bacillus subtilis*, *Sarcina lutea*, *Escherchia coli*, *Shigella sonnei*, and *Shigella shiga* and *Shigella dysenteriae* bacteria and *Aspergillus flavus* fungus [35]. The same group more recently explored the anticancer activity of this compound showing that it has dose-dependent antitumor activity against Ehrlich ascites carcinoma cells *in vivo* [36].

As it has been mentioned, some microorganisms have also produced this compound. Among them, actinomycetes, also called filamentous bacteria, is a group that produces this compound. *Streptomyces* sp. produced di-2-ethylhexyl phthalate which showed antibacterial activities against Gram-positive bacteria and fungi [37]. From *Streptomyces mirabilis* has also been isolated the same compound with antimicrobial activities mainly against Gram-positive bacteria and yeasts [38]. An actinomycete identified as *Nocardia levis* produced this phthalate that showed a moderate antimicrobial activity against some Gram-positive bacteria and some fungi [39]. There is no much literature reporting the production of this compound by other genera of bacteria. The only report is about *Bacillus cereus* bacteria symbiotically associated with an entomopathogenic nematode sp., namely, *Rhabditis (Oscheius)* sp. which secreted DEHP to the culture showing high-antimicrobial activity [40]. The fungi are the largest microbial group secreting this compound. *Penicillium* species produce a very diversified array of active secondary metabolites, including antibacterial, antifungal, immunosuppressant, and cholesterol-lowering agents. Recently, it has been reported that *Penicillium janthinellum* produced DEHP as a major bioactive compound, showing in this report a potent dose-dependent antitumor activity against Ehrlich cells in vivo [41]. Another fungus that produces DEHP is *Aspergillus fumigatus* which was shown to secrete this compound [42]. Very recently, the same compound was isolated from *Aspergillus awamori* showing activity against *Candida albicans* fungus and against Gram-positive bacteria *Sarcina lutea*, having at the same time, cytotoxic activity against some carcinoma cell lines [43].

5. Conclusions

Di-2-ethylhexylphtalate is used as plasticizer during industrial production of PVC, paints, adhesives, cosmetics, and food packaging, among others [15]. These compounds easily migrate to the environment; therefore, they are present in any environment such as plants, soils, water, and sediments [26]. This situation can create a human health problem because of the high exposing levels for humans. In this sense, there have been many studies about the human health problems associated with the exposure to this kind of compounds, probing that they can cause infertility and reproductive problems in males [27]. To solve this problem, many researchers have tried to remove these pollutants from the environment. Many study results indicate that biodegradation under aerobic conditions is the best option for the mineralization of phthalate esters in the environment. The use of microorganisms for this purpose has been extensively reported [6]. However, it has been found that DEHP compound is synthesized by many organisms such as plants, bacteria, or fungi, and several studies have demonstrated different biological activities for this compound [38]. These data show that many microorganisms produce di-2-ethylhexylphtalate but it is unclear if this compound is originated from secondary metabolisms of the living organisms or is absorbed from the accumulation of high levels of this compound from air and water. There is only one study with di-n-butyl phthalate (DBP) that demonstrates biosource production of this compound by filamentous fungi confirming that the fungal generation of DBP was largely through shikimic acid pathway, which was assembled by phthalic acid with butyl alcohol through esterification [44]. This study is very important because it confirmed the metabolic ability for biosynthesis of DBP. Until today, no studies have been done about other phthalates such as DEHP but it is very likely that the behavior is very similar, and the microorganisms can produce this compound through any of metabolic pathway. Therefore, this arises in the big question, is di-2-ethylhexylphthalate a natural product, rather than a pollutant? More research is needed to answer this question.

Conflicts of Interest

The authors declare that there are no conflicts of interest regarding the publication of this paper.

Acknowledgments

We thank VIEP (100518932 VIEP2018) and CONACYT (251512) for financial support.

References

[1] M. M. A. Daiem, J. Rivera-Utrilla, R. Ocampo-Perez, J. D. Mendez-Diaz, and M. Sanchez-Polo, "Environmental impact of phthalic acid esters and their removal from water and sediments by different technologies—a review," *Journal of Environmental Management*, vol. 109, pp. 164–178, 2012.

[2] D.-W. Gao and Z.-D. Wen, "Phthalate esters in the environment: a critical review of their occurrence, biodegradation, and removal during wastewater treatment processes," *Science of the Total Environment*, vol. 541, pp. 986–1001, 2016.

[3] S. Magdouli, R. Daghrir, S. K. Brar, P. Drogui, and R. D. Tyagi, "Di 2-ethylhexylphtalate in the aquatic and terrestrial environment: a critical review," *Journal of Enviromental Management*, vol. 127, pp. 36–49, 2013.

[4] S. Net, R. Sempere, A. Delmont, A. Paluselli, and B. Ouddane, "Occurrence, fate, behaviour and ecotoxicological state of phthalates in different environmental matrices," *Environmental Science and Technology*, vol. 49, no. 7, pp. 4019–4035, 2015.

[5] M. A. Park, K. A. Hwang, H. R. Lee, B. R. Yi, E. B. Jeung, and K. C. Choi, "Cell growth of BG-1 ovarian cancer cells is promoted by di-n-butyl phthalate and hexabromocyclododecane via upregulation of the cyclin D and cyclin-dependent kinase-4 genes," *Molecular Medicine Reports*, vol. 5, no. 3, pp. 761–766, 2012.

[6] D.-W. Liang, T. Zhang, H. H. P. Fang, and J. He, "Phthalates biodegradation in the environment," *Applied and environmental biotechnology*, vol. 80, no. 2, pp. 183–198, 2008.

[7] Y. Wang, B. Miao, D. Hou, X. Wu, and B. Peng, "Biodegradation of di-n-butyl phthalate and expression of the 3,4-

phthalate dioxygenase gene in *Arthrobacter* sp. ZH2 strain," *Process biochemistry*, vol. 47, no. 6, pp. 936–940, 2012.

[8] J. Wang, M.-Y. Zhang, T. Chen et al., "Isolation and identification of a di-(2-ethylhexyl) phthalate-degrading bacterium and its role in the bioremediation of a contaminated soil," *Pedosphere*, vol. 25, no. 2, pp. 202–211, 2015.

[9] H.-M. Zhao, H. Du, J. Lin et al., "Complete degradation of the endocrine disruptor di-(2-ethylhexyl) phthalate by a novel *Agromyces* sp. MT-O strain and its application to bioremediation of contaminated soil," *Science of the total Environment*, vol. 562, pp. 170–178, 2016.

[10] J. Xu, Q. Lu, R. Alves de Toledo, and H. Shim, "Degradation of di-2-ethylhexyl phthalate (DEHP) by an indigenous isolate *Acinetobacter* sp. SN13," *International Biodeterioration and Biodegradation*, vol. 117, pp. 205–214, 2017.

[11] M. Ahuactzin-Perez, J. L. Torres, B. R. Rodriguez-Pastrana et al., "Fungal biodegradation of dibutyl phthalate and toxicity of its breakdown products on the basis of fungal and bacterial growth," *World Journal of Microbiology and Biotechnology*, vol. 30, no. 11, pp. 2811–2819, 2014.

[12] S. Pradeep and S. Benjamin, "Mycelial fungi completely remediate di (2-ethylhexyl)phthalate, the hazardous plasticizer in PVC blood storage bag," *Journal of Hazardous Materials*, vol. 235-236, pp. 69–77, 2012.

[13] P. A. Przybylinska and M. Wyszkowski, "Environmental contamination with phthalates and its impact on living organisms," *Ecological Chemistry and Engineering S*, vol. 23, no. 2, pp. 347–356, 2016.

[14] U. Heudorf, V. Mersch-Sundermann, and J. Angerer, "Phthalates: Toxicology and exposure," *International journal of hygiene and environmental health*, vol. 210, no. 5, pp. 623–634, 2007.

[15] T. Fierens, K. Servaes, M. Van Holderbeke et al., "Analysis of phthalates in food products and packaging materials sold on the Belgian market," *Food and Chemical Toxicology*, vol. 50, no. 7, pp. 2575–2583, 2012.

[16] Y. Tsumura, S. Ishimitsu, A. Kaihara, K. Yoshii, Y. Nakamura, and Y. Tonogai, "Di (2-ethylhexyl) phthalate contamination of retail packed lunches caused by PVC gloves used in the preparation of foods," *Food Additives and Contaminants*, vol. 18, no. 6, pp. 569–579, 2001.

[17] F. T. Sejidov, Y. mansoori, and N. Goodarzi, "Esterification reaction using solid heterogeneous acid catalysts under solvent-less condition," *Journal of Molecular Catalysis A: Chemical*, vol. 240, no. 1-2, pp. 186–190, 2005.

[18] J. Skrzypek, M. Lachowska, M. Kulawska, and H. Moroz, "Synthesis of bis(2-ethylhexyl) phthalate over methane sulfonic acid catalyst kinetic investigations," *Reaction Kinetics and Catalysis Letters*, vol. 93, no. 2, pp. 281–286, 2008.

[19] C. R. Dias, M. F. Portela, and G. C. Bond, "Synthesis of phthalic anhydride: catalysts, kinetics, and reaction modeling," *Catalysis Reviews*, vol. 39, no. 3, pp. 169–207, 1997.

[20] B. Rindone, F. Saliu, and R. Suarez-Bertoa, "The synthesis of phthalic anhydride via ozonation of naphthalene," *Ozone: Science and Engineering*, vol. 32, no. 3, pp. 161–165, 2010.

[21] C. Kohlpaintner, M. Schulte, J. Falbe, P. Lappe, and J. Weber, "Aldehydes, Aliphatic," in *Ullmann's Encyclopedia of Industrial Chemistry*, Wiley-VCH, Weinheim, Germany, 2008.

[22] S. Miao, H. An, X. Zhao, and Y. Wang, "Catalytic performance of Cu–Mg–Al in the one-step synthesis of 2-ethylhexanol from *n*-butyraldehyde," *Reaction Kinetics, Mechanisms and Catalysis*, 2018, In press.

[23] P. A. Clausen, Z. Liu, V. Kofoed-Sorensen, J. Little, and P. Wolkoff, "Influence of temperature on the emission of di-(2-ethylhexyl)phthalate (DEHP) from PVC flooring in the emission cell FLEC," *Environmental Science & Technology*, vol. 46, no. 2, pp. 909–915, 2012.

[24] C. A. Staples, D. R. Peterson, T. F. Parkerton, and W. J. Adams, "The environmental fate of phthalate esters: a literature review," *Chemosphere*, vol. 35, no. 4, pp. 667–749, 1997.

[25] X. Y. Hu, B. Wen, and X. Q. Shan, "Survey of phthalate pollution in arable soils in China," *Journal of Environmental Monitoring*, vol. 5, no. 4, pp. 649–653, 2003.

[26] T. T. Ma, P. Christie, Y. M. Luo, and Y. Teng, "Phthalate esters contamination in soil and plants on agricultural land near an electronic waste recycling site," *Environmental Geochemistry and Health*, vol. 35, no. 4, pp. 465–476, 2013.

[27] A. P. Van Wezel, P. Van Vlaardingen, R. Posthumus, G. H. Crommentuijn, and D. T. H. M. Sijm, "Environmental risk limits for two phthalates, with special emphasis on endocrine disruptive properties," *Ecotoxicology and Environmental Safety*, vol. 46, no. 3, pp. 305–321, 2000.

[28] WHO, *Guidelines for Drinking Water Quality*, WHO, Geneva, Switzerland, 3rd edition, 2004.

[29] I. Rusyn and J. C. Corton, "Mechanistic considerations for human relevance of cancer hazard of di(2-ethylhexyl) phthalate," *Mutation Research*, vol. 750, no. 2, pp. 141–158, 2011.

[30] G. Pan, T. Hanaoka, M. Yoshimura et al., "Decreased serum free testosterone in workers exposed to high levels of di-*n*-butyl phthalate (DBP) and di- 2-ethylhexyl phthalate (DEHP): a cross-sectional study in China," *Environmental Health Perspectives*, vol. 114, no. 11, pp. 1643–1648, 2006.

[31] R. Lambrot, V. Muczynski, C. Lecureuil et al., "Phthalates impair germ cell development in the human fetal testis in vitro without change in testosterone production," *Environmental Health Perspectives*, vol. 117, no. 1, pp. 32–37, 2009.

[32] B. A. Jonsson, J. Richthoff, L. Rylander, A. Giwercman, and L. Hagmar, "Urinary phthalate metabolites and biomarkers of reproductive function in young men," *Epidemiology*, vol. 16, no. 4, pp. 487–493, 2005.

[33] T. T. H. Pham, R. D. Tyagi, S. K. Brar, and R. Y. Surampalli, "Effect of ultrasonication and Fenton oxidation on biodegradation of bis(2-ethylhexyl) phthalate (DEHP) in wastewater sludge," *Chemosphere*, vol. 82, no. 6, pp. 923–928, 2011.

[34] D.-W. Gao and Z.-D. Wen, "Phthalate esters in the environment: a critical reviewof their occurrence, biodegradation, and removal during wastewater treatment processes," *Science of the Total Environment*, vol. 541, pp. 986–1001, 2016.

[35] M. R. Habib and M. R. Karim, "Antimicrobial and cytotoxic activity of di-(2-ethylhexyl) phthalate and anhydrosophoradiol-3-acetate isolated from *Calotropis gigantea* (Linn.) Flower," *Mycobiology*, vol. 37, no. 1, pp. 31–36, 2009.

[36] M. R. Habib and M. R. Karim, "Antitumour evaluation of di-(2-ethylhexyl) phthalate (DEHP) isolated from *Calotropis gigantea* L. flower," *Acta Pharmaceutica*, vol. 62, no. 4, pp. 607–615, 2012.

[37] S. Smaoui, L. Mellouli, A. Lebrihi, Y. Coppel, L. F. B. Fguira, and F. Mathieu, "Purification and structure elucidation of three naturally bioactive molecules from the new terrestrial Streptomyces sp. TN17 strain," *Natural Product Research*, vol. 25, no. 8, pp. 806–814, 2011.

[38] M. H. El-Sayed, "Di-(2-ethylhexyl) phthalate, a major bioactive metabolite with antimicrobial and cytotoxic activity isolated from the culture filtrate of newly isolated soil

streptomyces (*Streptomyces mirabilis* Strain NSQu-25)," *World Applied Sciences Journal*, vol. 20, pp. 1202–1212, 2012.

[39] A. Kavitha, P. Prabhakar, M. Vijayalakshmi, and Y. Venkateswarlu, "Production of bioactive metabolites by Nocardia levis MK-VL_113," *Letters in Applied Microbiology*, vol. 49, no. 4, pp. 484–490, 2009.

[40] K. M. Anju, M. M. Archana, C. Mohandas, and B. Nambisan, "An antimicrobial phthalate derivative from bacillus cereus, the symbiotic bacterium associated with a novel entomopathogenic nematode, *Rhabditis (Oscheius)* Sp.," *International Journal of Pharmacy and Pharmaceutical Sciences*, vol. 7, pp. 238–242, 2015.

[41] O. H. El-Sayed, M. M. S. Asker, S. M. Shash, and S. R. Hamed, "Isolation, structure elucidation and biological activity of di-(2-ethylhexyl) phthalate produced by penicillium janthinellum 62," *International Journal of ChemTech Research*, vol. 8, pp. 58–66, 2015.

[42] M. S. Abdel-Aziz, M. A. Ghareeb, A. M. Saad, L. A. Refahy, and A. A. Hamed, "Chromatographic isolation and structural elucidation of secondary metabolites from the soil-inhabiting fungus aspergillus fumigatus 3T-EGY," *Acta Chromatographica*, 2017, In press.

[43] M. M. Lotfy, H. M. Hassan, M. H. Hetta, A. O. El-Gendy, and R. Mohammed, "Di-(2-ethylhexyl) Phthalate, a major bioactive metabolite with antimicrobial and cytotoxic activity isolated from River Nile derived fungus *Aspergillus awamori*," *Beni-Suef University Journal of Basic and Applied Sciences*, vol. 7, no. 3, pp. 263–269, 2018.

[44] C. Tian, J. Ni, F. Chang et al., "Bio-Source of di-n-butyl phthalate production by filamentous fungi," *Scientific Reports*, vol. 6, article 19791, 2016.

2

Cellular and Molecular Evidence of Acetaldehyde Elimination and Intracellular Environment Antioxidation by L-Cysteine

Zhenjing Li,[1,2] Xiaohong Zhang,[1,3] Yibin Xue,[1,2] Jingkai Zhang,[1] Meiling Li,[1] and Cheng Luo[1,2]

[1]*State Key Laboratory of Food Nutrition and Safety, Tianjin University of Science & Technology, Tianjin 300457, China*
[2]*Key Laboratory of Food Nutrition and Safety, Ministry of Education, School of Food Engineering and Biotechnology, Tianjin University of Science and Technology, Tianjin 300457, China*
[3]*China CFAPA Testing Technology Company Limited of Dalian, Dalian 116000, China*

Correspondence should be addressed to Cheng Luo; luo58@yahoo.com

Academic Editor: Fabio Polticelli

Acetaldehyde is a harmful metabolite of smoking and drinking. This study was initially intended to facilitate the understanding of the possible injury mechanism of A549 cells damaged by acetaldehyde and the possible protective mechanism of L-cysteine (L-Cys) by analyzing the oxidative damage indicators, as well as the changes in cell morphology and gene expression. Results from the dithiodimorpholine nitrobenzoic acid colorimetric determination for glutathione peroxidase (GSH-Px) activity in L-Cys groups were significantly higher ($P < 0.01$) than those in the acetaldehyde group in a dose-dependent manner. The expression of cytochrome *c* oxidase subunit II (COII) mRNA was significantly reduced compared with the control group ($P < 0.01$) and was noticeably restored in the L-Cys groups. Scanning electronic microscopy observation, DAPI staining, and flow cytometry also indicated that L-Cys could effectively attenuate the oxidative damage to A549 cells caused by acetaldehyde and reduces the rate of apoptosis. In conclusion, the protective effects of L-Cys on A549 cells against oxidative damage by acetaldehyde were dose-dependent within the range of 10 μmol/L to 160 μmol/L. Acetaldehyde damaged the mitochondria and resulted in the apoptosis of A549 cells by reactive oxygen species (ROS), e.g., free radicals, but L-Cys reversed the release of cytochrome *c* from the mitochondria, reduced the rate of apoptosis, and protected cells from ROS and oxidative stress.

1. Introduction

Acetaldehyde in the human body is primarily oxidized and metabolized by alcohol dehydrogenase in the liver. Acetaldehyde can also be increased by the uptake of beer, bread, and fruits containing this metabolite. Acetaldehyde can also be acquired from air or from tobacco smoking. Its low boiling point (20.2°C) causes acetaldehyde to evaporate easily. The perception limit of acetaldehyde in air ranges from 0.07 mg/L to 0.25 mg/L, which is within the normal range of concentrations for fruity odor [1]. However, higher concentrations will irritate the eyes, nose, throat, and lungs, resulting in cough and polypnea. The intake of high concentrations of acetaldehyde will make people sick, lose consciousness, contract emphysema, or even die from respiratory and cardiovascular failure [2]. Acetaldehyde has been shown to act on the liver by depressing mitochondrial function, decreasing fatty acid oxidation, enhancing glycogenolysis, and decreasing gluconeogenesis from glycerol during alcohol consumption [3]. Acetaldehyde reacts with DNA to form a variety of adducts that may cause polymerase errors and mutations in critical genes, frequently activating proto-oncogenes while inactivating tumor suppressor genes to result in tumorigenesis or carcinogenesis [4–6]. Moreover, acetaldehyde is also considered as a major cause of alcoholism and alcohol addiction [7, 8]. Therefore, the effective elimination of acetaldehyde is important not only in preventing cellular toxicity but also in the efficient removal of alcohol [9].

Acetaldehyde can be eliminated from the human body in three ways. The first way is to reduce the generation of acetaldehyde by avoiding alcohol drinking and smoking. The second way is achieved via accelerating the metabolism of acetaldehyde, such as by increasing the activity of acetaldehyde dehydrogenase (ALDH) in vivo or the content of ALDH to accelerate the conversion of acetaldehyde to acetic acid. The last way is to enable acetaldehyde to combine with another substance to become nontoxic and then be excreted [9].

The reaction of acetaldehyde with most amino acids generates relatively free and stable imines [10]. In particular, L-cysteine (L-Cys) can react with acetaldehyde to produce 2-methylthiazolidine-4-carboxylic acid, which is nontoxic to the human body and can be directly excreted [11]. Therefore, L-Cys can accelerate the consumption of acetaldehyde and effectively detoxify organisms subjected to acetaldehyde. Alternatively, L-Cys can also be converted into 2-methylthiazolidine-4-carboxylic acid (MTCA), a stable and nontoxic compound [12]. MTCA conversion has been regarded as a way to eliminate carcinogenic acetaldehyde in the digestion system of humans [13].

Methods used for acetaldehyde depletion by L-Cys have been increasingly recognized by different scientific and industrial communities [14, 15]. L-Cys, with its capability to chemically neutralize acetaldehyde, reportedly prevented ethanol- and acetaldehyde-induced conditioned place preference in the same strain of rats [16]. L-Cys also has a great affinity with acetaldehyde in air and therefore precipitates acetaldehyde [17]. Other studies have shown that acetaldehyde may possess reinforcing properties and that L-Cys can reduce the need to self-administer acetaldehyde [15]. Previous studies demonstrated that acetaldehyde has a significant damage effect on A549 cells in a concentration-dependent manner and L-Cys could protect A549 cells damaged by acetaldehyde and significantly increased its antioxidant capability [16, 18]. In the present study, the apoptosis-promoting mechanisms of acetaldehyde and the possible protective mechanism of L-Cys are further investigated.

2. Materials and Methods

2.1. Cell Culture. The human non-small-cell lung cancer A549 cell line (received as a gift from the Tianjin Institute of Lung Cancer, Tianjin University of Medicine) was routinely cultured in RPMI 1640 medium (Thermo Fisher Scientific Inc., Beijing, China) supplemented with 10% fetal bovine serum (FBS, Gibco) and 1× penicillin/streptomycin (100 U/mL $P + 0.1$ mg/mL·S, Solarbio, Beijing, China) in a humidified incubator (HERAcell 150i, Thermo fisher, USA) at 37°C under 5% CO_2. After growing to 80% confluence in a T25 tissue culture flask (TPP Biochrom AG, Trasadingen, Switzerland), the cells were plated at a density of 1×10^6 or 8×10^3 cells/mL in 6-well plates (Corning, New York, USA) for the following bioassays.

2.2. Cell Treatment. One control group (cells untreated with acetaldehyle), an acetaldehyde group (200 μmol/L), and three L-Cys groups with high, medium, and low concentrations of L-Cys (640, 160, and 40 μmol/L) were included in the study. Cells in the logarithmic phase were diluted to a unicellular suspension of 6×10^4 cells/mL and then inoculated to 6-well plates at 37°C under 5% CO_2 for 24 h. When the cellular confluence was 80%, RPMI 1640 medium without FBS was used to extend the culture for 24 h to maintain synchronization. When the cellular confluence was 90%, 640, 160, and 40 μmol/L, final concentrations of the L-Cys culture solutions (Solarbio, Beijing, China) were prepared for each L-Cys protection group, whereas the control and acetaldehyde groups continued being cultured after replacing with a fresh culture medium. The culture media of all groups were gathered and discarded after 2 h and then rinsed by phosphate-buffered saline (PBS, 0.2 M, pH 7.2). The acetaldehyde and L-Cys groups were prepared to the final concentration of 200 μmol/L of acetaldehyde culture medium. In the control group, an equivalent volume of fresh cultured medium was used. The cells were then chosen for further analysis after another 24 h of incubation. All the experiments were repeated thrice with six duplicate wells.

2.3. Cell Viability by MTT Assay. The cell viabilities of A549 cells treated with acetaldehyde or L-Cys at different concentrations were determined using 3-(4,5-dimethylthiazol-2-yl)-2,5-diphenyltetrazoliumbromide (MTT) assay to evaluate the possible cytotoxic effects of acetaldehyde or L-Cys.

For the toxicity analysis of acetaldehyde, the A549 cells were treated with different concentrations of acetaldehyde in 96-well plates for 26 h. For the toxicity analysis of L-Cys, the cells were incubated for 2 h with different concentrations of L-Cys in 96-well plates, and then 200 μmol/L of acetaldehyde was added for another 24 h of culture.

Aliquots of 20 μL of 5 mg/mL MTT (Solarbio, Beijing, China) in PBS were added to each well in the 96-well plate. The plates were incubated for another 4 h and then centrifuged. The culture medium was then discarded. The plates were carefully washed twice with PBS buffer. Aliquots of 150 μL of dimethyl sulfoxide (DMSO, Amresco, USA) were added to each well and oscillated for 30 min to extract the insoluble formazan that formed. The absorbance was measured at 570 nm on a plate reader (ELISA reader, ASYS Hitech, GmbH, Austria).

2.4. Intracellular Activity of GSH-Px. A549 cells were treated with acetaldehyde and L-Cys in a 24-well plate according to the method in the MTT assay. The cells were collected after 24 h, disrupted by ultrasound (Kunshan ultrasonic cleaning machine, 200 w) in 300 μL of PBS for 30 s, and centrifuged at 12,000 r/min at 4°C for 15 min. The supernatant was collected. A Glutathione peroxidase (GSH-Px) kit (Nanjing Jiancheng Bioengineering Institute, Nanjing, China) was used to detect the total protein concentration and GSH-Px activity. The experiment was repeated thrice.

2.5. Expression of COII mRNA by PCR. Approximately 2 million A549 cells were harvested after cultivation. Total RNA was extracted using the RNeasy Mini-Kit (Qiagen RNeasy Mini Kit, Germany) according to the manufacturer protocol. RNA integrity was evaluated through electrophoresis in 1.2% agarose gel. Nondegraded total RNA was displayed as 28S and 18S bands. First-strand cDNA was synthesized using the PrimeScript 1st Strand cDNA Synthesis kit (Fermentas First Strand cDNA Synthesis Kit, Fermentas) with the Oligo dT-Adaptor primer. PCR amplification was performed using gene-specific PCR primers. The primer sequences were designed by Invitrogen Trading (Shanghai, China) according to their mRNA sequences in the gene bank (accession number of COII: MH378688.1). The sequences of gene-specific PCR primers, the appropriate annealing temperature, and the length of production are summarized in Table 1. The amplification was performed in 20 μL reactions, which contained 2 μL of cDNA, 1 μL each of forward and reverse primers, 10 μL of PCR 2× buffer with $MgCl_2$, 10 mM of dNTP mixture, 2.5 U Taq polymerase, and 7 μL of sterile ddH_2O. Thermal cycling was performed under the following conditions: initial activation at 94°C for 180 s, followed by 35 cycles of denaturation at 94°C for 30 s, annealing at 58°C for 30 s, and an extension at 72°C for 45 s. Mouse β-actin was used as an external standard. A 5 μL portion of the RT-PCR product, added with 1 μL sample loading buffer, underwent electrophoresis in 1.5% of agarose gel (121 V, 25 min). Later, electrophoretic bands were scanned for optical density (OD) using a gel scanner (BIORAD Gel Doc XR). The bands were photographed, and the OD was analyzed. The grey density value of each amplified band was recorded, and the ratio of the OD value of each targeted amplified segment and that of the corresponding internal control β amplified segment was taken as the semiquantitative indicator of cytochrome *c* oxidase subunit II (COII) mRNA.

2.6. Cell Morphology Observation. Morphological changes of the cells were tracked using inverted microscopy (AMG EVOS), scanning electron microscopy (Hitachi, Tokyo, Japan), and fluorescence microscopy (GE Healthcare UK Ltd., Buckinghamshire, UK) in this study. For inverted and scanning electron microscopy test, A549 cells (1 × 106 cells/mL) were grown on cover slips in 6-well plates and were treated with L-Cys at different concentrations. The cells were then observed directly. For fluorescence microscopy test, A549 Cells were firstly seeded onto 6-well plates and cultured to adhere overnight. After 48 h of treatment with L-Cys, the cells were washed with PBS and stained with 2 μg/mL DAPI for 15 min. Images were then captured using the IN Cell analyzer 1000 (GE Healthcare UK Ltd., Buckinghamshire, UK). Each image shown is representative of 20 randomly observed fields.

2.7. Apoptosis Rates and Cell Cycles by Flow Cytometry. A549 cells (1×10^6 cells/mL) were grown on cover slips in 6-well plates. The cells were grouped and treated with acetaldehyde, L-Cys, and a combination of acetaldehyde and L-Cys. The cells were then washed in PBS, collected through trypsinization, fixed in 70% glacial ethanol, washed in PBS, resuspended in 1 mL of PBS containing 50 U/mL RNase and 50 μg/mL propidium iodide (PI), and then incubated for 40 min in the dark at 4°C. Cell cycle analysis was performed through flow cytometry (BD, Franklin Lakes, NJ, USA), and the population of cells in each phase was calculated using the ModFit LT software program. Each experiment was conducted thrice.

2.8. Statistical Analysis. Each experiment was repeated at least thrice. Numerical data are presented as mean ± SE. The difference between means was analyzed using one-way ANOVA, and LSD was used for multiple comparisons. All statistical analyses were performed using SPSS 17.0 software (SPSS, Chicago, IL, USA).

3. Results

3.1. Influence of Acetaldehyde and L-Cys on Cell Viabilities of A549 Cells. The cell viability of A549 cells decreased within the range of 0 μmol/L to 900 μmol/L of acetaldehyde (Figure 1(a)) in a dose-dependent manner, which indicates that acetaldehyde can inhibit cell proliferation. Results from Figure 1(b) suggest that L-Cys could protect cells from acetaldehyde-mediated damage in a concentration-dependent manner. Compared with the control group with acetaldehyde treated only, cell viability exhibited a statistically significant difference at the L-Cys concentrations of 40, 80, and 160 mol/L ($P < 0.01$). When the L-Cys concentration was greater than 160 mol/L (320, 640 mol/L), cell viability decreased; however, it is still statistically significant higher than the control group with acetaldehyde treated only ($P < 0.01$) (Figure 1(b)).

3.2. L-Cys Influences GSH-Px Activity in A549 Cell Damaged by Acetaldehyde. Similar to cell viability, the intracellular enzyme activity of GSH-Px was decreased with the increment of acetaldehyde concentration (Figure 2(a)), while pretreating cells with L-Cys could obviously weaken the acetaldehyde-mediated damage, and the effects were dose-dependent (Figure 2(b)). For example, GSP-Px activity was significantly lower than that of the non-acetaldehyde-treated control group when acetaldehyde concentration was up to 50 μmol/L ($P < 0.05$); pretreating cell with 40 μmol/L L-Cys significantly increased the GSH-Px activity when comparing with the group treated with 200 μmol/L acetaldehyde ($P < 0.01$).

3.3. Influence of L-Cys Treatment on COII mRNA Expression Damaged by Acetaldehyde. To uncover the possible mechanism of acetaldehyde-induced damage and L-Cys protection of A549 cells, COII mRNA was detected using RT-PCR. Electrophoretogram of RNA and influence of L-Cys on the COII mRNA expression of the A549 cells damaged by acetaldehyde are shown in Figures 3(a), 3(b), and 3(c), respectively. The expression of mitochondrial COII mRNA was significantly reduced compared with the control group and was noticeably restored in the L-Cys groups. This finding clearly showed that the oxidative stress of A549 cells mediated by acetaldehyde could

TABLE 1: Summary of the gene-specific PCR primer sequences, the length of amplificated product, the appropriate annealing temperature used in the experiments, and the number of the amplification cycles.

Gene	Sense primer (5′⟶3′)	Antisense primer (5′⟶3′)	Product size	T_m	Cycles
β-Actin	AAATCTGGCACCACACCTT	AGCACTGTGTTGGCGTAGAG	646 bp	58°C	35
COII	ATCTTAATGGCACATGCAGC	AGATCTCGGGTGACATTTCGA	726 bp	58°C	35

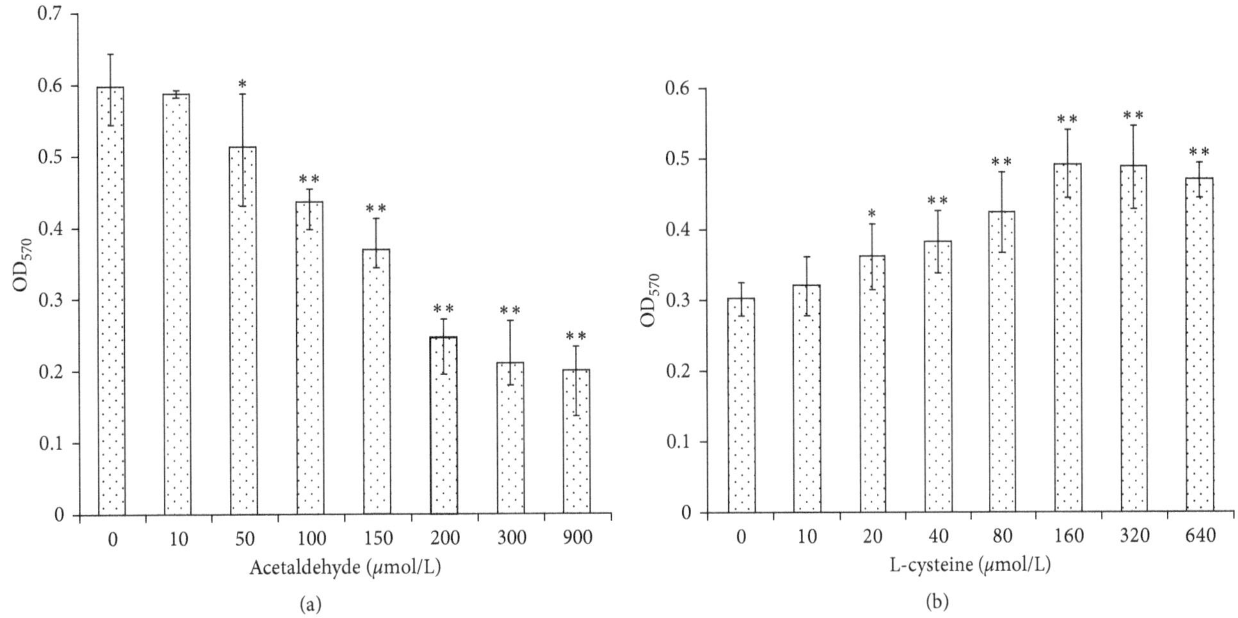

FIGURE 1: Cell viability of A549 cells influenced by acetaldehyde (a) and influence of L-Cys on cell viability of A549 cells damaged by 200 μmol/L acetaldehyde (b). Data are represented as the mean ± S.E. $n = 6 - 12$; $^{*}P < 0.05$ and $^{**}P < 0.01$ with respect to the control group.

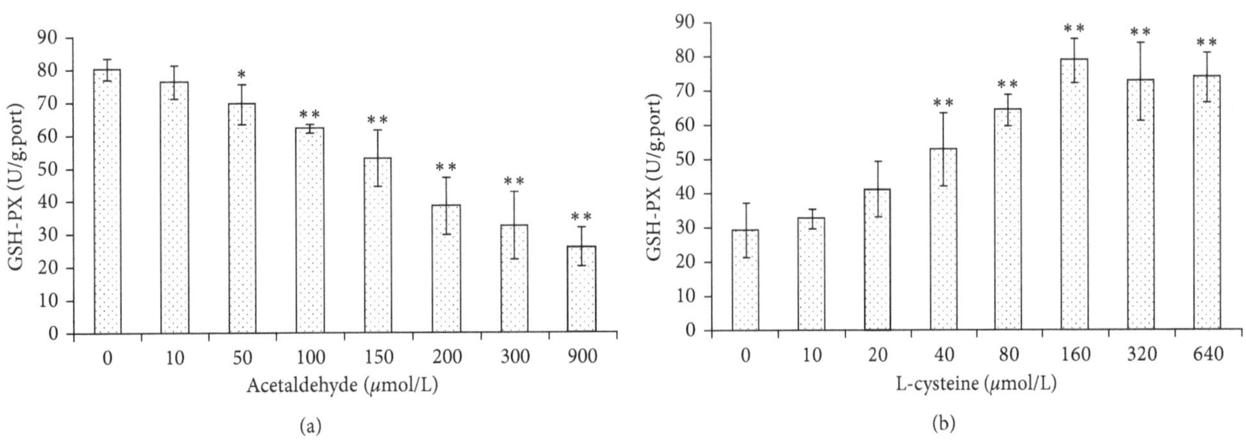

FIGURE 2: GSH-Px activity of A549 cells influenced by acetaldehyde (a) and influence of L-Cys on the GSH-Px activity of A549 cells damaged by 200 μmol/L acetaldehyde (b). Data are represented as the mean ± S.E. $n = 6 - 12$; $^{*}P < 0.05$ and $^{**}P < 0.01$ with respect to the related control group.

reduce COII expression, which was probably related to DNA damage of mitochondrion. Compared with the 200 μmol/L acetaldehyde group, a low concentration of L-Cys (40 μmol/L) increased the expression of COII mRNA ($P < 0.05$), whereas middle (160 μmol/L) and high concentrations (640 μmol/L) of L-Cys resulted in significantly higher expression ($P < 0.01$), which indicates the protective effect of L-Cys on A549 cells damaged by acetaldehyde (Figures 3(b) and 3(c)).

3.4. Cellular Morphology. Intact A549 cells of the control group were of sharp fusiform or polygonal shape (Figure 4(a)). Cellular density was significantly reduced, and cellular

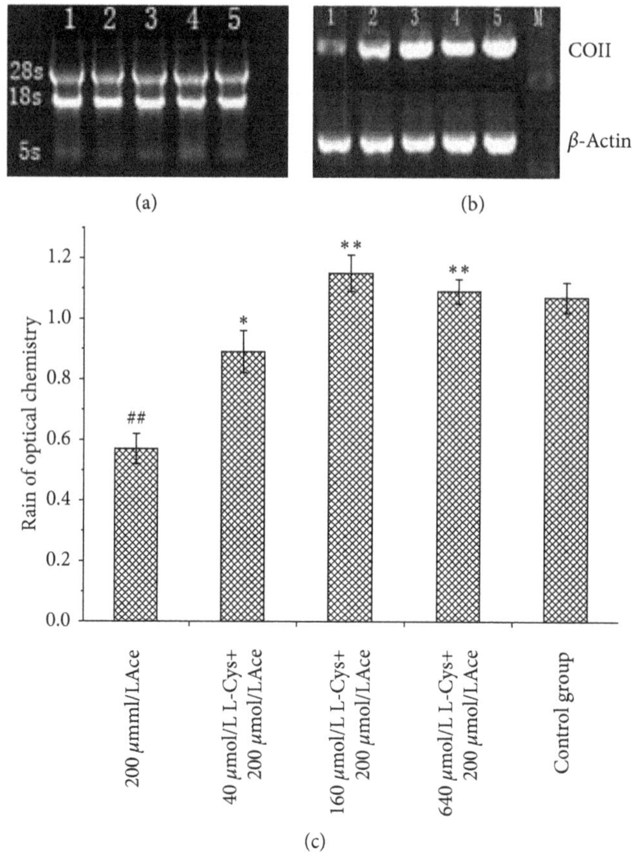

FIGURE 3: Electrophoretogram of RNA (a) and influence of L-Cys on the COII mRNA expression of the A549 cells damaged by acetaldehyde (b, c). Lanes 1 to 5 represent 200 μmol/L acetaldehyde, 40 μmol/L L-Cys + 200 μmol/L acetaldehyde, 160 μmol/L L-Cys + 200 μmol/L acetaldehyde, 640 μmol/L L-Cys + 200 μmol/L acetaldehyde, and 0 μmol/L L-Cys + 0 μmol/L acetaldehyde (control group), respectively. Data are presented as the mean ± S.E. $n = 6 - 12$; $^{\#\#}P < 0.01$, with respect to the control group; $^{*}P < 0.05$ and $^{**}P < 0.01$ with respect to the 200 μmol/L acetaldehyde group.

morphology became blurred after being treated by 200 μmol/L acetaldehyde (Figure 4(b)). L-Cys improved the cell morphology at low (Figure 4(c)), medium (Figure 4(d)), and high concentrations (Figure 4(e)).

The control group under SEM had clear cellular contour and smooth surfaces with microvilli (Figure 4(f)). Under 200 μmol/L acetaldehyde, the microvilli on the cellular surface disappeared and exhibited a blurred contour, an uneven surface, visible tooth-like projections, and cellular crenation (Figure 4(g)). Low, medium, and high L-Cys concentration groups had significantly improved cellular edges or surfaces and overall morphology when compared with the acetaldehyde group (Figures 4(h), 4(i), and 4(j)). This finding reveals that L-Cys had a protective effect on cells damaged by acetaldehyde.

A living cell nucleus generally has evenly dispersed blue fluorescence under fluorescence microscopy, whereas cell nuclei nuclear apoptosis has dense and massive fluorescence stained particles, and blue fluorescence is brighter than that of normal cells. In apoptotic cells, numerous fluorescent scattered DNA fragments are also seen. As indicated by Figures 4(k) to 4(o), A549 cells in the control group have evenly dispersed blue fluorescence under fluorescence microscopy (Figure 4(k)). After being treated with 200 μmol/L acetaldehyde, apoptotic cells exhibit morphological changes, such as nuclear fragmentation and pyknosis. Nuclear fragmentation, pyknosis, and other apoptotic manifestations are significantly reduced in low, medium, and high L-Cys concentration groups. The medium concentration group has shown a protective effect, which demonstrates that acetaldehyde could induce cellular apoptosis, whereas L-Cys protected cells to some degree (Figures 4(m), 4(n), and 4(o)).

3.5. Apoptosis of A549 Cells by Flow Cytometry. A549 cells with the same treatment were detected through PI single nucleus staining for flow cytometry to analyze cell counts in the different cell cycle phases and in apoptosis. As shown in Figure 5, the cells of the acetaldehyde group in the G1 phase increased from 77.14% to 82.43%, whereas those in the S phase decreased from 16.56% to 11.32% compared with the control group. This finding shows that acetaldehyde can block cells in the G1 phase from proceeding to the S phase, thus inhibiting cellular growth. Compared with the acetaldehyde group, low, medium, and high L-Cys concentration groups have significantly reduced percentages of cells in the G1 phase and increased the proportions of cells in the S phase. Therefore, L-Cys can protect cells from damage caused by acetaldehyde. The cells in the hypodiploid phase before G1 (Sub-G1) were

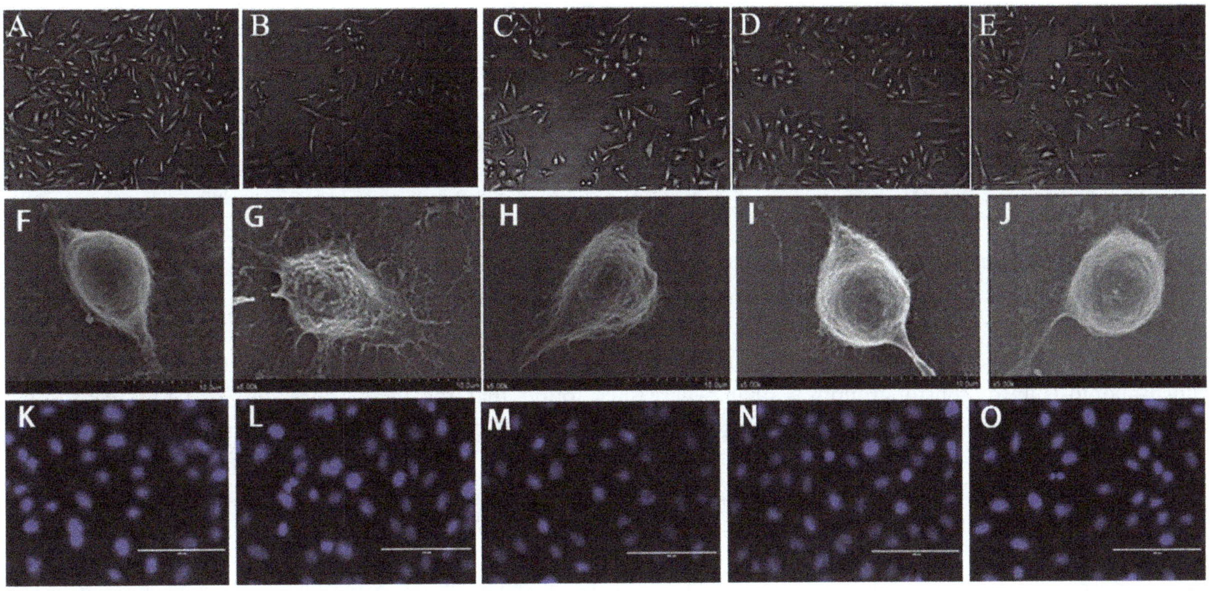

FIGURE 4: Observation of A549 cell morphology under inverted biological microscope (×100), scanning electron microscope (×5000), and fluorescence microscope (×400). (a, f, k) Control; (b, g, l) 200 μmol/L acetaldehyde; (c, h, m) 40 μmol/L L-Cys + 200 μmol/L acetaldehyde; (d, i, n) 160 μmol/L L-Cys + 200 μmol/L acetaldehyde; (e, j, o) 640 μmol/L L-Cys + 200 μmol/L acetaldehyde.

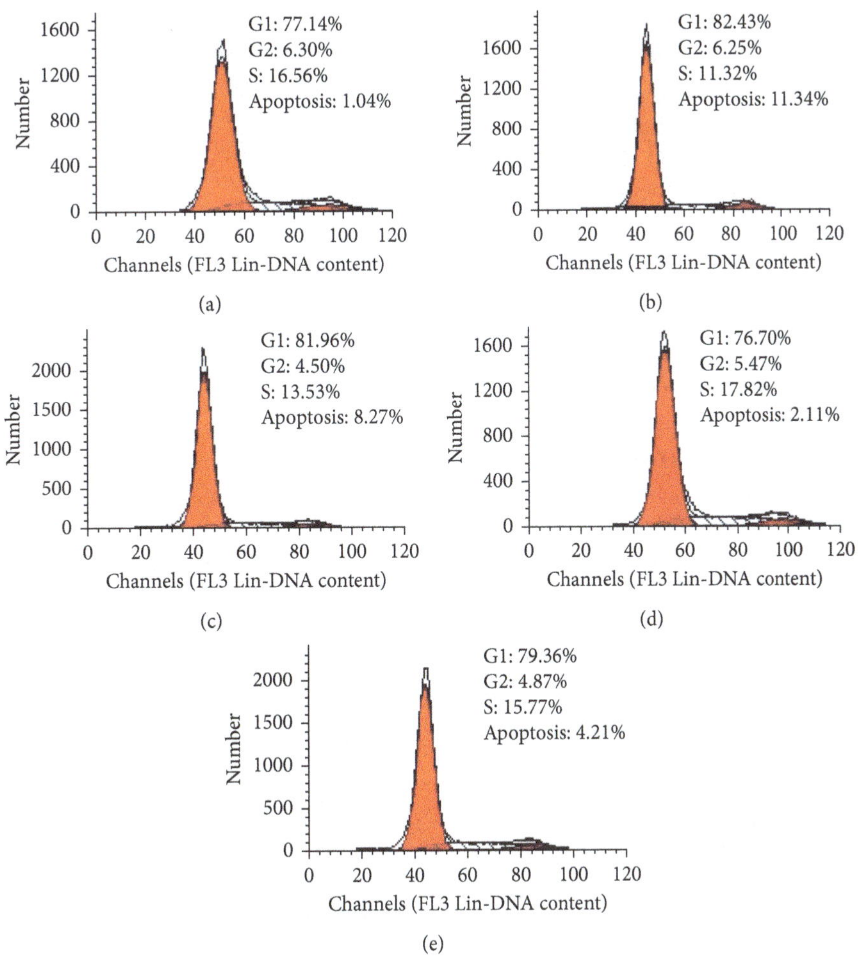

FIGURE 5: Influence of L-Cys on cell cycle and the apoptosis rate of A549 cell damaged by acetaldehyde. (a) Control; (b) 200 μmol/L acetaldehyde; (c) 40 μmol/L L-Cys + acetaldehyde; (d) 160 μmol/L L-Cys + acetaldehyde; (e) 640 μmol/L L-Cys + acetaldehyde.

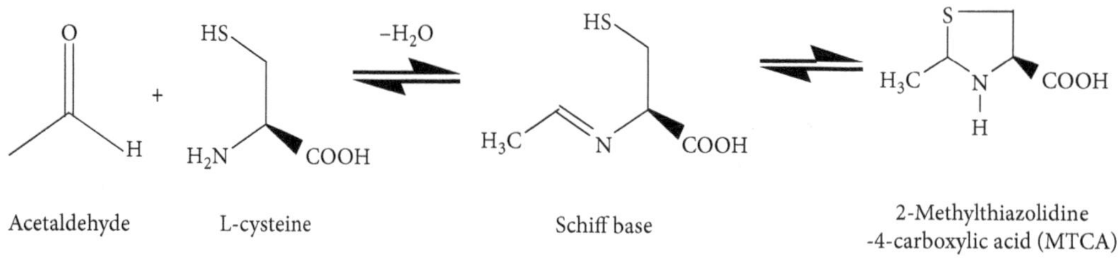

FIGURE 6: Chemical structural formula of L-Cys reacts with acetaldehyde to generate 2-methylthiazolidine-4-carboxylic acid (MTCA).

identified to be apoptotic cells. Results show that the number of cellular apoptosis in the control group was small, whereas the apoptotic rate of the cells in the acetaldehyde group increased from 1.04% to 11.34% (Figure 5(b)). The number of cellular apoptosis in the L-Cys groups declined the most with the medium L-Cys concentration group.

4. Discussion

Reactive oxygen species (ROS) are constantly produced and eliminated from various cells as products of metabolism. Organisms can normally regulate ROS to control cell apoptosis and proliferation [19, 20]. At low physiological levels, ROS function as "redox messengers" in intracellular signaling and regulation, whereas excess ROS induce the oxidative modification of cellular macromolecules, inhibit protein function, and promote cell death [21]. Cell growth by ROS and their role as a messenger in various signaling pathways is tightly regulated because once the amount of cellular ROS dramatically changes, the variation of cellular activity may be predicted [19, 21].

ROS are largely generated in the mitochondria. Bare mitochondrial DNA (mtDNA) is prone to attacks from free oxygen radicals. Consequently, structure and function can be decayed, which is why mitochondrial damage is always found in different degenerative chronic diseases [22, 23]. COII is the cytochrome *c* active center coded by mtDNA, and its structural and functional changes leading to damage of the mitochondrial respiratory chain are connected to the release of cytochrome *c* [24]. The expression of COII mRNA suggests that 200 μmol/L acetaldehyde can significantly reduce the expression of cellular COII mRNA. Two possible pathways exist relative to the damage to mitochondrial COII: one is to cause the COII structure to change and release cytochrome *c* to increase the oxidative damage of the mitochondrion and cause cells to continue apoptosis, and the other is to cause oxidative damage to mitochondrial DNA to reduce the expression of cytochrome oxidase.

L-Cys reacts with acetaldehyde to generate imine (Figure 6), which is harmless to the human body and can be directly excreted, making L-Cys protective to organisms [20, 25]. This research demonstrates that L-Cys significantly improves LDH activity, MDA contents, and other antioxidant indexes in acetaldehyde-damaging cells and significantly decreased the apoptosis rate [16]. In this article, we found that the gene expression of COII mRNA in low, medium, and high L-Cys concentration groups significantly increased compared with that of the 200 μmol/L acetaldehyde group. Thus, L-Cys is significantly protective to acetaldehyde-damaging cells.

Acetaldehyde, as an external and intracellular oxidant, can cause direct oxidative damage to biological macromolecules [26, 27]. Smoking, alcoholism, and air pollution increases the level of acetaldehyde in the human body. If a large amount of acetaldehyde produced in vivo cannot be removed promptly, the dynamic pro-oxidative system and antioxidant system of the organism lose balance, such that tissue or cells are frequently oxidatively damaged [28, 29]. Alcohol in the liver is rapidly metabolized into acetaldehyde, and acetate is the main source for carbon systemic circulation. So, there is no toxic effect if the metabolism functions normally. However, in the case of large consumption of alcohol, not only acetaldehyde can be accumulated but also acrolein through oxidation of lipid can be increased in liver. The acrolein has been recently regarded as a pathogenic mediator for many liver diseases [30]. Our results indicate that acetaldehyde can induce cellular damage, reduce GSH-Px activity, and cause apoptosis, whereas L-Cys increases GSH-Px activity. Combined with previous research results, we conclude that L-Cys is a powerful protective compound against the oxidative damage induced by acetaldehyde, which is possibly achieved by increasing the antioxidant environment in cells.

Conflicts of Interest

The authors declare that there are no conflicts of interest regarding the publication of this paper.

Authors' Contributions

Zhenjing Li and Xiaohong Zhang have equally contributed to this work.

Acknowledgments

This project was supported by an initial funding from the Tianjin city government for "1000 talents plan" program to C. L.

References

[1] W. Summer, "Odour Pollution of Air Causes and Control," in *Chemical and Process Engineering Series*, Leonard Hill, London, UK, 1971.

[2] M. J. Ellenhorn, *Ellenhorn's Medical Toxicology: Diagnosis and Treatment of Human Poisoning*, Williams and Wilkins, Baltimore, MD, USA, 2nd edition, 1997.

[3] E. L. William, "Acetaldehyde," *Journal of Chemical Health and Safety*, vol. 16, no. 5, pp. 43-44, 2009.

[4] K. Hori, S. Miyamoto, Y. Yukawa, M. Muto, T. Chiba, and T. Matsuda, "Stability of acetaldehyde-derived DNA adduct in vitro," *Biochemical and Biophysical Research Communications*, vol. 423, no. 4, pp. 642–646, 2012.

[5] H. S. Yu, T. Oyama, T. Isse et al., "Formation of acetaldehyde-derived DNA adducts due to alcohol exposure," *Chemico-Biological Interaction*, vol. 188, no. 3, pp. 367–375, 2010.

[6] C. J. Eriksson, "The role of acetaldehyde in the actions of alcohol (update 2000)," *Alcoholism Clinical Experimental Research*, vol. 25, no. 1, pp. 15S–32S, 2010.

[7] X. S. Deng and R. A. Deitrich, "Putative role of brain acetaldehyde in ethanol addiction," *Current Drug Abuse Reviewse*, vol. 1, no. 1, pp. 3–8, 2008.

[8] L. Tampier and M. E. Quintanilla, "Effect of acetaldehyde on acute tolerance and ethanol consumption in drinker and nondrinker rats," *Journal of Studies on Alcohol*, vol. 63, no. 3, pp. 257–262, 2002.

[9] A. I. Cederbaum, "Alcohol metabolism," *Clinics in Liver Disease*, vol. 16, no. 4, pp. 667–685, 2012.

[10] F. A. Carey and R. J. Sundberg, *Advanced Organic Chemistry Part A: Structure and Mechanisms*, Springer, Heidelberg, Germany, 2007.

[11] H. T. Nagasawa, D. J. Goon, W. P. vMuldoon, and R. T. Zera, "2-Substituted thiazolidine-4(R)-carboxylic acids as prodrugs of L-cysteine. Protection of mice against acetaminophen hepatotoxicity," *Journal of Medicinal Chemistry*, vol. 27, no. 5, pp. 591–596, 1984.

[12] A. Kartal, J. Hietala, I. Laakso et al., "Formulation and in-vivo evaluation of L-cysteine chewing gums for binding carcinogenic acetaldehyde in the saliva during smoking," *Journal of Pharmacy and Pharmacology*, vol. 59, no. 10, pp. 1353–1358, 2007.

[13] A. Kartal-Hodzic, T. Marvola, M. Schmitt, K. Harju, M. Peltoniemi, and M. Sivén, "Permeability and toxicity characteristics of L-cysteine and 2-methyl-thiazolidine-4-carboxylic acid in Caco-2 cells," *Pharmaceutical Development and Technology*, vol. 18, no. 6, pp. 1288–1293, 2013.

[14] V. Salaspuro, J. Hietala, P. Kaihovaara, L. Pihlajarinne, M. Marvola, and M. Salaspuro, "Removal of acetaldehyde from saliva by a slow-release buccal tablet of L-cysteine," *International Journal of Cancer*, vol. 97, no. 3, pp. 361–364, 2002.

[15] A. T. Peana, G. Muggironi, G. R. Fois, M. Zinellu, D. Sirca, and M. Diana, "Effect of (L)-cysteine on acetaldehyde self-administration," *Alcohol*, vol. 46, no. 5, pp. 489–497, 2012.

[16] A. T. Peana, A. R. Assaretti, G. Muggironi, P. Enrico, and M. Diana, "Reduction of ethanol-derived acetaldehyde induced motivational properties by L-cysteine," *Alcoholism Clinical and Experimental Research*, vol. 33, no. 1, pp. 43–48, 2009.

[17] K. Yamashita, M. Noguchi, A. Mizukoshi, and Y. Yanagisawa, "Acetaldehyde removal from indoor air through chemical absorption using L-cysteine," *International Journal of Environmental Research and Public Health*, vol. 7, no. 9, pp. 3489–3498, 2010.

[18] C. Luo and X. Zhang, "Elimination of acetaldehyde and antioxidation of A549 intracellular environment by L-cysteine," *Food Science*, vol. 34, pp. 247–250, 2013.

[19] P. D. Ray, B. W. Huang, and Y. Tsuji, "Reactive oxygen species (ROS) homeostasis and redox regulation in cellular signaling," *Cellular Signalling*, vol. 24, no. 5, pp. 981–990, 2012.

[20] S. Nakajima and M. Kitamura, "Bidirectional regulation of NF-κB by reactive oxygen species: a role of unfolded protein response," *Free Radical Biology and Medicine*, vol. 65, pp. 162–174, 2013.

[21] M. L. Circu and T. Y. Aw, "Reactive oxygen species, cellular redox systems, and apoptosis," *Free Radical Biology and Medicine*, vol. 48, no. 6, pp. 749–762, 2010.

[22] M. Filosto, M. Scarpelli, M. S. Cotelli et al., "The role of mitochondria in neurodegenerative diseases," *Journal of Neurology*, vol. 258, no. 10, pp. 1763–1774, 2011.

[23] A. J. Kowaltowski, R. F. Castilho, and A. E. Vercesi, "Mitochondria and reactive oxygen species," *Free Radical Biology and Medicine*, vol. 15, no. 4, pp. 333–343, 2009.

[24] P. Caroppi, F. Sinibaldi, L. Fiorucci, and R. Santucci, "Apoptosis and human diseases mitochondrion damage and lethal role of released cytochrome *c* as proapoptotic protein," *Current Medicinal Chemistry*, vol. 16, no. 31, pp. 4058–4065, 2009.

[25] H. Sprince, C. M. Parker, G. G. Smith, and L. J. Gonzales, "Protection against acetaldehyde toxicity in the rat by L-cysteine, thiamin and L-2-methylthiazolidine-4-carboxylic acid," *Agents Actions*, vol. 4, no. 2, pp. 125–130, 1974.

[26] B. E. Farfán Labonne, M. Gutiérrez, L. E Gómez-Quiroz et al., "Acetaldehyde-induced mitochondrial dysfunction sensitizes hepatocytes to oxidative damage," *Cell Biology and Toxicology*, vol. 25, no. 6, pp. 599–609, 2009.

[27] T. Uemura, Y. Tanaka, K. Higashi et al., "Acetaldehyde-induced cytotoxicity involves induction of spermine oxidase at the transcriptional level," *Toxicology*, vol. 310, pp. 1–7, 2013.

[28] J. B. Hoek and J. G. Pastorino, "Ethanol, oxidative stress, and cytokine-induced liver cell injury," *Alcohol*, vol. 27, no. 1, pp. 63–68, 2002.

[29] M. C. Gutierrez-Ruiz, L. E. Gomez Quiroz, E. Hernandez et al., "Cytokine response and oxidative stress produced by ethanol, acetaldehyde and endotoxin treatment in HepG2 cells," *Israel Medical Association Journal*, vol. 3, no. 2, pp. 131–136, 2001.

[30] W. Y. Chen, J. Zhang, S. Ghare, S. Barve, C. Mcclain, and S. Joshibarve, "Acrolein is a pathogenic mediator of alcoholic liver disease and the scavenger hydralazine is protective in mice," *Cellular and Molecular Gastroenterology and Hepatology*, vol. 2, no. 5, pp. 685–700, 2016.

3

Five Regioisomers of Dimethyl Dodecahedrane Derivatives: A Hybrid DFT B3LYP Study

Seol Lee, YongJae Cho, and Kee Hag Lee

Department of Chemistry and Nanoscale Sciences and Technology Institute, Wonkwang University, Iksan, Jeonbuk 570-749, Republic of Korea

Correspondence should be addressed to Kee Hag Lee; khlee@wku.ac.kr

Academic Editor: Arturo Espinosa Ferao

The hybrid density functional (B3LYP/6-31G(d, p)) method was used to understand why 1,2-dimethyl dodecahedrane has not been reported yet. From the direct dimethyl substitution of the dodecahedrane cage, we could have five $C_{20}H_{18}(CH_3)_2$ derivatives which were geometrically optimized without constraints. The results suggest that 1,16-dimethyl dodecahedrane derivative is the most stable, whereas the 1,2-dimethyl derivative is the most unstable; this implies that the distortion due to steric hindrance interactions would be a crucial effect on the relative energies of the dimethyl dodecahedranes. It would be disadvantaged thermodynamically for 1,2-dimethyl derivative that, in the experimental results, was not synthesized yet. The LUMO of each of the derivatives was equivalently delocalized over the void within the cage, implying that dimethyl derivatives are able to encapsulate atoms at the center of the cage. The HOMO was limitedly delocalized on the cage. The characteristics of the HOMO of derivatives show three patterns, implying that each derivative might undergo one of three entirely different sets of characteristic chemical reactions with electrophilic reagents.

Dedicated to Professor Roald Hoffmann on the occasion of his 80th birthday

1. Introduction

The geometry of fullerenes obeys the isolated-pentagon rule. However, this rule cannot be satisfied for the smallest fullerene C_{20}, which contains twelve 5-membered rings (MRs) and zero 6-MRs. Therefore, the C_{20} fullerene would be highly reactive owing to the fusion of strained 5-MRs. Furthermore, this strain could be removed through reactions of C_{20} that form sp^3 bonds. By using density functional theory method, the structures and electronic states of alkyl radicals (R, C_nH_{2n+1}, $n = 1–4$) added to C_{20} fullerenes were investigated [1]. Also it was proposed that thiyl radical-promoted polycyclization to form dodecahedrane would be exergonic [2].

Density functional calculations of the reactions CX_3 radical + $C_{20}H_{20}$ (X = H, F, Cl, and Br) based on two pathways (H-displacement and H-abstraction from $C_{20}H_{20}$) showed that CH_3 radical prefers H-abstraction from $C_{20}H_{20}$ while F-displacement is favorable for the reaction of CF_3 radical with the fullerene. Exothermic characters of H-abstraction with the lower potential barrier indicate that the H-abstraction would dominate the reaction $CCl_3^{\bullet} + C_{20}H_{20}$ while endothermic characters with high potential barrier heights for two reaction pathways of $C_{20}H_{20}$ with $CBr_3^{\bullet}$ indicate unfavorable thermodynamically and kinetically pathways for these reactions [3].

Using the PBE0/cc-pVDZ quantum chemical method, dipole moments, polarizabilities, the first hyperpolarizabilities for 1,2-diethynyldodecahedrane $C_{20}H_{18}(CCH)_2$ and diethynylmethanododecahedrane $C_{20}H_{18}C(CCH)_2$ including two neighboring ethynyl groups in vacuum and in tetrahydrofuran (THF) were reported [4].

A dodecahedrane derivative, namely, the $\mathbf{D_{3d}}$ symmetric 1,16-dimethyl derivative, has been synthesized experimentally [5–8]. Since then, the synthesis of methyl-substituted dodecahedrane was also confirmed by X-ray analysis. Moreover, Pólya's theorem was applied to the enumeration of substituted dodecahedrane isomers [9]. The total synthesis of dodecahedrane was confirmed by the single-peak nature of the ^{1}H and ^{13}C NMR spectra [10], as well as the X-ray spectra [11, 12]. Methyl dodecahedrane, 1,4-dimethyl ($\mathbf{C_s}$), 1,6-dimethyl ($\mathbf{C_{2v}}$), 1,16-dimethyl ($\mathbf{D_{3d}}$) dodecahedranes, and trimethyl dodecahedrane have subsequently been reported [13–15]. However, at present no 1,2-dimethyl ($\mathbf{C_{2v}}$) and 1,7-dimethyl ($\mathbf{C_2}$) dodecahedranes have been reported, yet.

The structures of dodecahedrane and its derivatives were studied at the level of molecular mechanics calculations [16, 17], INDO calculations [18], and hybrid density functional B3LYP calculations [19]. Recently the relative energies, electronic properties, and atomic structures of disubstituted $C_{20}H_{18}X_2$ regioisomers (X = F, Cl, Br, or OH) at the level of B3LYP/6-31G(d,p) have been determined [20]. Here the predominant electronic configurations of the regioisomers with all substituents X which are higher electron negativity than carbon are not so different, but the HOMO and LUMO maps for five regioisomers are not similar, implying that different exohedral complexes may undergo a distinct set of characteristic chemical reactions.

However, to the best of our knowledge, no the first principle calculations of the full optimization of the five dimethyl dodecahedranes, $C_{20}H_{18}(CH_3)_2$ derivatives using B3LYP/6-31G(d,p), have been reported. Such calculations would be required in order to understand, in the point of energetics, why only three derivatives have been synthesized. Accordingly, in this study we determine the relative energies, electronic properties, and atomic structures of five derivatives with the same spin state, following the methods of Hwang et al. 2012 [20]. Also it is interesting to see the effect of the methyl substituents on the frontier orbitals, because carbon atoms have the same electronegativity in the substituents and the cage.

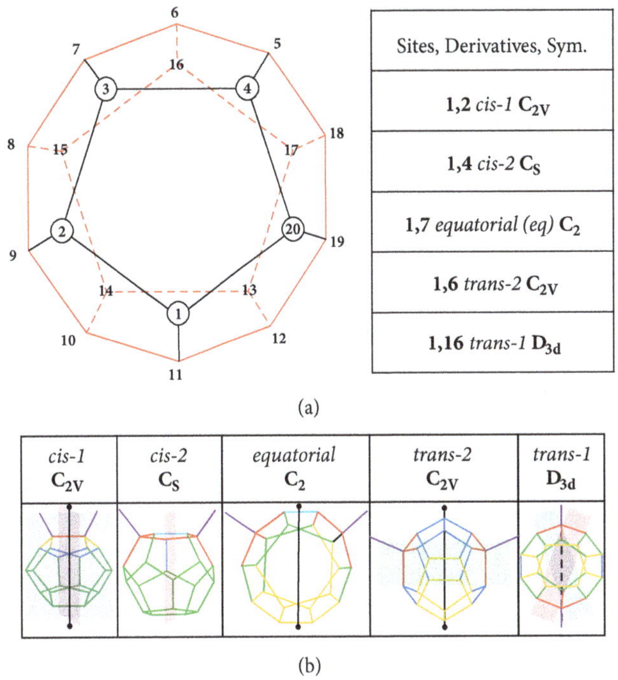

FIGURE 1: (a) The dodecahedrane and its numbering system (hydrogen atoms deleted for clarity). (b) The dimethyl substituted dodecahedrane derivatives $C_{20}H_{18}(CH_3)_2$.

2. Calculations

In this study, hybrid density functional theory (DFT) with Becke's three-parameter hybrid method and the Lee-Yang-Parr exchange-correlation functional theory (B3LYP) [21–28] were used to optimize the geometries of the $C_{20}H_{20}$ and five dimethyl derivatives $C_{20}H_{18}(CH_3)_2$, as shown in Figure 1. Furthermore, the electron basis set 6-31G(d,p) was used [25]. We fully optimized the geometries of all the $C_{20}H_{18}(CH_3)_2$ derivatives using the Gaussian 2003 B.04 and Gaussian 2009 D01 package suites [26, 27]. We obtain highly accurate geometries, by using the convergence criterion with tight optimization and an ultrafine pruned (99,590) grid. Here, we analyzed the relative energies as well as the HOMO and LUMO orbitals of the derivatives [20, 28].

To attest the veracity of the B3LYP/6-31G(d,p) relative energy values, as shown in Supplementary Material (see Supplementary Material available online at https://doi.org/10.1155/2017/5494038), we calculated single-point energies at the levels of M06-2X [29], PBE [30], B98 [31], B97D3 [32, 33], B3LYPD3 [21, 22, 33], PBED3 [30, 33], and M06-2XD3 [29, 33] with 6-31G(d,p) basis set on the optimized geometries and using the Zero Point Energy-Correction (ZPEC) already obtained at the B3LYP/6-31G(d) level, using Gaussian 2009 D01 package suites [27]. Here D3 represents the dispersion-corrected functional with the D3 version of Grimme's dispersion with Becke-Johnson damping [33]. Also using the post-HF MP2 [34]/def2-TZVPP [35]//B3LYP/6-31G(d) level, we computed the energies as single-point calculations with the ZPEC already obtained at the B3LYP/6-31G(d) level.

3. Results and Discussion

By performing calculations for the full optimization of the atomic structures of the five $C_{20}H_{18}(CH_3)_2$ derivatives using B3LYP/6-31G(d,p) without any constraints, we determined the relative energies of the derivatives in Table 1 with the symmetries as shown in Figure 1.

The geometrical structures and the frontier orbitals (HOMOs and LUMOs) of the five $C_{20}H_{18}(CH_3)_2$ derivatives are shown in Figure 2. The calculated C-C bond length in the cage of neutral $C_{20}H_{20}$ with I_h symmetry is 1.556 Å, which is in the range of the experimental result (i.e., 1.555 ± 0.001 Å) [36].

In the five dimethyl derivatives whose geometries were fully optimized, the relative Zero Point Energy-Correction (ZPEC) energies of dimethyl dodecahedranes are arranged in the following order: *1,16-dimethyl derivative* (*trans*-1, $\mathbf{D_{3d}}$) ≈ *1,7-dimethyl derivative* (*equatorial*, $\mathbf{C_2}$) ≈ *1,6-dimethyl derivative* (*trans*-2, $\mathbf{C_{2v}}$) ≈ *1,4-dimethyl derivative* (*cis*-2, $\mathbf{C_s}$)

TABLE 1: The relative energies (rel. E), relative ZPEC energies (rel. ZPEC), relative enthalpy (rel. ΔH^0), and relative Gibbs free energies (rel. ΔG^0) in units of meV and distance (**d**, Å) between the two methyls of the five $C_{20}H_{18}(CH_3)_2$ derivatives using B3LYP/6-31G(d, p) calculations[a].

Derivative	Rel. E	Rel. ZPEC	ΔH^0	ΔG^0	**d**
***1,2**-cis-1*	112.7	121.6	3	9	2.88
***1,4**-cis-2*	2.9	2.5	−3	−5	4.32
***1,7**-eq*	0.8	0.4	−2	−2	6.11
***1,6**-trans-2*	1.9	0.2	−4	−7	6.99
***1,16**-trans-1*	0.0	0.0	0	0	7.48

[a]The relative value of each derivative is the energy minus the energy of *trans-1* isomer.

< *1,2-dimethyl derivative* (*cis*-1, $\mathbf{C_{2v}}$). The distance between dimethyl groups in the 1,16-dimethyl derivative with the lowest energy is the longest among these. In 1,2-dimethyl derivative, the two substituents are located on the surface of the $C_{20}H_{20}$ fullerene within a distance that is shorter than the van der Waals distance (i.e., 3.4 Å).

The increasing order of relative ZPEC energies in them is the same to the decreasing order of the distance between the two methyl groups of derivatives. As shown in Figure 3 and Supplementary Materials, the veracity of the B3LYP/6-31G(d,p) relative energy values is attested at the levels of M06-2X, PBE, B98, B97D3, B3LYPD3, PBED3, and M06-2XD3, and MP2 in the point of view that the relative energy of *cis-1* is about 10 kJ/mol higher than the others. The difference of relative energies of four isomers except *cis-1* is less than 0.7 kJ/mol in the levels of DFT but within 2 kJ/mol in the MP2 method.

Our result suggests that both the strength of the cage distortion and the van der Waals force between the pair of methyl groups of derivatives are important factors that affect the position of exohedral dimethyl substitution on the dodecahedrane cage. And it could be suggested that in 1,2-dimethyl derivative, steric impedance brought on by the nearby two methyl groups is very crucial. It is interesting to compare the results above with the case of $C_{20}H_{18}X_2$ (X = F, Cl, Br, OH). In this case, except for the dihydroxy derivative, 1,2-disubstituted derivatives have the highest energy whereas 1,16-derivatives have the lowest energy; however, in the dihydroxy derivatives, the increasing order is reversed [20, 37].

Here our calculations concerning the dimethyl dodecahedrane derivatives could explain the fact why three dimethyl dodecahedrane derivatives were reported in X-ray and NMR experiments [7, 15], but the 1,2-dimethyl dodecahedrane derivative has not been synthesized yet. And it could be suggested that the more stable derivative in the viewpoint of energetics could be synthesized easier than others.

In the following substitution reaction, the thermal reaction energies are endothermic:

$$C_{20}H_{20} + 2CH_3 \longrightarrow C_{20}H_{18}\left(CH_3\right)_2 + 2H \quad (1)$$

where 2.48, 2.86, 2.56, 3.78, and 5.79 eV of the thermal reaction energies at 298.15 K and 1 atm are absorbed for *1,16-dimethyl derivative* (*trans*-1, $\mathbf{D_{3d}}$), *1,7-dimethyl derivative* (*equatorial*, $\mathbf{C_2}$), *1,6-dimethyl derivative* (*trans-2*, $\mathbf{C_{2v}}$), *1,4-dimethyl derivative* (*cis*-2, $\mathbf{C_s}$), and *1,2-dimethyl derivative* (*cis*-1, $\mathbf{C_{2v}}$), respectively. The Gibbs free energies are 10.285, 10.278, 10.283, 10.280, and 10.294 eV for *1,16-dimethyl derivative* (*trans*-1, $\mathbf{D_{3d}}$), *C 1,6-dimethyl derivative* (*trans*-2, $\mathbf{C_{2v}}$), *1,7-dimethyl derivative* (*equatorial*, $\mathbf{C_2}$), *1,4-dimethyl derivative* (*cis*-2, $\mathbf{C_s}$), and *1,2-dimethyl derivative* (*cis*-1, $\mathbf{C_{2v}}$), respectively.

Even though the very small variation of Gibbs free energy is not sufficient to decide which structure should be totally experimentally disfavored, it could suggest that, in the point of thermodynamics, a synthesis of *cis-1* isomer might be a little difficult compared to the others. In the experimental results, dodecahedrane, methyl, and dimethyl derivatives are synthesized [12], but *cis-1* and equatorial isomers are not synthesized yet.

Figure 2 shows the HOMOs and LUMOs for the directed dimethyl derivatives of the dodecahedrane cage. When we calculate the energy gaps from the HOMO/LUMO energy levels of the derivatives, $C_{20}H_{20}$ and $C_{20}H_{18}(CH_3)_2$ have slightly different energy gaps. Here the HOMOs of the $C_{20}H_{18}(CH_3)_2$ derivatives with the same number of symmetry elements are very similar, but the HOMOs of derivatives with the different number of symmetry elements are different, despite the fact that the LUMOs among different geometric structures are identical. This suggests that each derivative undergoes one of three distinct sets of characteristic chemical reactions only for the HOMO. It is interesting to see that they are different to the case of disubstituted $C_{20}H_{18}X_2$ regioisomers (X = F, Cl, Br, or OH) which have five distinct set of reactions for both the HOMO and the LUMO [20].

Figure 2 shows the effect of the disubstituents on the frontier orbitals of $C_{20}H_{20}$. The HOMO and LUMO energies of five $C_{20}H_{18}(CH_3)_2$ derivatives are higher than the HOMO and LUMO levels of $C_{20}H_{20}$ (HOMO: −7.09 eV; LUMO: 0.90 eV). Also the band gaps of dimethyl substituted derivatives are wider than the gaps of $C_{20}H_{20}$ and C_{20} (HOMO: −5.03 eV; LUMO: −3.10 eV). Moreover, the LUMO energies are positive and higher than that of $C_{20}H_{20}$, which implies that these could function less efficiently as electron acceptors than $C_{20}H_{20}$.

4. Conclusion

We obtained the fully optimized geometries of the $C_{20}H_{18}(CH_3)_2$ derivatives (the smallest fullerene disubstituents) without any constraints using the hybrid B3LYP/6-31G(d,p) density functional method. We observed that the relative energy of the $C_{20}H_{20}$ disubstituted derivatives is mostly dependent on the distance between the pair of two methyl groups. It is interesting to note that the 1,2-dimethyl derivatives is less stable than the 1,16-dimethyl derivative. This implies that, in the $C_{20}H_{18}(CH_3)_2$ derivatives, the effect of distortion due to the steric hindrance interactions on the relative energy would be crucial. Therefore, our calculations in the point of thermodynamic view could support the experimental results, wherein three dimethyl dodecahedrane derivatives except the 1,2-dimethyl dodecahedrane derivative

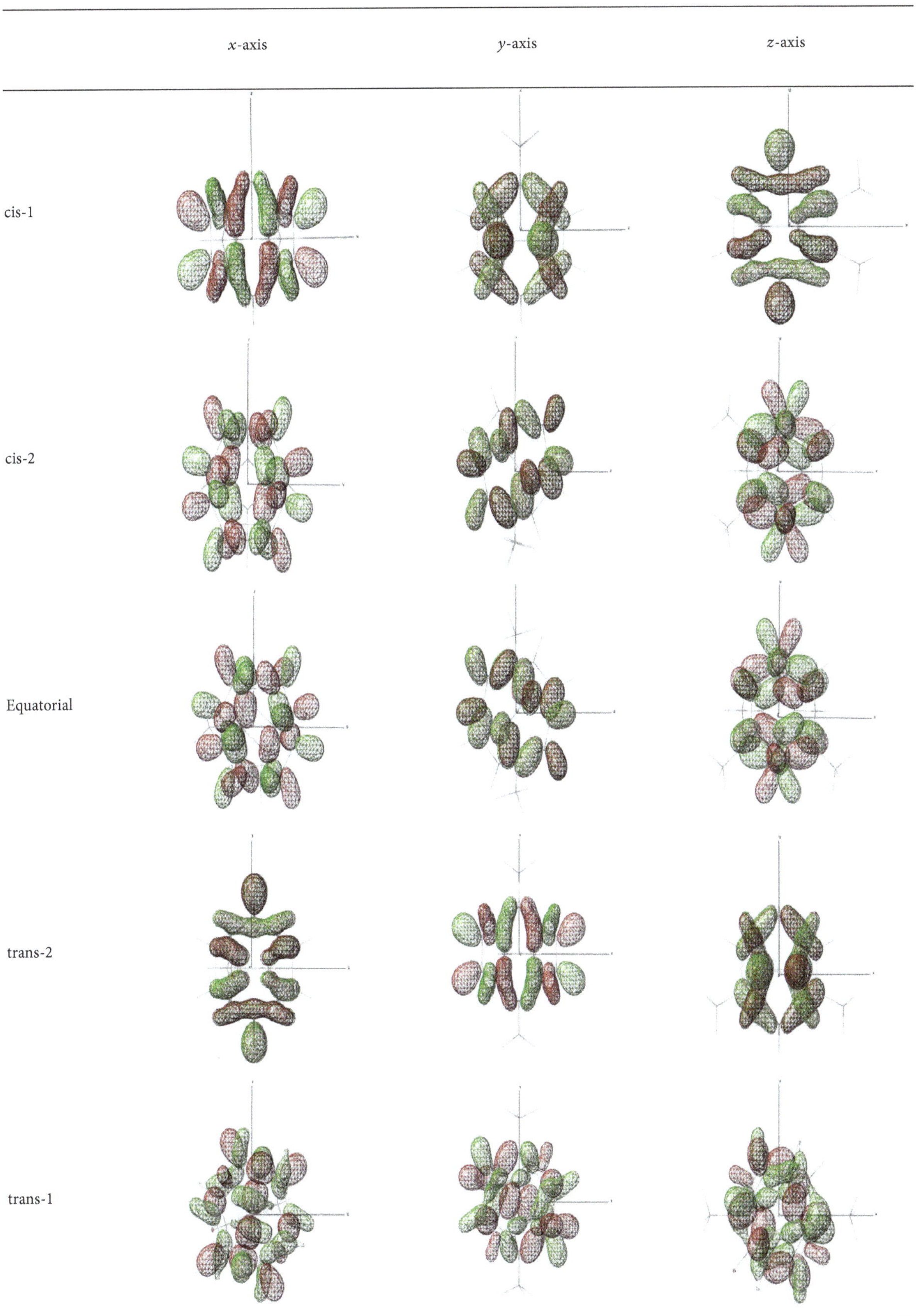

(a)

FIGURE 2: Continued.

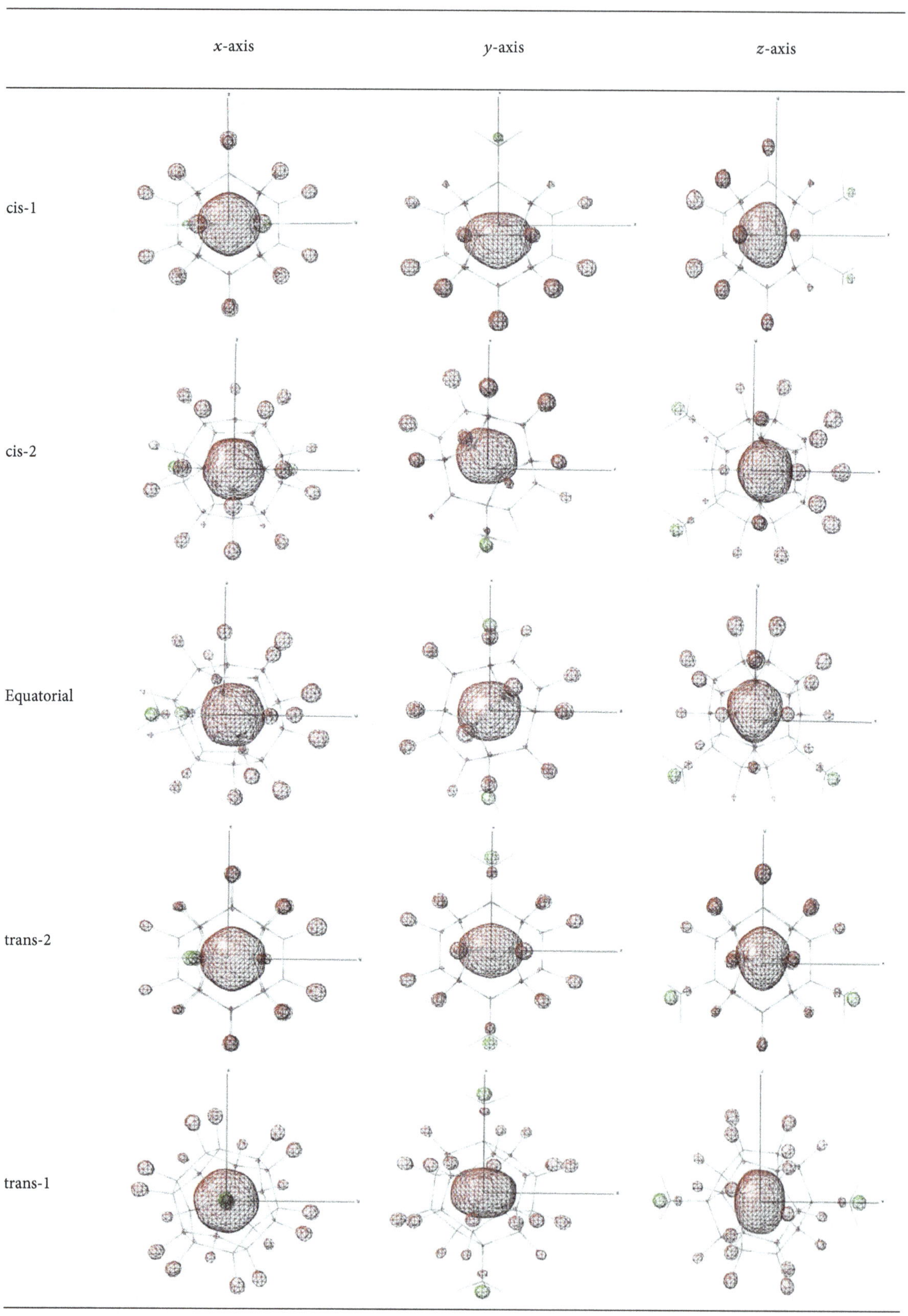

(b)

FIGURE 2: The (a) HOMOs and (b) LUMOs of the fully optimized geometries of the five $C_{20}H_{18}(CH_3)_2$ derivatives using B3LYP/6-31G(d,p) calculations. Here X-, Y-, and Z-axis represent the frontier orbitals, looking from x, y, and z direction, respectively. The HOMOs of $C_{20}H_{18}(CH_3)_2$ derivatives (isovalue = 0.05). *trans-2* (HOMO: −7.08 eV) and the others (HOMO: −7.05); $C_{20}H_{20}$ (HOMO: −7.09 eV). The LUMOs of $C_{20}H_{18}(CH_3)_2$ derivatives (isovalue = 0.05). Five dimethyl derivatives (LUMO: 0.98 eV); $C_{20}H_{20}$ (LUMO: 0.90 eV).

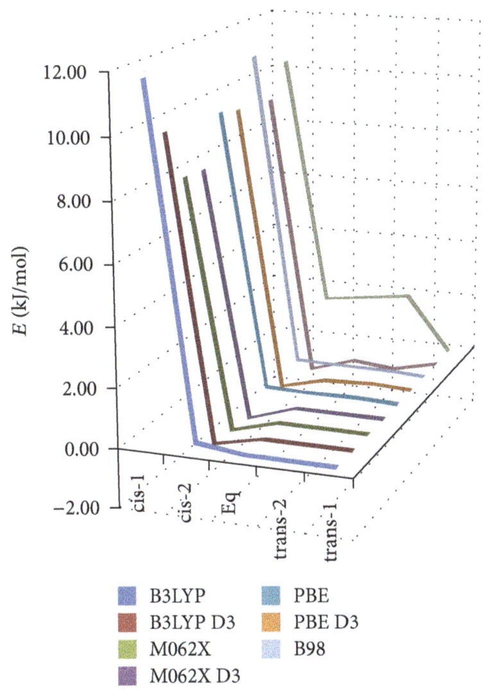

FIGURE 3: Trends in relative energies (unit: kJ/mol) for five isomers computed as single-point calculations at all different levels of DFT/6-31G(d, p)//B3LYP/6-31G(d, p) and MP2/def2-TZVPP//B3LYP/6-31G(d, p).

were preparative-synthetic and suggest that *1,7-dimethyl derivative* (*equatorial*, C_2) would be synthesized in the near future.

Conflicts of Interest

The authors declare that there are no conflicts of interest regarding the publication of this paper.

Acknowledgments

This study was supported by Wonkwang University in 2015.

References

[1] S. Abe, S. Kawano, Y. Toida et al., "Electronic states of alkyl-radical-functionalized C20 fullerene using density functional theory," *Japanese Journal of Applied Physics*, vol. 55, no. 3, Article ID 03DD03, 2016.

[2] S. C. Wang and D. J. Tantillo, "Viability of dodecahedrane-forming radical polycyclizations," *Organic & Biomolecular Chemistry*, vol. 15, no. 9, pp. 1976–1979, 2017.

[3] R. Ghafouri and F. Ektefa, "Theoretical study on the mechanism of reactions of CX3 radicals (X = H, F, Cl and Br) with C20H20 and C20F20 fullerenes," *Journal of Molecular Structure*, vol. 1127, pp. 296–302, 2017.

[4] S. G. Semenov and M. V. Makarova, "A quantum chemical study of diethynyl derivatives of dodecahedrane and buckminsterfullerene in vacuum and in tetrahydrofuran," *Optics and Spectroscopy*, vol. 118, no. 1, pp. 46–49, 2015.

[5] L. A. Paquette, D. W. Balogh, R. Usha, D. Kountz, and G. G. Christoph, "Crystal and molecular structure of a pentagonal dodecahedrane," *Science*, vol. 211, no. 4482, pp. 575-576, 1981.

[6] L. A. Paquette and D. W. Balogh, "An expedient synthesis of 1,16-dimethyldodecahedrane," *Journal of the American Chemical Society*, vol. 104, no. 3, pp. 774–783, 1982.

[7] G. G. Christoph, P. Engel, R. Usha, D. W. Balogh, and L. A. Paquette, "Comparative X-ray crystal structure analysis of C2v-dimethylmonosecododecahedrane and D3d-dimethyldodecahedrane. The remarkable consequences of a missing framework bond on molecular topology," *Journal of the American Chemical Society*, vol. 104, no. 3, pp. 784–791, 1982.

[8] L. A. Paquette, "Dodecahedrane—The chemical transliteration of Plato's universe (A Review)," *Proceedings of the National Academy of Sciences of the United States of America*, vol. 79, no. 14, pp. 4495–4500, 1982.

[9] L. A. Paquette, R. J. Ternansky, D. W. Balogh, and W. J. Taylor, "Total synthesis of a monosubstituted dodecahedrane. The methyl derivative," *Journal of the American Chemical Society*, vol. 105, no. 16, pp. 5441–5446, 1983.

[10] L. A. Paquette, R. J. Ternansky, D. W. Balogh, and G. Kentgen, "Total synthesis of dodecahedrane," *Journal of the American Chemical Society*, vol. 105, no. 16, pp. 5446–5450, 1983.

[11] H. Prinzbach and W.-D. Fessener, *Organic Synthesis*, Blackwell, Oxford, UK, 1987.

[12] M. Bertau, F. Wahl, A. Weiler et al., "From pagodanes to dodecahedranes - Search for a serviceable access to the parent (C20H20) hydrocarbon," *Tetrahedron*, vol. 53, no. 29, pp. 10029–10040, 1997.

[13] L. A. Paquette, Y. Miyahara, and C. W. Doecke, "Preparatively useful dehydrogenative method for dodecahedrane synthesis," *Journal of the American Chemical Society*, vol. 108, no. 7, pp. 1716–1718, 1986.

[14] I. Santas, D. W. Balogh, C. W. Doecke, A. G. Marshall, and L. A. Paquette, "Gas-phase basicities of cubane, dodecahedrane, and methyl-and 1,16-dimethyldodecahedrane as measured by Fourier transform ion cyclotron resonance mass spectrometry," *Journal of the American Chemical Society*, vol. 108, no. 26, pp. 8183–8185, 1986.

[15] L. A. Paquette and Y. Miyahara, "Directed alkyl substitution of the dodecahedrane nucleus. The 1,4-dimethyl, 1,6-dimethyl, and 1,4,16-trimethyl derivatives. Indanododecahedrane by stepwise dehydrogenation of a benzylated seco derivative," *The Journal of Organic Chemistry*, vol. 52, no. 7, pp. 1265–1272, 1987.

[16] N. L. Allinger, H. J. Geise, W. Pyckhout, L. A. Paquette, and J. C. Gallucci, "Structures of norbornane and dodecahedrane by molecular mechanics calculations (MM3), X-ray crystallography, and electron diffraction," *Journal of the American Chemical Society*, vol. 111, no. 3, pp. 1106–1114, 1989.

[17] F. Wahl, J. Wörth, and H. Prinzbach, "The Pagodane Route to Dodecahedranes: An Improved Approach to the $C_{20}H_{20}$ Parent Framework; Partial and Total Functionalizations—Does C_{20}-Fullerene Exist?" *Angewandte Chemie International Edition*, vol. 32, no. 12, pp. 1722–1726, 1993.

[18] J. M. Schulman and R. L. Disch, "Theoretical studies of dodecahedrane. 2. Dodecahedrane, inclusion compounds, and fluorine derivatives," *Journal of the American Chemical Society*, vol. 100, no. 18, pp. 5677–5681, 1978.

[19] D. Moran, F. Stahl, E. D. Jemmis, H. F. Schaefer III, and P. V. R. Schleyer, "Structures, stabilities, and ionization potentials of dodecahedrane endohedral complexes," *Journal of Physical Chemistry A*, vol. 106, no. 20, pp. 5144–5154, 2002.

[20] Y. G. Hwang, S. Lee, and K. H. Lee, "DFT Study for Substitution Patterns of $C_{20}H_{18}X_2$ Regioisomers (X = F, Cl, Br, or OH)," *Bulletin of the Korean Chemical Society*, vol. 33, no. 2, pp. 641–646, 2012.

[21] A. D. Becke, "Density-functional thermochemistry. III. The role of exact exchange," *The Journal of Chemical Physics*, vol. 98, no. 7, p. 5648, 1993.

[22] C. Lee, W. Yang, and R. G. Parr, "Development of the Colle-Salvetti correlation-energy formula into a functional of the electron density," *Physical Review B*, vol. 37, no. 2, pp. 785–789, 1988.

[23] P. J. Stephen, F. J. Devlin, C. F. Chablowski, and M. J. Frisch, "Ab Initio Calculation of Vibrational Absorption and Circular Dichroism Spectra Using Density Functional Force Fields," *The Journal of Physical Chemistry*, vol. 98, no. 45, pp. 11623–11627, 1994.

[24] R. H. Herteing and W. Koch, "On the parameterization of the local correlation functional. What is Becke-3-LYP?" *Chemical Physics Letters*, vol. 268, no. 5-6, pp. 345–351, 1997.

[25] W. J. Hehre, K. Ditchfield, and J. A. Pople, "Self-consistent molecular orbital methods. XII. Further extensions of gaussian-type basis sets for use in molecular orbital studies of organic molecules," *The Journal of Chemical Physics*, vol. 56, no. 5, pp. 2257–2261, 1972.

[26] M. J. Frisch, G. W. Trucks, H. B. Schlegel et al., *A Gaussian 03, B.04*, Gaussian, Inc., Pittsburgh, Pa, USA, 2003.

[27] M. J. Frisch, G. W. Trucks, H. B. Schlegel et al., *Gaussian 09W, Rev: D01*, Gaussian, Inc., Pittsburgh, Pa, USA, 2009.

[28] T. Clark and R. Koch, *The Chemist's Electronic Book of Orbitals*, Springer Berlin Heidelberg, Berlin, Germany, 1999.

[29] Y. Zhao and D. G. Truhlar, "The M06 suite of density functionals for main group thermochemistry, thermochemical kinetics, noncovalent interactions, excited states, and transition elements: two new functionals and systematic testing of four M06-class functionals and 12 other functionals," *Theoretical Chemistry Accounts*, vol. 120, no. 1, pp. 215–241, 2008.

[30] J. P. Perdew, K. Burke, and M. Ernzerhof, "Generalized gradient approximation made simple [Phys. Rev. Lett. 77, 3865 (1996)]," *Physical Review Letters*, vol. 78, no. 7, p. 1396, 1997.

[31] H. L. Schmider and A. D. Becke, "Optimized density functionals from the extended G2 test set," *The Journal of Chemical Physics*, vol. 108, no. 23, p. 9624, 1998.

[32] A. D. Beck, "Density-functional thermochemistry. V. Systematic optimization of exchange-correlation functionals," *The Journal of Chemical Physics*, vol. 107, no. 20, p. 8554, 1997.

[33] S. Grimme, S. Ehrlich, and L. Goerigk, "Effect of the damping function in dispersion corrected density functional theory," *Journal of Computational Chemistry*, vol. 32, no. 7, pp. 1456–1465, 2011.

[34] M. Head-Gordon, J. A. Pople, and M. J. Frisch, "MP2 energy evaluation by direct methods," *Chemical Physics Letters*, vol. 153, no. 6, pp. 503–506, 1988.

[35] F. Weigend and R. Ahlrichs, "Balanced basis sets of split valence, triple zeta valence and quadruple zeta valence quality for H to Rn: design and assessment of accuracy," *Physical Chemistry Chemical Physics*, vol. 7, no. 18, pp. 3297–3305, 2005.

[36] B. S. Hudson, D. G. Allis, S. F. Parker, A. J. Ramirez-Cuesta, H. Herman, and H. Prinzbach, "Infrared, Raman, and inelastic neutron scattering spectra of dodecahedrane: An I h molecule in T h site symmetry," *Journal of Physical Chemistry A*, vol. 109, no. 15, pp. 3418–3424, 2005.

[37] F. Wahl, A. Weiler, P. Landenberger et al., "Towards perfunctionalized dodecahedranes-en route to C20 fullerene," *Chemistry - A European Journal*, vol. 12, no. 24, pp. 6255–6267, 2006.

4

Effects of Neutralization, Decoloration, and Deodorization on Polycyclic Aromatic Hydrocarbons during Laboratory-Scale Oil Refining Process

Yuxiang Ma, Longkai Shi, Yulan Liu, and Qiyu Lu

College of Food Science and Technology, Henan University of Technology, Zhengzhou 450001, China

Correspondence should be addressed to Qiyu Lu; qiyulu7120@vip.sina.com

Academic Editor: Feng Xu

The influence of technological operations during oil refining process on polycyclic aromatic hydrocarbons (PAHs) in neutralized, bleached, and deodorized oils was investigated on the basis of laboratory-scale study. Under the best experimental conditions, benzo[a]pyrene decreased by 85.1%, 99.7%, and 40.8% in neutralized, bleached, and deodorized oils, respectively. Total of 16 analytes decreased by 55.7%, 87.5%, and 47.7%, respectively. Bleaching with activated charcoal was the most efficient procedure to reduce PAHs in crude oil. Neutralization had a modest influence on sixteen analytes; however, deodorization was only responsible for a slight decrease in the light PAHs and heavy PAHs contents. Data obtained in this study suggest that the use of activated carbon during oil refining process is highly recommended; moreover, these results provide a useful guidance for oil refining plant to reduce security risk and ensure the quality of the vegetable oil products.

1. Introduction

Edible vegetable oils consist of a complex mixture of glycerides (mainly triacylglycerols and partially diacylglycerols and monoacylglycerols), hydrocarbons, alcohols, phospholipids, free fatty acids, pigments, sterols, tocopherols, and volatile compounds, and so on [1, 2]. Usually, oil refining process is done to eliminate the unwanted minor components, which make oils not suitable for sale and consumption. Furthermore, the purification process is trying to reduce possible damage to the neutral oil as well as minimize refining loss [3, 4]. All compounds, which are detrimental to flavour, color, stability, and safety of the refined final oil products will be removed. They are primarily phospholipids, free fatty acids, pigments, trace metals, volatiles, persistent organic pollutants (POPs), and other impurities [1, 5].

Among various POPs, polycyclic aromatic hydrocarbons (PAHs) are a class of ubiquitous chemical contaminations in vegetable oils. Academically, they are divided into two parts, light PAHs (LPAHs, up to four fused rings) and heavy PAHs (HPAHs, more than four fused rings). The persistent nature of PAHs related to their great chemical stability leads to their significant accumulation in the food chain; moreover, their high liposolubility results in their occurrence in oils and fatty foods, particularly in edible oils. In 2005, China has set a limit of 10 μg/kg for benzo[a]pyrene (BaP) in edible vegetable oils for the sake of dietary safety associated with oil consumption [6]. Meanwhile, Commission Regulation Number 835/2011 legislated the maximum levels for PAHs in fats and oils for direct human consumption or ingredient for food products [7]. It established 2 μg/kg for BaP and 10 μg/kg for summed four PAHs (called PAH_4, including BaP, benzo[b]fluoranthene (BbFlu), benz[a]anthracene (BaA), and chrysene (Chr)) as the most suitable indicator of PAHs in foodstuffs. However, no countries or organizations worldwide have set the regulations and limits for the content of total PAHs in commercial oils, partly due to the fact that there are hundreds of these sorts of pollutants in the world, thus resulting in difficulty in separation, determination, and calculation. Previously, we developed and validated a liquid chromatography-atmospheric pressure photoionization-tandem mass spectrometry method for analyzing United

States Environmental Protection Agency (EPA) 16 PAHs in edible oils [8], and the occurrence as well as concentration of 16 analytes in eighty-five brands of vegetable oils in China was detected and statistically analyzed [9]. Now, investigations on technologies for the removal of PAHs from edible oils reach considerable importance for reducing health risks to consumers and the efficiency related to PAHs reduction is still questioned due to the fact that available information regarding specific process conditions is limited.

Normally, the amount of PAHs in crude vegetable oils can be eliminated during oil refining process, thus generating low levels of contamination in the final oil products compared to in the crude ones [10]. In China, conventional oil refining process consists of four steps (i.e., degumming, neutralization, decoloration, and deodorization), and the last three procedures were reported with good removing effect on PAHs in edible vegetable oils [10–14]. Neutralization, also known as alkali refining, is used to eliminate free fatty acids that can promote lipid oxidation by forming soapstock. Adsorbents, such as activated clay and activated carbon, can be used to remove oil-solubility pigments (e.g., carotene and chlorophyll), metals, and oxidation products as well as remaining phospholipids and soaps, and this procedure is defined as decoloration or bleaching. Deodorization, or sometimes called steam distillation, removes volatile components and small molecular decomposed peroxides at elevated temperature under vacuum to improve flavour quality, shelf life, and stability of the final oil products.

The aim of this study is to investigate the effects of three main steps, namely, neutralization, decoloration, and deodorization, on PAHs in crude vegetable oils, to examine the respective influence of parameters employed on their efficiencies in reducing of PAHs, and to characterize the distribution and evaluate the behaviour of PAHs during the whole technological procedure adopted. This is a comprehensive effort to report on the levels of selected PAHs measured in different stages of the oil refining. Meanwhile, the inclusion of multiple PAHs provides a full-scale representation of the entire group of pollutants in edible oils during the refining process.

2. Materials and Methods

2.1. Materials and Chemicals. Mixture of EPA 16 PAHs was provided by O2si (Charleston, South Carolina, USA) with the concentration of 200 μg/mL (dissolved in acetonitrile); the full names and abbreviations of 16 analytes related to this study are shown in Table 1. Sixteen corresponding isotopical analogues were obtained from Dr. Ehrenstorfer GmbH (Augsburg, Germany) and C/D/N Isotopes (Pointe-Claire, Quebec, Canada). Purities of individual standard components were guaranteed to be above 97.9%.

Acetonitrile and toluene (HPLC grade) were purchased from Sigma (St. Louis, MO, USA). Deionized water was obtained from a Millipore Milli-Q water purification system (Billerica, MA, USA). Degummed peanut oil sample with a high contamination level of PAHs, which was used for neutralization, bleaching, and deodorization, was prepared in our own laboratory. Potassium hydroxide (KOH) used for alkali refining was purchased from Merck Chemicals Co., Ltd. (Shanghai, China). In this study, three absorbents were chosen as they represent some of the types used by refineries in China. Activated clay (AC) was purchased from Maige Adsorbents Co., Ltd. (Jiangsu, China). Notit-8015 activated charcoal (NACc) was obtained from Zhongji Chemicals Import & Export Co., Ltd. (Shanghai, China). WY activated charcoal (WYACc) was obtained from Chongqing Beyond Ocean Carbon Manufacture Co., Ltd. (Chongqing, China).

TABLE 1: The full names and abbreviations of EPA 16 PAHs related to this study.

PAH	Abbreviation
Naphthalene	Nap
Acenaphthylene	Acy
Acenaphthene	Ace
Fluorene	Fl
Phenanthrene	Phe
Anthracene	Ant
Fluoranthene	Flu
Pyrene	Pyr
Benz[a]anthracene	BaA
Chrysene	Chr
Benzo[b]fluoranthene	BbFlu
Benzo[k]fluoranthene	BkFlu
Benzo[a]pyrene	BaP
Dibenz[a,h]anthracene	DBahA
Benzo[ghi]perylene	BghiP
Indeno[1,2,3-cd]pyrene	IP

2.2. Oil Refining Process. Fifty grams of prepared oils was used for each experiment. Three oil refining procedures were respectively operated with the prepared degummed oil sample; thus the efficiency of individual steps for removing PAHs could be obtained and compared. A Flowchart for experimental design and all parameters is shown in Figure 1. In this study, orthogonal experiments were employed to select the best removal conditions for the two former processes (neutralization and decoloration). Nine experiments were performed for each alkali refining and bleaching based on the standard L_9 orthogonal arrays, whereas single factor experiment was used to optimize the counterpart for deodorization.

2.2.1. Neutralization. The indicated concentration of KOH was add to the prepared oil sample to neutralize the free fatty acids and the desired concentration of additional amount of KOH was infused to ensure the complete formation of soaps. This was proceeded with agitation at 40°C originally, and once the KOH was added, the temperature was increased to the indicated temperature immediately, maintaining the temperature for the desired time. Then the soaps were separated by centrifugation at 5400 r/min. To remove the traces of soaps dissolved in the treated oil and the remains of the basic reagents added, the oils obtained in the previous

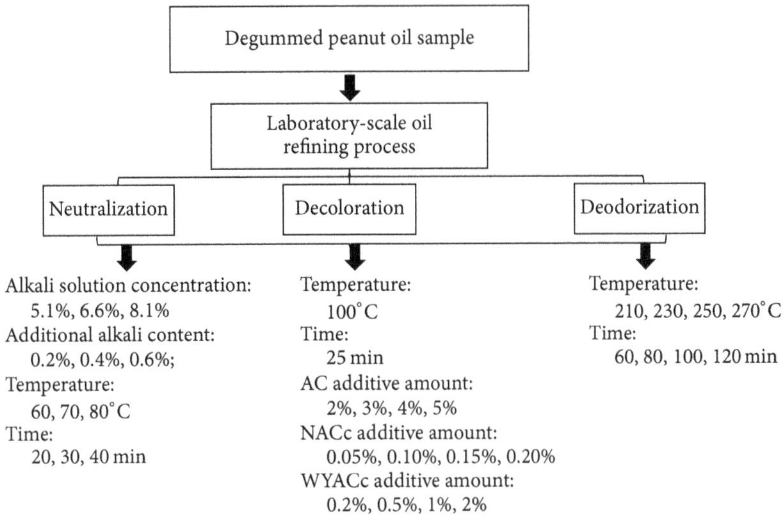

FIGURE 1: Flowchart for experimental design and all parameters.

stages were washed with three successive additions of 10% of 95°C pure water. Then removing the washing water by centrifuging and the neutralized oil was obtained.

2.2.2. Decoloration. The removal of PAHs by the oil bleaching procedure was compared using three commercial sorbents: AC, NACc, and WYACc. According to the actual amount of these three kinds of processing agents used in the oil refinery, the proportions of AC, NACc, and WYACc were optimized and set between 2 and 4%, 0.05 and 0.2%, and 0.2 and 2%, respectively. Vacuum bleaching was performed in 30–35 mbar by using a vacuum pump and the treatment process was as follows: the adopted adsorbent was add to the prepared oil sample, and the mixture was maintained by agitation for the desired time at the indicated temperature in a vacuum atmosphere. Then the adsorbents were separated by centrifugation at 5400 r/min, and the bleached oil was obtained.

2.2.3. Deodorization. The deodorization was accomplished at the indicated temperature and 80 Pa with the use of direct stream (20 kg/T·h) in order to remove volatile compounds and thermolabile pigments. Under the high temperature and high vacuum conditions, tested oil was maintained for the desired time and then cooled down to ambient temperature and the deodorized oil was obtained.

2.3. Oil PAHs Determination and Analytical Conditions. Oil PAHs determination scheme and analytical means were described in detail in a previous published paper [8]; all of the oil samples were prepared and determined in triplicate. A flowchart describing the sample preparation and purification procedure is provided in Figure 2. Ion pairs of 16 analytes and their analogous are given in Table S1 (Supplementary Material, available online at https://doi.org/10.1155/2017/7824761); data of the method validation are provided in Tables S2 and S3. Briefly, the calibration curves of each analyte showed a linear response with correlation coefficients between 0.9992 and 0.9999. The recoveries of 16 analytes at three spiking levels were ranging from 77.8 to 106.4%, and the relative standard deviations were lower than 10%. The limits of quantitation for 16 PAHs were in the range of 0.02–0.43 μg/kg. Table S4 shows the LC conditions and all of the required instrumental parameters. Data analysis was performed using Origin 8.0 and Microsoft Office Excel 2015.

3. Results and Discussion

3.1. Effects of Neutralization on PAHs. As it is known, parameters (alkali concentration, amount of additional alkali, temperature, and time) of the alkali refining process have a significant impact on the quantity and composition of the neutralized oil. An L_9 (3^4) orthogonal matrix with four factors, each factor containing three levels, was selected to arrange the experiments. Alkali solution concentrations were 5.1%, 6.6%, and 8.1%; additional alkali contents were 0.2%, 0.4%, and 0.6%; temperatures were 60, 70, and 80°C; times were 20, 30, and 40 min. The levels for each parameter variables were selected from a series of preliminary trials without using particular experimental designs (data were not shown).

Results of experiments in various conditions are displayed in Table 2. Optimum condition of each parameter was shown as the highest value of the removal rate. The optimum conditions for elimination of BaP, LPAHs, and PAH_{16} were achieved at alkali concentration of 5.1%, additional alkali amount of 0.6%, temperature of 80°C, and time of 40 min. For HPAHs, the best results were achieved at alkali concentration of 6.6%, additional alkali amount of 0.6%, temperature of 70°C, and time of 20 min. And for PAH_4, the best conditions were alkali concentration of 5.1%, additional alkali amount of 0.4%, temperature of 70°C, and time of 30 min. In addition to the main effect, the signal-to-noise (S/N) ratio analysis was used to evaluate the quality of the results obtained. S/N

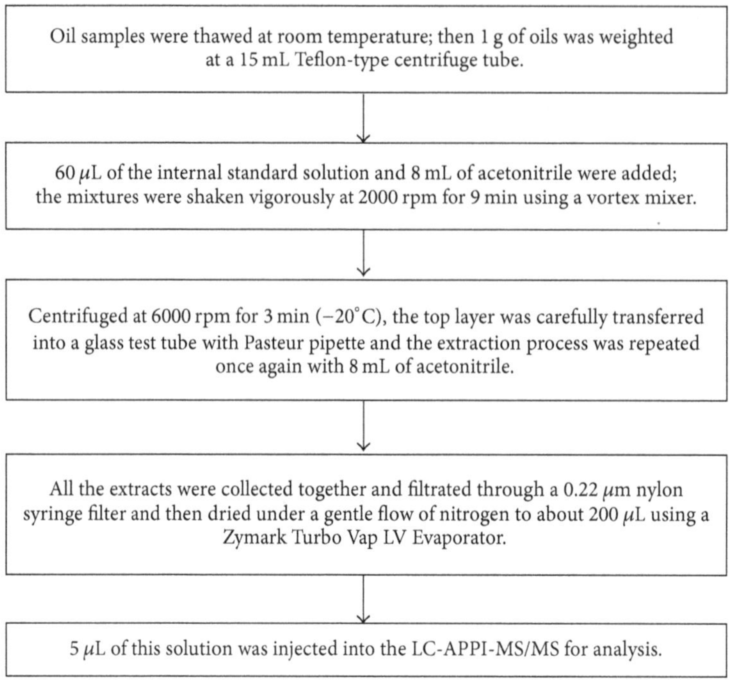

FIGURE 2: Flowchart for oil sample preparation and purification procedure.

ratio with "higher is better" goal was selected to optimize the process. Table 2 also shows the result of S/N ratio analysis, and S/N ratio analysis revealed that the optimum conditions for the five indicators were not similar to the optimum conditions obtained using the main effect plot. S/N ratio analysis showed that time is the highest influence parameter on BaP and HPAHs, whereas alkali solution concentration is the highest influence parameter on PAH_4, LPAHs, and PAH_{16}. Taking everything into consideration, the optimized conditions obtained were alkali concentration of 5.1%, additional alkali amount of 0.2%, temperature of 70°C, and time of 40 min.

3.2. Effects of Decoloration on PAHs. The effects of three absorbents on PAHs were compared in this study. The temperature and time for single factor experiments were set at 100°C and 25 min, respectively. AC additive amounts were 2%, 3%, 4%, and 5%; NACc additive amounts were 0.05%, 0.10%, 0.15%, and 0.20%; WYACc additive amounts were 0.2%, 0.5%, 1%, and 2%. It was observed that increase in added contents of absorbent had a favourable effect on removing concentration of PAHs in crude oils. The removal rates of all three absorbents increase with amount rise at constant time and constant temperature. NACc was the best sorbent (additive amount was 0.2%) with the removal rates of 99.1, 78.9%, 99.1%, 98.5%, and 81.5 for BaP, LPAHs, HPAHs, PAH_4, and PAH_{16}, respectively. These could be explained by the NACc characteristics and the molecular conformation, namely, the large surface area of NACc that provides suitable pore volume and pore size for adequate adsorption. The data we observed demonstrated that almost all the 4–6 rings analytes can be eliminated by using 0.2% of NACc. Moreover, the removal efficiencies for analytes less than 4 rings were quite satisfactory (approximately 80% for individual object), thus proving its high ability to adsorb PAHs in edible oils. However, AC was the weakest one, even if at the biggest quantity (5%); the removal rates of BaP, LPAHs, HPAHs, PAH_4, and PAH_{16} were 79.7%, 43.9%, 83.5%, 81.6%, and 48.9%, respectively. Especially in the LPAHs and PAH_{16}, low removal efficiencies (less than 50%) were observed. In terms of WYACc, comparable results were obtained when 1% of WYACc was added (compared with 0.2% of NACc). Individually, the removal rates of BaP, LPAHs, HPAHs, PAH_4, and PAH_{16} were 99.84%, 74.61%, 99.38%, 98.85%, and 77.74% when 1% of WYACc was employed. It is worth mentioning that when 2% of WYACc was used, almost all the 16 PAHs can be reduced, even the 2-3-ring PAHs. The removal rates of LPAHs and PAH_{16} were 93.1% and 94.0% for 2% of WYACc.

NACc was the best choice in this study, due to its high efficiency at extremely low content of addition. However, the expensive price of this material is one of the limits to its application in oil bleaching process. For this reason, WYACc, the one possessing the comparative PAHs removal efficiency, was selected to proceed with the orthogonal experiment.

In orthogonal test, parameters including WYACc amount, temperature, and time were optimized. An L_9 (3^4) orthogonal matrix with three factors, each factor containing three levels, was selected to arrange the experiments. WYACc amounts were 0.2%, 0.5%, and 1.0%; temperatures were 100, 110, and 120°C; times were 15, 25, and 35 min. Results of experiments in various conditions are shown in Table 3. The optimum conditions for elimination of BaP, HPAHs, LPAHs, and PAH_{16} were achieved at WYACc amount of 1.0%, temperature of 100°C, and time of 35 min. For PAH_4, the best conditions were achieved at WYACc amount of

Table 2: Results of neutralization on PAHs.

	Run	Parameter				Removal rate (%)				
		Alkali concentration (%)	Temperature (°C)	Additional alkali (%)	Time (min)	BaP	HPAHs[a]	PAH_4[b]	LPAHs[c]	PAH_{16}
	1	5.1	60	0.2	20	67.9	47.5	42.6	53.6	52.8
	2	5.1	70	0.4	30	64.2	43.3	47.2	52.7	51.5
	3	5.1	80	0.6	40	80.4	43.4	40.1	56.9	55.2
	4	6.6	60	0.4	40	80.3	41.7	38.7	48.4	47.6
	5	6.6	70	0.6	20	77.1	47.8	40.7	50.4	50.0
	6	6.6	80	0.2	30	75.4	45.2	42.0	50.0	49.4
	7	8.1	60	0.6	30	38.8	33.6	30.3	40.7	39.8
	8	8.1	70	0.2	40	77.1	44.8	39.8	55.1	53.8
	9	8.1	80	0.4	20	71.3	41.8	37.8	48.4	47.6
Item	Level	Alkali concentration (%)	Temperature (°C)	Additional alkali (%)	Time (min)					
BaP	1	70.8	62.4	73.5	72.1					
	2	77.6	72.8	72.0	59.5					
	3	62.4	75.7	65.4	79.3					
	Δ	15.2	13.3	8.1	19.8					
	Rank	2	3	4	1					
HPAHs[a]	1	44.7	40.9	45.8	45.7					
	2	44.9	45.3	42.3	40.7					
	3	40.1	43.5	41.6	43.3					
	Δ	4.8	4.4	4.2	5					
	Rank	2	3	4	1					
PAH_4[b]	1	43.3	37.2	41.5	40.4					
	2	40.5	42.6	41.2	39.8					
	3	36.0	40.0	37.0	39.5					
	Δ	7.3	5.4	4.5	0.9					
	Rank	1	2	3	4					
LPAHs[c]	1	54.4	47.6	52.9	50.8					
	2	49.6	52.7	49.8	47.8					
	3	48.1	51.8	49.3	53.5					
	Δ	6.3	5.1	3.6	5.7					
	Rank	1	3	4	2					
PAH_{16}	1	53.2	46.7	52.0	50.1					
	2	49.0	51.8	48.9	46.9					
	3	47.1	50.7	48.3	52.2					
	Δ	6.1	5.1	3.7	5.3					
	Rank	1	3	4	2					

[a]HPAHs, including BbFlu, BkFlu, BaP, DBahA, BghiP, and IP; [b]PAH4, including BaA, Chr, BbFlu, and BaP; [c]LPAHs, including Nap, Acy, Ace, Fl, Phe, Ant, Flu, Pyr, BaA, and Chr.

1.0%, temperature of 120°C, and time of 25 min (see Table 3). S/N ratio analysis revealed that the optimum conditions for the five indicators were similar to the optimum conditions obtained by using the main effect plot. S/N ratio analysis showed that WYACc amount is the highest influence parameter on all the five indicators, whereas temperature is the lowest influence parameter. Taking everything into consideration, the optimized conditions chosen were WYACc amount of 1.0%, temperature of 110°C, and time of 35 min.

3.3. Effects of Deodorization on PAHs. Steam distillation of prepared oil sample was accomplished using the single factor experiment. temperature and time were tested; temperatures were 210, 230, 250, and 270°C and times were 60, 80, 100, and 120 min. To study the effect of time on the deodorization efficiency, the related test was performed at different times ranging from 60 to 120 min; similarly, to study the effect of temperature on the deodorization efficiency, the related test was performed at different temperatures ranging from 210 to 270°C. Figures 3 and 4 show the effects of deodorization time and temperature on BaP and PAH_{16}, respectively. Increase in temperature had a favourable effect on removal of BaP at short time of operation (60 min), both at 210 and 230°C. Independently of the temperature conditions, longer time of deodorization had a more positive effect on the amount

Table 3: Results of decoloration on PAHs.

Run	Parameter			Removal rate (%)				
	Absorbent amount (%)	Temperature (°C)	Time (min)	BaP	HPAHs[a]	PAH_4[b]	LPAHs[c]	PAH_{16}
1	0.2	100	15	66.5	67.9	60.3	11.5	18.6
2	0.2	110	25	78.3	77.0	65.8	20.8	27.9
3	0.2	120	35	87.9	87.6	80.4	17.2	26.1
4	0.5	100	25	96.6	96.1	92.9	54.8	60.0
5	0.5	110	35	97.8	96.9	94.9	56.2	61.3
6	0.5	120	15	84.3	78.1	76.0	42.3	46.8
7	1.0	100	35	99.7	99.1	97.7	80.7	83.0
8	1.0	110	15	98.8	97.6	97.0	74.8	77.7
9	1.0	120	25	98.9	98.4	97.8	74.9	77.9

Item	Level	Absorbent amount (%)	Temperature (°C)	Time (min)
BaP	1	77.6	87.6	83.2
	2	92.9	91.6	91.3
	3	99.1	90.4	95.1
	Δ	21.5	4.0	11.9
	Rank	1	3	2
HPAHs[a]	1	68.8	83.6	77.8
	2	87.9	85.9	85.5
	3	97.5	84.7	91.0
	Δ	28.7	2.3	13.2
	Rank	1	3	2
PAH_4[b]	1	77.5	87.7	81.2
	2	90.4	90.5	90.5
	3	98.4	88.1	94.6
	Δ	20.9	2.8	13.4
	Rank	1	3	2
LPAHs[c]	1	16.5	49.0	42.9
	2	51.1	50.6	50.2
	3	76.8	44.8	51.4
	Δ	60.3	5.8	8.5
	Rank	1	3	2
PAH_{16}	1	24.2	53.9	47.7
	2	56.1	55.6	55.3
	3	79.5	50.3	56.8
	Δ	55.3	5.3	9.1
	Rank	1	3	2

[a]HPAHs, including BbFlu, BkFlu, BaP, DBahA, BghiP, and IP; [b]PAH4, including BaA, Chr, BbFlu, and BaP; [c]LPAHs, including Nap, Acy, Ace, Fl, Phe, Ant, Flu, Pyr, BaA, and Chr.

of PAHs removed, and vice versa. The residue amounts of BaP, LPAHs, HPAHs, PAH_4, and PAH_{16} in the deodorized oil decrease with temperature and time increase. The optimum results were achieved at temperature of 270°C and time of 120 min, and the removal rates obtained were 40.8%, 48.4%, 42.7%, 42.2%, and 47.7%.

3.4. Comparison of Effects of Three Refining Stages on PAHs. Optimum conditions obtained above from means calculation and orthogonal method must be evaluated through real experiment to check the accuracy of proposed optimum condition. Confirmation experiments (for neutralization and decoloration processes) were individually performed using the proposed optimum conditions from the orthogonal experiments. Concentrations of individual PAHs as well as LPAHs, HPAHs, PAH_4, and PAH_{16} in crude oil and each step of the oil refining process are presented in Table 4. All the analyses were carried out in triplicate and the experimental results obtained were expressed as means ± SD.

The amounts of BaP, PAH_4, and PAH_{16} in crude oil were 24.00 ± 1.51, 129.20 ± 8.11, and 763.94 ± 16.97 μg/kg, respectively. Meanwhile, LPAHs and HPAHs were 667.44 ± 15.96 and 96.50 ± 7.34 μg/kg, respectively. Individual levels of PAHs ranged from 2.84 ± 0.16 to 212.74 ± 8.66 μg/kg, showing

TABLE 4: PAHs content (mean ± SD (n = 3)) in crude oil and each step of the oil refining process.

PAH	Mean ± SD (n = 3)			
	Crude oil	Neutralized oil	Bleached oil	Deodorized oil
Nap	34.20 ± 2.28	10.75 ± 0.78	3.17 ± 1.32	4.25 ± 0.63
Acy	5.20 ± 0.45	2.74 ± 0.14	1.06 ± 0.01	1.38 ± 0.05
Ace	22.38 ± 1.95	4.80 ± 0.52	5.80 ± 0.28	3.61 ± 0.11
Fl	67.85 ± 5.35	44.40 ± 2.22	11.05 ± 1.00	19.61 ± 1.28
Phe	212.74 ± 8.66	47.73 ± 2.39	44.00 ± 2.54	134.40 ± 3.32
Ant	41.67 ± 2.56	27.00 ± 1.42	4.62 ± 0.05	18.11 ± 1.22
Flu	135.80 ± 9.01	64.60 ± 4.03	14.48 ± 0.63	77.88 ± 6.00
Pyr	71.60 ± 5.48	39.60 ± 2.58	9.52 ± 0.11	40.95 ± 2.33
BaA	38.80 ± 2.22	24.11 ± 0.98	0.54 ± 0.01	22.19 ± 0.98
Chr	37.20 ± 2.63	22.60 ± 0.82	0.87 ± 0.01	22.09 ± 1.36
BbFlu	29.20 ± 1.98	23.10 ± 1.13	0.27 ± 0.01	16.15 ± 1.11
BkFlu	12.36 ± 0.63	4.24 ± 0.51	0.16 ± 0.01	7.10 ± 0.58
BaP	24.00 ± 1.51	3.58 ± 0.37	0.06 ± 0.01	14.21 ± 1.08
DBahA	2.84 ± 0.16	0.69 ± 0.01	0.05 ± 0.01	1.09 ± 0.02
BghiP	14.10 ± 1.13	9.91 ± 1.28	0.12 ± 0.01	8.06 ± 0.05
IP	14.00 ± 1.18	8.56 ± 0.80	0.03 ± 0.00	8.74 ± 1.08
LPAHs[a]	667.44 ± 15.96	288.33 ± 7.28	95.11 ± 3.33	344.46 ± 8.57
HPAHs[b]	96.50 ± 7.34	50.08 ± 3.00	0.69 ± 0.02	55.34 ± 2.22
PAH_4[c]	129.20 ± 8.11	73.39 ± 4.82	1.74 ± 0.01	74.64 ± 2.54
PAH_{16}	763.94 ± 16.97	338.41 ± 7.54	95.80 ± 4.51	399.8 ± 9.22

[a]LPAHs, including Nap, Acy, Ace, Fl, Phe, Ant, Flu, Pyr, BaA, and Chr; [b]HPAHs, including BbFlu, BkFlu, BaP, DBahA, BghiP, and IP; [c]PAH4, including BaA, Chr, BbFlu, and BaP.

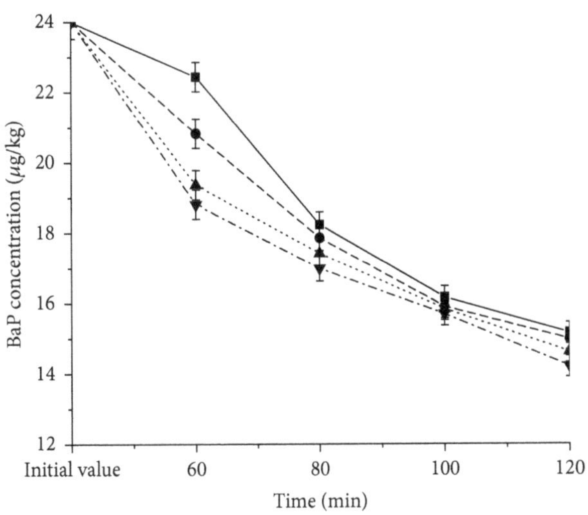

FIGURE 3: Effect of deodorization on BaP.

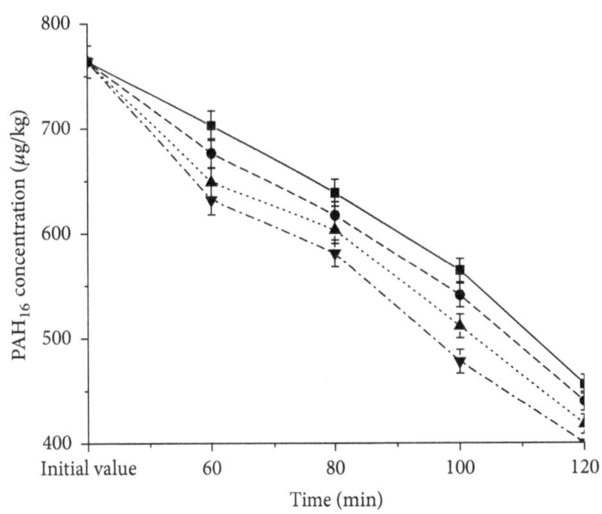

FIGURE 4: Effect of deodorization on PAH_{16}.

a broad range of PAHs concentrations in crude vegetable oil. Phe, Flu, and Pyr were the three kinds of compounds with the highest concentrations; levels of 3 analytes were 212.74 ± 8.66, 135.80 ± 9.01, and 71.60 ± 5.48 μg/kg, respectively. DBahA was the lowest one oppositely, with the concentration of 2.84 ± 0.16 μg/kg.

Determination of other intermediary oils (neutralized, bleached, and deodorized oils) revealed that all analytes showed reduction in their levels (see Table 4). Percent reductions of intermediary oils were relative to the initial crude oil levels. As a whole, bleaching with WYACc was the most efficient means to eliminate PAHs in crude oil, while stream distillation was the worst. BaP contents decreased by 85.1%, 99.7%, and 40.8% in neutralized, bleached, and deodorized oils, respectively. Similarly, PAH_4 contents diminished by 43.2%, 98.7%, and 42.2%, respectively, and PAH_{16} contents decreased by 55.7%, 87.5%, and 47.7%, respectively. Previous literatures reported the decrease of PAHs during oil refining process was 27%–82% [11], 82%–84% [12], and 72%–87% [14], and all the oil samples (crude, neutralized, bleached, and deodorized oils) were sampled in the vegetable oil refinery. The differences among technological parameters used in the oil refining process, and the situations of samples or the laboratory conditions in refined process, may result in the small variances among the data obtained by this study and other authors.

A decrease of PAHs amounts is clearly observed from both of the LPAHs and HPAHs, in higher or lower percentage, independent of the treatments adopted. Neutralization and, chiefly, decoloration with WYACc were the more effective steps contributing to the LPAHs decrease, with the removal rates being 56.8% and 85.8%, respectively. For HPAHs, decoloration was the best, with the removal rate of 99.3%. Meanwhile, neutralization was also involved in the reduction of the 5-6 rings analytes, with the HPAHs contents diminished by 48.1%. However, deodorization was only responsible for a slight decrease in the LPAHs and HPAHs contents, with the removal rates being 48.4% and 42.7%, respectively. In comparison, other results published previously showed a more modest reduction in the HPAHs content during oil refining process: 41% [11] and 22% [12]. Meanwhile, activated carbon was not used in any of these studies. The differences of removal rates of LPAHs during deodorization reported by this study (48.4%) and by Cejpek et al. [11] and Larsson et al. [12] (varied between 57% and 92%) may be resulting from the discrepancy of operating conditions applied. Because LPAHs were in greater abundance than HPAHs in tested oil samples, the former had a higher relative contribution to the summed total PAHs. Thus, the PAH_{16} profile and removal behaviour observed in this study were similar to LPAHs (see Table 4).

4. Conclusion

Effects of neutralization, decoloration, and deodorization on PAHs during oil refining process were investigated in this study. The results of the experiment showed that the optimum conditions of neutralization were alkali concentration of 5.1%, additional alkali amount of 0.2%, temperature of 70°C, and time of 40 min. The optimum conditions of decoloration were WYACc amount of 1.0%, temperature of 110°C, and time of 35 min and the optimum conditions of deodorization were temperature of 270°C and time of 120 min. An evident decrease of PAHs contents during oil refining process was observed, and bleaching with activated charcoal was the best to reduce PAHs in crude oil. Under the optimum experimental conditions, BaP contents decreased by 85.1%, 99.7%, and 40.8% in neutralized, bleached, and deodorized oils, respectively. PAH_4 contents diminished by 43.2%, 98.7%, and 42.2%, respectively, and PAH_{16} contents decreased by 55.7%, 87.5%, and 47.7%, respectively. Bleaching with WYACc was the most efficient method in the comparison study to absorb PAHs in crude oil, while stream distillation was the worst. Neutralization had a modest influence on 16 analytes, with the removal rate of 55.7%. However, deodorization was only responsible for a slight decrease in the LPAHs and HPAHs contents (removal rates were 48.4% and 42.7%, resp.). It is worth mentioning that the high efficiency of activated charcoal indicates that the use of this material to eliminate PAHs during oil refining process is highly recommended.

Conflicts of Interest

No conflicts of interest exist in the submission of this manuscript, and the manuscript is approved by all authors for publication.

Acknowledgments

This work was supported by the Research Fund of National 13th Five-Year Plan of China (2016YFD0401405-2) and the earmarked fund for Modern Agro-industry Technology Research System (CARS15-1-10). Longkai Shi and Yulan Liu must thank Hongwei Hou from China National Tobacco Quality Supervision & Test Center (Zhengzhou, China) for providing the LC-APPI-MS/MS system.

References

[1] S. Chumsantea, K. Aryusuk, S. Lilitchan, N. Jeyashoke, and K. Krisnangkura, "Reducing oil losses in alkali refining," *Journal of the American Oil Chemists' Society*, vol. 89, no. 10, pp. 1913–1919, 2012.

[2] M. V. Ruiz-Méndez, "Contribution of denaturing and deodorization processes of oils to toxic oil syndrome," *Chemico-Biological Interactions*, vol. 192, no. 1-2, pp. 142–144, 2011.

[3] J. Ortega-García, N. Gámez-Meza, J. A. Noriega-Rodriguez et al., "Refining of high oleic safflower oil: effect on the sterols and tocopherols content," *European Food Research and Technology*, vol. 223, no. 6, pp. 775–779, 2006.

[4] E. Sabah and M. Majdan, "Removal of phosphorus from vegetable oil by acid-activated sepiolite," *Journal of Food Engineering*, vol. 91, no. 3, pp. 423–427, 2009.

[5] R. Ajemba and O. Onukwuli, "Adsorptive removal of colour pigment from palm oil using acid activated Nteje clay. Kinetics, equilibrium and thermodynamics. Physicochem," *Problem Miner. Process*, vol. 49, no. 1, pp. 369–381, 2013.

[6] GB 2716-2005, "Hygienic standard for edible vegetable oil," (In Chinese).

[7] "Commission Regulation (EU) No. 835/2011 of 19 August 2011 amending Regulation EC No. 1881/2006 as regards maximum levels for foodstuffs," 2011.

[8] L.-K. Shi, Y.-L. Liu, H.-M. Liu, and M.-M. Zhang, "One-step solvent extraction followed by liquid chromatography-atmospheric pressure photoionization tandem mass spectrometry for the determination of polycyclic aromatic hydrocarbons in edible oils," *Analytical and Bioanalytical Chemistry*, vol. 407, no. 13, pp. 3605–3616, 2015.

[9] L.-K. Shi, D.-D. Zhang, and Y.-L. Liu, "Incidence and survey of polycyclic aromatic hydrocarbons in edible vegetable oils in China," *Food Control*, vol. 62, pp. 165–170, 2016.

[10] M. C. Rojo Camargo, P. R. Antoniolli, and E. Vicente, "Evaluation of polycyclic aromatic hydrocarbons content in different stages of soybean oils processing," *Food Chemistry*, vol. 135, no. 3, pp. 937–942, 2012.

[11] K. Cejpek, J. Hajslova, V. Kocourek, M. Tomaniova, and J. Cmolik, "Changes in PAH levels during production of rapeseed oil," *Food Additives and Contaminants*, vol. 15, no. 5, pp. 563–574, 1998.

[12] B. K. Larsson, A. T. Eriksson, and M. Cervenka, "Polycyclic aromatic hydrocarbons in crude and deodorized vegetable oils," *Journal of the American Oil Chemists' Society*, vol. 64, no. 3, pp. 365–370, 1987.

[13] M. León-Camacho, I. Viera-Alcaide, and M. V. Ruiz-Méndez, "Elimination of polycyclic aromatic hydrocarbons by bleaching of olive pomace oil," *European Journal of Lipid Science and Technology*, vol. 105, no. 1, pp. 9–16, 2003.

[14] V. H. Teixeira, S. Casal, and M. B. P. P. Oliveira, "PAHs content in sunflower, soybean and virgin olive oils: evaluation in commercial samples and during refining process," *Food Chemistry*, vol. 104, no. 1, pp. 106–112, 2007.

Synthesis and Pharmacological Activities of Pyrano[2,3-d]pyrimidine and Pyrano[2,3-d]pyrimidine-5-one Derivatives as New Fused Heterocyclic Systems

Naglaa F. H. Mahmoud, Eman A. El-Bordany, and Galal A. Elsayed

Chemistry Department, Faculty of Science, Ain Shams University, Abbassia, Cairo 11566, Egypt

Correspondence should be addressed to Naglaa F. H. Mahmoud; naglaa.fawzy@yahoo.com

Academic Editor: Orazio Nicolotti

Various fused oxazine such as 4-(4-methoxyphenyl)-3,7-dimethyl-1,4-dihydro-5H-pyrazolo [4′,3′:5,6]pyrano[2,3-d][1,3]oxazin-5-one **2** has been prepared and utilized as a starting material for novel pyrazolopyranopyrimidinones **3**, **5**, **6**, and **7a–c** and pyrazolopyranopyrimidines **4**, **9**, **10**, and **11** which are expected to possess considerable chemical and pharmacological activities. The structures of the new compounds have been elucidated by spectroscopic data and elemental analysis. The antioxidant and anticancer activities of synthesized products have been evaluated.

1. Introduction

Heterocyclic compounds are one of the most important molecules in the synthesis of pharmacologically active compounds [1–4]. Enaminonitriles have been utilized as starting materials for synthesis of huge number of heterocyclic compounds containing pyranopyrimidine [5–9] and pyridopyrimidine [10–13] moieties. Oxazines [14] and quinazolines [15] are natural products which exhibit a widespread spectrum of biological activity including antitumor [16–18], anti-inflammatory [19], and antiplatelet [20] properties. In view of these reports and as a continuation of our previous work on heterocyclic chemistry [21–27], we aimed to incorporate the pyrano moiety into the 1,5-position of oxazinone and pyrimidinone ring systems to obtain new heterocyclic system which is anticipated to possess notable chemical and pharmacological activities. So we utilized enaminonitrile in the synthesis of pyrano[2,3-d][1,3]oxazin-5-one derivative and pyrano[2,3-d]pyrimidine and pyrano[2,3-d]pyrimidine-5-one derivatives in addition to studying their antioxidant activities.

2. Results and Discussion

2.1. Chemistry. Our present investigation aimed to synthesize of 6-amino-4-(4-methoxyphenyl)-3-methyl-1,4-dihydropyrano[2,3-c]pyrazole-5-carbonitrile **1** [28], which was used as a precursor for the synthesis of previously unreported 4-(4-methoxyphenyl)-3,7-dimethyl-1,4-dihydro-5H-pyrazolo[4′,3′:5,6]pyrano[2,3-d][1, 3]oxazin-5-one **2** as new heterocyclic compound containing oxazine and pyranopyrazole moieties. Compound **1** has been prepared by four-component one-pot reaction [29] and three-component one-pot reaction [30] (Scheme 1).

Treatment of **1** by freshly distilled acetic anhydride afforded **2** as a sole product. The structure of **2** was elucidated by spectral analysis. The IR spectrum showed a strong absorption band at 1732 cm^{-1} corresponding to lactonic carbonyl group and the disappearance of the cyano group. Also the structure was confirmed by elemental analysis; ^{1}H-NMR spectrum showed a singlet at δ 12.52 ppm of NH, a doublet at δ 7.162–7.136 ppm and δ 6.84–6.81 ppm of aromatic protons, a singlet at δ 4.7 ppm of benzylic proton, and a singlet at δ 3.68 ppm of three protons for OCH_3 group.

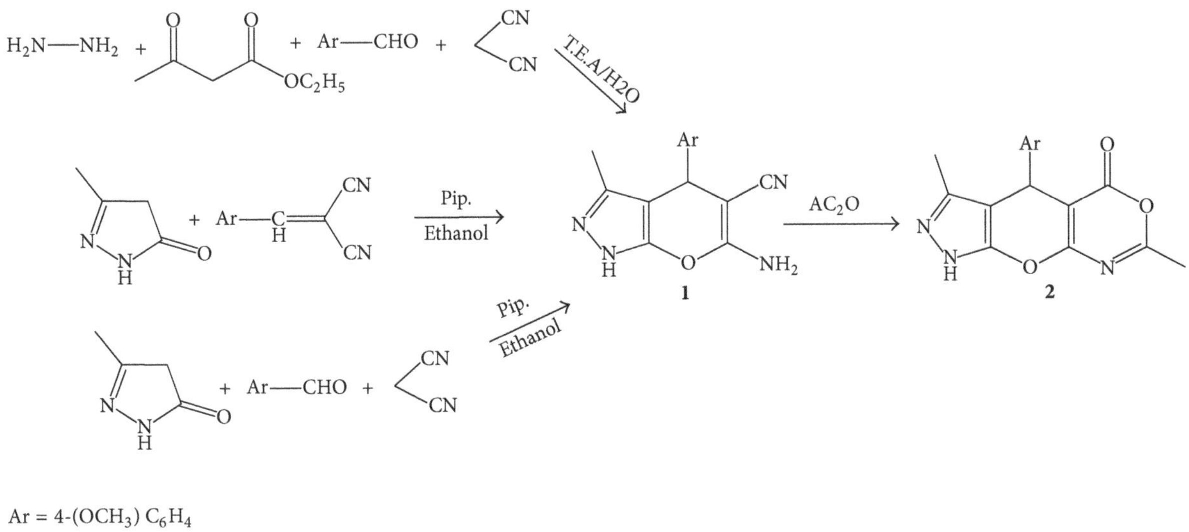

SCHEME 1

Ammonolysis of **2** by fusion with ammonium acetate and/or formamide afforded 4-(4-methoxyphenyl)-3,7-dimethyl-4,6-dihydro-pyrazolo[4′,3′:5,6]pyrano[2,3-d]pyrimidine-5(1H)-one **3**. The structure of **3** was confirmed by the IR spectrum which exhibited strong absorption bands at 3200, 3281, 1664, and 1610 cm^{-1} corresponding to ν_{NH}, $\nu_{C=O}$, and $\nu_{C=N}$, respectively and lacked a band corresponding to lactonic carbonyl group and also the presence of acidic hydrogen in the ^{1}H-NMR spectrum at δ 12.38. The electrophilicity of the lactonic carbonyl group of compound **2** was studied by nucleophilic reaction with 1,2-diaminobenzene and thiosemicarbazide, as 1,4-dinucleophile, to afford 4-(4-methoxyphenyl)-3,7-dimethyl-4-hydrobenzimidazolo[1,2,c]pyrazolo[4′3′:5,5]pyrano[2,3-d]pyrimidine **4** and N-[4-(4-methoxyphenyl)-3,7-dimethyl-5-oxo-1,4-dihydro-pyrazolo[4′,3′:5,6]pyrano[2,3-d]pyrimidin-6(5H)-yl]thiourea **5**, as sole products. The pyrazolo-pyranopyrimidine derivative **4** was formed via ring opening of the oxazine ring by an amino group followed by ring closer and then dehydration with the other amino group. The IR spectrum of compound **4** showed strong absorption bands at 3180 and 1609 cm^{-1} corresponding to ν_{NH} and $\nu_{C=N}$. The IR spectrum of compound **5** showed well defined absorption bands at 3251, 3158, 3121, 1665, 1619, and 1254 cm^{-1} corresponding to ν_{NH2}, ν_{NH}, $\nu_{C=O}$, $\nu_{C=N}$, and $\nu_{C=S}$. Compound **2** was allowed to react with primary amines, namely, hydrazine hydrate, 4-methylaniline, 4-methoxyaniline, and/or 4-aminoaniline, to afford the pyrazolopyranopyrimidine derivatives **6** and **7a–c**, respectively (Scheme 2).

4-(4-Methoxyphenyl)-3,7dimethy-l-4,6-dihydropyrazolo[4′,3′:5,6]pyrano[2,3-d]pyrimidin-5(1H)-thione **8** was synthesized by sulfurization of **3** using Lawson's reagent or by reaction of thiourea with 5-chloro-4-(4-methoxyphenyl)-3,7-dimethyl-1,4-dihydropy-razolo[4′,3′:5,6]pyrano[2,3-d]pyrimidine **9**. Compound **9** was prepared by chlorination of **3** by $POCl_3$ and PCl_5 mixture. Hydrazinolysis of **9** afforded 5-hydrazino-4-(4-methoxyphenyl)-3,7-dimethyl-1,4-dihydropyrazolo[4′,3′:5,6]pyrano[2,3-d]pyrimidine **10**.

IR spectrum of compound **10** showed strong absorption bands at 3215, 3156, 3161, and 1612 cm^{-1} corresponding to ν_{NH2}, ν_{NH}, and $\nu_{C=N}$. Also the structure was confirmed by ^{1}H-NMR spectrum that showed a singlet at δ 12.01 ppm of NH, multiplet at δ 7.83 ppm of aromatic protons, a singlet at δ 4.8 ppm of benzylic proton, a singlet at δ 3.7 ppm of three protons for OCH_3 group, and a singlet at δ 5.354 of NH_2.

Acetylation of **10** by refluxing with acetic anhydride gave the diacetyl derivative **11** and not the expected triazolo derivative **12**. The structure of **11** was confirmed by the spectral data such as the IR which showed the absorbance for the carbonyl group at 1732 and 1695 cm^{-1} and the two methyl groups signals at δ 2.678 ppm for $2CH_3$ in ^{1}H-NMR spectrum (Scheme 3).

2.2. Pharmacological Activity. Antioxidant activity using 2,2′-azinobis(3-ethylbenzothiazoline-6-sulfonic acid) (ABTS) inhibition. The prepared compounds were tested for antioxidant activity as reflected in their ability to inhibit oxidation in rat brain and kidney homogenates, from the result in Table 1. Compounds **1**, **4**, **7a**, and **10** exhibit a similar antioxidant activity to ascorbic acid, used as standard. Compounds **8** and **11** exhibited moderate antioxidant activity, while **2**, **3**, **7b**, **5**, and **9** showed lower activity.

Bleomycin Dependent DNA Damage. The bleomycins are a family of glycopeptides antibiotics [31] that are routinely used as antitumor agents. The bleomycin assay has been adopted for assessing the prooxidant activity of food antioxidants. The antitumor antibiotic bleomycin binds iron ions and DNA. If the samples are able to reduce bleomycin–Fe^{3+} to bleomycin–Fe, DNA degradation in the system will be stimulated, resulting in a positive test for prooxidant activity. DNA

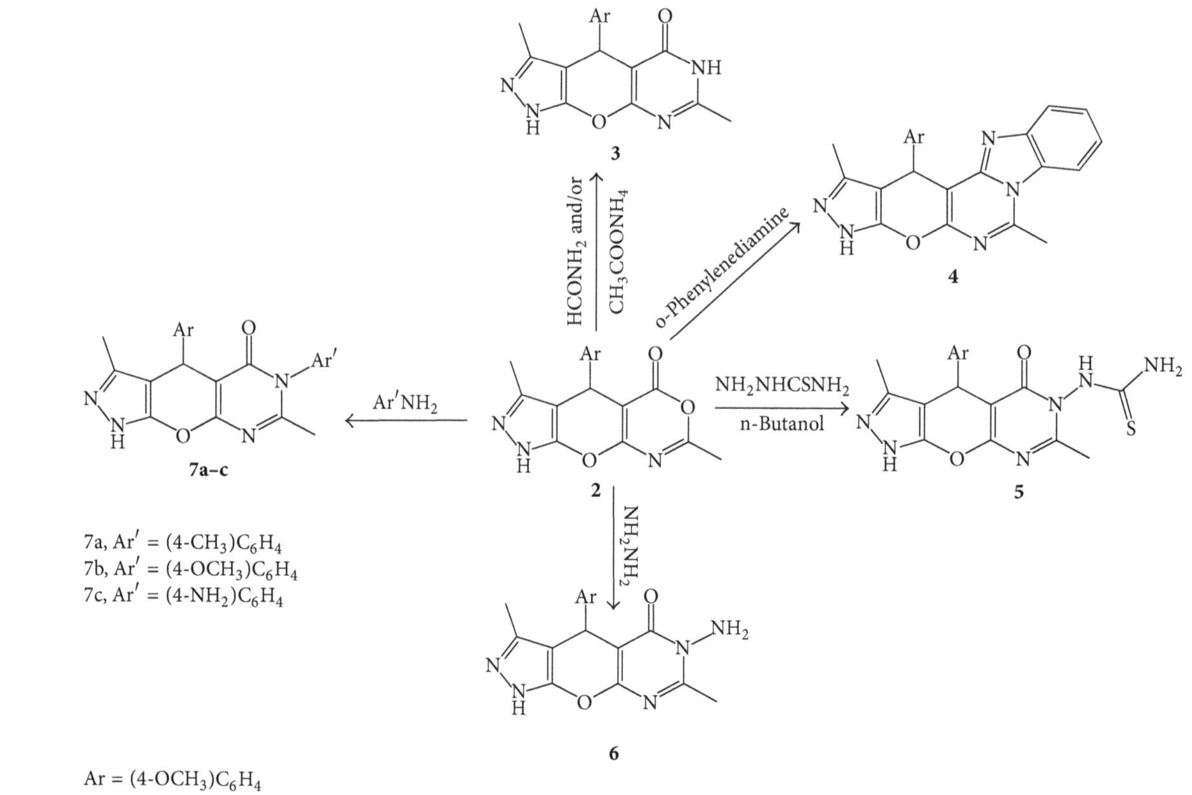

Scheme 2

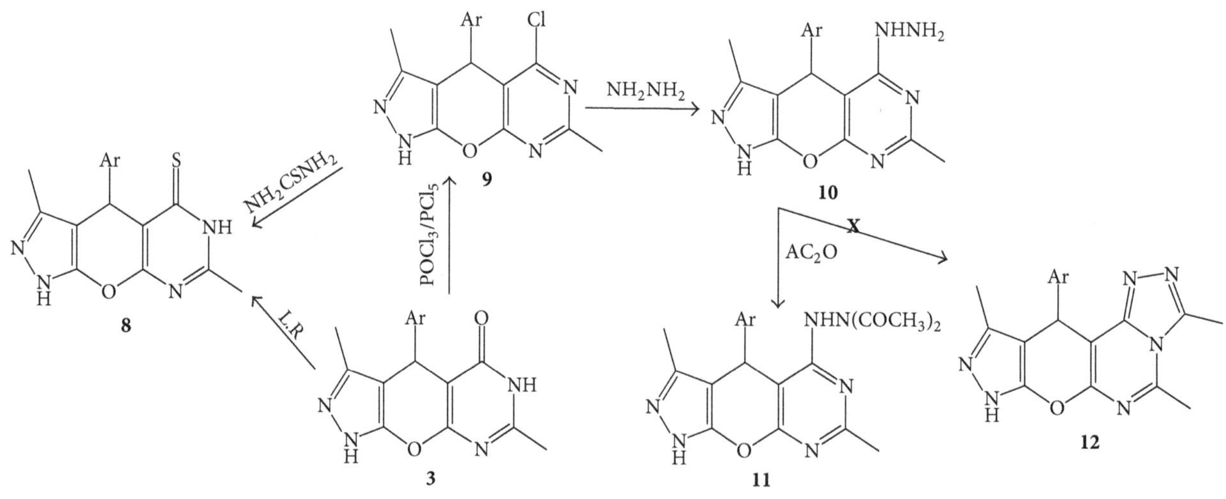

Scheme 3

degradation is accompanied by the formation of a product similar to malondialdehyde.

L-Ascorbic acid was used as the reducing agent to reduce Fe^{3+} to Fe^{2+}. The synthesized compounds were selected for bleomycin dependent DNA damage testing (Table 1).

Results in Table 1 showed that compounds **1**, **4**, **7a**, **8**, and **10** have the ability to protect DNA from the damage induced by bleomycin. On the other hand, the rest of the compounds exhibited weak activities.

Cytotoxic Activity of Some Compounds against Human Tumor Cells. The prepared compounds were tested for cytotoxic activity against four human tumor cells. Best results were observed for compounds **1**, **4**, **7a**, and **10** which were found

TABLE 1: The antioxidant activities of the synthesized compounds.

Compound number	Antioxidant activity		Bleomycin dependent DNA damage
	Absorbance	Inhibition %	
1	0.111	78.6	0.098
2	0.427	17.9	0.127
3	0.419	19.4	0.135
4	0.062	88.1	0.083
5	0.393	24.4	0.138
7a	0.083	84.0	0.091
7b	0.428	17.7	0.151
8	0.274	47.3	0.089
9	0.446	14.2	0.142
10	0.060	88.5	0.066
11	0.355	31.7	0.112
Ascorbic acid	0.065	89.2	0.076
ABTS	0.520	—	—

TABLE 2: The cytotoxic activity of some compounds against human tumor cells.

Compounds	In vitro cytotoxicity IC50 (μg/ml)*			
	HePG2	HCT-116	PC3	MCF-7
5-FU	7.9 ± 0.12	5.3 ± 0.14	8.3 ± 0.25	5.4 ± 0.21
1	9.4 ± 0.27	13.5 ± 1.65	16.9 ± 1.43	26.8 ± 1.52
2	82.1 ± 5.34	95.9 ± 6.34	>100	>100
3	45.6 ± 3.21	51.1 ± 3.50	61.1 ± 4.96	37.2 ± 2.07
4	8.4 ± 0.30	9.0 ± 0.41	16.0 ± 1.30	14.6 ± 1.36
5	36.2 ± 3.22	52.2 ± 3.67	41.3 ± 2.65	40.1 ± 3.22
7a	8.7 ± 0.26	13.61 ± .48	17.9 ± 1.57	19.5 ± 1.81
7b	63.5 ± 3.98	>100	85.2 ± 6.31	85.5 ± 5.62
8	16.4 ± 1.04	31.4 ± 2.64	19.9 ± 1.68	26.9 ± 1.91
9	52.0 ± 2.83	69.2 ± 4.05	47.2 ± 2.54	58.5 ± 3.54
10	8.3 ± 0.41	11.2 ± 0.86	8.9 ± 0.50	9.1 ± 0.65
11	22.1 ± 1.56	37.8 ± 1.93	26.4 ± 1.79	30.0 ± 2.80

*IC50 (μg/ml): 1–10 (very strong), 11–20 (strong), 21–50 (moderate), 51–100 (weak), and above 100 (noncytotoxic); 5-FU = 5-fluorouracil.

to very strong cytotoxic compounds. Compound **8** was also considered strongly cytotoxic. Finally, compounds **3**, **5**, and **11** were moderate in their cytotoxic activity and compounds **2**, **7b**, and **9** were weakly cytotoxic (Table 2).

2.3. Experimental. All melting points were measured on a Gallenkamp electric melting point apparatus and are uncorrected. The infrared spectra were recorded in potassium bromide disks on a pyeUnicam SP-3-300 and Shimdazu FTIR 8101 PC infrared spectrophotometers.

The ^{1}H-NMR and ^{13}C-NMR were recorded at a Varian Mercury VX-300 MHz and 75 MHz, respectively, using TMS as internal standard in deuterated chloroform ($CDCl_3$) or deuterated dimethyl sulfoxide (DMSO-d6). Chemical shifts are measured in ppm. The mass spectra were recorded on a Shimadzu GCMS-QP-1000EX mass spectrometer at 70 eV. Elemental analyses were carried out at the Microanalytical Center of Cairo University. All the reactions and the purity of the new compounds were followed and checked by TLC using TLC aluminum sheets silica gel F_{254}.

*2.4. Synthesis of 6-Amino-4-(4-methoxyphenyl)-3-methyl-1,4-dihydropyrano[2,3-c]pyrazole-5-carbonitrile **(1)***

Method 1. An equimolar mixture of anisaldehyde (10 mmol, 1.36 mL), ethyl acetoacetate (10 mmol, 1.3 mL), hydrazine hydrate (0.5 mL), and malononitrile (10 mmol, 0.66 g) was stirred for 30 min at room temperature in the presence of triethyl amine and water (5 ml). The precipitate was filtered off and recrystallized from ethanol, yield (81%).

Method 2. A mixture of 3-methyl-5-pyrazolone (10 mmol, 0.98 g) and arylidine (10 mmol, 1.1 g) was refluxed in (30 mL) ethanol and 5 drops of piperidine for 2 h. The precipitate was filtrated off and recrystallized from ethanol to give solid 1, yield (87%).

Method 3. A mixture of 3-methyl-5-pyrazolone (10 mmol, 0.98 g), anisaldehyde (10 mmol, 1.36 mL), and malononitrile (10 mmol, 0.66 g) was refluxed in (30 mL) ethanol in presence of 5 drops of piperidine for 2 h. The precipitate was filtrated off and recrystallized from ethanol to give solid 1, yield (80%)

*2.5. Synthesis of 4-(4-Methoxyphenyl)-3,7-dimethyl-1,4-dihydro-5H-pyrazolo[4′,3′:5,6]pyrano[2,3-d][1,3]oxazin-5-one **(2)**.* A mixture of **1** (10 mmol, 2.82 g) and freshly distilled acetic anhydride (20 mL) was heated under reflux for 3 h. The solid that deposited after distilling the excess solvent was collected and recrystallized from ethanol to give **(2)**.

m.p. over 300°C. Anal. Calcd. for $C_{17}H_{15}N_3O_4$ (325.32): C, 62.76; H, 4.65; N, 12.92. Found: C, 63.07; H, 4.29; N, 13.10. FTIR (KBr) (cm^{-1}): 3157 (NH), 1732 (C=O), 1624 (C=N), 1606 (C=C). ^{1}H-NMR (DMSO-d_6): δ_H (ppm) 12.52 (s, 1H, NH, D_2O exchangeable), 7.162–7.136 (d, J = 7.8, 2H, Ar-H), 6.84–6.81 (d, J = 9, H, Ar-H), 4.7 (s, 1H, benzylic proton), 3.68 (s, 3H, OCH_3), 2, 16 (s, 3H, CH_3) and 1.831 (s, 3H, CH_3). ^{13}C-NMR (75 MHz, DMSO-d_6): δ 173.7, 165, 163.4, 161.9, 157.2, 138.6, 134.2, 130.3, 130.3, 114.4, 114.5, 112.9, 99.8, 55.5, 34.4, 15.7, 13.4. MS, m/z (%): 325 ($M^{\bullet+}$, 6), 281 (27), 241 (13), 218 (100), 107 (28).

*2.6. Synthesis of 4-(4-Methoxyphenyl)-3,7-dimethyl-4,6-dihydropyrazolo[4′,3′:5,6]pyrano[2,3-d]pyrimidine-5-(1H)-one **(3)**.* A mixture of **2** (10 mmol, 3.25 g) and formamide (10 mL) or excess ammonium acetate was refluxed for 4 h. After cooling white solid was collected, washed with water, dried, and recrystallized from benzene: ethanol mixture (1 : 1) to give **(3)**.

m.p. over 300°C Anal. Calcd. for $C_{17}H_{16}N_4O_3$ (324.33): C, 62.95; H, 4.97; N, 17.27. Found: C, 62.07; H, 4.92; N, 17.10. FTIR (KBr) (cm^{-1}): 3200, 3281 (NH), 1664 (C=O), 1610 (C=N) ^{1}H-NMR (DMSO-d_6): δ_H (ppm) 12.381 (s, 1H, NH, D_2O exchangeable), 12.01 (s, 1H, NH, D_2O exchangeable), 7.359–7.334 (d, J = 7.5, 2H, Ar-H), 6.790–6.762 (d, J = 8.4, 2H, Ar-H), 4.899 (s, 1H, benzylic proton), 3.668 (s, 3H, OCH_3), 2.176 (s, 3H, CH_3) and 1.791 (s, 3H, CH_3). MS, m/z (%): 324 ($M^{\bullet+}$, 16), 281 (19), 240 (24), 217 (100), 107 (51).

*2.7. Synthesis of 4-(4-Methoxyphenyl)-3,7-dimethyl-4-hydrobenzoimidazolo[1,2-c]pyrazolo[4′,3′:5,5]pyrano[3,2-d]pyrimidine **(4)**.* A mixture of 2 (10 mmol, 3.25 g) and 2-amino aniline (10 mmol, 1.08 g) in n-butanol (20 mL) was heated under reflux for 4 h. The solid that separated after concentration and cooling was filtered off and recrystallized from petroleum ether at 60–80°C.

m.p. 102–104°C Anal. Calcd. For $C_{23}H_{19}N_5O_2$ (397): C, 69.51; H, 4.82; N, 17.62. Found: C, 69.26; H, 4.78; N, 17.23. FTIR (KBr) (cm^{-1}): 3180 (NH), 1609 (C=N) ^{1}H-NMR (DMSO-d_6): δ_H (ppm) 11.252 (s, 1H, NH, D_2O exchangeable), 8.023–7.215 (m, 8H, Ar-H), 5.099 (s, 1H, benzylic proton), 3.212 (s, 3H, OCH_3), 2.801 (s, 3H, CH_3) and 2,176 (s, 3H, CH_3). MS, m/z (%): 395 ($M^{\bullet+}$, 7), 317 (9), 290 (12), 267 (10), 181 (25), 80 (100).

*2.8. Synthesis of N-[4-(4-Methoxyphenyl)-3,7-dimethyl-5-oxo-1,4-dihydropyrazolo[4′,3′:5,6]pyrano[2,3-d]pyrimidine-6(5H)-yl]thiourea **(5)**.* A mixture of **2** (10 mmol, 3.25 g) and thiosemicarbazide (10 mmol, 0.91 g) in dimethylformamide (20 mL) was heated under reflux for 6 h. The solid that was separated after concentration and cooling was filtered off, washed with water, and recrystallized from ethanol to give **(5)**.

m.p. 265–267°C Anal. Calcd. for $C_{18}H_{18}N_6O_3S$ (398.44): C, 54.26; H, 4.55; N, 21.09; S, 8.05. Found: C, 53.70; H, 4.19; N, 20.83. FTIR (KBr) (cm^{-1}): 3251, 3158, 3121 (NH_2, NH), 1665 (CO), 1619 (C=N) and 1254 (C=S). ^{1}H-NMR (DMSO-d_6): δ_H (ppm) 11.912 (s, 1H, NH, D_2O exchangeable), 7.923–7.751 (m, 4H, Ar-H), 8.213 (s, 1H, NH, D_2O exchangeable), 6.754 (s, 2H, NH_2, D_2O exchangeable), 5.013 (s, 1H, benzylic proton), 3.934 (s, 3H, OCH_3), 2.178 (s, 3H, CH_3) and 1.987 (s, 3H, CH_3). ^{13}C-NMR (75 MHz, DMSO-d_6): δ 182.7, 163.2, 161.9, 160.3, 158.5, 157.1, 139.8, 135.2, 130.3, 130.3, 114.6, 114.6, 113.8, 102.9, 55.9, 35.7, 22.3, 13.3,. MS, m/z (%): 398 ($M^{\bullet+}$, 23), 323 (20), 291 (18), 107 (100), 75 (21).

*2.9. Synthesis of 6-Amino-4-(4-methoxyphenyl)-3,7-dimethyl-4,6-dihydropyrazolo[4′,3′:5,6]pyrano[2,3-d]pyrimidin-5(1H)-one **(6)**.* To a solution of **2** (10 mmol, 3.25 g) in 20 mL of ethanol, hydrazine hydrate (0.5 mL) was added. The reaction mixture was refluxed for 4 h. The separated solid was filtered, dried, and recrystallized from ethanol to give **(6)**.

m.p. 268–290°C Anal. Calcd. for $C_{17}H_{17}N_5O_3$ (339.35): C, 60.17; H, 5.05; N, 20.64. Found: C, 60.07; H, 4.94; N, 20.53. FTIR (KBr) (cm^{-1}): 3325, 3210, 3154 (NH_2, NH), 1669 (CO), 1609 (C=N). ^{1}H-NMR (DMSO-d_6): δ_H (ppm) 12.012 (s, 1H, NH, D_2O exchangeable), 7.632–7.115 (m, 4H, Ar-H), 6.845 (s, 2H, NH_2, D_2O exchangeable), 5.013 (s, 1H, benzylic proton), 3.798 (s, 3H, OCH_3), 2.867 (s, 3H, CH_3) and 1.872 (s, 3H, CH_3). MS, m/z (%): 322 ($M^{\bullet+}$ -NH_3, 23), 282 (19), 240 (10), 80 (100).

*2.10. General Procedure for the Synthesis of Pyrazolopyranopyrimidine **(7a–c)**.* A solution of **2** (10 mmol, 3.25 g) in dimethylformamide (20 mL) and (10 mmol) of 4-methyl aniline, 4-methoxyaniline, and/or 4-amino aniline was heated under reflux for 5 h. The solid that was separated out after concentration and cooling was recrystallized from the proper solvent to give **(7a–c)**.

*2.11. 4-(4-Methoxyphenyl)-3,7-dimethyl-6-(p-tolyl)-4,6-dihydropyrazolo[4′,3′:5,6]pyrano[2,3-d]pyrimidin-5(1H)-one **(7a)**.* Solvent: ethanol m.p. 278–280°C. Anal. Calcd. for $C_{24}H_{22}N_4O_3$ (414.46): C, 69.55; H, 5.35; N, 13.52. Found: C, 68.67; H, 4.89; N, 12.98. FTIR (KBr) (cm^{-1}): 3191 (NH), 1682 (CO), 1608 (C=N). ^{1}H-NMR (DMSO-d_6): δ_H (ppm) 10.972 (s, 1H, NH, D_2O exchangeable), 7.963–7.785 (m, 8H, Ar-H), 4.876 (s, 1H, benzylic proton), 3.912 (s, 3H, OCH_3), 2.911 (s, 3H, CH_3) and 2.176 (s, 3H, CH_3) and 1.987 (s, 3H, CH_3). MS, m/z (%): 414 ($M^{\bullet+}$, 17), 386 (8), 307 (19), 107 (100).

*2.12. 4,6-Bis(4-methoxyphenyl)-3,7-dimethyl-4,6-dihydropyrazolo[4′,3′:5,6]pyrano[2,3-d]pyrimidin-5(1H)-one **(7b)**.* Solvent: ethanol. m.p. over 300°C Anal. C; alcd. for $C_{24}H_{22}N_4O_4$ (430.46): C, 66.97; H, 5.15; N, 13.02. Found: C, 66.17; H, 4.98; N, 13.41 FTIR (KBr) (cm^{-1}): 3154 (NH), 1685 (CO), 1611 (C=N). ^{1}H-NMR (DMSO-d_6): δ_H (ppm) 11.241

(s, 1H, NH, D_2O exchangeable), 8.054–7.815 (m, 8H, Ar-H), 5.045 (s, 1H, benzylic proton), 3.912 (s, 6H, 2 OCH_3), 2.832 (s, 3H, CH_3), 2.092 (s, 3H, CH_3) and 1.898 (s, 3H, CH_3). MS, *m/z* (%): 430 ($M^{\bullet+}$, 8), 402 (7), 295 (19), 80 (100).

2.13. 6-(4-Aminophenyl)-4-(methoxyphenyl)-3,7-dimethyl-4,6-dihydropyrazolo[4′,3′:5,6]pyrano[2,3-d]pyrimidin-5(1H)-one ***(7c)***. Solvent: dioxane. m.p. over 300°C Anal. Calcd. for $C_{23}H_{21}N_5O_3$ (415.44): C, 66.49; H, 5.09; N, 16.86. Found: C, 66.16; H, 4.92; N, 16.16. FTIR (KBr) (cm^{-1}): 3312, 3215, 3145 (NH_2, NH), 1679 (CO), 1611 (C=N). ^{1}H-NMR (DMSO-d_6): δ_H (ppm) 9.471 (s, 1H, NH, D_2O exchangeable), 7.439–7.005 (m, 8H, Ar-H), 6.450 (s, 2H, NH_2, D_2O exchangeable), 4.713 (s, 1H, benzylic proton), 3.803 (s, 3H, OCH_3), 2.367 (s, 3H, CH_3) and 1.912 (s, 3H, CH_3). MS, *m/z* (%): 415 ($M^{\bullet+}$, 10), 398 (9), 372 (20), 77 (100).

2.14. Synthesis of 4-(4-Methoxyphenyl)-3,7-dimethyl-4,6-dihydropyrazolo[4′,3′:5,6]pyrano[2,3-d]pyrimidine-5(1H)-thione ***(8)***. A mixture of **9** (10 mmol, 3.4 g) and thiourea (10 mmol, 0.76 g) in ethanol (20 mL) was heated under reflux for 6 h. The separated solid was filtered, dried, and recrystallized from ethanol to give **(8)**.

m.p. 218–220°C. Anal. Calcd. for $C_{17}H_{16}N_4O_2S$ (340): C, 59.98; H, 4.74; N, 16.46. Found: C, 60.03; H, 4.91; N, 16.10. FTIR (KBr) (cm^{-1}): 3158, 3121 (NH), and 1225 (C=S). ^{1}H-NMR (DMSO-d_6): δ_H (ppm) 11.912 (s, 1H, NH, D_2O exchangeable), 10.112 (s, 1H, NH, D_2O exchangeable), 7.923–7.751 (m, 4H, Ar-H), 4.731 (s, 1H, benzylic proton), 3.876 (s, 3H, OCH_3), 2.112 (s, 3H, CH_3) and 1.831 (s, 3H, CH_3). MS, *m/z* (%): 340 ($M^{\bullet+}$, 16), 281 (19), 241 (23), 233 (100), 107 (18).

2.15. Synthesis of 5-Chloro-4-(4-methoxyphenyl)-3,7-dimethyl-1,4-dihydropyrazolo[4′,3′:5,6]pyrano[2,3-d]pyrimidine ***(9)***. A mixture of **3** (10 mmol, 3.24 g), phosphorus oxychloride (15 mL), and phosphorus pentachloride (150 mmol, 3.13 g) was heated in water bath at 80°C for 8 h. The reaction mixture after cooling was poured onto crushed ice and the solid obtained was filtered off and recrystallized from ethanol to give pale brown solid **(9)**.

m.p. 198–200°C. Anal. Calcd. for $C_{17}H_{15}ClN_4O_2$ (342): C, 59.57; H, 4.41; N, 10.34. Found: C, 58.97; H, 3.95; N, 10.10. FTIR (KBr) (cm^{-1}): 3123 (NH), 1620 (C=N) ^{1}H-NMR (DMSO-d_6): δ_H (ppm) 12.012 (s, 1H, NH, D_2O exchangeable), 7.832–7.615 (m, 4H, Ar-H), 4.832 (s, 1H, benzylic proton), 3.843 (s, 3H, OCH_3), 2.546 (s, 3H, CH_3) and 1.687 (s, 3H, CH_3). MS, *m/z* (%): 342 ($M^{\bullet+}$, 10), 344 (25), 235 (12), 237 (30), 214 (28), 201 (100), 107 (30).

2.16. Synthesis of 5-Hydrazino-4-(4-methoxyphenyl)-3,7-dimethyl-1,4-dihydropyrazolo[4′,3′:5,6]pyrano[2,3-d]pyrimidine ***(10)***. A mixture of **9** (10 mmol, 3.4 g) and hydrazine hydrate (0.5 mL) in (30 mL) ethanol was heated under reflux for 5 h. The solid separated after concentration and cooling was filtered off and recrystallized from ethanol to give **(10)**.

m.p. 243–245°C Anal. Calcd. for $C_{17}H_{18}N_6O_2$ (338.36): C, 60.34; H, 5.36; N, 24.84. Found: C, 61.07; H, 5.23; N, 24.73. FTIR (KBr) (cm^{-1}): 3215, 3156, 3161 (NH_2, NH), 1612 (C=N) and. ^{1}H-NMR (DMSO-d_6): δ_H (ppm) 12.012 (s, 1H, NH, D_2O exchangeable), 8.123–7.836 (m, 4H, Ar-H), 6.013 (s, 1H, NH, D_2O exchangeable), 5.354 (s, 2H, NH_2, D_2O exchangeable), 4.803 (s, 1H, benzylic proton), 3.734 (s, 3H, OCH_3), 2,576 (s, 3H, CH_3) and 2.087 (s, 3H, CH_3). MS, *m/z* (%): 338 ($M^{\bullet+}$, 12), 307 (21), 244 (11), 201 (100), 107 (18).

2.17. Synthesis of 5-(2N,N-diacetylhydrazino)-4-(4-methoxyphenyl)-3,7-dimethyl-1,4-dihydropyrazolo[4′,3′:5,6]pyrano[2,3-d]pyrimidine ***(11)***. A mixture of **10** (10 mmol, 3.3 g) was heated under reflux with 30 mL of fresh distilled acetic anhydride for 8 h. The excess acetic anhydride was removed by distillation and the separated product was filtered off and recrystallized from ethanol to give **(11)**.

m.p. 278–280°C Anal. Calcd. for $C_{21}H_{22}N_6O_4$ (422.44): C, 59.71; H, 5.25; N, 19.81. Found: C, 60.17; H, 4.96; N, 19.23. FTIR (KBr) (cm^{-1}): 3211, 3198 (NH), 1732, 1695 (CO), and 1619 (C=N). ^{1}H-NMR (DMSO-d_6): δ_H (ppm) 12.901 (s, 1H, NH, D_2O exchangeable), 7.902–7.714 (m, 4H, Ar-H), 6.213 (s, 1H, NH, D_2O exchangeable), 5.630 (s, 1H, benzylic proton), 3.919 (s, 3H, OCH_3), 2.013 (s, 3H, CH_3), 2.678 (s, 6H, 2CH_3) and 1.987 (s, 3H, CH_3). ^{13}C-NMR (75 MHz, DMSO-d_6): δ 176.6, 174.6, 173.8, 171.8, 163.5, 160.2, 157.2, 138.7, 130.7, 130.0, 130.0, 114.6, 114.6, 113.1, 98.2, 56.0, 39.3, 25.2, 21.4, 21.4, 12.8. MS, *m/z* (%): 422 ($M^{\bullet+}$, 13), 322 (17), 307 (100), 201 (23), 107 (28), 101 (16).

Conflicts of Interest

The authors declare that they have no conflicts of interest.

Acknowledgments

The authors would like to express their appreciation for Ain Shams University.

References

[1] T. Echer and S. Hauptmann, *The Chemistry of Heterocycles Structure, Reaction, Syntheses, and Applications*, Wiley-VCH, Weinheim, 2003.

[2] C. G. Hartung, A. C. Backes, B. Felber, A. Missio, and A. Philipp, “Efficient microwave-assisted synthesis of highly functionalized pyrimidine derivatives,” *Tetrahedron*, vol. 62, no. 43, pp. 10055–10064, 2006.

[3] C. J. Shishoo, V. S. Shirsath, I. S. Rathod, and V. D. Yande, “Design, synthesis and antihistaminic (H1) activity of some condensed 3-aminopyrimidin-4(3H)-ones,” *European Journal of Medicinal Chemistry*, vol. 35, no. 3, pp. 351–358, 2000.

[4] R. V. Chambhare, B. G. Khadse, A. S. Bobde, and R. H. Bahekar, “Synthesis and preliminary evaluation of some N-[5-(2-furanyl)-2-methyl-4-oxo-4H-thieno[2,3-d]pyrimidin-3-yl]-carboxamide and 3-substituted-5-(2-furanyl)-2-methyl-3H-thieno[2,3-d]pyrimidin-4-ones as antimicrobial agents,” *European Journal of Medicinal Chemistry*, vol. 38, no. 1, pp. 89–100, 2003.

[5] M. Svobodová, P. Šimůnek, V. MacHáček, L. Štruncová, and A. Růžička, “Four-coordinate organoboron compounds from β-enaminonitriles and diazonium salts,” *Tetrahedron*, vol. 68, no. 8, pp. 2052–2060, 2012.

[6] D. Villemin, Z. Belhadj, N. Cheikh, N. Choukchou-Braham, N. Bar, and J. Lohier, "Solventless convenient synthesis of new cyano-2-aminopyridine derivatives from enaminonitriles," *Tetrahedron Letters*, vol. 54, no. 13, pp. 1664–1668, 2013.

[7] H. M. Aly and M. M. Kamal, "Efficient one-pot preparation of novel fused chromeno[2,3-d]pyrimidine and pyrano[2,3-d]pyrimidine derivatives," *European Journal of Medicinal Chemistry*, vol. 47, no. 1, pp. 18–23, 2012.

[8] N. R. Kamdar, D. D. Haveliwala, P. T. Mistry, and S. K. Patel, "Design, synthesis and in vitro evaluation of antitubercular and antimicrobial activity of some novel pyranopyrimidines," *European Journal of Medicinal Chemistry*, vol. 45, no. 11, pp. 5056–5063, 2010.

[9] R. K. Rad, J. Azizian, and H. Kefayati, "Electrogenerated acetonitrile anions/tetrabutylammonium cations: an effective catalytic system for the synthesis of novel chromeno[3′,4′:5, 6]pyrano[2,3-d]pyrimidines," *Tetrahedron Letters*, vol. 55, no. 50, pp. 6887–6890, 2014.

[10] K. Horváti, B. Bacsa, N. Szabó et al., "Antimycobacterial activity of peptide conjugate of pyridopyrimidine derivative against Mycobacterium tuberculosis in a series of in vitro and in vivo models," *Tuberculosis*, vol. 95, no. 1, pp. S207–S211, 2015.

[11] F. Buron, J. Y. Mérour, M. Akssira, G. Guillaumet, and S. Routier, "Recent advances in the chemistry and biology of pyridopyrimidines," *European Journal of Medicinal Chemistry*, vol. 95, pp. 76–95, 2015.

[12] J. Zhang, J. Huang, C. Liu et al., "Discovery of a series of pyridopyrimidine derivatives as potential topoisomerase I inhibitors," *Chinese Chemical Letters*, vol. 25, no. 7, pp. 1025–1028, 2014.

[13] S. Labadie, K. Barrett, W. S. Blair et al., "Design and evaluation of novel 8-oxo-pyridopyrimidine Jak1/2 inhibitors," *Bioorganic and Medicinal Chemistry Letters*, vol. 23, no. 21, pp. 5923–5930, 2013.

[14] J. Ilaš, P. Š. Anderluh, M. S. Dolenc, and D. Kikelj, "Recent advances in the synthesis of 2H-1,4-benzoxazin-3-(4H)-ones and 3,4-dihydro-2H-1,4-benzoxazines," *Tetrahedron*, vol. 61, no. 31, pp. 7325–7348, 2005.

[15] D. J. Connolly, D. Cusack, T. P. O'Sullivan, and P. J. Guiry, "Synthesis of quinazolinones and quinazolines," *Tetrahedron*, vol. 61, no. 43, pp. 10153–10202, 2005.

[16] A. M. Alanazi, A. A. M. Abdel-Aziz, I. A. Al-Suwaidan, S. G. Abdel-Hamide, T. Z. Shawer, and A. S. El-Azab, "Design, synthesis and biological evaluation of some novel substituted quinazolines as antitumor agents," *European Journal of Medicinal Chemistry*, vol. 79, pp. 446–454, 2014.

[17] M. N. Noolvi, H. M. Patel, V. Bhardwaj, and A. Chauhan, "Synthesis and in vitro antitumor activity of substituted quinazoline and quinoxaline derivatives: search for anticancer agent," *European Journal of Medicinal Chemistry*, vol. 46, no. 6, pp. 2327–2346, 2011.

[18] A. S. El-Azab, M. A. Al-Omar, A. A. M. Abdel-Aziz et al., "Design, synthesis and biological evaluation of novel quinazoline derivatives as potential antitumor agents: molecular docking study," *European Journal of Medicinal Chemistry*, vol. 45, no. 9, pp. 4188–4198, 2010.

[19] K. M. Amin, M. M. Kamel, M. M. Anwar, M. Khedr, and Y. M. Syam, "Synthesis, biological evaluation and molecular docking of novel series of spiro [(2H,3H) quinazoline-2,1′- cyclohexan]-4(1H)- one derivatives as anti-inflammatory and analgesic agents," *European Journal of Medicinal Chemistry*, vol. 45, no. 6, pp. 2117–2131, 2010.

[20] K. M. Pritchard, J. Al-Rawi, and C. Bradley, "Synthesis, identification and antiplatelet evaluation of 2-morpholino substituted benzoxazines," *European Journal of Medicinal Chemistry*, vol. 42, no. 9, pp. 1200–1210, 2007.

[21] N. Fawzy, M. O. I. Ghobashy, and A. K. El-Ziaty, "Synthesis and antimicrobial evaluation of some new quinazolin-4(3H)-one derivatives," *European Journal of Chemistry*, vol. 3, no. 4, pp. 437–441, 2012.

[22] M. R. Mahmoud, E. A. A. El-Bordany, N. F. Hassan, and F. S. M. A. El-Azm, "New 2,3-disubstituted quinazolin-4(3H)-ones from 2-undecyl-3,1-benzoxazin- 4-one," *Journal of Chemical Research*, no. 9, pp. 541–544, 2007.

[23] M. El-Hashash, J. Morsy, M. Azab, and N. Mahmoud, "Design, synthesis and anticancer activity of novel 2,3-and 2,4-disubstituted quinazoline and quinazolinone derivatives," *Heterocycles*, vol. 92, no. 2, pp. 316–329, 2016.

[24] A. M. M. El-Saghier and N. Mahmoud, "One-pot Synthesis and An Antibacterial Activity of 2-(2-hydroxyphenyl)-4-oxothiazolidin-3-yl)alkyl/ phenylthiazolidin-4-one derivatives and reaction with aromatic aldehydes," *Ciencia e Tecnica Vitivinicola*, vol. 31, 2016.

[25] M. A. El-hashash, A. M. El-Naggar, E. A. El-Bordany, M. I. Marzouk, and T. M. S. Nawar, "Regioselectivity and regiospecificity of benzoxazinone (2-isopropyl-4H-3,1-benzoxazinone) derivatives toward nitrogen nucleophiles and evaluation of antimicrobial activity," *Synthetic Communications*, vol. 46, no. 14, pp. 1230–1241, 2016.

[26] M. A. El-Hashash, A. M. El-Naggar, E. A. El-Bordany, M. I. Marzouk, and T. M. S. Nawar, "6-Iodo-2-isopropyl-4H-3,1-benzoxazin-4-one as building block in heterocyclic synthesis," *Synthetic Communications*, vol. 46, no. 24, pp. 2009–2021, 2016.

[27] H. M. F. Madkour, A. S. El-Sayed, A. I. S. Mounir, and E. A. A. El- Bordany, "Synthesis of some stable 4H-3, 1-benzoxazin-4-ones and their behavior toward nucleophiles," *Bulletin of the polish*, vol. 47, no. 3, pp. 217–229, 1999.

[28] A. K. Elziaty, O. E. A. Mostafa, E. A. ELBordany, M. Nabil, and H. M. F. Madkour, "Access tonew pyranopyrazoles and related heterocycles," *International Journal of Scientific & Engineering Research*, vol. 5, no. 1, pp. 727–735, 2014.

[29] H. Kiyani, H. A. Samimi, F. Ghorbani, and S. Esmaieli, "One-pot, four-component synthesis of pyrano[2,3-c]pyrazoles catalyzed by sodium benzoate in aqueous medium," *Current Chemistry Letters*, vol. 2, no. 4, pp. 197–206, 2013.

[30] M. M. Mashaly and M. Hammouda, "New simple and one-pot synthetic routes to polyfunctionally substituted pyridines; 1,4-dihydropyridazines and 4H-1,2-oxazine," *Zeitschrift fur Naturforschung - Section B Journal of Chemical Sciences*, vol. 54, no. 9, pp. 1205–1209, 1999.

[31] J. Gutteridge, D. Rowley, and B. Halliwell, "Superoxide-dependent formation of hydroxyl radicals in the presence of iron salts. detection of free iron in biological systems by using bleomycin-dependent degradation of DNA," *Biochemical Journal*, vol. 199, no. 1, pp. 263–265, 1981.

6

Advances in the Preparation of Fluorinated Isoquinolines: A Decade of Progress

Joseph C. Sloop

School of Science and Technology, Georgia Gwinnett College, 1000 University Center Lane, Lawrenceville, GA 30043, USA

Correspondence should be addressed to Joseph C. Sloop; jsloop@ggc.edu

Academic Editor: Esteban P. Urriolabeitia

Heterocyclic molecules incorporating fluorinated isoquinoline components are found in many medicinally and agriculturally important bioactive products as well as industrially impactful materials. Within the past decade, a variety of isoquinolinic ring assembly techniques has enabled the introduction of diverse fluorine-containing functionalities which can enhance potential bioactivity and industrial utility. This review examines recent noncatalyzed and transition metal catalyzed synthetic approaches to the assembly of isoquinoline derivatives that are ring-fluorinated and/or result in the incorporation of fluorine-containing functional groups. Specifically, efficient synthetic methods and regioselectivity in the incorporation of functional groups into isoquinoline ring systems are examined.

1. Introduction

1.1. Discovery of Isoquinoline and Early Synthesis Efforts. Isoquinoline, the bicyclic aromatic heterocycle shown in Figure 1, was first isolated as the sulfate salt from coal tar in 1885 by Hoogewerf and van Dorp [1]. By 1893, several syntheses of isoquinolinoid compounds were published by Pomeranz, Fritsch, Bischler, and Napieralski, reactions which bear their names [2–4]. Early in the 20th century, Pictet, Gams, and Spengler developed slightly different approaches to prepare isoquinoline derivatives [5, 6].

While the synthetic methods shown in Figure 1 enabled construction of different parts of the "A" ring component of the isoquinoline heterocycle, most early preparations required strong acids and refluxing conditions or dehydrating agents and dehydrogenation catalysts which limited functional group survival during the reactions.

1.2. Medicinal and Industrial Uses of Isoquinolines. Since its discovery, isoquinoline has remained an aromatic heterocycle of broad appeal to the synthetic organic, materials science, pharmaceutical, and agrochemical communities. Isoquinoline derivatives have important industrial applications, where they serve as solvents for aromatic molecules, fluorosensors, as components in paints, dyes, and electronic devices [7–9]. Isoquinoline derivatives such as papaverine have been isolated from natural sources and the isoquinolinoid pharmacore can be found in numerous drugs that possess antitumor, anesthetic, and antibiotic properties [10–12]. Figure 2 shows some representative examples of important isoquinoline-based molecules.

1.3. Effects of Fluorine Incorporation on Molecular Properties. When fluorine and fluorine-containing groups are incorporated into molecules, dramatic shifts of molecular properties occur in many instances. The inductive effect brought about by fluorine's high electronegativity and its small van der Waals radius of 1.47 Å changes molecular structural and stereoelectronic properties such as conformation, pK_a, polarity, solubility, and hydrogen-bonding capacity [13–18]. An example of how these alterations bear directly on materials science applications is found in the fluorinated isoquinoline-based electrophosphorescent iridium complexes used in color displays [9]. See Figure 3.

For a number of years, the pharmaceutical industry has leveraged these fluorine-induced molecular property modulations in drug discovery strategies when constructing

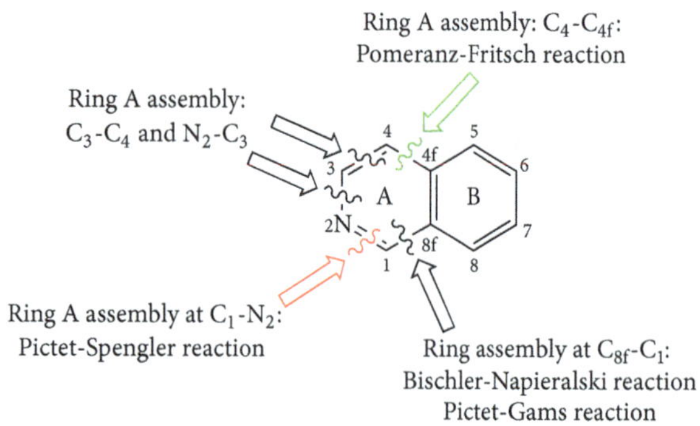

FIGURE 1: Isoquinoline ring assembly methods.

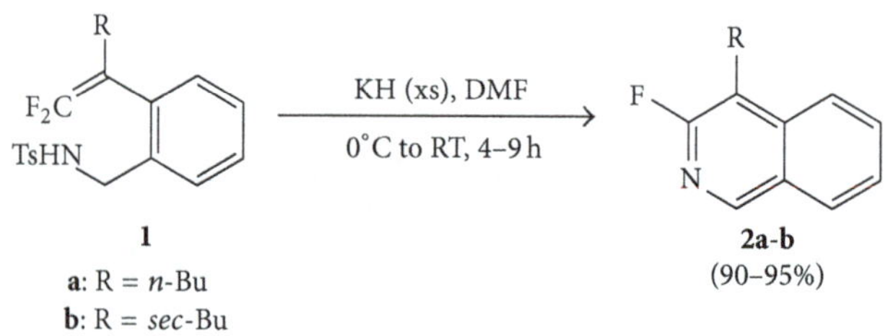

SCHEME 1: 1,1-Difluorostyrene aminotosylate cyclization.

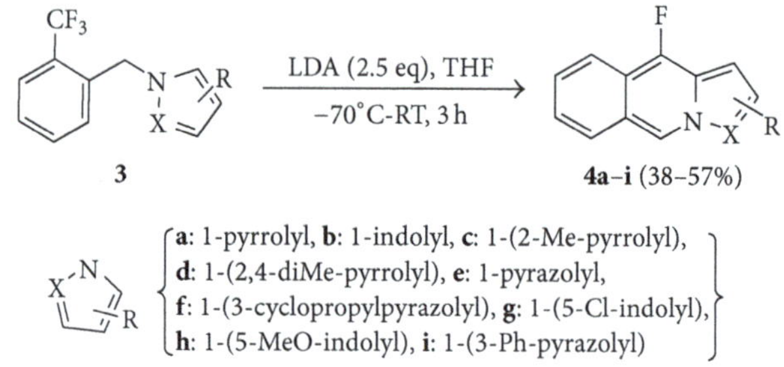

SCHEME 2: Fused-ring 4-fluoroisoqinoline derivative synthesis.

heteroaromatic pharmacores. Today, nearly 20% of FDA-approved drugs contain fluorine [19]. The very strong σ_{sp3} C-F bond (110 kcal/mol) increases the metabolic stability of drugs, enabling better bioavailability and binding affinity [19–22]. Isoquinolines functionalized with fluorine and fluorine-containing groups, the focus of this review, are key pharmacores with many applications. For example, isoquinoline carboxamides labeled with ^{18}F have found use as radiolabeling ligands for positron emission tomography [22]. Additionally, the preparation of fluorinated, fluoroalkylated, and fluoroarylated isoquinoline variants with antibacterial and antiparasitic properties continues to be an area of significant interest to the medicinal chemistry community. Figure 3 contains several examples of medicinally important isoquinoline derivatives which bear fluorine in their structure.

1.4. Recent Synthesis Efforts in the Preparation of Fluorinated Isoquinolines. This review examines a number of fluorinated isoquinoline preparations which capitalize on noncatalyzed reactions and cyclizations as well as those which employ transition metal catalysis to achieve the isoquinoline architecture. As Figure 4 shows, multiple methods that permit construction of the isoquinoline ring A and ring B components have been advanced and will be examined.

More specifically, we will consider those processes shown in Figure 4 which result in isoquinoline ring fluorinations, di- and trifluoromethylations, fluoroarylations, and trifluoromethylarylations. As Figure 5 indicates, fluorine functionality may be introduced at every carbon center of the isoquinoline core, and more than fifty-five examples will be explored in this review.

We will first examine nontransition metal mediated processes and then discuss the scope of transition metal catalysis in the preparation of fluorine-containing isoquinolines.

2. Ring-Fluorinated, Di- and Trifluoromethylated, Perfluoroalkylated, and Fluoroarylated Isoquinoline Synthesis via Nontransition Metal Mediated Processes

This portion of the review catalogues a broad cross section of isoquinoline preparations reported in the last decade which are not catalyzed by transition metal complexes. Methods examined include both intramolecular and intermolecular cycloadditions, tandem reactions, and multicomponent and single-pot processes which produce a wide array of A and B ring-fluorinated, di- and trifluoromethylated, perfluoroalkylated, and fluoroarylated polyfunctional isoquinolines. In addition, several trifluoromethylation and perfluoroalkylation reactions are shown that produce polyfunctional isoquinolines with a high degree of R_f substitution site regioselectivity.

2.1. 3- and 4-Fluoroisoquinolines via Intramolecular Ring A Cyclization at N_2-C_3 and C_3-C_4. Ichikawa and coworkers cyclized the difluoroalkene aminotosylate **1** (Scheme 1) under basic conditions in a 6-endo-trig fashion at N_2-C_3 to obtain the 4-butylated 3-fluoroisoquinoline series **2** in excellent yield [23].

Kiselyov reported a base-promoted intramolecular ring closure of *ortho*-trifluoromethyl benzyl heterocycles **3** en route to nine tricyclic 4-fluoroisoquinoline derivatives **4**. See Scheme 2. The reaction likely proceeds through a C_3-C_4 cyclization of a quinone methide intermediate arising from elimination of fluoride anion [24].

2.2. 6- and 7-Fluoroisoquinolines via Intermolecular Ring A Cyclization at C_{8f}-C_1, N_2-C_3, C_1-N_2, and C_3-C_4. In a recent *Nature* communication, Xie et al. noted that assembly of isoquinoline ring systems at C_{8f}-C_1 and N_2-C_3 by metal-free [4 + 2] cycloadditions was rare [25]. An example of their efforts is shown in Scheme 3, where a microwave-mediated, Brønsted acid-catalyzed [4 + 2] cycloaddition of the *p*-fluorophenyl ynamide **5** and nitriles **6a-b** produces 7-fluoroisoquinolines **7a-b** in good yield. Conversely, Feng and Wu reported in 2016 the base-promoted C_1-N_2 and C_3-C_4

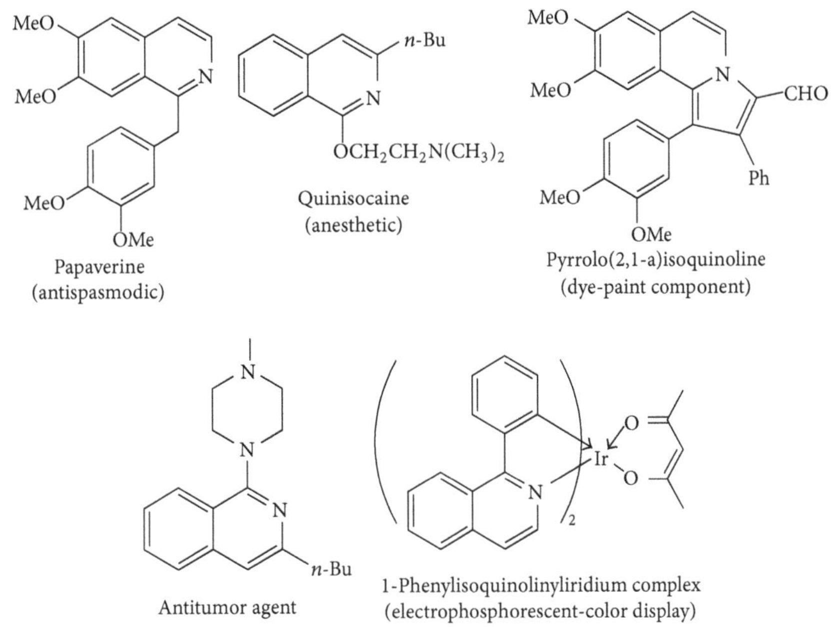

FIGURE 2: Selected medicinal, agricultural, and industrial isoquinoline derivatives.

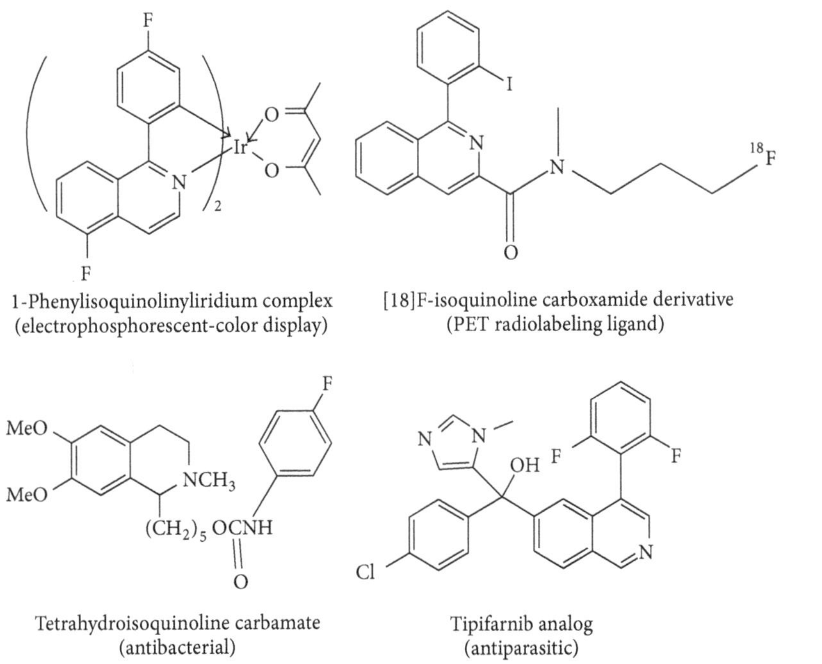

FIGURE 3: Selected fluorinated isoquinoline derivatives with medicinal and industrial applications.

cycloaddition of benzonitriles **8** and **9** to prepare the 1-amino-isoquinolines **10a-b** which are fluorinated at the 6- and 7-positions in moderate yields [26].

Yang et al. employed a tandem C_{8f}-C_1 and C_1-N_2bond-forming process with azidoacrylate **11** and asymmetric diazo diketone **12** (Scheme 4) to obtain a 1:1 ratio of 7-fluoroisoquinoline isomers **13a** and **13b** in good yield [27]. This interesting reaction cascade begins with the in situ phosphazene formation of **11** via the Staudinger-Meyer reaction with concomitant ketene formation from **12** via a

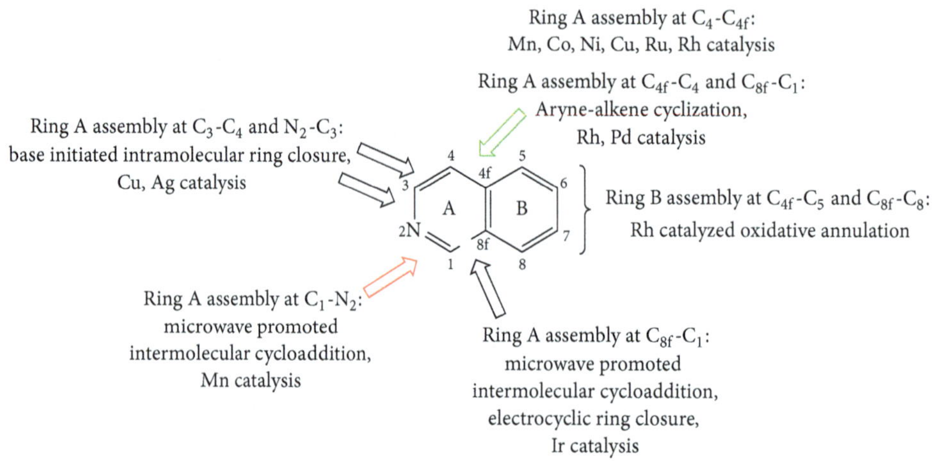

FIGURE 4: Recent isoquinoline synthesis methods.

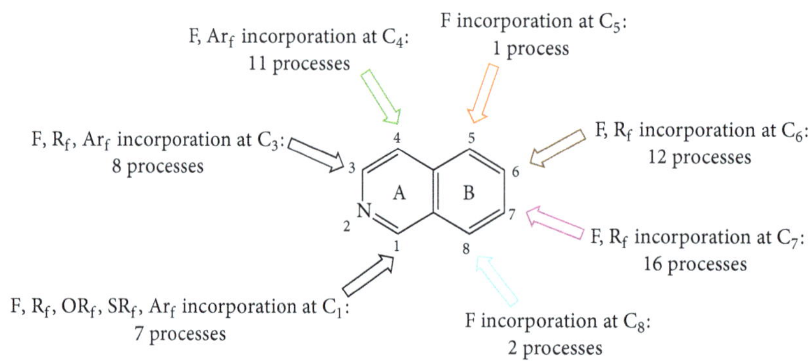

FIGURE 5: Isoquinoline fluorination, fluoroalkylation, and fluoroarylation sites.

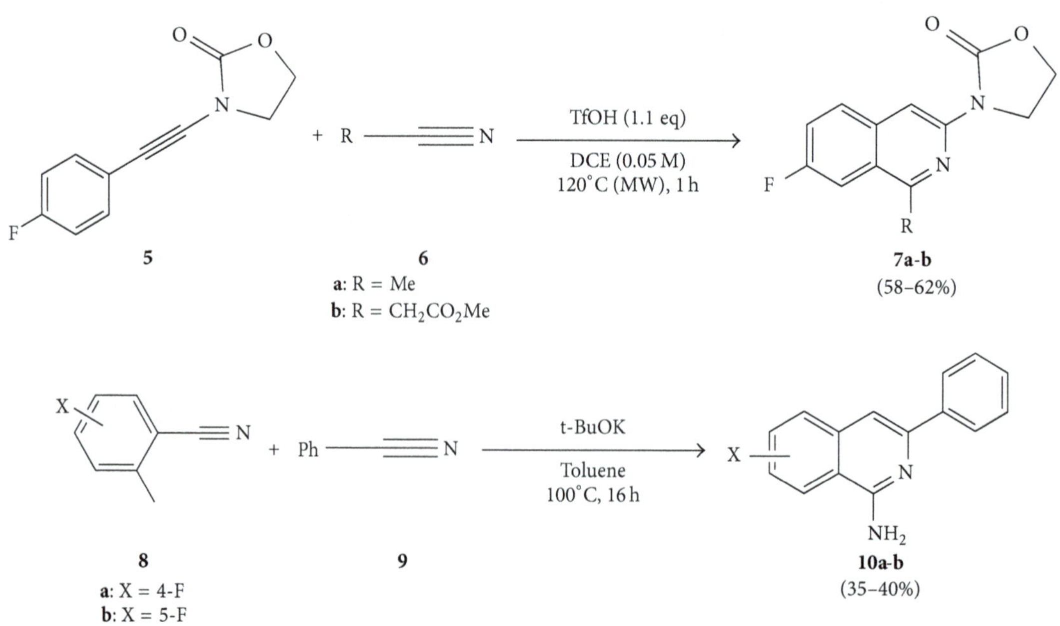

SCHEME 3: Preparation of 6- and 7-fluoroisoquinolines via ynamide-nitrile and nitrile-nitrile cycloadditions.

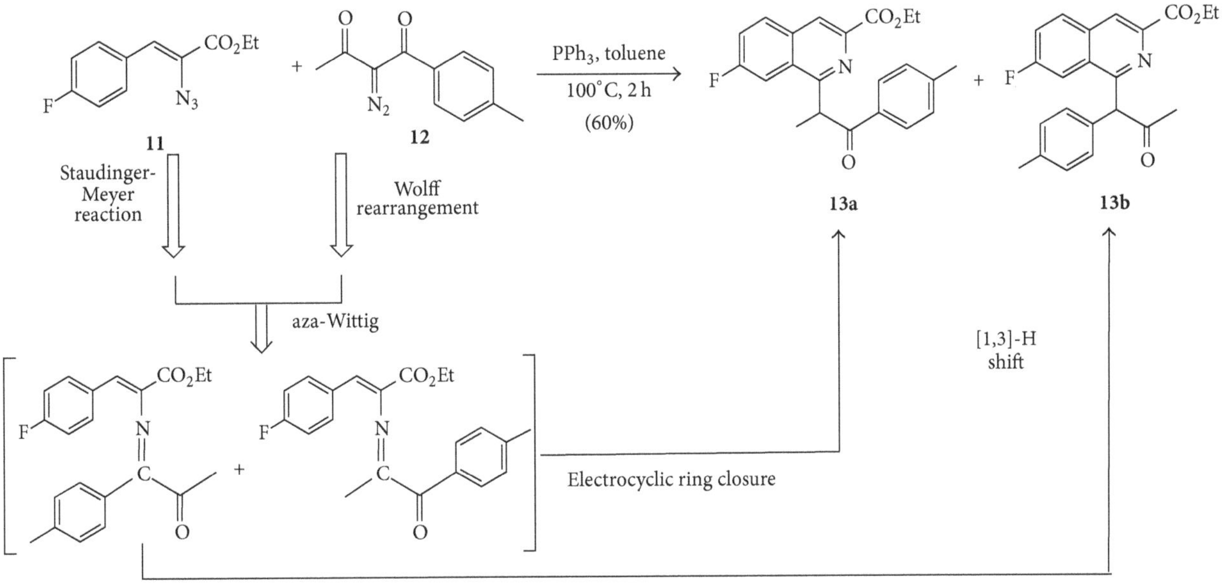

SCHEME 4: Triphenylphosphine-mediated 7-fluoroisoquinoline synthesis.

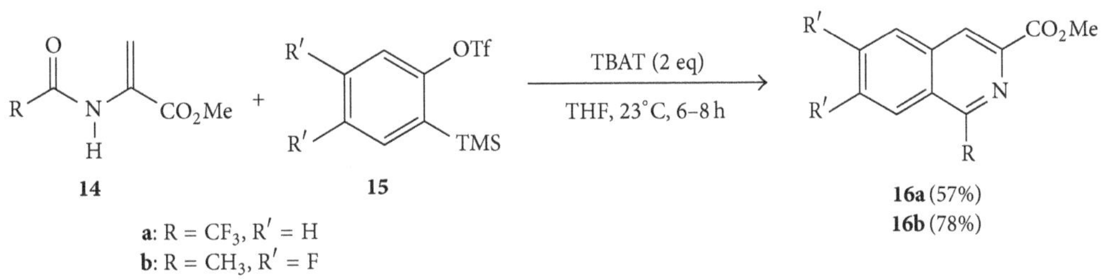

SCHEME 5: 6,7-Difluoro- and 1-trifluoromethylisoquinolines via aryne annulation.

Wolff rearrangement. An aza-Wittig reaction between the phosphazene and ketene yields the isomeric N-vinylic ketene imines in brackets. Finally, electrocyclic ring closure and a subsequent [1,3]-H migration leads to the formation of isoquinoline isomers.

2.3. Trifluoromethylated, Ring-Fluorinated, and Perfluorinated Isoquinolines via Inter- and Intramolecular Ring A Cyclizations at C_{8f}-C_1 and C_{4f}-C_4. Stoltz's team leveraged their success in C-C bond insertion reactions to develop a $Bu_4NPh_3SiF_2$ (TBAT) promoted N-acyl dehydroamino ester-(**14**) aryne (**15**) C_{8f}-C_1/C_{4f}-C_4 annulation that resulted in the successful synthesis of a series of polysubstituted isoquinolines and indoles [28]. TBAT serves as the fluoride anion source that desilylates **15** to generate the aryne upon loss of the triflate. Scheme 5 depicts the preparation of methyl-3-carboxy-1-trifluoromethylisoquinoline (**16a**) and methyl-3-carboxy-6,7-difluoro-1-methylisoquinoline (**16b**) in good overall yield.

Zhang and Studer [29] recently reported the C_{8f}-C_1 intramolecular cyclization of β-aryl-α-isocyano-acrylates **17** and radical perfluoroalkylation using Togni reagents **18** as the R_f source to produce eighteen examples of highly functionalized 1-trifluoromethyl-, 1-pentafluoroethyl-, and 1-heptafluoropropylisoquinolines **19** in good overall yields. See Scheme 6.

2.4. Preparation of Selectively Trifluoromethylated and Difluoromethylated Isoquinolines at C_1 and C_4. Kuninobu et al. successfully conducted a benzylic trifluoromethylation of the isoquinolinium N-oxide **20** using KBF_3CF_3 in BF_3OEt_2 to produce the 1-(2,2,2-trifluoroethyl)isoquinoline **21** in 71% yield [30]. See Scheme 7. As shown in Scheme 8, Ichikawa employed a dehydrogenation treatment of dihydroisoquinoline **22** to achieve the 4-trifluoromethylisoquinoline **23** in high yield. The same substrate was subjected to dehydrofluorination (with subsequent alkene isomerization) to obtain 4-difluoromethylisoquinoline **24** in good yield [31].

2.5. Preparation of 4-Fluoroarylated Isoquinolines. During an examination of tipifarnib analogs for bioactivity, Chennamaneni and coworkers coupled the trityl-protected, tetrahydroisoquinoline **25** with the heterocyclic ketone **26** (Scheme 9) en route to the 4-(2,6-difluorophenyl) isoquinoline **27** in 60% yield [32].

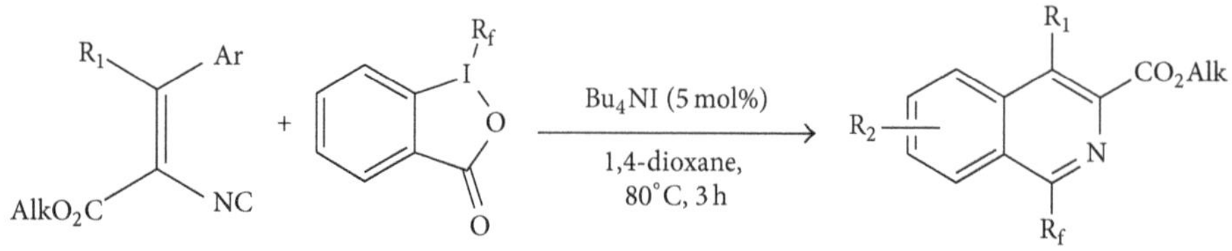

17 18 19a–r (53–80%)

Alk = Me, $R_f = CF_3$

a: R_1 = Ph, R_2 = H
b: R_1 = Ph, R_2 = 6-OMe, 7-OMe
c: R_1 = Ph, R_2 = 7-Cl
d: R_1 = 4-MeC_6H_4, R_2 = 7-Me
e: R_1 = 4-$MeOC_6H_4$, R_2 = 7-OMe
f: R_1 = 4-ClC_6H_4, R_2 = 7-Cl
g: R_1 = 4-BrC_6H_4, R_2 = 7-Br
h: R_1 = H, R_2 = 7-OMe
i: R_1 = Me, R_2 = H
j: R_1 = Me, R_2 = 7-Me
k: R_1 = *n*-Pr, R_2 = H
l: R_1 = *i*-Pr, R_2 = H
m: R_1 = 4-ClC_6H_4, R_2 = H
n: R_1 = 4-BrC_6H_4, R_2 = H
o: R_1 = cyclohexyl, R_2 = H

Alk = Me, $R_f = CF_2CF_3$

p: R_1 = Ph, R_2 = H

Alk = Me, $R_f = (CF_2)_2CF_3$

q: R_1 = Ph, R_2 = H

Alk = *t*-Bu, $R_f = CF_3$

r: R_1 = Ph, R_2 = H

Scheme 6: 1-Perfluoroalkylisoquinolines via isonitrile cyclization and radical fluoroalkylation.

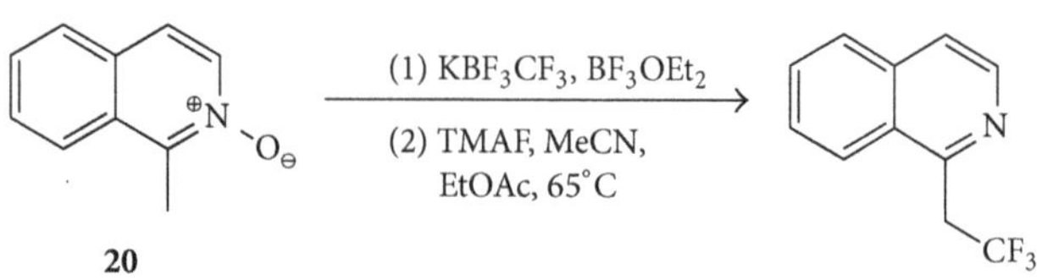

20 21 (71%)

Scheme 7: Benzylic trifluoromethylation of 1-methylisoquinolinium *N*-oxide.

Using amine **28** and 4-fluorobenzaldehyde shown in Scheme 10, Awuah and Capretta combined microwave conditions and acid-catalysis to condense and cyclize the imine adduct at C_{8f}-C_1 followed by dehydrogenation to produce the 2-(4-fluorophenyl) isoquinoline **29a** in excellent overall yield. In a second reaction, **28** was reacted with 4-fluorophenylacetic acid under microwave conditions in the presence of phosphoryl chloride to prepare the 1-(4-fluorobenzoyl)isoqionoline **29b** in good yield [33].

3. Fluorinated, Trifluoromethylated, Fluoroarylated, and Trifluoromethylarylated Isoquinoline Synthesis via Transition Metal Mediated Processes

This section catalogues a representative sample of isoquinoline preparations catalyzed by periods 4, 5, and 6 transition metal complexes. Methods examined include those which produce a wide array of A and B ring-fluorinated, trifluoromethylated, and fluoroarylated polyfunctional isoquinolines. In addition, several selective fluorination and trifluoromethylation reactions, done in concert with transition metal catalysts, are shown that produce polyfunctional isoquinolines with a high degree of fluorination and trifluoromethylation site regiocontrol.

3.1. Period 4 Transition Metal Catalysis: Ring-Fluorinated, Trifluoromethylated, Fluoroarylated, and Trifluoromethylarylated Isoquinolines via Inter- and Intramolecular Ring A Cyclization at N_2-C_3, C_{4f}-C_4, and C_1-C_4

3.1.1. Manganese Catalysis. He et al. researchers utilized the manganese catalyst $MnBn(CO)_5$ in a [4 + 2] annulation at N_2-C_3 and C_{4f}-C_4 of imines **30** and alkynes **31** to prepare six fluorinated isoquinolines **32** in good to excellent yields shown in Scheme 11 [34]. Their strategy affords three ring B fluoroisoquinoline regioisomers (**32a–c**) and a trifluoromethylisoquinoline (**32d**) as well as a ring A fluorophenylisoquinoline (**32e**) and a ring A trifluoromethylphenylisoquinoline (**32f**). Mao's group recently reported the manganese catalyzed coupling and cyclization of vinyl isocyanides **33** and assorted arylhydrazines **34** en route to three new examples of 7-fluoroisoqinoline **35a**, the 1-(4-fluorophenyl)-isoquinoline **35b**, and the 1-(4-trifluoromethylphenyl)-isoquinoline **35c** in good overall yields [35].

3.1.2. Cobalt Catalysis. Cobalt (III) catalysis has been effectively utilized in N_2-C_3 and C_{4f}-C_4 bond formation/cyclizations of oximes, amidines, and hydrazones to prepare a broad array of isoquinolines bearing fluorine and fluorine-containing groups. Examples are depicted in Scheme 12. Sun and coworkers used $Cp{*}CoI_2(CO)$ to catalyze the regioselective cyclization of the acyl oxime series **36** with both internal and terminal alkynes **37** en route to a diverse set

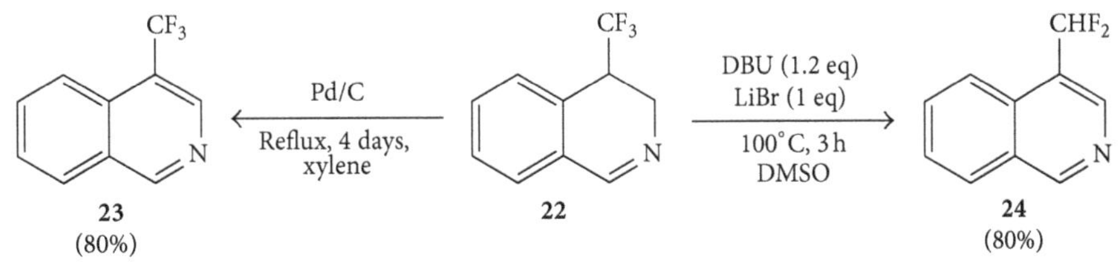

SCHEME 8: Preparation of 4-trifluoromethyl- and 4-difluoromethylisoquinolines.

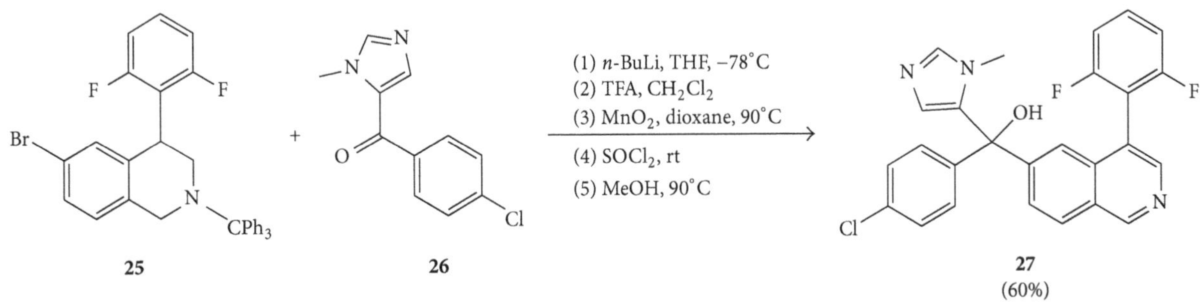

SCHEME 9: Preparation of difluoroarylated tipifarnib analog.

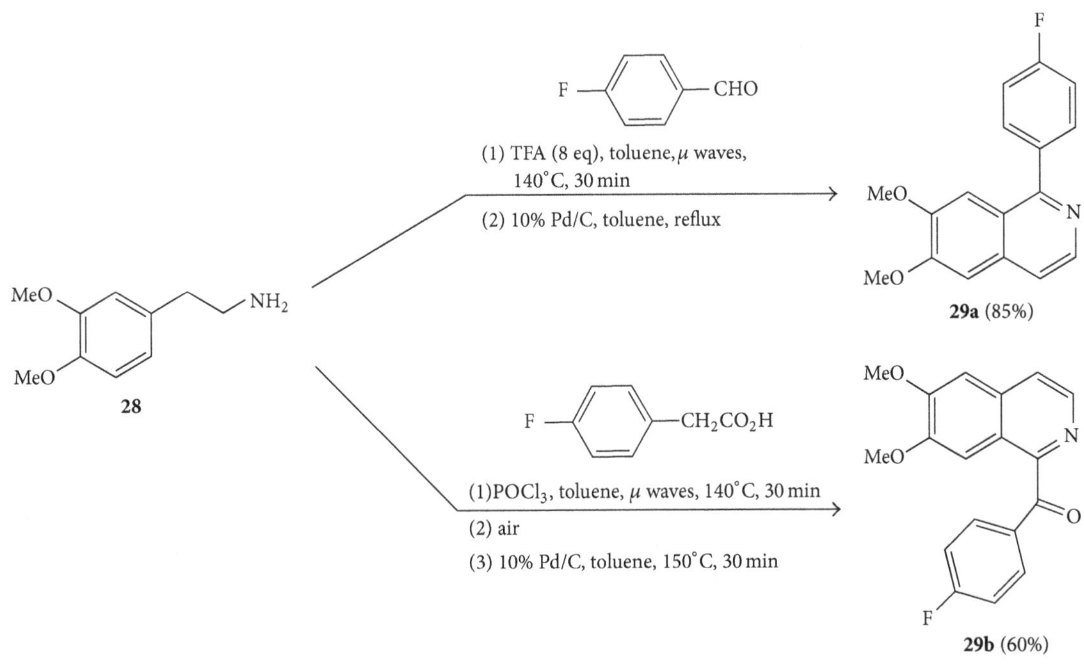

SCHEME 10: Synthesis of fluorophenyl- and fluorobenzoylisoquinolines.

of eleven fluorine-containing isoquinolines **38** [36]. Li's team used the same catalyst system with a slightly different base to couple aryl amidines **39** and diazo compounds **40** to deliver five examples of monofluorinated and trifluoromethylated 1-aminoisoquinolines **41** in fair to excellent yields [37]. Pawar et al. just released a study of Co(III)-catalyzed C–H/N–N bond functionalization of arylhydrazones **42** with internal alkynes **43** for the synthesis of three ring-fluorinated and fluoroarylated isoquinoline derivatives **44** in very good yields [38], while a 2016 report by Yu et al. catalogued the cycloaddition of oxadiazolones **45** with alkynes **46** to prepare eight ring-fluorinated, trifluoromethylated, and fluoroarylated isoquinolines **47** in yields ranging from 45 to 85% [39]. The relatively mild reaction conditions used in these processes permit toleration of a wide scope of substrates.

3.1.3. Nickel Catalysis. Yoshida's group conducted a nickel catalyzed, N_2-C_3/C_{4f}-C_4 bond formation/cycloaddition of

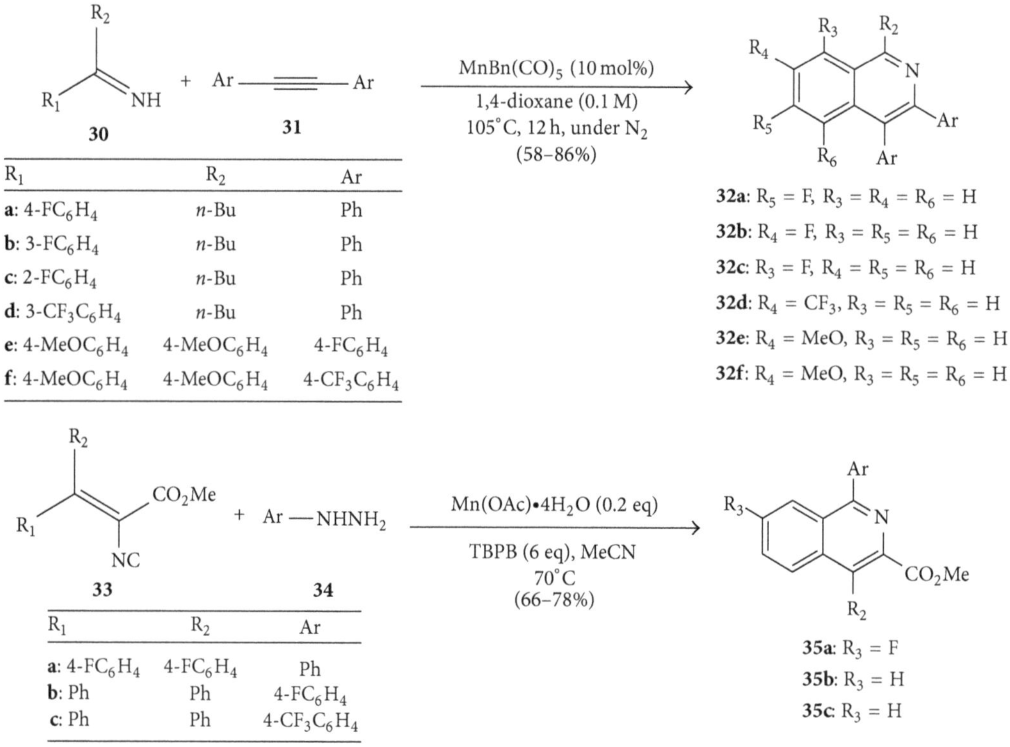

R_1	R_2	Ar
a: 4-FC_6H_4	*n*-Bu	Ph
b: 3-FC_6H_4	*n*-Bu	Ph
c: 2-FC_6H_4	*n*-Bu	Ph
d: 3-$CF_3C_6H_4$	*n*-Bu	Ph
e: 4-$MeOC_6H_4$	4-$MeOC_6H_4$	4-FC_6H_4
f: 4-$MeOC_6H_4$	4-$MeOC_6H_4$	4-$CF_3C_6H_4$

R_1	R_2	Ar
a: 4-FC_6H_4	4-FC_6H_4	Ph
b: Ph	Ph	4-FC_6H_4
c: Ph	Ph	4-$CF_3C_6H_4$

Scheme 11: Manganese catalyzed fluoroisoquinoline synthesis.

aromatic ketoximes **48** with 4-octyne **49** to prepare fluorine-containing isoquinolines **50a** and **50b** in good overall yield [40]. See Scheme 13.

3.1.4. Copper Catalysis. Ohto et al. developed a copper(I)-catalyzed four-component coupling reaction whereby 2-ethynylbenzaldehydes **51**, paraformaldehyde, diisopropylamine, and *t*-$BuNH_2$ were cyclized at N_2-C_3 to give 6-fluoro- and 7-fluoroisoquinolines **52** [41]. See Scheme 14. Fan et al. used isoquinoline-*N*-oxides **53** and the Togni reagent **54** catalyzed by copper(II) triflate, to prepare sixteen 1-(trifluoromethyl)isoquinolines **55** [42]. Mormino et al. subjected 3-bromoisoquinoline **56** to the phenanthroline-ligated copper trifluoromethylating reagent (phen)$CuCF_3$ **57**, preparing the 4-trifluoromethylisoquinoline **58** via a radical substitution [43]. These processes provide the target isoquinolines in good overall yields.

3.2. *Period 5 Transition Metal Catalysis: Ring-Fluorinated, Trifluoromethylated, Fluoroarylated, and Trifluoromethylarylated Isoquinolines via Inter- and Intramolecular Ring A Cyclization at N_2-C_3, C_{4f}-C_4, C_{8f}-C_8, and C_1-C_4*

3.2.1. Ruthenium Cataliysis. Scheme 15 depicts two ruthenium catalyzed cyclizations to produce fluorine-containing isoquinolines via N_2-C_3 and C_{4f}-C_4 bond formation. He et al. cyclized several diaryl imines **59** with diarylacetylenes **60** with a ruthenium catalyst to prepare six tetrasubstituted, ring-fluorinated, trifluoromethylated, fluoroarylated, and trifluoromethylarylated isoquinolines **61** in yield ranging from 40 to 86% [44]. Chinnagolla and coworkers used a similar methodology and cyclized several diaryl imine chlorides **62** with diarylacetylenes **63** with a ruthenium catalyst to prepare a series of six tetrasubstituted trifluoroethoxyisoquinolines **64** in yield ranging from 61 to 81% [45].

3.2.2. Rhodium Catalysis. Two rhodium-catalyzed preparations of isoquinolines that incorporate fluorine and trifluoromethyl groups are shown in Scheme 16. Qian's group conducted a rhodium-catalyzed C_{8f}-C_8/C_{4f}-C_4 annulation of fluoropicolinamide **65** with diphenylacetylene **66** to produce the tetraphenyl 4-fluoroisoquinoline **67** in excellent yield [46]. Guimond and Fagnou performed a highly efficient rhodium-catalyzed oxidative cross-coupling and C_{4f}-C_4/N_2-C_3 cycloaddition of aldimines **68** and 4-octyne **69** en route to 6-fluoro- and 6-trifluoromethylisoquinolines **70a** and **70b** [47]. These catalyzed processes are believed to occur via a sequence involving Rh (III) insertion followed by reductive elimination/electrocyclization to produce the isoquinolines.

3.2.3. Palladium Catalysis. Scheme 17 depicts work conducted by Pilgrim and coworkers in which the protected aryl bromide series **71**, acetophenone **72**, an electrophile, and ammonium chloride were combined in a palladium catalyzed, three-step, one-pot C_{4f}-C_4/N_2-C_3 cycloaddition process to furnish ring-fluorinated isoquinoline **73a**, a trifluoromethylated isoquinoline **73b,** and trifluoromethylarylated isoquinolines **73c–d** in overall yields ranging from 46 to 73% [48]. This procedure begins with the arylation reaction of an enolate, followed by reaction with Selectfluor II,

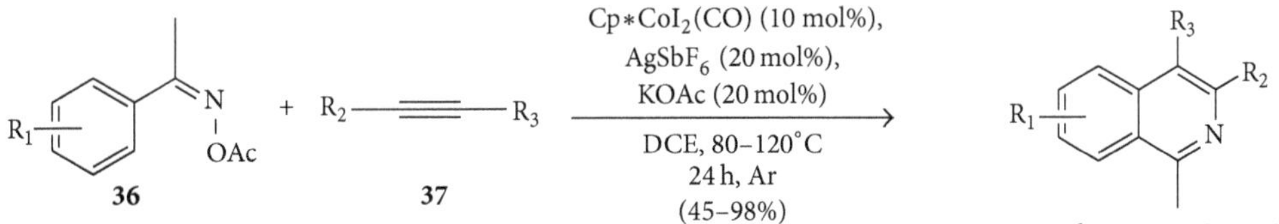

a: R_1 = 3-Cl, R_2 = 4-FC_6H_4, R_3 = H
b: R_1 = 3-Br, 4-F, R_2 = 2-MeC_6H_4, R_3 = H
c: R_1 = 3-CF_3, R_2 = 2-MeC_6H_4, R_3 = H
d: R_1 = 3-CF_3, R_2 = $(CH_2)_2Ph$, R_3 = H
e: R_1 = 3-CF_3, R_2 = $(CH_2)_2Me$, R_3 = H
f: R_1 = 3-CF_3, R_2 = $(CH_2)_4Me$, R_3 = H
g: R_1 = 3-CF_3, R_2 = cyclohexyl, R_3 = H
h: R_1 = 3-CF_3, R_2 = R_3 = Ph
i: R_1 = 4-Cl, R_2 = 4-FC_6H_4, R_3 = H
j: R_1 = 4-F, R_2 = Ph, R_3 = H
k: R_1 = 4-CF_3, R_2 = Ph, R_3 = H

38a: R_1 = 7-Cl, R_2 = 4-FC_6H_4, R_3 = H
38b: R_1 = 7-Br, 4-F, R_2 = 2-MeC_6H_4, R_3 = H
38c: R_1 = 7-CF_3, R_2 = 2-MeC_6H_4, R_3 = H
38d: R_1 = 7-CF_3, R_2 = $(CH_2)_2Ph$, R_3 = H
38e: R_1 = 7-CF_3, R_2 = $(CH_2)_2Me$, R_3 = H
38f: R_1 = 7-CF_3, R_2 = $(CH_2)_4Me$, R_3 = H
38g: R_1 = 7-CF_3, R_2 = cyclohexyl, R_3 = H
38h: R_1 = 7-CF_3, R_2 = R_3 = Ph
38i: R_1 = 6-Cl, R_2 = 4-FC_6H_4, R_3 = H
38j: R_1 = 6-F, R_2 = Ph, R_3 = H
38k: R_1 = 6-CF_3, R_2 = Ph, R_3 = H

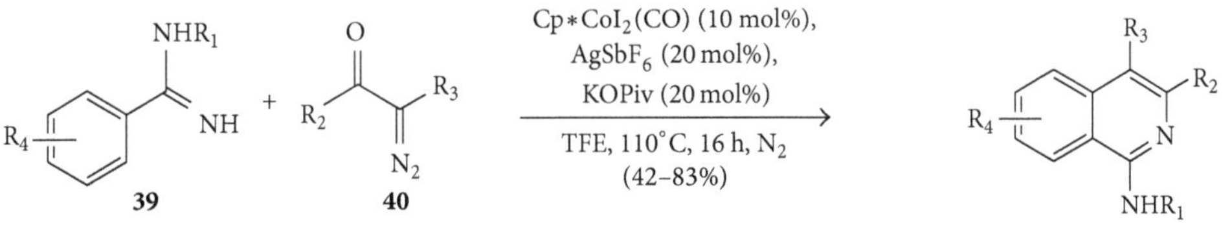

a: R_1 = t-Bu, R_2 = Me, R_3 = CO_2Et, R_4 = 4-F
b: R_1 = t-Bu, R_2 = Me, R_3 = CO_2Et, R_4 = 3-F
c: R_1 = t-Bu, R_2 = Me, R_3 = CO_2Et, R_4 = 3-CF_3
d: R_1 = t-Bu, R_2 = Me, R_3 = CO_2Et, R_4 = 2-F
e: R_1 = t-Bu, R_2 = H, R_3 = Ph, R_4 = 4-F

41a: R_1 = t-Bu, R_2 = Me, R_3 = CO_2Et, R_4 = 6-F
41b: R_1 = t-Bu, R_2 = Me, R_3 = CO_2Et, R_4 = 7-F
41c: R_1 = t-Bu, R_2 = Me, R_3 = CO_2Et, R_4 = 7-CF_3
41d: R_1 = t-Bu, R_2 = Me, R_3 = CO_2Et, R_4 = 8-F
41e: R_1 = t-Bu, R_2 = H, R_3 = Ph, R_4 = 6-F

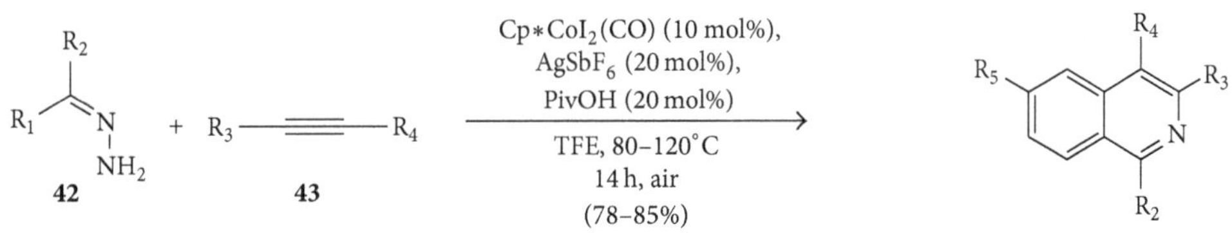

a: R_1 = R_2 = Ph, R_3 = R_4 = 4-$CF_3C_6H_4$
b: R_1 = R_2 = 4-FC_6H_4, R_3 = R_4 = Ph
c: R_1 = R_2 = Ph, R_3 = R_4 = 4-FC_6H_4

44a: R_2 = Ph, R_3 = R_4 = 4-$CF_3C_6H_4$, R_5 = H
44b: R_2 = 4-FC_6H_4, R_3 = R_4 = Ph, R_5 = F
44c: R_2 = Ph, R_3 = R_4 = 4-FC_6H_4, R_5 = H

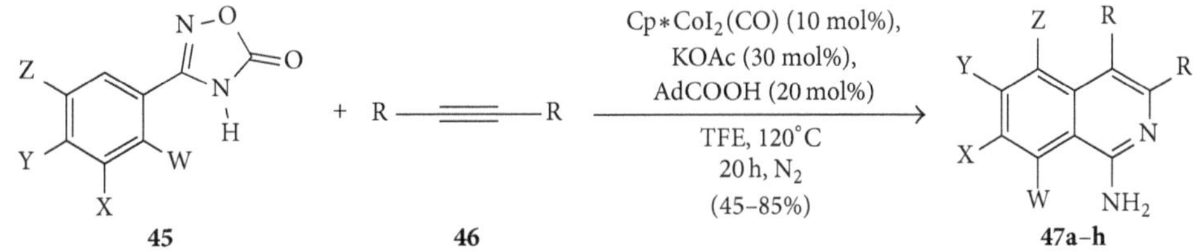

a: Y = F, W = X = Z = H — R = Ph
b: Y = 4-CF_3, W = X = Z = H — R = Ph
c: W = F, X = Y = Z = H — R = Ph
d: Y = 4-CF_3, W = X = Z = H — R = Ph
e: W = H, X = Y = Z = F — R = Ph
f: Y = Cl, W = X = Z = H — R = 4-FC_6H_4
g: Y = Cl, W = X = Z = H — R = 2-FC_6H_4
h: Y = Cl, W = X = Z = H — R = 3-FC_6H_4

Scheme 12: Cobalt catalyzed fluoroisoquinoline syntheses.

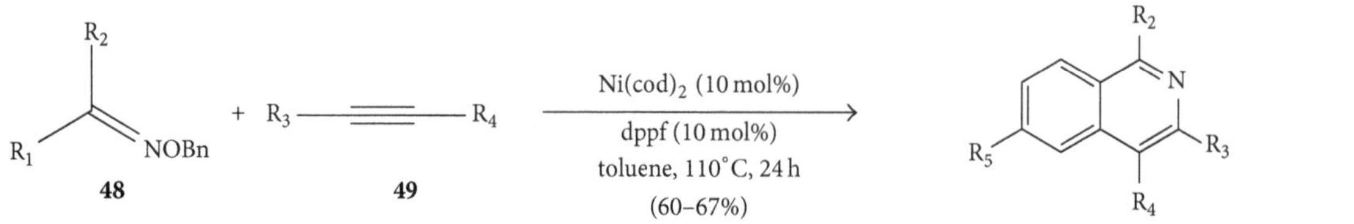

48 **49**

a: $R_1 = R_2 = 4\text{-}FC_6H_4$, $R_3 = R_4 = Pr$
b: $R_1 = 4\text{-}FC_6H_4$, $R_2 = 4\text{-}MeOC_6H_4$, $R_3 = R_4 = Pr$

50a: $R_2 = 4\text{-}FC_6H_4$, $R_3 = R_4 = Pr$, $R_5 = F$
50b: $R_2 = 4\text{-}MeOC_6H_4$, $R_3 = R_4 = Pr$, $R_5 = F$

SCHEME 13: Nickel catalyzed 6-fluoro and 1-(4-fluorophenyl)isoquinoline synthesis.

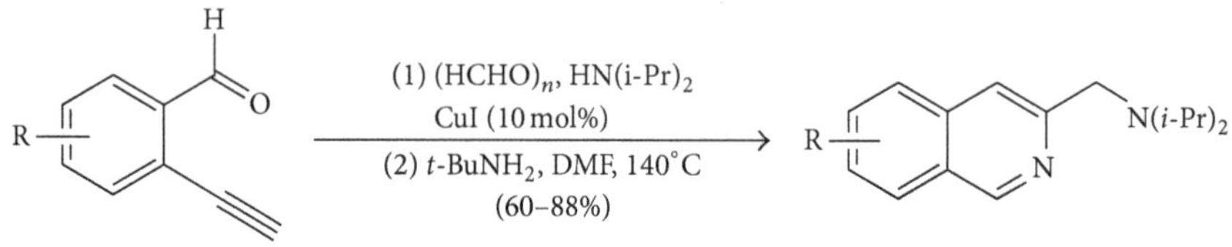

51

52a: R = 6-F
52b: R = 7-F

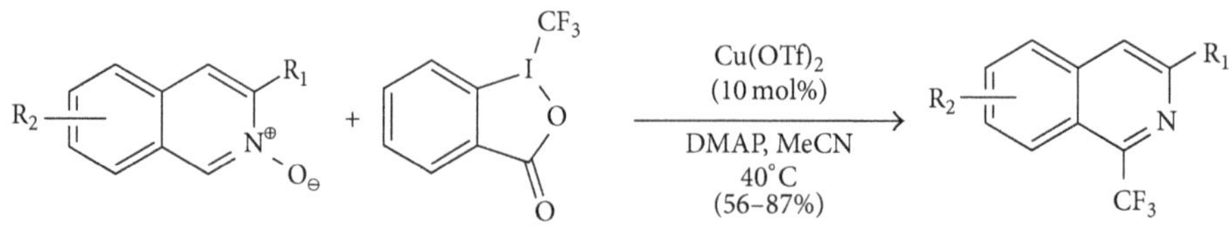

53 **54** **55a–p**

a: R_1 = Ph, R_2 = H
b: R_1 = Ph, R_2 = 6-OMe, 7-OMe
c: R_1 = Ph, R_2 = 7-Cl
d: R_1 = Ph, R_2 = 7-Me
e: R_1 = *t*-Bu, R_2 = H
f: R_1 = *n*-Bu, R_2 = 7-Cl
g: R_1 = Ph, R_2 = 7-F
h: R_1 = Ph, R_2 = 6-F
i: R_1 = Ph, R_2 = 5-OMe
j: R_1 = *t*-Bu, R_2 = H
k: R_1 = *n*-Bu, R_2 = 7-Cl
l: R_1 = *p*-Anisyl, R_2 = H
m: R_1 = *p*-ClC_6H_4, R_2 = H
n: R_1 = cyclopropyl, R_2 = H
o: R_1 = cyclopropyl, R_2 = 7-F
p: R_1 = cyclopropyl, R_2 = 7-Cl

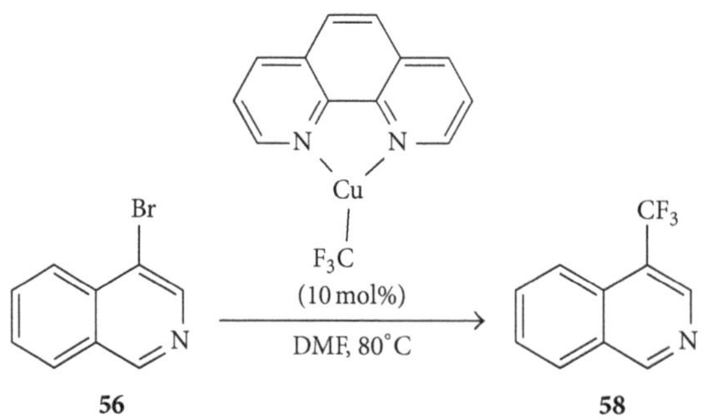

56 **58** (67%)

SCHEME 14: Copper catalyzed fluoroisoquinoline synthesis.

allyl bromide, or 4-bromobenzotrifluoride, respectively, and aromatization with ammonium chloride.

Scheme 18 shows two $Pd(OAc)_2$ catalyzed isoquinoline producing cyclizations. Tian et al. developed an efficient, Pd-catalyzed Heck reaction in which the 2-triflate substituted aryl ketone series **74** was coupled and intramolecularly cyclized with enamine **75** at C_{4f}-C_4/N_2-C_3 and isomerized to produce four examples of ring-fluorinated and fluoroarylated isoquinolines **76** [49]. Yang's group conducted a palladium catalyzed, microwave-assisted, one-pot reaction via a C_{4f}-C_4/N_2-C_3 cyclization for the synthesis of isoquinolines [50]. The coupling-imination-annulation sequence of reactions using *o*-bromoarylaldehydes **77** and terminal alkynes **78** with ammonium acetate produced three disubstituted

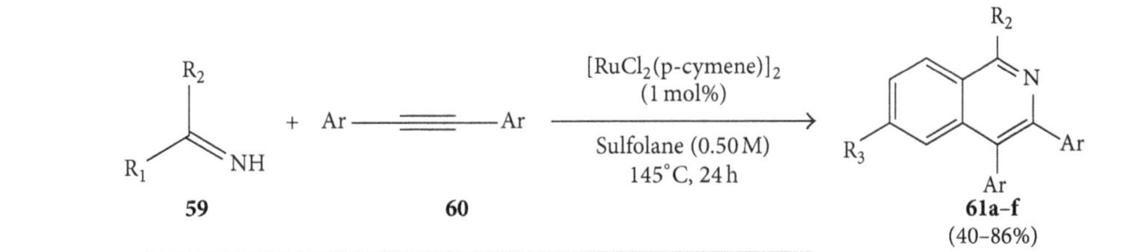

R_1	R_2	Ar	R_3
a: $4\text{-}FC_6H_4$	$4\text{-}FC_6H_4$	Ph	F
b: $4\text{-}CF_3C_6H_4$	$4\text{-}CF_3C_6H_4$	Ph	CF_3
c: $4\text{-}MeOC_6H_4$	$4\text{-}MeOC_6H_4$	$4\text{-}FC_6H_4$	MeO
d: $4\text{-}MeOC_6H_4$	$4\text{-}MeOC_6H_4$	$4\text{-}CF_3C_6H_4$	MeO
e: $4\text{-}MeOC_6H_4$	$4\text{-}MeOC_6H_4$	Ph, $4\text{-}CF_3C_6H_4$	MeO
f: $4\text{-}MeOC_6H_4$	$4\text{-}MeOC_6H_4$	$4\text{-}CF_3C_6H_4$, $4\text{-}CH_3C_6H_4$	MeO

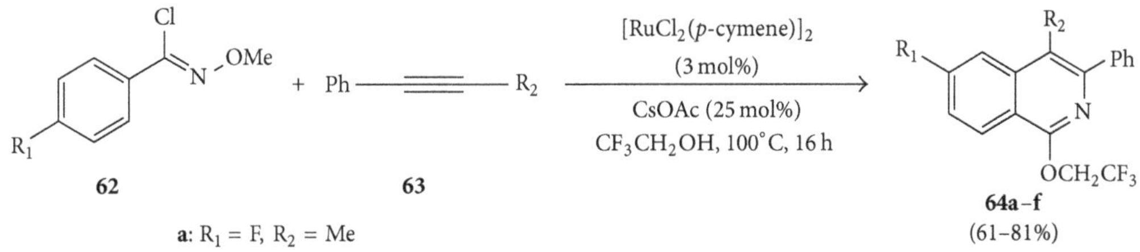

a: R_1 = F, R_2 = Me
b: R_1 = Cl, R_2 = Et
c: R_1 = Cl, R_2 = *n*-Bu
d: R_1 = Cl, R_2 = allyl
e: R_1 = Cl, R_2 = Ph
f: R_1 = Br, R_2 = Ph

Scheme 15: Ruthenium catalyzed fluoroisoquinoline and trifluoromethylisoquinoline synthesis.

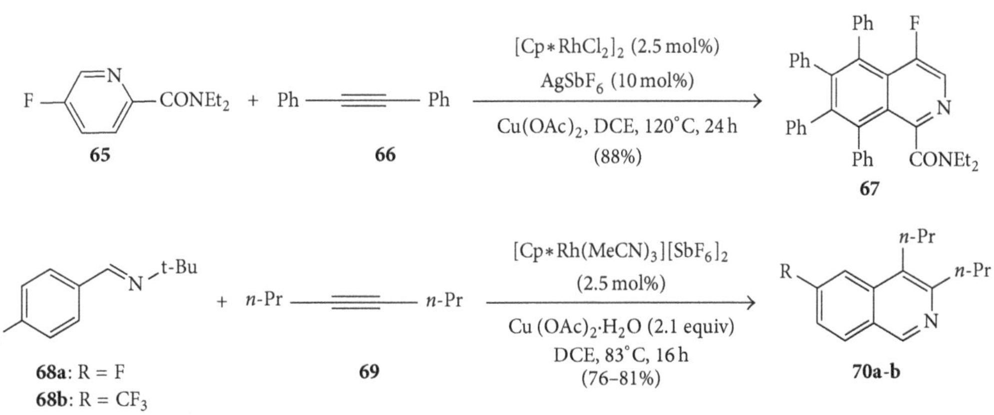

Scheme 16: Rhodium-catalyzed fluoroisoquinoline and trifluoroisoquinoline synthesis.

6-fluoroisoquinolines **79**. These reactions feature a wide range of substrates with various functional groups, and the corresponding products were obtained in good yields.

3.2.4. Silver Catalysis. Several silver (I) catalyzed N_2-C_3 cyclizations of *o*-iminylaryl alkynes have been employed successfully to prepare a broad selection of fluorinated, trifluoromethylthiolated, fluoroarylated, and trifluoromethylarylated isoquinolines. See Scheme 19. Xiao et al. used a silver(I)-catalyzed reaction of 2-alkynylbenzaldoximes **80** with silver (trifluoromethyl)thiolate **81** in the presence of *p*-methoxybenzenesulfonyl chloride to synthesize eleven new 1-[(trifluoromethyl)thio]isoquinolines **82** in yields ranging from 36 to 85% [51]. Xu and Lui utilized an Ag(I)-catalyzed, NFSI aminofluorination of aryliminoalkynes **83** in the development of an efficient synthesis of thirteen examples of 4-fluoroisoquinolines **84** [52]. Jeganathan and Pitchumani used a Ag(I)-exchanged K10-montmorillonite clay catalyzed ring

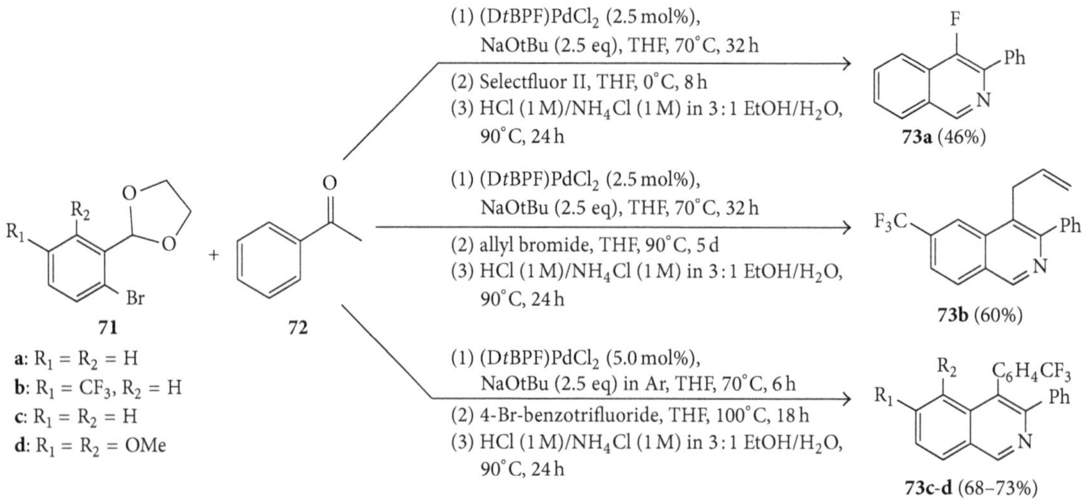

SCHEME 17: Di-t-butylphosphinoferrocene-based palladium catalyzed fluoroisoquinoline and trifluoromethylisoquinoline syntheses.

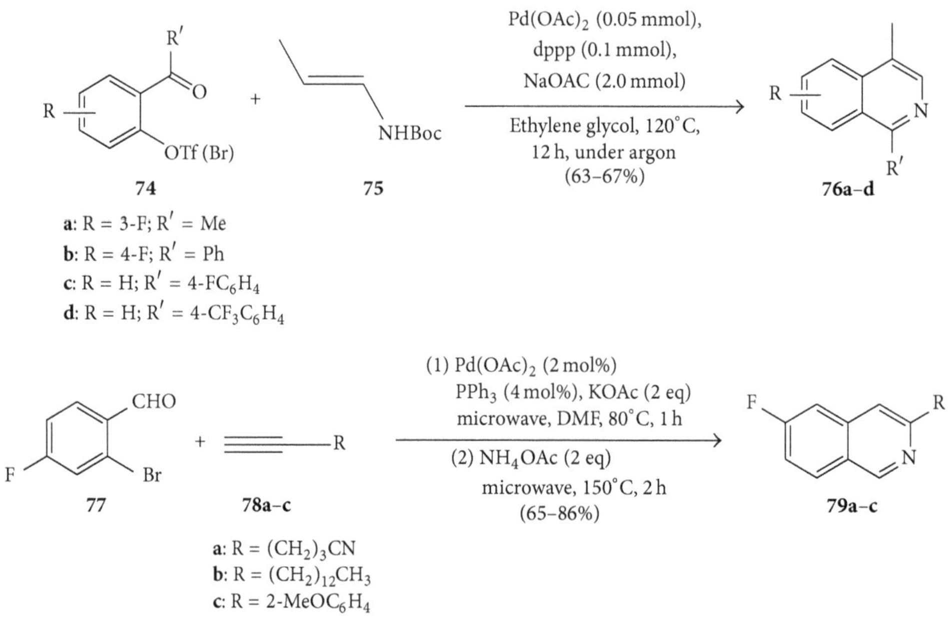

SCHEME 18: $Pd(OAc)_2$ catalyzed fluoroisoquinoline syntheses.

closure of iminoalkynes **85** to prepare three fluoroarylated isoquinolines **86** in good overall yield [53]. This effective cyclization accommodates a variety of iminoalkynes and is noteworthy for its catalyst reusability, simplicity, and environmentally responsible process.

3.3. Period 6 Transition Metal Catalyzed Ring-Fluorinated, Trifluoromethylated, and Fluoroarylated Isoquinolines via Intermolecular Ring A Cyclization at C_{8f}-C_1

3.3.1. Iridium Catalysis. Our final entry highlighted in Scheme 20 is the iridium catalyzed fluoroisoquinoline synthesis via a C_{8f}-C_1 bond-forming/cyclization reaction, conducted by Jiang et al. [54]. This synthesis uses a visible light-promoted insertion of vinyl isocyanides **87** with diphenyliodonium tetrafluoroborate **88** at room temperature to prepare six multisubstituted ring-fluorinated, trifluoromethylated, and fluoroarylated isoquinoline derivatives **89**.

4. Conclusion

The past decade has been a period of intense exploration of new approaches to the preparation of industrially and medicinally important isoquinolines which incorporate fluorine and fluorine-containing functional groups. This review has examined both nonmetal catalyzed and transition metal catalyzed processes that lead to a wide array of isoquinolines that have fluorine and fluorine-containing groups installed at nearly every position on the fused-ring isoquinoline heterocycle. The reactions reviewed span processes which construct the

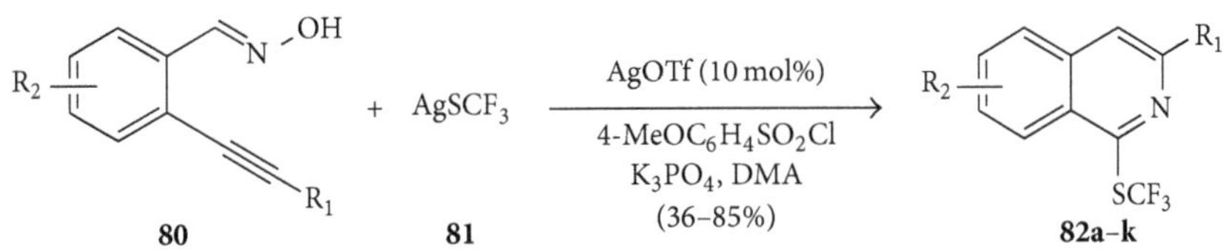

80 + **81** → **82a–k**

AgOTf (10 mol%), 4-$MeOC_6H_4SO_2Cl$, K_3PO_4, DMA (36–85%)

a: R_1 = Ph, R_2 = H
b: R_1 = 4-MeC_6H_4, R_2 = H
c: R_1 = 4-$MeOC_6H_4$, R_2 = H
d: R_1 = 4-FC_6H_4, R_2 = H
e: R_1 = 4-ClC_6H_4, R_2 = H
f: R_1 = cyclopropyl, R_2 = H
g: R_1 = *n*-butyl, R_2 = H
h: R_1 = Ph, R_2 = Me
i: R_1 = Ph, R_2 = OMe
j: R_1 = Ph, R_2 = F
k: R_1 = Ph, R_2 = Cl

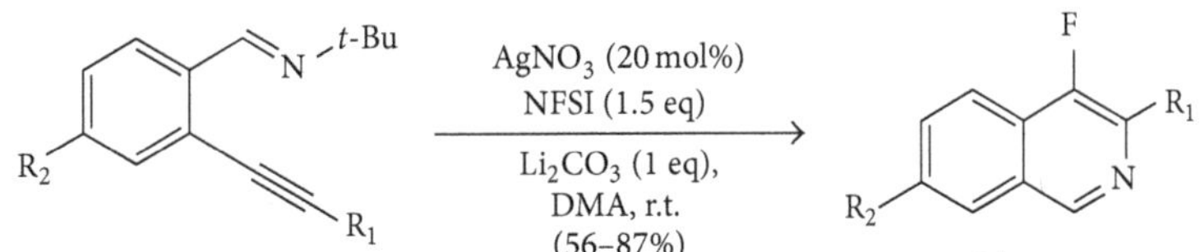

83 → **84a–m**

$AgNO_3$ (20 mol%), NFSI (1.5 eq), Li_2CO_3 (1 eq), DMA, r.t. (56–87%)

a: R_1 = Bu, R_2 = H
b: R_1 = Bu, R_2 = Cl
c: R_1 = Bu, R_2 = F
d: R_1 = Bu, R_2 = OMe
e: R_1 = *t*-Bu, R_2 = H
f: R_1 = Ph, R_2 = H
g: R_1 = $CH_2NTsBoc$, R_2 = H
h: R_1 = CH_2CH_2OAc, R_2 = H
i: R_1 = $(CH_2)_3CO_2Me$, R_2 = H
j: R_1 = *p*-Anisyl, R_2 = H
k: R_1 = *p*-FC_6H_4, R_2 = H
l: R_1 = cyclohexenyl, R_2 = H
m: R_1 = cyclopropyl, R_2 = H

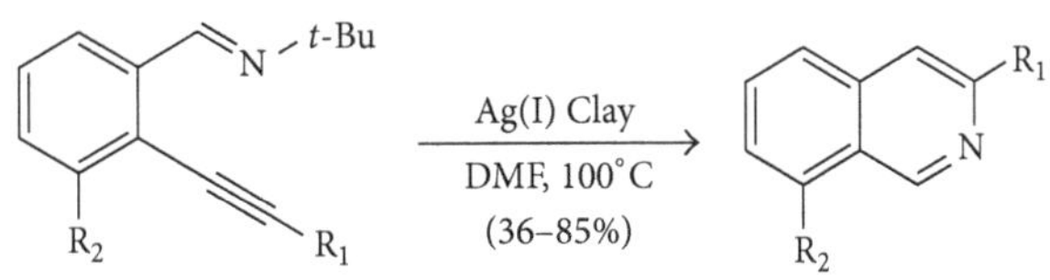

85 → **86a–c**

Ag(I) Clay, DMF, 100°C (36–85%)

a: R_1 = 4- $CF_3C_6H_4$, R_2 = H
b: R_1 = 3-$CF_3C_6H_4$, R_2 = H
c: R_1 = 3-FC_6H_4, R_2 = OMe

Scheme 19: Silver catalyzed fluoroisoquinoline syntheses.

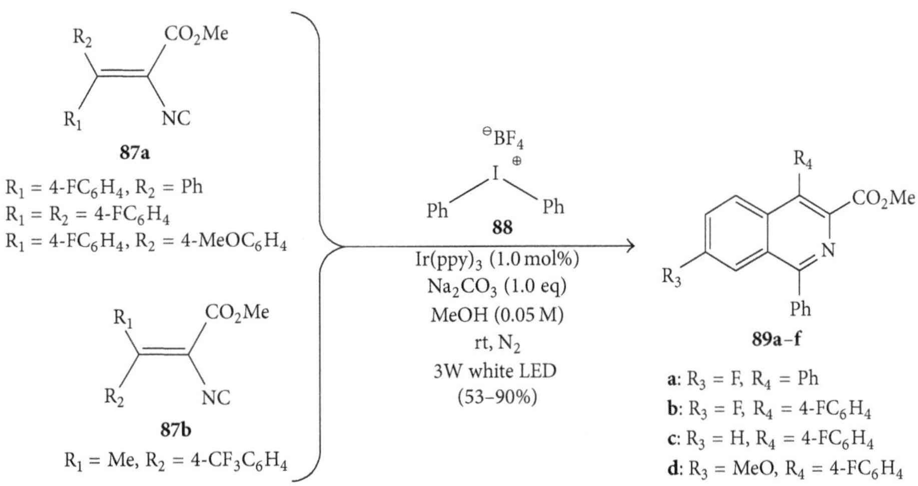

87a
R_1 = 4-FC_6H_4, R_2 = Ph
R_1 = R_2 = 4-FC_6H_4
R_1 = 4-FC_6H_4, R_2 = 4-$MeOC_6H_4$

87b
R_1 = Me, R_2 = 4-$CF_3C_6H_4$

88

Ir(ppy)$_3$ (1.0 mol%), Na_2CO_3 (1.0 eq), MeOH (0.05 M), rt, N_2, 3W white LED (53–90%)

89a–f
a: R_3 = F, R_4 = Ph
b: R_3 = F, R_4 = 4-FC_6H_4
c: R_3 = H, R_4 = 4-FC_6H_4
d: R_3 = MeO, R_4 = 4-FC_6H_4
e: R_3 = F, R_4 = 4-$MeOC_6H_4$
f: R_3 = CF_3, R_4 = Me

Scheme 20: Iridium catalyzed fluoroisoquinoline and trifluoromethylisoquinoline syntheses.

isoquinoline core via both A ring and B ring cyclizations. Additionally, these investigations have led to the discovery of milder reaction conditions, improved yields, enhanced regioselectivity, and site-specific ring monofluorination as well as difluoromethylation, trifluoromethylation, perfluoroalkylation, fluoroarylation, and trifluoromethylarylation examples. In all, more than 30 processes producing over 160 new isoquinoline examples have been highlighted.

Competing Interests

The author declares that there are no competing interests regarding the publication of this paper.

References

[1] S. Hoogewerf and W. van Dorp, "Sur un isomére de la quinoléine," *Recueil des Travaux Chemiques des Pays-Bas*, vol. 4, no. 4, pp. 125–129, 1885.

[2] C. Pomeranz, "Über eine neue Isochinolinsynthese," *Monatshefte für Chemie*, vol. 14, no. 1, pp. 116–119, 1893.

[3] P. Fritsch, "Synthesen in der Isocumarin– und Isochinolinreihe," *Berichte der Deutschen Chemischen Gesellschaft*, vol. 26, no. 1, pp. 419–422, 1893.

[4] A. Bischler and B. Napieralski, "Zur Kenntniss einer neuen Isochinolinsynthese," *Berichte der deutschen chemischen Gesellschaft*, vol. 26, no. 2, pp. 1903–1908, 1893.

[5] A. Pictet and A. Gams, "Über eine neue Methode zur synthetischen Darstellung der Isochinolinbasen," *Berichte der Deutschen Chemischen Gesellschaft*, vol. 43, no. 2, pp. 2384–2391, 1910.

[6] A. Pictet and T. Spengler, "Über die Bildung von Isochinolin–derivaten durch Einwirkung von Methylal auf Phenyl–äthylamin, Phenyl–alanin und Tyrosin," *Berichte der Deutschen Chemischen Gesellschaft*, vol. 44, no. 3, pp. 2030–2036, 1911.

[7] D. Collado, E. Perez-Inestrosa, R. Suau, J.-P. Desvergne, and H. Bouas-Laurent, "Bis(isoquinoline N-oxide) pincers as a new type of metal cation dual channel fluorosensor," *Organic Letters*, vol. 4, no. 5, pp. 855–858, 2002.

[8] G. Cuny, Minnesota Mining and Manufacturing Company. Patent, Pyrrolo(2,1-a)isoquinoline dyes, EP 0780443 A2, June 1997.

[9] Y. Kuramochi and O. Ishitani, "Iridium(III) 1-phenylisoquinoline complexes as a photosensitizer for photocatalytic CO_2 reduction: a mixed system with a Re(I) catalyst and a supramolecular photocatalyst," *Inorganic Chemistry*, vol. 55, no. 11, pp. 5702–5709, 2016.

[10] E. Schlittler and J. Müller, "Eine neue Modifikation der Isochinolinsynthese nach *Pomeraz–Fritsch*," *Helvetica Chimica Acta*, vol. 31, no. 3, pp. 914–924, 1948.

[11] A. L. Smith, F. F. DeMorin, N. A. Paras et al., "Selective inhibitors of the mutant B-Raf pathway: discovery of a potent and orally bioavailable aminoisoquinoline," *Journal of Medicinal Chemistry*, vol. 52, no. 20, pp. 6189–6192, 2009.

[12] A. Galán, L. Moreno, J. Párraga et al., "Novel isoquinoline derivatives as antimicrobial agents," *Bioorganic and Medicinal Chemistry*, vol. 21, no. 11, pp. 3221–3230, 2013.

[13] D. O'Hagan, "Understanding organofluorine chemistry. An introduction to the C–F bond," *Chemical Society Reviews*, vol. 37, no. 2, pp. 308–319, 2008.

[14] S. Purser, P. R. Moore, S. Swallow, and V. Gouverneur, "Fluorine in medicinal chemistry," *Chemical Society Reviews*, vol. 37, no. 2, pp. 320–330, 2008.

[15] W. K. Hagmann, "The many roles for fluorine in medicinal chemistry," *Journal of Medicinal Chemistry*, vol. 51, no. 15, pp. 4359–4369, 2008.

[16] S. Lee, J.-S. Park, and T. R. Lee, "Analysis of the wettability of partially fluorinated polymers reveals the surprisingly strong acid-base character of poly(vinylidene fluoride)," *Bulletin of the Korean Chemical Society*, vol. 32, no. 1, pp. 41–48, 2011.

[17] J. Yun, S. Park, J. Heo et al., "Enhancement of charge transport properties of small molecule semiconductors by controlling fluorine substitution and effects on photovoltaic properties of organic solar cells and perovskite solar cells," *Chemical Science*, vol. 7, pp. 6649–6661, 2016.

[18] T. Fujita and J. Ichikawa, "Fluorinated isoquinolines: synthesis, properties and applications," in *Fluorine in Heterocyclic Chemistry Volume 2: 6-Membered Heterocycles*, V. Nenajdenko, Ed., pp. 181–210, Springer International Publishing, Cham, Switzerland, 2014.

[19] Z. Y. Li and Z.-X. Jiang, "Fluorine is flourishing in pharmaceuticals," *Journal of Biomolecular Research & Therapeutics*, vol. 1, article 2, 2012.

[20] H.-J. Böhm, D. Banner, S. Bendels et al., "Fluorine in medicinal chemistry," *ChemBioChem*, vol. 5, no. 5, pp. 637–643, 2004.

[21] E. P. Gillis, K. J. Eastman, M. D. Hill, D. J. Donnelly, and N. A. Meanwell, "Applications of fluorine in medicinal chemistry," *Journal of Medicinal Chemistry*, vol. 58, no. 21, pp. 8315–8359, 2015.

[22] W. Yu, E. Wang, R. J. Voll, A. H. Miller, and M. M. Goodman, "Synthesis, fluorine-18 radiolabeling, and in vitro characterization of 1-iodophenyl-N-methyl-N-fluoroalkyl-3-isoquinoline carboxamide derivatives as potential PET radioligands for imaging peripheral benzodiazepine receptor," *Bioorganic and Medicinal Chemistry*, vol. 16, no. 11, pp. 6145–6155, 2008.

[23] J. Ichikawa, K. Sakoda, H. Moriyama, and Y. Wada, "Syntheses of ring-fluorinated isoquinolines and quinolines via intramolecular substitution: cyclization of 1,1-difluoro-1-alkenes bearing a sulfonamide moiety," *Synthesis*, no. 10, pp. 1590–1598, 2006.

[24] A. S. Kiselyov, "A convenient procedure for the synthesis of fused fluoro isoquinolines," *Tetrahedron*, vol. 62, no. 4, pp. 543–548, 2006.

[25] L.-G. Xie, S. Niyomchon, A. J. Mota, L. González, and N. Maulide, "Metal-free intermolecular formal cycloadditions enable an orthogonal access to nitrogen heterocycles," *Nature Communications*, vol. 7, Article ID 10914, pp. 1–9, 2016.

[26] J.-B. Feng and X.-F. Wu, "Potassium *tert*-butoxide-promoted synthesis of 1-aminoisoquinolines from 2-methylbenzonitriles and benzonitriles under catalyst-free conditions," *Advanced Synthesis and Catalysis*, vol. 358, no. 13, pp. 2179–2185, 2016.

[27] Y.-Y. Yang, W.-G. Shou, Z.-B. Chen, D. Hong, and Y.-G. Wang, "A tandem approach to isoquinolines from 2-azido-3-arylacrylates and α-diazocarbonyl compounds," *The Journal of Organic Chemistry*, vol. 73, no. 10, pp. 3928–3930, 2008.

[28] C. D. Gilmore, K. M. Allan, and B. M. Stoltz, "Orthogonal synthesis of indolines and isoquinolines via aryne annulation," *Journal of the American Chemical Society*, vol. 130, no. 5, pp. 1558–1559, 2008.

[29] B. Zhang and A. Studer, "1-Trifluoromethylated isoquinolines *via* radical trifluoromethylation of isonitriles," *Organic and Biomolecular Chemistry*, vol. 12, pp. 9895–9898, 2014.

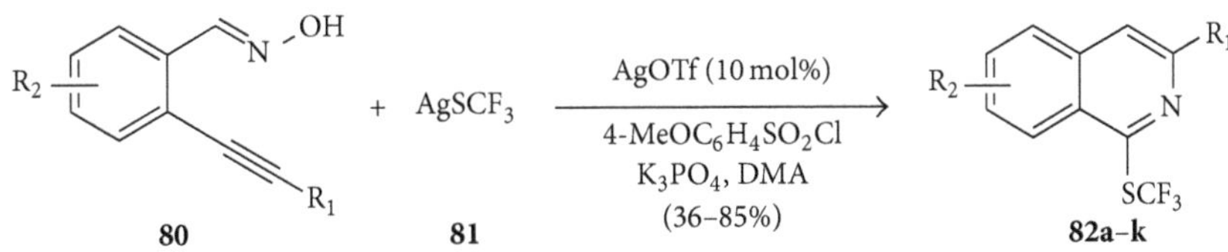

a: R_1 = Ph, R_2 = H
b: R_1 = 4-MeC_6H_4, R_2 = H
c: R_1 = 4-$MeOC_6H_4$, R_2 = H
d: R_1 = 4-FC_6H_4, R_2 = H
e: R_1 = 4-ClC_6H_4, R_2 = H
f: R_1 = cyclopropyl, R_2 = H
g: R_1 = *n*-butyl, R_2 = H
h: R_1 = Ph, R_2 = Me
i: R_1 = Ph, R_2 = OMe
j: R_1 = Ph, R_2 = F
k: R_1 = Ph, R_2 = Cl

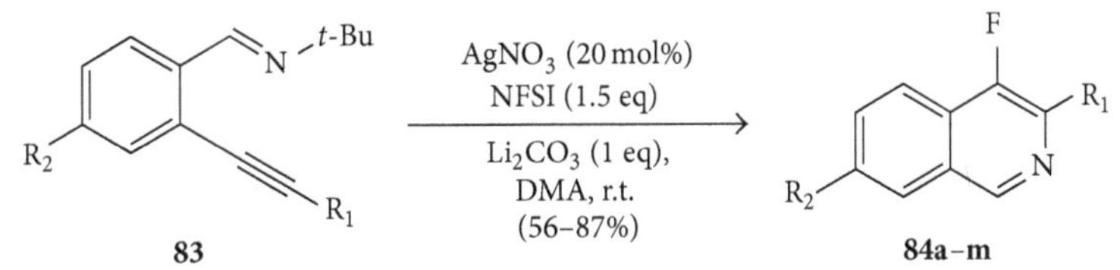

a: R_1 = Bu, R_2 = H
b: R_1 = Bu, R_2 = Cl
c: R_1 = Bu, R_2 = F
d: R_1 = Bu, R_2 = OMe
e: R_1 = *t*-Bu, R_2 = H
f: R_1 = Ph, R_2 = H
g: R_1 = $CH_2NTsBoc$, R_2 = H
h: R_1 = CH_2CH_2OAc, R_2 = H
i: R_1 = $(CH_2)_3CO_2Me$, R_2 = H
j: R_1 = *p*-Anisyl, R_2 = H
k: R_1 = *p*-FC_6H_4, R_2 = H
l: R_1 = cyclohexenyl, R_2 = H
m: R_1 = cyclopropyl, R_2 = H

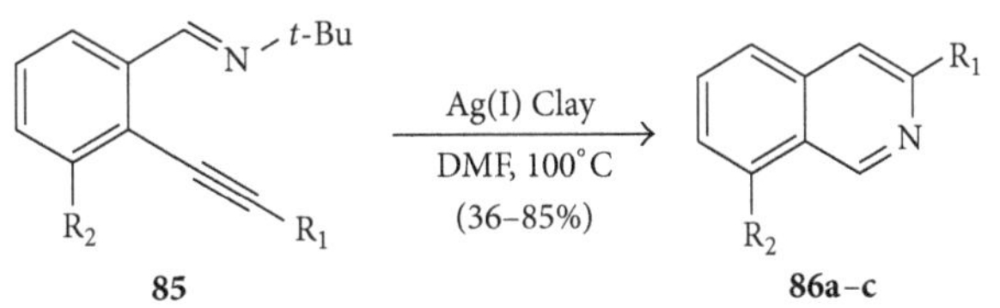

a: R_1 = 4- $CF_3C_6H_4$, R_2 = H
b: R_1 = 3-$CF_3C_6H_4$, R_2 = H
c: R_1 = 3-FC_6H_4, R_2 = OMe

SCHEME 19: Silver catalyzed fluoroisoquinoline syntheses.

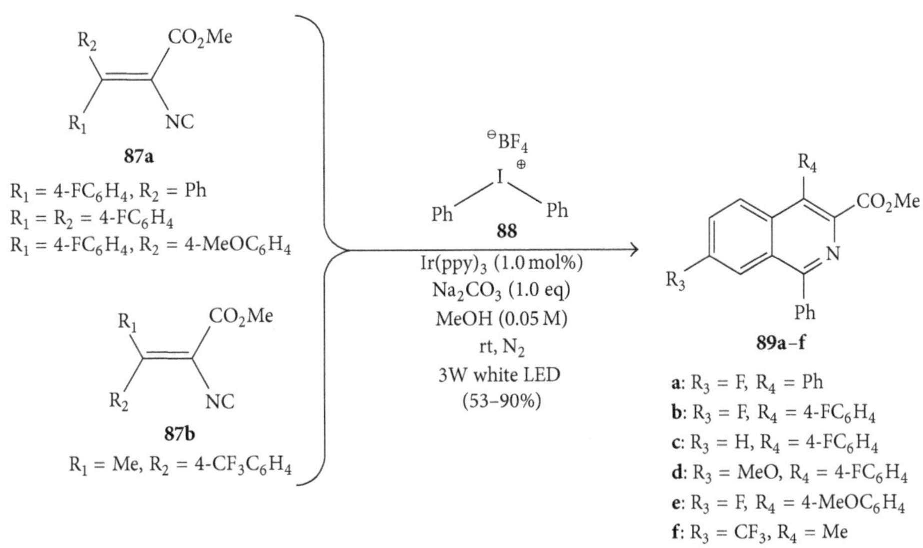

SCHEME 20: Iridium catalyzed fluoroisoquinoline and trifluoromethylisoquinoline syntheses.

isoquinoline core via both A ring and B ring cyclizations. Additionally, these investigations have led to the discovery of milder reaction conditions, improved yields, enhanced regioselectivity, and site-specific ring monofluorination as well as difluoromethylation, trifluoromethylation, perfluoroalkylation, fluoroarylation, and trifluoromethylarylation examples. In all, more than 30 processes producing over 160 new isoquinoline examples have been highlighted.

Competing Interests

The author declares that there are no competing interests regarding the publication of this paper.

References

[1] S. Hoogewerf and W. van Dorp, "Sur un isomére de la quinoléine," *Recueil des Travaux Chemiques des Pays-Bas*, vol. 4, no. 4, pp. 125–129, 1885.

[2] C. Pomeranz, "Über eine neue Isochinolinsynthese," *Monatshefte für Chemie*, vol. 14, no. 1, pp. 116–119, 1893.

[3] P. Fritsch, "Synthesen in der Isocumarin– und Isochinolinreihe," *Berichte der Deutschen Chemischen Gesellschaft*, vol. 26, no. 1, pp. 419–422, 1893.

[4] A. Bischler and B. Napieralski, "Zur Kenntniss einer neuen Isochinolinsynthese," *Berichte der deutschen chemischen Gesellschaft*, vol. 26, no. 2, pp. 1903–1908, 1893.

[5] A. Pictet and A. Gams, "Über eine neue Methode zur synthetischen Darstellung der Isochinolinbasen," *Berichte der Deutschen Chemischen Gesellschaft*, vol. 43, no. 2, pp. 2384–2391, 1910.

[6] A. Pictet and T. Spengler, "Über die Bildung von Isochinolin–derivaten durch Einwirkung von Methylal auf Phenyl–äthylamin, Phenyl–alanin und Tyrosin," *Berichte der Deutschen Chemischen Gesellschaft*, vol. 44, no. 3, pp. 2030–2036, 1911.

[7] D. Collado, E. Perez-Inestrosa, R. Suau, J.-P. Desvergne, and H. Bouas-Laurent, "Bis(isoquinoline N-oxide) pincers as a new type of metal cation dual channel fluorosensor," *Organic Letters*, vol. 4, no. 5, pp. 855–858, 2002.

[8] G. Cuny, Minnesota Mining and Manufacturing Company. Patent, Pyrrolo(2,1-a)isoquinoline dyes, EP 0780443 A2, June 1997.

[9] Y. Kuramochi and O. Ishitani, "Iridium(III) 1-phenylisoquinoline complexes as a photosensitizer for photocatalytic CO_2 reduction: a mixed system with a Re(I) catalyst and a supramolecular photocatalyst," *Inorganic Chemistry*, vol. 55, no. 11, pp. 5702–5709, 2016.

[10] E. Schlittler and J. Müller, "Eine neue Modifikation der Isochinolinsynthese nach *Pomeraz–Fritsch*," *Helvetica Chimica Acta*, vol. 31, no. 3, pp. 914–924, 1948.

[11] A. L. Smith, F. F. DeMorin, N. A. Paras et al., "Selective inhibitors of the mutant B-Raf pathway: discovery of a potent and orally bioavailable aminoisoquinoline," *Journal of Medicinal Chemistry*, vol. 52, no. 20, pp. 6189–6192, 2009.

[12] A. Galán, L. Moreno, J. Párraga et al., "Novel isoquinoline derivatives as antimicrobial agents," *Bioorganic and Medicinal Chemistry*, vol. 21, no. 11, pp. 3221–3230, 2013.

[13] D. O'Hagan, "Understanding organofluorine chemistry. An introduction to the C–F bond," *Chemical Society Reviews*, vol. 37, no. 2, pp. 308–319, 2008.

[14] S. Purser, P. R. Moore, S. Swallow, and V. Gouverneur, "Fluorine in medicinal chemistry," *Chemical Society Reviews*, vol. 37, no. 2, pp. 320–330, 2008.

[15] W. K. Hagmann, "The many roles for fluorine in medicinal chemistry," *Journal of Medicinal Chemistry*, vol. 51, no. 15, pp. 4359–4369, 2008.

[16] S. Lee, J.-S. Park, and T. R. Lee, "Analysis of the wettability of partially fluorinated polymers reveals the surprisingly strong acid-base character of poly(vinylidene fluoride)," *Bulletin of the Korean Chemical Society*, vol. 32, no. 1, pp. 41–48, 2011.

[17] J. Yun, S. Park, J. Heo et al., "Enhancement of charge transport properties of small molecule semiconductors by controlling fluorine substitution and effects on photovoltaic properties of organic solar cells and perovskite solar cells," *Chemical Science*, vol. 7, pp. 6649–6661, 2016.

[18] T. Fujita and J. Ichikawa, "Fluorinated isoquinolines: synthesis, properties and applications," in *Fluorine in Heterocyclic Chemistry Volume 2: 6-Membered Heterocycles*, V. Nenajdenko, Ed., pp. 181–210, Springer International Publishing, Cham, Switzerland, 2014.

[19] Z. Y. Li and Z.-X. Jiang, "Fluorine is flourishing in pharmaceuticals," *Journal of Biomolecular Research & Therapeutics*, vol. 1, article 2, 2012.

[20] H.-J. Böhm, D. Banner, S. Bendels et al., "Fluorine in medicinal chemistry," *ChemBioChem*, vol. 5, no. 5, pp. 637–643, 2004.

[21] E. P. Gillis, K. J. Eastman, M. D. Hill, D. J. Donnelly, and N. A. Meanwell, "Applications of fluorine in medicinal chemistry," *Journal of Medicinal Chemistry*, vol. 58, no. 21, pp. 8315–8359, 2015.

[22] W. Yu, E. Wang, R. J. Voll, A. H. Miller, and M. M. Goodman, "Synthesis, fluorine-18 radiolabeling, and in vitro characterization of 1-iodophenyl-N-methyl-N-fluoroalkyl-3-isoquinoline carboxamide derivatives as potential PET radioligands for imaging peripheral benzodiazepine receptor," *Bioorganic and Medicinal Chemistry*, vol. 16, no. 11, pp. 6145–6155, 2008.

[23] J. Ichikawa, K. Sakoda, H. Moriyama, and Y. Wada, "Syntheses of ring-fluorinated isoquinolines and quinolines via intramolecular substitution: cyclization of 1,1-difluoro-1-alkenes bearing a sulfonamide moiety," *Synthesis*, no. 10, pp. 1590–1598, 2006.

[24] A. S. Kiselyov, "A convenient procedure for the synthesis of fused fluoro isoquinolines," *Tetrahedron*, vol. 62, no. 4, pp. 543–548, 2006.

[25] L.-G. Xie, S. Niyomchon, A. J. Mota, L. González, and N. Maulide, "Metal-free intermolecular formal cycloadditions enable an orthogonal access to nitrogen heterocycles," *Nature Communications*, vol. 7, Article ID 10914, pp. 1–9, 2016.

[26] J.-B. Feng and X.-F. Wu, "Potassium *tert*-butoxide-promoted synthesis of 1-aminoisoquinolines from 2-methylbenzonitriles and benzonitriles under catalyst-free conditions," *Advanced Synthesis and Catalysis*, vol. 358, no. 13, pp. 2179–2185, 2016.

[27] Y.-Y. Yang, W.-G. Shou, Z.-B. Chen, D. Hong, and Y.-G. Wang, "A tandem approach to isoquinolines from 2-azido-3-arylacrylates and α-diazocarbonyl compounds," *The Journal of Organic Chemistry*, vol. 73, no. 10, pp. 3928–3930, 2008.

[28] C. D. Gilmore, K. M. Allan, and B. M. Stoltz, "Orthogonal synthesis of indolines and isoquinolines via aryne annulation," *Journal of the American Chemical Society*, vol. 130, no. 5, pp. 1558–1559, 2008.

[29] B. Zhang and A. Studer, "1-Trifluoromethylated isoquinolines *via* radical trifluoromethylation of isonitriles," *Organic and Biomolecular Chemistry*, vol. 12, pp. 9895–9898, 2014.

[30] Y. Kuninobu, M. Nagase, and M. Kanai, "Benzylic C(sp^3)-H perfluoroalkylation of six-membered heteroaromatic compounds," *Angewandte Chemie—International Edition*, vol. 54, no. 35, pp. 10263–10266, 2015.

[31] J. Ichikawa, "Synthetic methods for heterocycles and carbocycles bearing fluorinated one-carbon units (=CF_2, CHF_2, or CF_3): intramolecular reaction of 2-Trifluoromethyl-1-alkenes," *Journal of Synthetic Organic Chemistry, Japan*, vol. 68, no. 11, pp. 1175–1184, 2010.

[32] N. K. Chennamaneni, J. Arif, F. S. Buckner, and M. H. Gelb, "Isoquinoline-based analogs of the cancer drug clinical candidate tipifarnib as anti-Trypanosoma cruzi agents," *Bioorganic and Medicinal Chemistry Letters*, vol. 19, no. 23, pp. 6582–6584, 2009.

[33] E. Awuah and A. Capretta, "Strategies and synthetic methods directed toward the preparation of libraries of substituted isoquinolines," *Journal of Organic Chemistry*, vol. 75, no. 16, pp. 5627–5634, 2010.

[34] R. He, Z.-T. Huang, Q.-Y. Zheng, and C. Wang, "Manganese-catalyzed dehydrogenative [4 + 2] annulation of N–H imines and alkynes by C–H/N–H activation," *Angewandte Chemie—International Edition*, vol. 53, no. 19, pp. 4950–4953, 2014.

[35] H. Mao, M. Gao, B. Liu, and B. Xu, "Manganese(II)-catalyzed modular synthesis of isoquinolines from vinyl isocyanides and hydrazines," *Organic Chemistry Frontiers*, vol. 3, no. 4, pp. 516–521, 2016.

[36] B. Sun, T. Yoshino, M. Kanai, and S. Matsunaga, "Cp∗CoIII catalyzed site-selective C-H activation of unsymmetrical O-acyl oximes: synthesis of multisubstituted isoquinolines from terminal and internal alkynes," *Angewandte Chemie—International Edition*, vol. 54, no. 44, pp. 12968–12972, 2015.

[37] J. Li, M. Tang, L. Zang, X. Zhang, Z. Zhang, and L. Ackermann, "Amidines for versatile cobalt(III)-catalyzed synthesis of isoquinolines through C–H functionalization with diazo compounds," *Organic Letters*, vol. 18, no. 11, pp. 2742–2745, 2016.

[38] A. Pawar, D. Agarwal, and D. Lade, "Cp*Co(III)-Catalyzed C–H/N–N functionalization of arylhydrazones for the synthesis of isoquinolines," *The Journal of Organic Chemistry*, vol. 81, no. 22, pp. 11409–11415, 2016.

[39] X. Yu, K. Chen, F. Yang, S. Zha, and J. Zhu, "Oxadiazolone-enabled synthesis of primary azaaromatic amines," *Organic Letters*, vol. 18, no. 20, pp. 5412–5415, 2016.

[40] Y. Yoshida, T. Kurahashi, and S. Matsubara, "Nickel-catalyzed cycloaddition of aromatic (O-benzyl)ketoximes with alkynes to produce isoquinoline and isoquinoline N-oxide derivatives," *Chemistry Letters*, vol. 40, no. 10, pp. 1140–1142, 2011.

[41] Y. Ohta, S. Oishi, N. Fujii, and H. Ohno, "Facile synthesis of 3-(aminomethyl)isoquinolines by copper-catalysed domino four-component coupling and cyclisation," *Chemical Communications*, no. 7, pp. 835–837, 2008.

[42] C. Fan, J. Song, G. Qiu, G. Liu, and J. Wu, "Generation of 1-(trifluoromethyl)isoquinolines via a copper-catalyzed reaction of isoquinoline-N-oxide with Togni reagent," *Organic Chemistry Frontiers*, vol. 1, no. 8, pp. 924–928, 2014.

[43] M. G. Mormino, P. S. Fier, and J. F. Hartwig, "Copper-mediated perfluoroalkylation of heteroaryl bromides with (phen)CuR_F," *Organic Letters*, vol. 16, no. 6, pp. 1744–1747, 2014.

[44] K.-H. He, W.-D. Zhang, M.-Y. Yang et al., "Redox-divergent hydrogen-retentive or hydrogen-releasing synthesis of 3,4-dihydroisoquinolines or isoquinolines," *Organic Letters*, vol. 18, no. 12, pp. 2840–2843, 2016.

[45] R. K. Chinnagolla, S. Pimparkar, and M. Jeganmohan, "A regioselective synthesis of 1-haloisoquinolines via ruthenium-catalyzed cyclization of O-methylbenzohydroximoyl halides with alkynes," *Chemical Communications*, vol. 49, no. 35, pp. 3703–3705, 2013.

[46] Z.-C. Qian, J. Zhou, B. Li, and B.-F. Shi, "Efficient synthesis of isoquinolines via Rh(III)-catalyzed oxidative annulation of picolinamides with alkynes," *Synlett*, vol. 25, no. 7, pp. 1036–1040, 2014.

[47] N. Guimond and K. Fagnou, "Isoquinoline synthesis via rhodium-catalyzed oxidative cross-coupling/cyclization of aryl aldimines and alkynes," *Journal of the American Chemical Society*, vol. 131, no. 34, pp. 12050–12051, 2009.

[48] B. S. Pilgrim, A. E. Gatland, C. T. McTernan, P. A. Procopiou, and T. J. Donohoe, "Modular isoquinoline synthesis using catalytic enolate arylation and in situ functionalization," *Organic Letters*, vol. 15, no. 24, pp. 6190–6193, 2013.

[49] Y. Tian, J. Qi, C. Sun, D. Yin, X. Wang, and Q. Xiao, "One-pot synthesis of 4-methylisoquinolines via a sequential Pd-catalyzed Heck reaction and intramolecular cyclization," *Organic and Biomolecular Chemistry*, vol. 11, no. 42, pp. 7262–7266, 2013.

[50] D. Yang, S. Burugupalli, D. Daniel, and Y. Chen, "Microwave-assisted one-pot synthesis of isoquinolines, furopyridines, and thienopyridines by palladium-catalyzed sequential coupling-imination-annulation of 2-bromoarylaldehydes with terminal acetylenes and ammonium acetate," *Journal of Organic Chemistry*, vol. 77, no. 9, pp. 4466–4472, 2012.

[51] Q. Xiao, J. Sheng, Q. Ding, and J. Wu, "Facile assembly of 1-[(trifluoromethyl)thio]isoquinolines through reaction of 2-alkynylbenzaldoxime with silver (trifluoromethyl)thiolate," *European Journal of Organic Chemistry*, vol. 2014, no. 1, pp. 217–221, 2014.

[52] T. Xu and G. Liu, "Ag(I)-catalyzed aminofluorination of alkynes: efficient synthesis of 4-fluoroisoquinolines and 4-fluoropyrrolo[α]isoquinolines," *Organic Letters*, vol. 14, no. 21, pp. 5416–5419, 2012.

[53] M. Jeganathan and K. Pitchumani, "Synthesis of substituted isoquinolines via iminoalkyne cyclization using Ag(i) exchanged K10-montmorillonite clay as a reusable catalyst," *RSC Advances*, vol. 4, no. 73, pp. 38491–38497, 2014.

[54] H. Jiang, Y. Cheng, R. Wang, Y. Zhang, and S. Yu, "Synthesis of isoquinolines via visible light-promoted insertion of vinyl isocyanides with diaryliodonium salts," *Chemical Communications*, vol. 50, no. 46, pp. 6164–6167, 2014.

7

Discotic Liquid Crystals Based on Cu(I) Complexes with Benzoylthiourea Derivatives Containing a Perfluoroalkyl Chain

Monica Iliş and Viorel Cîrcu

Department of Inorganic Chemistry, University of Bucharest, 23 Dumbrava Rosie St., Sector 2, Bucharest 020464, Romania

Correspondence should be addressed to Viorel Cîrcu; viorel.circu@chimie.unibuc.ro

Academic Editor: Pedro M. Mancini

Mesomorphic three-coordinate copper(I) complexes ($[Cu(BTU)_2X]$, where X = Cl or Br) based on a new *N*-benzoylthiourea (BTU) ligand with two decyloxy and one perfluorooctyl groups at its periphery were designed and prepared. The BTU ligand coordinates via the S atom in a neutral monodentate fashion as confirmed by IR and NMR spectroscopy data. The liquid crystalline behavior of these copper(I) complexes was investigated by a combination of polarized optical microscopy (POM), differential scanning calorimetry (DSC), and X-ray diffraction analysis (XRD), while their thermal stability was studied by thermogravimetric analysis (TGA). These new copper(I) complexes have mesomorphic properties and exhibit a hexagonal columnar mesophase over a large temperature range, more than 100°C.

1. Introduction

Liquid crystalline compounds (LC) based on metal complexes (metallomesogens) represent a special class of liquid crystals with interesting properties related to metal ions employed, such as luminescence, magnetism, redox properties, and so on [1, 2]. The unique properties of thermotropic liquid crystals, in particular their electrooptical properties [3–7], have led to various applications (LCD, molecular sensors and detectors, optical switches, spatial light modulator, etc.) [8–10]. Metallomesogens based on copper(I) complexes are quite rare, even if the copper(I) can form complexes with low coordination number suitable for the stabilization of LC phases. The copper(I) complexes with liquid crystalline properties reported so far are restricted to few classes of ligands related to alkylthiolates [11], isocyanides [12–16], phenantroline [17], azamacrocycles [18], or Schiff bases derived from 2-iminopyridines [19, 20], giving rise to either mono- or binuclear two- or tetra-coordinate complexes. Ionic columnar metallomesogens formed by three-coordinate copper(I) complexes with bis(1-pyrazolyl)ethyl ether ligands were reported by Lin and Lai [21]. In this study, we report a new class of copper(I) metallomesogens based on neutral copper(I) three-coordinate complexes with benzoylthiourea (BTU) ligands. These benzoylthiourea derivatives (BTU) proved to have excellent properties to bind to various metal ions due to the very strong donor groups (carbonyl and thioamide) that yield various transition metal complexes. The BTU ligands can give neutral homoleptic or heteroleptic complexes either with *S*- and *O*-coordination, when the ligand is coordinated in monoanionic bidentate form by deprotonation or with only *S*-coordination if the ligand is bound in neutral form to the metal center [22–32]. The reaction between copper(II) salts and, more generally, the versatile acylthiourea compounds leads to a variety of copper(I) complexes, including tetra- and three-coordinate mononuclear complexes, halide-bridged dinuclear complexes, or cluster-type polynuclear complexes, depending on the halide type and reaction conditions [33–39]. The complexity of the redox reaction between the copper(II) salts and thiourea compounds results from the combination between the versatility of the acylthiourea ligands and the coordination flexibility of the copper(I) ion. The three-coordinate copper(I) complexes, which possess two molecules of the BTU ligand linked via the sulfur atom and one halide ion, have approximately an overall planar shape

[37–39]. The molecular shape arguments indicate that such complexes could be potential candidates for new liquid crystalline materials after careful functionalization of BTU derivatives with suitable mesogenic groups. The BTU compounds having terminal alkoxy groups at the both ends of the molecule show interesting mesogenic behavior, displaying nematic, smectic A, and smectic C phases, depending on the position and length of the alkyl chains [40]. On the contrary, partially fluorinated BTU compounds show comparable mesogenic properties, with an increased stability of lamellar phases and higher transition temperatures [41]. It is well known that the alkyl chains substitution with semi- or perfluoroalkylated chains leads to a serious increasing of both the transition temperatures and mesophases thermal ranges [42, 43]. Moreover, the stability of layered phases for such compounds with perfluoroalkyl groups is significantly increased due to the incompatibility between the aliphatic and perfluoroalkylated chains as well as the rigidity of the latter [44–50].

This paper reports the preparation and mesogenic behavior of a new class of copper(I) metallomesogens based on copper(I) halide complexes (chloride or bromide) with a novel benzoylthiourea ligand that have at one side two decyloxy chains and, at the other side, a perfluorooctyl group. The liquid crystalline properties of the new three-coordinate copper(I) complexes were studied by a combination of differential scanning calorimetry (DSC) and polarizing optical microscopy (POM). Moreover, these new complexes have mesomorphic properties and exhibit a columnar mesophase over a large temperature range, more than 100°C, as shown by DSC.

2. Experimental

2.1. Characterization Methods. All the chemicals were used as supplied. C, H, and N analyses were carried out with an EuroEA 3300 instrument. IR spectra were recorded on a Bruker spectrophotometer using KBr discs. ^{1}H and ^{13}C NMR spectra were recorded on a Bruker spectrometer operating at 500 MHz, using $CDCl_3$ as solvent. ^{1}H chemical shifts were referenced to the solvent peak position, δ 7.26 ppm. The phase assignments for the BTU ligand and its copper(I) complexes were evaluated by polarizing optical light microscopy (POM), placed on untreated glass slides, using a Nikon 50iPol microscope equipped with a Linkam THMS600 hot stage and TMS94 control processor. Temperatures and enthalpies of transitions were recorded by using the differential scanning calorimetry (DSC) technique employing a Diamond DSC PerkinElmer instrument. The BTU ligand and its copper(I) complexes were studied at 10°/min of scanning rate after being encapsulated in aluminium pans. Three heating/cooling cycles were performed on each sample. Thermogravimetric analysis for all samples was performed on a TA Q50 TGA instrument using alumina crucibles and nitrogen as purging gas. The heating rate employed was 10°C·min^{-1} from room temperature to 550°C. The powder XRD measurements were made on a D8 Advance diffractometer (Bruker AXS GmbH, Germany), in parallel beam setting, with monochromatized Cu-Kα1 radiation ($\lambda = 1.5406$ Å), scintillation detector, and horizontal sample stage. The measurements were performed in symmetric (θ–θ) geometry in the 2θ range from 1.5° to 30° in steps of 0.02°, with measuring times per step in the 5–40 s range. The temperature control of the samples during measurements was achieved by adapting a home-made heating stage to the sample stage of the diffractometer.

*2.2. Preparation of N-(4-Perfluorooctylphenylcarbamothioyl)-3,4-didecyloxybenzamide (**1**).* The synthesis of the BTU ligand **1** followed the procedure described elsewhere [40, 41]. In the first step, the acid chlorides were prepared by reacting the 3,4-didecyloxybenzoic acid (5 mmol) with an excess of thionyl chloride (20 mmol) in freshly distilled dichloromethane (25 ml) for 3 hours and heating under reflux. Then, the excess of thionyl chloride and the solvent were removed under reduced pressure. The resulting acid chloride was taken to the next step without further purification. The acid chloride was dissolved in dry acetone (20 ml), and a solution of NH_4SCN (5 mmol) in acetone (15 mL) was added dropwise for a period of 15 min under nitrogen. The mixture was heated under reflux for a period of 30 min. During the addition of the ammonium thiocyanate solution, the formation of a cloudy white precipitate was observed. The mixture was cooled down to room temperature, and a solution of *p*-perfluorooctylaniline (4.6 mmol) in acetone (10 mL) was added dropwise for a period of 30 min. The mixture was further stirred for 2 hours at room temperature, and then it was poured in 100 ml of deionized water. The resulting precipitate was filtered off and washed several times with water and ethanol followed by two times of recrystallization from a mixture of dichloromethane/ethanol to yield a white solid.

2.2.1. Compound 1. White crystalline solid. Yield: 77%. Anal. Calcd. For $C_{42}H_{51}F_{17}N_2O_3S$ (%): C, 51.11; H, 5.21; and N, 2.84; Found: C, 51.59; H, 5.35; and N, 2.65. ^{1}H NMR (500 MHz, $CDCl_3$) δ 12.99 (s, 1H), 9.05 (s, 1H), 7.98 (d, $J = 8.6$ Hz, 2H), 7.74 (d, $J = 8.6$ Hz, 2H), 7.60–7.30 (m, 2H), 6.93 (d, $J = 8.2$ Hz, 1H), 4.24–3.71 (m, 4H), 1.93–1.80 (m, 4H), and 1.55–1.20 (m, 28H), 0.88 (m, 6H). ^{13}C NMR (126 MHz, $CDCl_3$, ppm) δ: 178.5, 166.4, 154.2, 149.6, 141.1, 127.7, 127.6, 126.5, 123.3, 123.0, 121.1, 119.6, 112.4, 112.1, 69.6, 69.3, 31.8, 29.6, 29.4, 29.3, 29.1, 29.0, 26.1, 25.9, 22.7, and 14.1. IR (KBr, cm^{-1}): 3266 (ν_{NH}), 2924, 2853 (ν_{CH_2}), 1665 ($\nu_{C=O}$), 1503 ($\nu_{C\text{-}N}$), 1350 ($\nu_{C=S}$), and 1144 ($\nu_{C\text{-}N}$).

*2.3. Preparation of Copper(I) Complexes (**2a** and **2b**).* The synthesis of copper(I) complexes with benzoylthiourea ligand **1** is similar to the preparation of such copper(I) complexes described elsewhere [36, 38, 39]. A solution of the corresponding halide copper(II) salt (CuX_2, where X = Cl or Br; 0.5 mmol) dissolved in ethanol (5 ml) was added dropwise over a period of 5 min to a hot solution of the ligand (1 mmol) in ethanol (10 ml). A bright-yellow precipitate was formed during the addition of the solution of copper(II) salt. The mixture was further stirred and heated under reflux for 2 hours. After this period of time, the hot mixture was filtered and the resulting precipitate was washed several times with ethanol and dried in vacuum to yield a yellow solid. The yields were calculated based on copper salts. The ^{1}H NMR spectroscopy and the elemental analysis confirmed the formation and purity of these copper(I) complexes.

2.3.1. Compound ***2a***. Yellow solid. Yield: 74%. Anal. Calcd. For $C_{84}H_{102}ClCuF_{34}N_4O_6S_2$(%): C, 48.67; H, 4.96; and N, 2.70; Found: C, 49.23; H, 4.84; and N, 2.74. ^{1}H NMR (500 MHz, $CDCl_3$, ppm) δ 13.35 (s, 2H), 11.06 (s, 2H), 8.16–7.92 (m, 2H), 7.70–7.62 (m, 10H), 6.89 (d, $J = 8.6$ Hz, 2H), 4.09 (t, $J = 6.5$ Hz, 4H), 4.03 (t, $J = 6.5$ Hz, 4H), 2.05–1.69 (m, 8H), 1.51–1.20 (m, 56H), and 0.88 (m, 12H).

^{13}C NMR (126 MHz, $CDCl_3$) δ 179.1, 169.0, 154.6, 148.9, 139.9, 127.8, 125.4, 123.9, 122.2, 113.4, 111.7, 69.5, 69.1, 31.9, 30.9, 29.6, 29.5, 29.3, 29.2, 29.1, 26.0, 22.7, 22.3, and 14.1.

IR (KBr, cm^{-1}): 3274 (ν_{NH}), 2927, 2856 (ν_{CH_2}), 1664 ($\nu_{C=O}$), 1512 ($\nu_{C\text{-}N}$), and 1152 ($\nu_{C\text{-}N}$).

2.3.2. Compound ***2b***. Yellow solid. Yield: 78%. Anal. Calcd. For $C_{84}H_{102}BrCuF_{34}N_4O_6S_2$(%): C, 47.65; H, 4.86; and N, 2.65; Found: C, 48.51; H, 4.64; and N, 2.66. ^{1}H NMR (500 MHz, $CDCl_3$, ppm) δ 13.27 (s, 2H), 10.66 (s, 2H), 7.98 (d, 7.8 Hz, 2H), 7.68 (m, 4H), 7.59 (m, 6H), 6.88 (d, $J = 8.6$ Hz, 2H), 4.09 (t, $J = 6.5$ Hz, 4H), 4.01 (t, $J = 6.5$ Hz, 4H), 1.97–1.65 (m, 8H), 1.50–1.11 (m, 56H), and 0.88 (m, 12H).

^{13}C NMR (126 MHz, $CDCl_3$) δ 178.4, 168.6, 154.7, 149.0, 139.9, 127.8, 125.1, 123.8, 122.1, 113.3, 111.7, 69.6, 69.1, 31.9, 29.6, 29.5, 29.4, 29.3, 29.2, 29.0, 26.0, 22.7, and 14.1.

IR (KBr, cm^{-1}): 3275 (ν_{NH}), 2927, 2856 (ν_{CH_2}), 1665 ($\nu_{C=O}$), 1510 ($\nu_{C\text{-}N}$), and 1152 ($\nu_{C\text{-}N}$).

3. Results and Discussions

3.1. Synthesis. The novel BTU ligand having two terminal decyloxy chains on the benzoyl unit and a perfluorooctyl group on the aniline fragment was prepared through the reaction of 3,4-didecyloxybenzoyl isothiocyanate with 4-perfluorooctylaniline in high yield (77%) (Scheme 1) [40, 41, 51]. The copper(I) complexes were obtained by reaction of the BTU ligand with copper(II) halide salts in hot ethanol in a 2 : 1 molar ratio to give yellow solids. The reaction can be also performed by employing copper(I) salts giving identical products. The new BTU derivative and the copper(I) complexes were fully characterized by ^{1}H NMR and ^{13}C NMR and IR spectroscopies as well as by elemental analyses. The ^{1}H NMR spectrum of BTU ligand **1** contains two singlets, one at 9.05 ppm and a second one at 12.99 ppm, which can be assigned to the two NH groups. The first singlet can be assigned to the NH proton located between the carbonyl and thiocarbonyl groups, while the second singlet was assigned to the NH proton between the thiocarbonyl group and the benzene ring. On the contrary, the chemical shift values for the carbonyl and thiocarbonyl carbons were found at about 166.5 ppm and around 178.5 ppm, respectively, as reported for some other fluorinated BTU compounds [52, 53].

The ^{1}H NMR spectra of copper(I) complexes recorded in $CDCl_3$ solvent show the expected signals assigned to ligand with several significant shifts resulted from its coordination to the copper(I) center (Figure 1). The most significant changes were seen for the two NH singlets. Both signals were downfield shifted with a more pronounced shift for the signal assigned to the NH group located between the carbonyl and thiocarbonyl groups of the BTU ligand.

While the first singlet at 12.99 ppm in the ^{1}H NMR spectrum of uncoordinated ligand downfield shifts to 13.35 ppm for **2a** and to 13.27 ppm for **2b**, the second NH signal at 8.99 ppm for **1** is downfield shifted to 11.06 ppm for **2a** and to 10.66 ppm for **2b** as a consequence of the intramolecular hydrogen bonding between the halide ion and the second NH group. These chemical shift changes are attributed to the copper(I) complexes formation, and similar results were found for related three-coordinate copper(I) complexes [33–35]. Intramolecular hydrogen-bonding between the coordinated halide ions and NH group was evidenced by X-ray structural analysis for several related complexes [37–39].

Several regions of the FTIR spectra, recorded in KBr discs, are of interest for supporting the coordination via the sulfur atom: 3200–3400 cm^{-1} for $\nu_{N\text{-}H}$ frequencies, 1500–1700 cm^{-1} for strong $\nu_{C=O}$ and $\nu_{C\text{-}N}$ frequencies, and finally, 1100–1250 cm^{-1} where strong absorption bands assigned to a combination of $\nu_{C\text{-}N}$ and ν_{phenyl} stretching frequencies can be found [54–58]. For example, the intensity and the position of the carbonyl stretching $\nu_{C=O}$ at ~1665 cm^{-1} are essentially the same in the IR spectra of **2a** and **2b** as in the free ligand **1**. The medium intense band at 1350 cm^{-1} in the IR of ligand **1**, assigned to $\nu_{C=S}$, is absent in the IR spectra of copper(I) complexes [34]. Additionally, the very strong band at 1144 cm^{-1} for **1**, corresponding to $\nu_{C\text{-}N}$ of the thioamide group, shifts to higher wavenumber at 1152 cm^{-1} for **2a** and **2b**.

3.2. Liquid Crystalline Properties. The new copper(I) complexes and the benzoylthiourea ligand were investigated for their liquid crystalline properties by a combination of hot-stage polarizing optical microscopy (POM) and differential scanning calorimetry (DSC). The thermal parameters for all compounds are presented in Table 1.

The heating run of the DSC curve recorded for ligand **1** shows only one transition assigned to a crystal-to-isotropic state transition (Table 1). On cooling from the isotropic state down to 0°C, three transitions were observed, an isotropic to a crystalline phase (Cr_1) transition followed by other two transitions between crystalline states (Cr_2 and Cr_3). No LC phase was evidenced for the BTU ligand. On the contrary, it was found that both copper(I) complexes show mesogenic properties.

Two transitions were detected in the DSC traces both during the heating and cooling runs for the two copper(I) complexes: a first transition from the isotropic state to a columnar phase followed by a second transition assigned to a glass transition around 50°C (Figures 2 and 3). All these

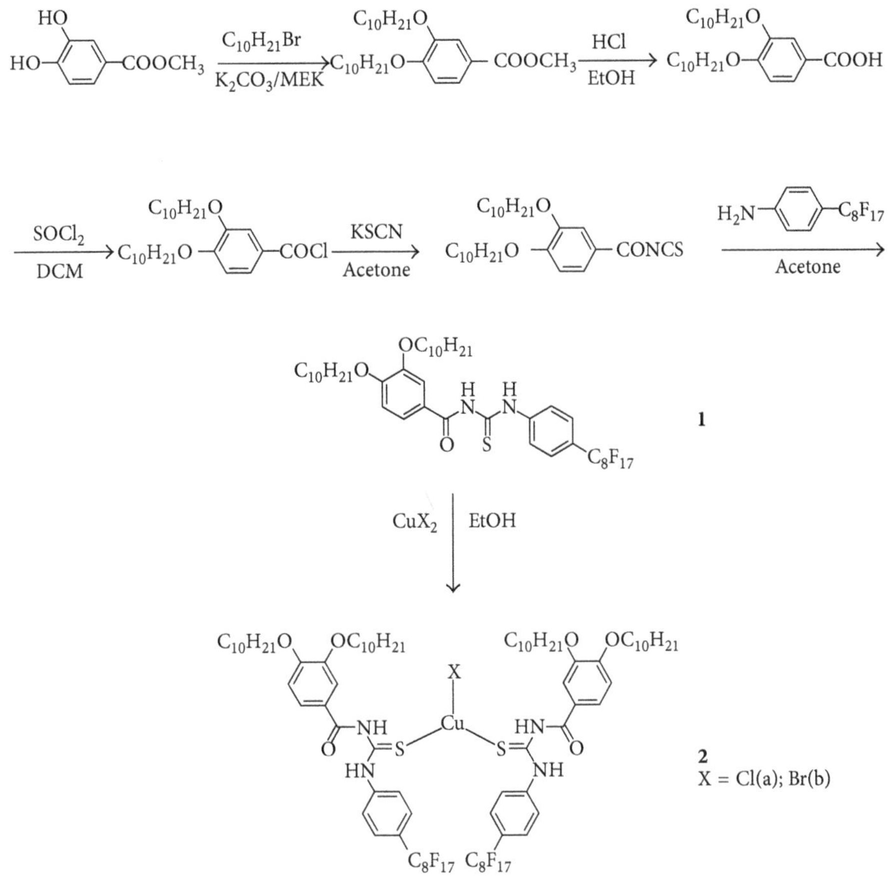

SCHEME 1: Synthesis of *N*-benzoyl-*N*′-arylthiourea perfluoroalkyl terminated BTU and of copper(I) complexes.

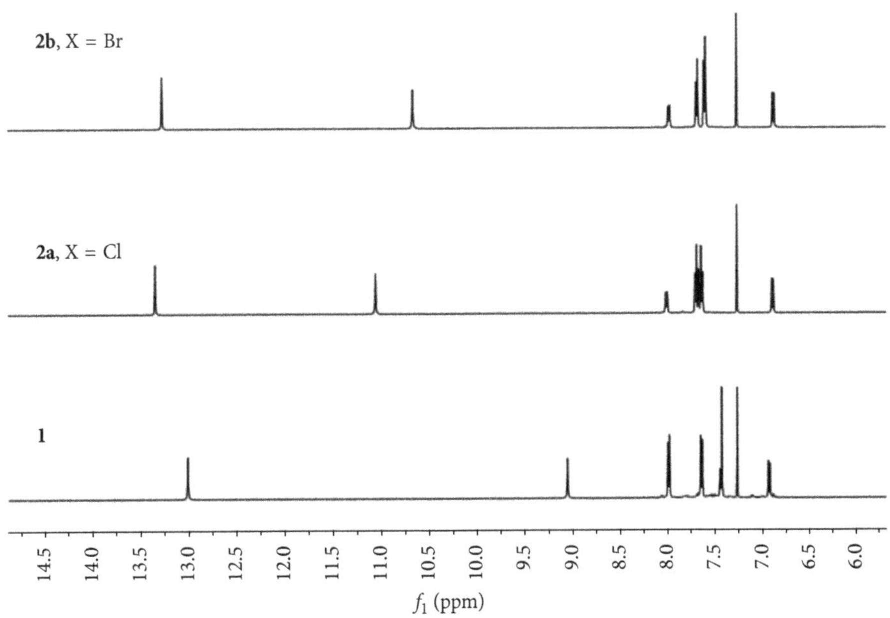

FIGURE 1: The low field region of the ^{1}H NMR spectra of the benzoylthiourea derivative **1** and its copper(I) complexes **2a** and **2b**.

transitions were detected by POM as well, and an example of textural changes for compound **2a** is presented in Figure 4. These results are consistent with the POM studies, where typical optical textures of a hexagonal columnar (Col_h) phase were observed for both copper(I) complexes (Figures 4 and 5). The clearing temperatures of the two copper(I) complexes (Table 1, measured during the first heating run) depend significantly on the size of the halide ion.

TABLE 1: Transition temperatures (in °C) and enthalpies (kJ·mol^{-1}) for ligand **1** and its copper(I) complexes.

Compound	Transitions, T/°C (ΔH/kJ·mol^{-1})[a]
1	Cr 110 (58.7) Iso 99 (−35.0) Cr_1 87 (−7.1) Cr_2 17 (−7.7) Cr_3
2a	g 50[b] Col_h 162 (13.7) Iso 162 (−13.0) Col_h 50[b] g
2b	g 52 Col_h 173 (18.2) Iso 170 (−16.3) Col_h 47 g

[a] Cr_1, Cr_2, and Cr_3 = crystalline phases; Iso = isotropic phase; Col_h = hexagonal columnar phase; g = glassy state. [b] Transition detected by POM.

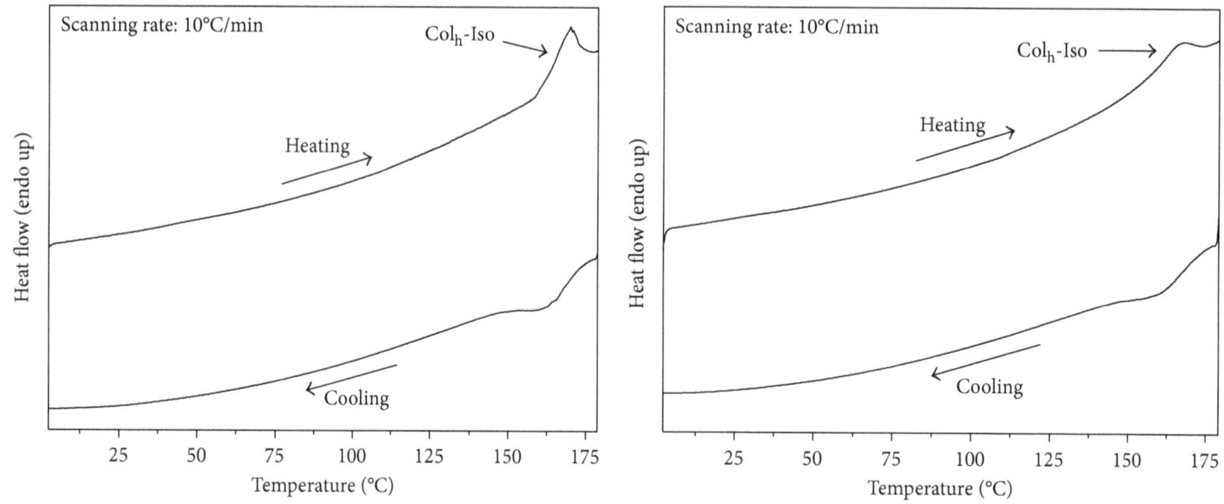

(a) (b)

FIGURE 2: DSC traces for **2a**: first (a) and second heating-cooling cycles (b).

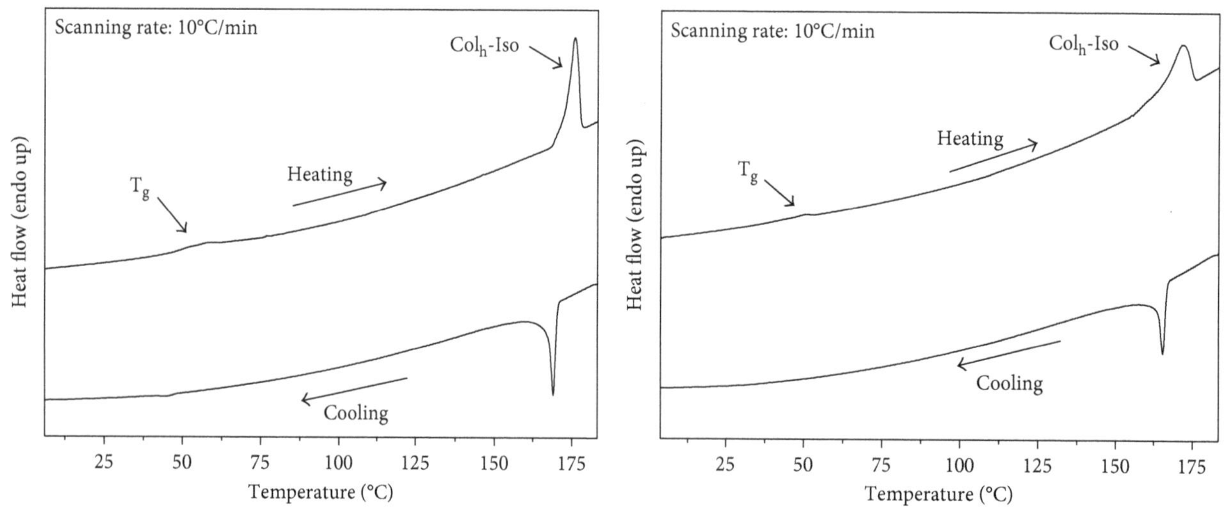

(a) (b)

FIGURE 3: DSC traces for **2b**: first (a) and second heating-cooling cycles (b).

The lowest clearing temperature was recorded for chlorocomplex **2a** (162°C), while a higher clearing temperature was measured for the bromocomplex **2b** (173°C). Interestingly, these two copper(I) compounds are stable in the glassy state at room temperature as no crystalline phases were seen during the first and subsequent heating-cooling cycles.

The mesophase assignment for the two Cu(I) complexes was confirmed further by powder X-ray diffraction (XRD) measurements (Table 2). The samples were heated up to 125°C in the mesomorphic domain, and typical patterns that are characteristic of a hexagonal packing were obtained in each case.

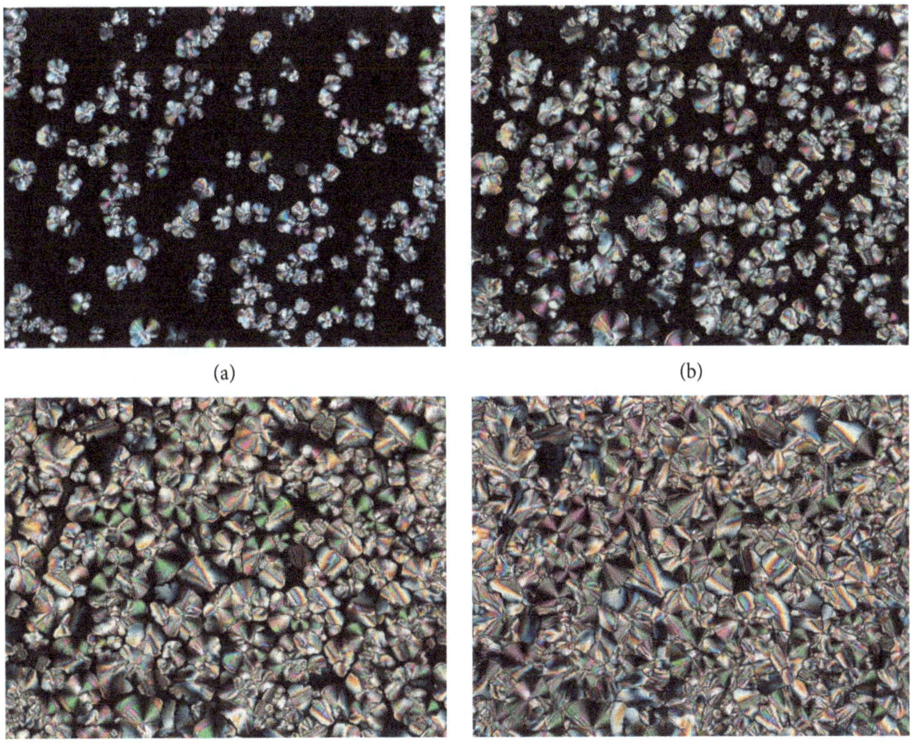

(a) (b) (c) (d)

FIGURE 4: Pictures taken at the polarizing optical microscope showing the optical textures of **2a** at 170°C (a), 160°C (b), 85°C (c), and 35°C (d).

(a) (b)

FIGURE 5: Pictures taken at the polarizing optical microscope showing the optical textures of **2b** at 175°C (a) and 35°C (b).

The diffractograms of **2a** and **2b** show a strong maximum from the (100) reflection followed by a series of four weaker sharp peaks with a d-spacing ratio of $d/\sqrt{3}$, $d/2$, $d/\sqrt{7}$, and $d/4$ from the (110), (200), (210), and (400) reflections for **2a** or $d/\sqrt{3}$, $d/2$, $d/\sqrt{7}$, and $d/3$ from the (110), (200), (210), and (300) reflections for **2b**, respectively (Figure 6).

It is important to mention that the transition temperatures recorded both during the heating and cooling runs by DSC were slightly shifted towards lower values on subsequent heating-cooling cycles for complexes **2a** and **2b**. These observations indicate a possible slight decomposition of the copper(I) complexes on heating above the clearing points. Therefore, a study regarding the thermal stability of these compounds was performed in the next step.

3.3. Thermal Stability of the Ligands and Copper(I) Complexes. The two copper(I) complexes and the BTU ligand were investigated by TG analysis in nitrogen, in the 25–550°C temperature range, and the decomposition curves are presented in Figure 7. These TG curves do not contain steps or other indication of mass loss in the 25–160°C region, and therefore, the complexes do not contain small molecules (water or ethanol solvent). On the contrary, the results of the thermogravimetric analyses show that the copper(I) complexes **2a** and **2b** have a higher thermal stability compared to the BTU ligand **1**, and their decomposition started at temperatures around 180°C. For exemplification, Figure 7 gives the TG curves for the BTU ligand **1** and its copper(I) complexes **2a** and **2b**. The BTU ligand starts decomposing around 160°C.

TABLE 2: Temperature dependent X-ray powder diffraction for Cu(I) compounds.

Compound	Mesophase	T (°C)	Indexation	d-spacing observed (Å)	d-spacing calculated (Å)	Lattice parameters (Å)[a]
2a	Col_h	125	100	26.91	26.91	$a = 31.07$
			110	15.49	15.54	
			200	13.42	13.46	
			210	10.04	10.17	
			400	6.66	6.73	
			Broad[b]	4.6	—	
2b	Col_h	125	100	26.51	26.51	$a = 30.61$
			110	15.17	15.30	
			200	13.18	13.26	
			210	10.00	10.02	
			300	8.80	8.84	
			Broad[b]	4.6	—	

[a]The Col_h lattice parameter $a = 2\langle d_{100}\rangle/\sqrt{3}$. [b]Broad peak assigned to molten alkyl chains.

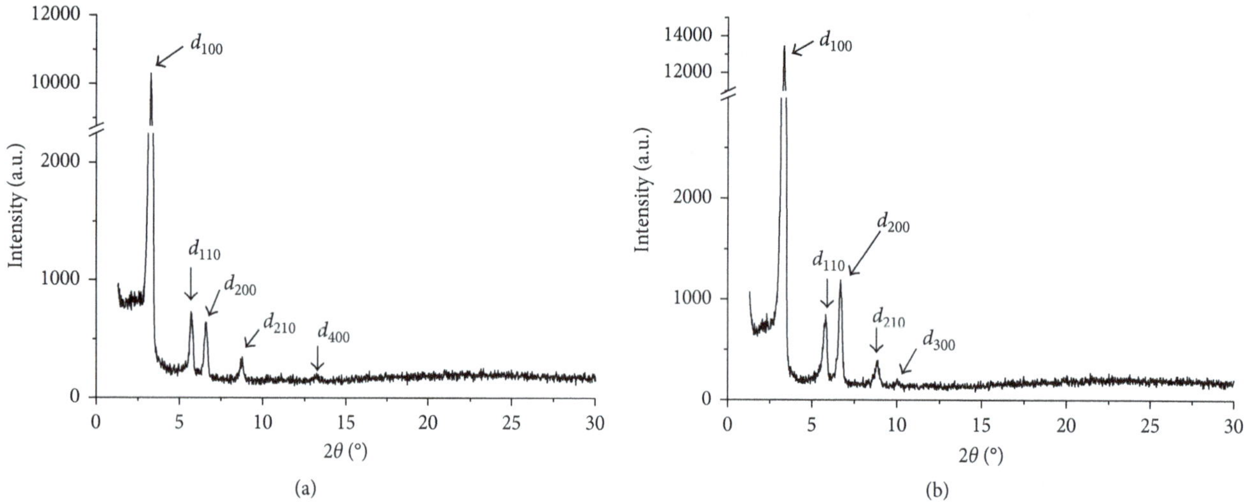

FIGURE 6: X-ray powder diffractograms for Cu(I) compounds: **2a** at 125°C (a) and **2b** at 125°C (b).

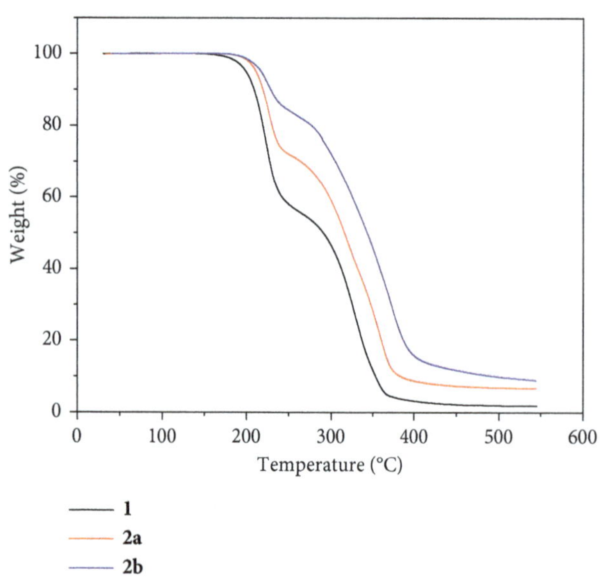

FIGURE 7: TG curves for ligand **1** and its copper(I) complexes **2a** and **2b**.

As it can be seen from Figure 7, the two copper(I) complexes undergo complete decomposition up to 500°C. A small weight loss, less than 2%, was recorded on heating the copper(I) complexes **2a** and **2b** above the clearing points around 180°C. The slight decomposition observed at temperatures higher than the clearing point could explain the transition temperatures shifting recorded by DSC. Similar observations were found for some palladium(II) complexes with mixed ligands, Schiff bases, and BTU derivatives, reported previously [59, 60].

4. Conclusions

Three-coordinate copper(I)-containing liquid crystalline materials with columnar mesophases were prepared by using an alkoxy-substituted *N*-benzoylthiourea derivative having a perfluorooctyl group at one end. The new copper(I) complexes exhibit a hexagonal columnar phase over a large temperature range, more than 100°C, as evidenced by DSC, POM observations, and powder X-ray diffraction analysis. The thermal stability of these copper(I) metallomesogens is limited by a slight decomposition on heating above the clearing points (~180°C).

Conflicts of Interest

The authors declare that they have no conflicts of interest.

Acknowledgments

The authors would like to thank to Dr. Iuliana Pasuk (National Institute of Materials Physics, Magurele, Romania) for her support with XRD measurements.

References

[1] R. W. Date, E. F. Iglesias, K. E. Rowe, J. M. Elliott, and D. W. Bruce, "Metallomesogens by ligand design," *Dalton Transactions*, no. 10, pp. 1914–1931, 2003.

[2] J. Torroba and D. W. Bruce, "Metallomesogens," in *Comprehensive Inorganic Chemistry II: From Elements to Applications*, vol. 8, pp. 837–917, Elsevier Ltd., New York, NY, USA, 2nd edition, 2013.

[3] D. Manaila-Maximean, "New grafted ferrite particles/liquid crystal composite under magnetic field," *Journal of Magnetism and Magnetic Materials*, vol. 452, pp. 343–348, 2017.

[4] J. M. Gilli, S. Thiberge, and D. Manaila-Maximean, "New aspect of the voltage/confinement ratio phase diagram for a confined homeotropic cholesteric," *Molecular Crystals and Liquid Crystals*, vol. 417, no. 1, pp. 207–213, 2004.

[5] D. Manaila-Maximean, R. Bena, and A. M. Albu, "Electric and electro-optic investigations on polymer dispersed liquid crystal films," *Modern Physics Letters B*, vol. 11, pp. 431–440, 1997.

[6] C. Rosu, D. M. Maximean, S. Kundu, P. L. Almeida, and O. Danila, "Perspectives on the electrically induced properties of electrospun cellulose/liquid crystal devices," *Journal of Electrostatics*, vol. 69, no. 6, pp. 623–630, 2011.

[7] D. Manaila-Maximean and C. Rosu, "Influence of polarizing electric field on electrical and optical properties of PDLC films," *Molecular Crystals and Liquid Crystals*, vol. 413, no. 1, pp. 9–19, 2004.

[8] E.-K. Fleischmann and R. Zentel, "Liquid-crystalline ordering as a concept in materials science: from semiconductors to stimuli-responsive devices," *Angewandte Chemie International Edition*, vol. 52, no. 34, pp. 8810–8827, 2013.

[9] S. Laschat, A. Baro, N. Steinke et al., "Discotic liquid crystals: from tailor-made synthesis to plastic electronics," *Angewandte Chemie International Edition*, vol. 46, no. 26, pp. 4832–4887, 2007.

[10] C. Tschierske, "Development of structural complexity by liquid-crystal self-assembly," *Angewandte Chemie International Edition*, vol. 52, no. 34, pp. 8828–8878, 2013.

[11] P. Espinet, M. C. Lequerica, and J. M. Martín-Alvarez, "Synthesis, structural characterization and mesogenic behavior of copper(I) n-alkylthiolates," *Chemistry-A European Journal*, vol. 5, no. 7, pp. 1982–1986, 1999.

[12] M. Benouazzane, S. Coco, P. Espinet, and J. Barbera, "Supramolecular organization in copper(I) isocyanide complexes: copper(I) liquid crystals from a simple molecular structure," *Journal of Materials Chemistry*, vol. 11, no. 6, pp. 1740–1744, 2001.

[13] M. Benouazzane, S. Coco, and P. Espinet, "Binuclear mesogenic copper(I) isocyanide complexes with an unusual inorganic core formed by two tetrahedra sharing an edge," *Inorganic Chemistry*, vol. 41, no. 22, pp. 5754–5759, 2002.

[14] S. Coco, C. Cordovilla, B. Donnio, P. Espinet, M. J. Garcia-Casas, and D. Guillon, "Self-organization of dendritic supermolecules, based on isocyanide-gold(I), -copper(I), -palladium(II), and -platinum(II) complexes, into micellar cubic mesophases," *Chemistry-A European Journal*, vol. 14, no. 12, pp. 3544–3552, 2008.

[15] R. Dembinski, P. Espinet, S. Lentijo et al., "Fluorophobic effect in metallomesogens-the synthesis and mesomorphism of Ag, Au, Cu, Fe, Pd, and Pt fluorous isocyanide complexes," *European Journal of Inorganic Chemistry*, vol. 2008, no. 10, pp. 1565–1572, 2008.

[16] R. Chico, E. de Domingo, C. Dominguez et al., "High one-dimensional charge mobility in semiconducting columnar mesophases of isocyano-triphenylene metal complexes," *Chemistry of Materials*, vol. 29, no. 17, pp. 7587–7595, 2017.

[17] R. Ziessel, G. Pickaert, F. Camerel et al., "Tuning organogels and mesophases with phenanthroline ligands and their copper complexes by inter- to intramolecular hydrogen bonds," *Journal of the American Chemical Society*, vol. 126, no. 39, pp. 12403–12413, 2004.

[18] F. Neve, M. Ghedini, A. M. Levelut, and O. Francescangeli, "Ionic metallomesogens. Lamellar mesophases in copper(I) azamacrocyclic complexes," *Chemistry of Materials*, vol. 6, no. 1, pp. 70–76, 1994.

[19] L. Douce, A. El-ghayoury, A. Skoulios, and R. Ziessel, "Columnar mesophases from tetrahedral copper(I) cores and Schiff-base derived polycatenar ligands," *Chemical Communications*, no. 20, pp. 2033-2034, 1999.

[20] A. El-ghayoury, L. Douce, A. Skoulios, and R. Ziessel, "Cation-induced macroscopic ordering of non-mesomorphic modules—a new application for metallohelicates," *Angewandte Chemie International Edition*, vol. 37, no. 16, pp. 2205–2208, 1998.

[21] H.-D. Lin and C. K. Lai, "Ionic columnar metallomesogens formed by three-coordinated copper(I) complexes," *Journal of the Chemical Society, Dalton Transactions*, no. 16, pp. 2383–2387, 2001.

[22] A. Saeed, U. Floerke, and M. F. Erben, "A review on the chemistry, coordination, structure and biological properties of 1-(acyl/aroyl)-3-(substituted) thioureas,," *Journal of Sulfur Chemistry*, vol. 35, no. 3, pp. 318–355, 2014.

[23] L. Beyer, E. Hoyer, J. Liebscher, and H. Hartmann, "Formation of complexes with *n*-acyl-thioureas," *Zeitschrift für Physikalische Chemie*, vol. 21, pp. 81–91, 1981.

[24] P. Muhl, K. Gloe, F. Dietze, E. Hoyer, and L. Beyer, "*N*-acylthioureas-effective extractants for liquid-liquid-extraction of metal-ions," *Zeitschrift für Physikalische Chemie*, vol. 26, pp. 81–94, 1986.

[25] J. Duque, O. Estevez-Hernandez, E. Reguera, J. Ellena, and Rodrigo S. Corrêa, "Synthesis, characterization, and single crystal X-ray structure of the 1-furoyl-3-cyclohexylthiourea cadmium chloride complex, $CdC_4H_3OC(O)NHC(S)NHC_6H_{11}]_4Cl_2$," *Journal of Coordination Chemistry*, vol. 62, no. 17, pp. 2804–2813, 2009.

[26] O. Estévez-Hernández, J. Duque, J. Rodríguez-Hernández, and E. Reguera, "Dinuclear and polymeric Hg(II) complexes with 1-(2-furoyl)thiourea derivatives: their crystal structure

and related properties," *Polyhedron*, vol. 97, pp. 148–156, 2015.

[27] T. Yesilkaynak, "2-Chloro-*N*-((5-chloropyridine-2-yl) carbamothioyl)benzamide and its Co(II), Ni(II) and Cu (II) metal complexes," *Journal of Thermal Analysis and Calorimetry*, vol. 124, no. 2, pp. 1029–1037, 2016.

[28] R. S. Correa, K. M. de Oliveira, F. G. Delolo et al., "Ru(II)-based complexes with *N*-(acyl)-*N′*,*N′*-(disubstituted)thiourea ligands: Synthesis, characterization, BSA- and DNA-binding studies of new cytotoxic agents against lung and prostate tumour cells," *Journal of Inorganic Biochemistry*, vol. 150, pp. 63–71, 2015.

[29] M. Sternberg, J. Rust, C. W. Lehmann, and F. Mohr, "Synthesis and metal complexes of thiourea ligands containing carbohydrate-derived substituents," *Helvetica Chimica Acta*, vol. 96, no. 2, pp. 280–288, 2013.

[30] S. S. Hassan, M. M. Shoukry, R. N. Shallan, and R. van Eldik, "Synthesis, characterization, speciation and biological studies on metal chelates of 1-benzoyl(1,2,4-triazol-3-yl)thiourea," *Journal of Coordination Chemistry*, vol. 70, no. 10, pp. 1761–1775, 2017.

[31] A. Molter, S. Kathrein, B. Kircher, and F. Mohr, "Anti-tumour active gold(I), palladium(II) and ruthenium(II) complexes with thio- and selenoureato ligands: a comparative study," *Dalton Transactions*, vol. 47, no. 14, pp. 5055–5064, 2018.

[32] C. Roşu, D. Manaila-Maximean, V. Cîrcu, Y. Molard, and T. Roisnel, "Differential negative resistance in the current–voltage characteristics of a new palladium (II) metallomesogen," *Liquid Crystals*, vol. 38, no. 6, pp. 757–765, 2011.

[33] X.-Y. Zhao, C.-B. Zhu, H.-P. Li, Y. Yang, and H. W. Roesky, "Synthesis and characterization of copper(I) halide complexes with *N*-(2,6-diisopropylphenyl)-*N′*-benzoylthiourea: monomeric, dimeric, and cage structures," *Zeitschrift für anorganische und allgemeine Chemie*, vol. 640, no. 8-9, pp. 1614–1621, 2014.

[34] Y. F. Yuan, J. T. Wang, M. C. Gimeno, A. Laguna, and P. G. Jones, "Synthesis and characterisation of copper complexes with *N*-ferrocenoyl-*N′*-aryl(alkyl)thioureas," *Inorganica Chimica Acta*, vol. 324, no. 1-2, pp. 309–317, 2001.

[35] D. Wang, S.-Y. Wu, H.-P. Li, Y. Yang, and H. W. Roesky, "Synthesis and characterization of copper complexes with the *N*-(2, 6-diisopropylphenyl)-*N′*-acylthiourea ligands," *European Journal of Inorganic Chemistry*, vol. 2017, no. 10, pp. 1406–1413, 2017.

[36] A. F. ElHusseiny, A. Eldissouky, A. M. Al-Hamza, and H. H. Hassan, "Structure–property relationship studies of copper(I) complexes of nanosized hypodentate ligands and evaluation of their antitumor and antimicrobial activities," *Journal of Coordination Chemistry*, vol. 68, no. 2, pp. 241–260, 2015.

[37] G. Li, D.-J. Che, Z.-F. Li, Y. Zhu, and D.-P. Zou, "Versatile coordination patterns in the reaction system of N-benzoyl-N'-(2 pyridyl)thiourea with $CuCl_2$. Their reaction conditions, systematic isolation and crystal structures," *New Journal of Chemistry*, vol. 26, no. 11, pp. 1629–1633, 2002.

[38] D. P. Singh, S. Pratap, S. K. Pandey, R. J. Butcher, and G. Marverti, "*N*-(naphthyl)-*N′*-(methoxy carbonyl)thiocarbamide and its Cu(I) complex: synthesis, spectroscopic, X-ray, DFT and in vitro cytotoxicity study," *Journal of Coordination Chemistry*, vol. 68, no. 2, pp. 261–276, 2015.

[39] L. Xian, T.-B. Wei, and Y.-M. Zhang, "Synthesis and crystal structure of bis(*N*-o-methylphenyl-*N′*-ethoxycarbonylthiourea) copper(I) chloride," *Journal of Coordination Chemistry*, vol. 57, no. 6, pp. 453–457, 2004.

[40] M. Ilis, M. Bucos, F. Dumitrascu, and V. Cîrcu, "Mesomorphic behaviour of *N*-benzoyl-*N′*-aryl thioureas liquid crystalline compounds," *Journal of Molecular Structure*, vol. 987, no. 1–3, pp. 1–6, 2011.

[41] M. Iliş, M. Micutz, I. Pasuk, T. Staicu, and V. Cîrcu, "Synthesis and liquid crystalline properties of novel fluorinated *N*-benzoyl thiourea compounds. Effect of perfluoroalkyl chains on the thermal behavior and smectic phases stability," *Journal of Fluorine Chemistry*, vol. 204, pp. 84–89, 2017.

[42] M. Hird, "Fluorinated liquid crystals–properties and applications," *Chemical Society Reviews*, vol. 36, no. 12, pp. 2070–2095, 2007.

[43] M. Spengler, R. Y. Dong, C. A. Michal, M. Pfletscher, and M. Giese, "Fluorination of supramolecular liquid crystals–tuning tool and analytical probe," *Journal of Materials Chemistry C*, vol. 5, no. 9, pp. 2235–2239, 2017.

[44] S. Marzouk, A. Khalfallah, B. Heinrich, J. E. Khiari, A. Kriaa, and S. Mérya, "Synthesis and mesomorphic properties of liquid crystals containing a perfluorinated segment via different linkers," *Journal of Fluorine Chemistry*, vol. 197, pp. 15–23, 2017.

[45] J. Y. Jang and Y. W. Park, "Smectic C mesogens with terminal perfluoroalkyl chains," *Molecular Crystals and Liquid Crystals*, vol. 606, no. 1, pp. 90–100, 2015.

[46] J. Y. Jang and Y. W. Park, "Synthesis and structural studies of smectic C mesogens with terminal perfluoroalkyl chains," *Liquid Crystals*, vol. 40, no. 4, pp. 511–515, 2013.

[47] B. Bilgin-Eran, H. Ocak, C. Tschierske, and U. Baumeister, "Synthesis and mesomorphism of fluoroalkylated platinum complexes," *Liquid Crystals*, vol. 39, no. 4, pp. 467–476, 2012.

[48] A. P. Piccionello, A. Calabrese, I. Pibiri, V. Giacalone, A. Pace, and S. Buscemia, "Synthesis of fluorinated bent-core mesogens (BCMs) containing the 1,2,4-oxadiazole ring," *Journal of Heterocyclic Chemistry*, vol. 53, no. 6, pp. 1935–1940, 2016.

[49] A. Iwan, "Characterisation and mesomorphic behavior of rod-shaped unsymmetrical imine with a fluorinated chain and a carboxylic group," *Molecular Crystals and Liquid Crystals*, vol. 528, no. 1, pp. 156–162, 2010.

[50] H. Ocak, B. Bilgin-Eran, C. Tschierske, U. Baumeister, and G. Pelzl, "Effect of fluorocarbon chains on the mesomorphic properties of chiral imines and their complexes with copper(II)," *Journal of Materials Chemistry*, vol. 19, no. 38, pp. 6995–7001, 2009.

[51] I. B. Douglass and F. B. Dains, "Some derivatives of benzoyl and furoyl isothiocyanates and their use in synthesizing heterocyclic compounds," *Journal of the American Chemical Society*, vol. 56, no. 3, pp. 719–721, 1934.

[52] J. Imrich, T. Bušová, P. Kristián, and J. Džara, "Synthesis and the 13C NMR spectra of *N*,*N*–disubstituted benzoylthioureas and their seleno and oxo analogues," *Chemical Papers*, vol. 48, pp. 42–46, 1998.

[53] F. Leßmann, L. Beyer, R. Richter, and R. Meusinger, "*N*-benzoylthiocarbamoyl amino acid ethyl esters as chelate ligands for transition metal ions," *Zeitschrift für Naturforschung*, vol. 53b, pp. 981–990, 1998.

[54] A. Saeed, Z. Ashraf, M. F. Erben, and J. Simpson, "Vibrational spectra and molecular structure of isomeric 1-(adamantan-1-ylcarbonyl)-3-(dichlorophenyl)thioureas," *Journal of Molecular Structure*, vol. 1129, pp. 283–291, 2017.

[55] L. Qiao, Y. Zhang, W. Hu et al., "Synthesis, structural characterization and quantum chemical calculations on 1-(isomeric methylbenzoyl)-3-(4-trifluoromethylphenyl)thioureas," *Journal of Molecular Structure*, vol. 1141, pp. 309–321, 2017.

[56] M. E. D. Lestard, D. M. Gil, O. Estévez-Hernández, M. F. Erben, and J. Duque, "Structural, vibrational and electronic characterization of 1-benzyl-3-furoyl-1-phenylthiourea: an experimental and theoretical study," *New Journal of Chemistry*, vol. 39, no. 9, pp. 7459–7471, 2015.

[57] O. Estévez-Hernández, F. F. Salomón, J. Duque, J. Rodríguez-Hernández, and D. M. Gil, "Synthesis, structural and spectroscopic studies of 1-(2-furoyl)-3-phenylthiourea: a theoretical and experimental approach," *Journal of Sulfur Chemistry*, vol. 37, no. 5, pp. 555–579, 2016.

[58] D. M. Gil, M. E. Defonsi Lestard, O. Estévez-Hernández, J. Duque, and E. Reguera, "Quantum chemical studies on molecular structure, spectroscopic (IR, Raman, UV–Vis), NBO and Homo–Lumo analysis of 1-benzyl-3-(2-furoyl) thiourea," *Spectrochimica Acta Part A: Molecular and Biomolecular Spectroscopy*, vol. 145, pp. 553–562, 2015.

[59] M. Ilis, D. Batalu, I. Pasuk, and V. Cîrcu, "Cyclometalated palladium(II) metallomesogens with Schiff bases and N-benzoyl thiourea derivatives as co-ligands," *Journal of Molecular Liquids*, vol. 233, pp. 45–51, 2017.

[60] M. Ilis, M. Micutz, and V. Cîrcu, "Luminescent palladium(II) metallomesogens based on cyclometalated Schiff bases and N-benzoyl thiourea derivatives as co-ligands," *Journal of Organometallic Chemistry*, vol. 836-837, pp. 81–89, 2017.

8

A Chemical Synthesis of Benzo[*a*]pyrene Hapten and Its Application in a Streptavidin-Horseradish Peroxidase-Based Enzyme-Linked Immunological Analysis

Zhen Ma and Huisheng Zhuang

School of Environmental Science and Engineering, Shanghai Jiao Tong University, Shanghai 200240, China

Correspondence should be addressed to Huisheng Zhuang; huishengzhuang@126.com

Academic Editor: Jorge F. Fernandez-Sanchez

A specific BaP hapten was synthesized and BaP antigens (immunogen, coating antigen) were prepared using diverse methods. Based on these works, a streptavidin-horseradish peroxidase-based enzyme-linked immunological analysis was developed and firstly used to detect BaP. Several physiochemical factors that may influence the assay performance were optimized in the assay. Under optimal conditions, a detection limit of 0.0094 ng mL^{-1} was obtained and good linearity was achieved within a range of 0.03–35.40 ng mL^{-1}. Satisfactory recovery of spiked samples (91.12–109.23%) was obtained and the coefficient of variation was acceptable. Finally, the detection results of BaP in environmental and food samples were consistent with those obtained using high-performance liquid chromatography. The proposed immunoassay is reliable and has great potential for detecting trace amounts of BaP.

1. Introduction

Polycyclic aromatic hydrocarbons (PAHs) are persistent organic pollutants that contain 2–8 aromatic rings in linear, angular, or cluster arrangements [1]. Due to their toxicity and carcinogenicity, PAHs have attracted the attention of environmental chemists and regulatory agencies for several decades [2, 3]. Among the PAHs, benzo[*a*]pyrene (BaP) has been widely studied. In recent years, much attention has been devoted to the existence of BaP in the environment. BaP is considered the most dangerous of PAHs because of its chemical and physical characteristics and ease of accumulation in the food chain [4, 5]. BaP is usually found in exhaust gas produced by the incomplete combustion of fuels, in all smoke resulting from the combustion of organic material and in charbroiled food [6–8]. BaP can enter into water bodies mainly via atmospheric fallout, municipal/industrial effluents, urban runoff, and oil spillage [9, 10]; it is emitted to the atmosphere from different sources, including industrial processes, combustion of fossils and fuels, and irrational application of pesticide [11].

BaP is a great potential hazard to human health. It is a type of carcinogenic compound, which can cause lung, gastric, bladder, and gastrointestinal cancers [12, 13]. Recent studies of BaP have also demonstrated its involvement in the development of cardiovascular diseases [14]. BaP is mutagenic and teratogenic; it can even pass through the placenta to endanger the embryogenesis [15]. Dietary intake of BaP is a major source of human exposure [16]. Related studies have shown that people in the general population without substantial environmental and occupational exposure had a higher exposure to BaP by food ingestion [17]. Probable sources of BaP in food are environmental contamination from soil, water, and the thermal treatment used for the preparation of food, such as roasting.

Up to now, many different methods have been described for the determination of BaP residues. The main methods are gas chromatography and mass spectrometry (GC-MS) and high-performance liquid chromatography (HPLC); however, these chromatographic methods usually require long periods of detection processes that are time-consuming and the follow-up maintenance is expensive. Compared

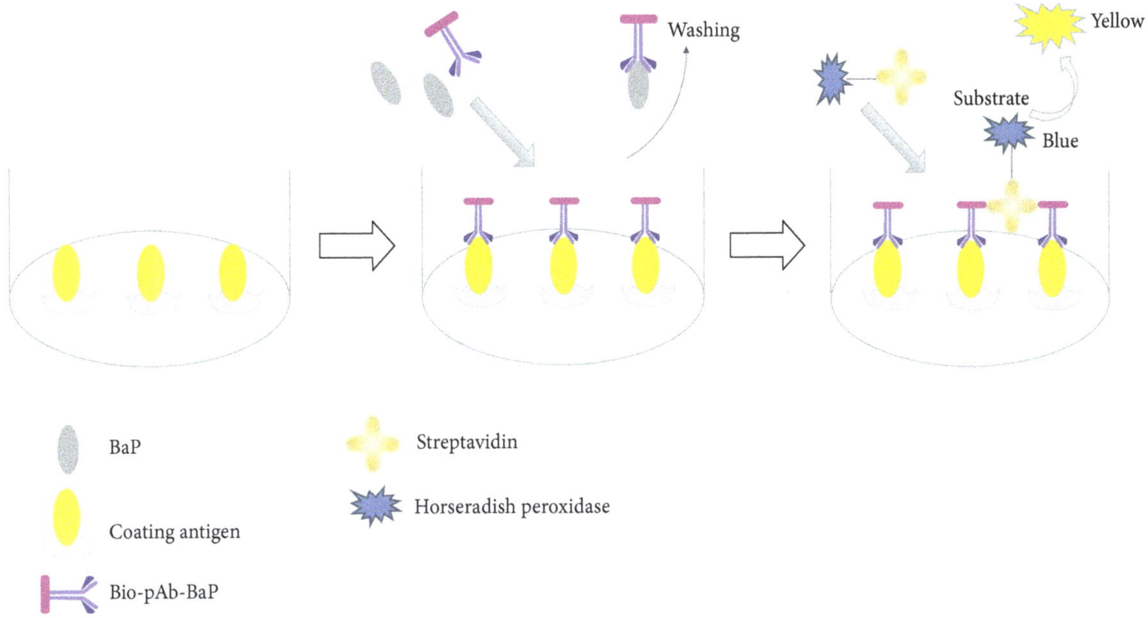

FIGURE 1: Principle of the SA-ELISA method. BaP coating antigens which are covered on the surface of the tubes competed with BaP analytes to combine with the bio-pAb-BaP. The horseradish peroxidase-conjugated streptavidin (SA-HRP) is fixed on the wall through the reaction with bio-pAb-BaP. As the increase of BaP concentration, less bio-pAb-BaP can be attached to the solid carrier. In this case, the amount of SA-HRP is less, and then quantitative analysis could be performed after sufficient colour development.

with chromatographic techniques, immunoassays are cost-effective, with adequate sensitivity and high selectivity [18]. Therefore, immunochemical techniques have been increasingly considered as complementary methods for residue analysis. Among these immunoassays, enzyme-linked immunosorbent assay (ELISA) is well suited for detecting trace amounts of pollutants. Some improved immunoassays were researched based on the conventional ELISA, such as fluorescence-enzyme immunoassay (FL-ELISA) and chemiluminescence immunoassay (CL-ELISA) [19, 20]. Accompanying the improved immunoassays above, the streptavidin-enzyme based ELISA can reduce nonspecific reactions with reagents due to the higher specificity and affinity between streptavidin and biotin, and it has great potential for detecting trace amounts of BaP. However, as we know, no information on detecting BaP by streptavidin-enzyme based ELISA has been reported.

In this research, a specific BaP hapten was synthesized. Diverse BaP antigens and biotinylated polyclonal BaP antibody (bio-pAb-BaP) were prepared. Based on these works, a streptavidin-horseradish peroxidase-based enzyme-linked immunoassay was developed (SA-ELISA) and firstly used for the sensitive detection of BaP. Some environmental and food samples were collected in Shanghai, China. Several physiochemical factors that may influence the assay performance were optimized, and the accuracy and reliability of the immunoassay were validated. Figure 1 shows the detection process of SA-ELISA method.

2. Materials and Methods

2.1. Buffers and Solutions. Phosphate-buffer saline (PBS: NaCl [137 mmol L^{-1}], KH_2PO_4 [2 mmol L^{-1}], Na_2HPO_4 [10 mmol L^{-1}] and KCl [2.7 mmol L^{-1}], pH 7.40), PBST (PBS with 0.05% Tween 20, pH 7.40), and carbonate buffer (CBS: Na_2CO_3 [15 mmol L^{-1}] and $NaHCO_3$ [34.9 mmol L^{-1}], pH 9.60) were used in the assay.

2.2. Materials and Chemicals. The BaP standard, other types of PAHs, chemicals for BaP hapten and antigen synthesis, 25% glutaraldehyde solution, polyethylene glycol 20,000 (PEG 20,000), N,N-dimethylformamide (DMF), and dimethyl sulfoxide (DMSO) were purchased from J&K Chemical (Beijing, China). 3,3′,5,5′-Tetramethyl benzidine (TMB), hydrogen peroxide (H_2O_2), keyhole limpet hemocyanin (KLH), poly-L-Lysine (PLL), polyvinyl alcohol (PVA), skimmed milk powder, H_2SO_4, HCl, gelatin, Tween 20, Coomassie Brilliant Blue G250, and inorganic salts were purchased from Sinopharm Chemical Reagent Co. Ltd. (Shanghai, China). Bovine serum albumin (BSA), ovalbumin (OVA), dialysis bags, biotinylated N-hydroxysuccinimide ester (BNHS), and SA-HRP were all purchased from Sangon Co. Ltd. (Shanghai, China). BaP hapten was purified through column chromatography using silica gel (40 μm average particle size) obtained from Shanghai Sanpont Co. Ltd. (Shanghai, China).

2.3. Synthesis of BaP Hapten. BaP molecule does not contain functional groups that can connect with proteins. Therefore, a specific BaP hapten was synthesized (Figure 2). Then, the BaP antigens and bio-pAb-BaP were prepared. For the preparation of BaP hapten, antigens, and pAb-BaP, we followed the methods of Ma and Zhuang 2017 [21].

BaP (200 mg), n-iodosuccinimid (NIS, 210 mg), and acidic alumina (5 g) were dissolved in toluene (15 mL); the mixture was allowed to react with stirring for 4 days

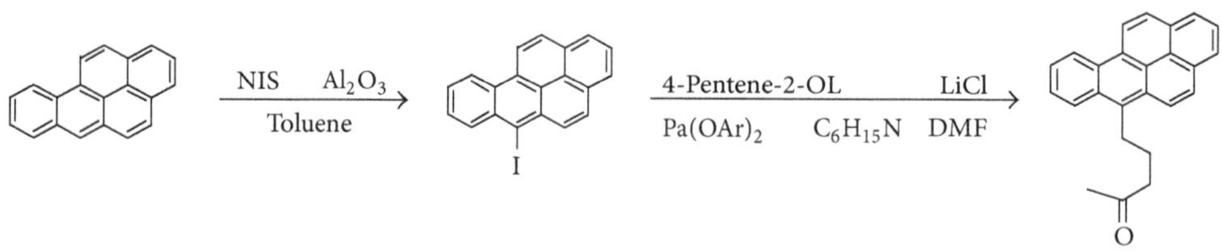

FIGURE 2: Synthesis of BaP hapten.

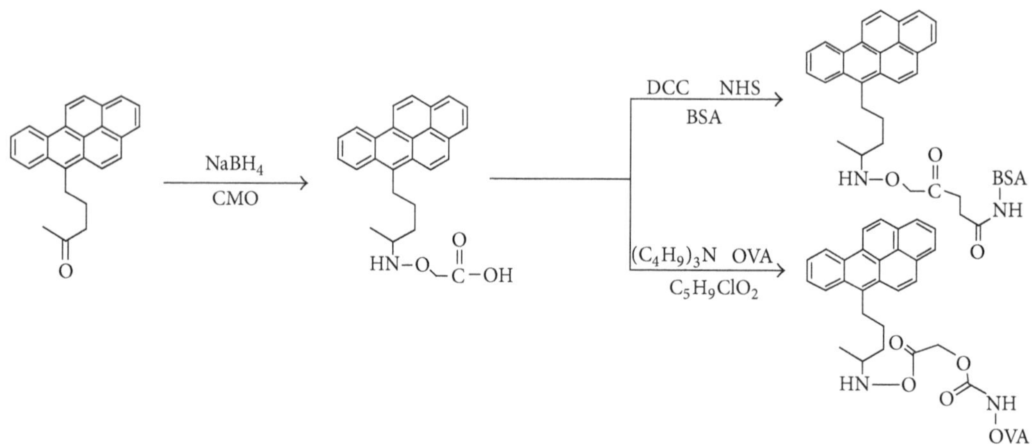

FIGURE 3: Synthesis of BaP immunogen and coating antigen.

at 20°C. The residual acidic alumina was washed with dichloromethane (25 mL × 4). Afterwards, the organic phase was collected and extracted by saturated sodium sulphite, saturated salt water, and double-distilled water (ddH_2O) successively. After being dried with $MgSO_4$, the mixture was concentrated down to near dryness; then, the yellow powder (6-I-BaP, 225 mg, yield: 75.3%) was obtained by recrystallization with dichloromethane and petroleum ether.

Next, 6-I-BaP (135 mg), palladium acetate (5 mg), and LiCl (10 mg) were dissolved in DMF (10 mL). Then, 4-pentene-2-alcohol (35 μL) and triethylamine (100 μL) were added. The mixture was allowed to react for 15 h at 120°C. After a completed reaction, toluene (70 mL) was added. The reaction mixture was extracted by saturated salt water and ddH_2O, respectively. After being dried with $MgSO_4$, the volatile was removed under reduced pressure. The crude product was purified through chromatography column filled with silica gel (leaching solution: tert-butyl methyl ether/petroleum ether = 1/2 (v/v)), and then the yellow powder BaP hapten (BaP = O 102 mg, yield: 87.1%) was obtained.

2.4. Immunogen and Coating Antigen. As a contact portion between carrier protein and BaP hapten, the linking arm can become an antigenic determinant to affect the specificity for conjugating with antibodies [22]. In order to reduce specific binding caused by the linking arm, BSA-BaP (immunogen) and OVA-BaP (coating antigen) were prepared by the activated ester method and mixed anhydride method, respectively (Figure 3).

BaP hapten (8 mg) and carboxymethyl amine hydrochloride (16 mg) were dissolved in pyridine (5 mL); the mixture was allowed to react in absence of light. After a completed reaction, sodium borohydride (36 mg) dissolved in ddH_2O (300 μL) was added to the reaction mixture and allowed to react for 5 h at ambient temperature. After the organic phase was evaporated to dryness, the crude product was washed with HCl solution (pH 2.0) and ddH_2O successively. After being dried, the yellow product BaP-COOH was obtained.

BaP-COOH (0.155 g) was dissolved in DMF (1 mL). N-Hydroxysuccinimide (NHS) and N, N′-dicyclohexylcarbodiimide (DCC) were dissolved in DMF (1 mL), respectively and added to the solution above successively. The mixture was allowed to react with stirring for 8 h at room temperature. After centrifugal separation, the obtained supernate was added dropwise into 10 mL of bovine serum albumin (BSA) solution (12 mg mL^{-1}, in PBS) and then stirred for 6 h at 4°C. The suspension was dialyzed against PBS for 3 days. After another centrifugal separation, the supernate (immunogen, BSA-BaP) was stored at −20°C.

BaP-COOH (0.155 g) was dissolved in DMF (1 mL), and then n-butylamine (50 μL) and isobutyl chloroformate (62.5 μL) were added sequentially. The mixture was stirred magnetically for 3 h at 4°C. Afterwards, the mixture was added dropwise into 10 mL of egg albumin (OVA) solution (12 mg mL^{-1}, in PBS) and allowed to react with stirring for 5 h at 4°C. The suspension was dialyzed against PBS for 3 days. After centrifugal separation, the supernate (coating antigen, OVA-BaP) was stored at −20°C.

2.5. Preparation of Bio-pAb-BaP. In this research, polyclonal anti-BaP antibody (pAb-BaP) was prepared by taking healthy rabbits as immune objects. Two white rabbits were immunized with BSA-BaP through subcutaneous and intramuscular injections. The immunization lasted for 4 months. pAb-BaP was separated from rabbit serum through ammonium sulphate precipitation method, dialyzed against PBS for 5 days, and stored at –20°C.

The prepared pAb-BaP was diluted with CBS buffer at a concentration of 2.0 mg mL^{-1}. Then, the solution was mixed with 1.0 mg mL^{-1} BNHS (in DMSO) in the mass ratio of 1 : 10. The mixture was stirred for 4 h, and dialyzed against PBS for 3 days. The obtained bio-pAb-BaP was stored at –20°C.

The specificity of pAb-BaP was evaluated by testing cross-reactivity (CR) using other types of PAHs. The CR values were calculated according to the formula CR (%) = (IC_{50} of BaP)/(IC_{50} of analogues) × 100%. The chemical structures of BaP analogues and CR results are shown in Table 1. In all cases, there was a low CR (below 7%) between BaP and other structurally similar compounds, indicating that the prepared pAb-BaP was suitable for detecting trace amount of BaP.

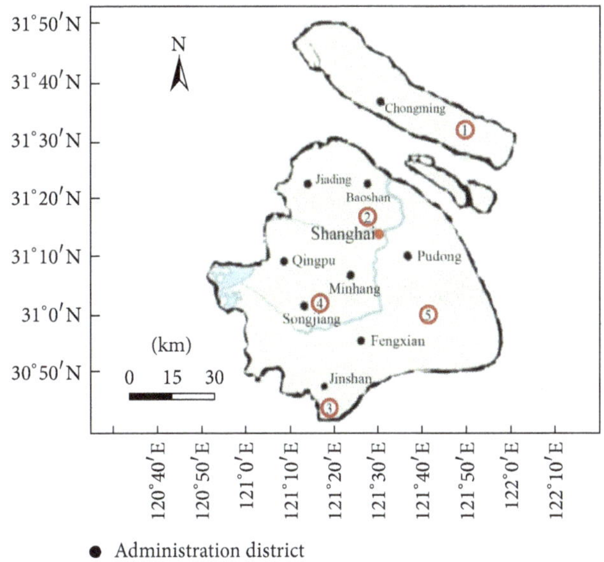

FIGURE 4: Locations of sampling sites in Shanghai, China. $PM_{2.5}$ samples were collected at sampling sites (1), (2), and (3). (1) Chongming Island; (2) central area of Hongkou; (3) Jinshan District. Agricultural soils were collected at sampling sites (4) and (5) (local vegetable bases).

2.6. Preparation of BaP Samples. $PM_{2.5}$ (particles in the air, the aerodynamic diameter of which is less than or equal to 2.5 μm) and agricultural soil samples were collected in Shanghai, China; the locations of sampling sites are shown in Figure 4. Vegetables and cereals were purchased from a local food market; roast lamb and duck were obtained from a barbecue shop. The treatment of environmental samples was similar to the description of the standard of People's Republic of China HJ 784-2016. First, $PM_{2.5}$ sampling membranes were broken down into pieces; agricultural soil samples were ground into powder after being freeze-dried. Then, each $PM_{2.5}$ sampling membrane (in pieces) and agricultural soil (10 g) were extracted for 16 h by soxhlet extraction with acetone/hexane (1 : 1, v/v, 50 mL) and centrifuged at 4500 rpm for 10 min. The organic layer was dried with Na_2SO_4 and then concentrated to nearly 2 mL.

The food samples were treated as described in the standard of People's Republic of China GB/T 5009. 27-2003. The vegetable samples (100 g) were grinded into pieces; the cereal samples (50 g) and barbecue food (50 g) were crushed. Next, all the samples were extracted by ultrasonic processing with hexane/acetone (1 : 1, v/v, 100 mL × 2) for 0.5 h. For the treatment of vegetables, the extract was washed with cyclohexane (100 mL × 2) and ddH_2O (100 mL × 2) successively. After being dried with Na_2SO_4, the cyclohexane layer was condensed to about 2 mL. In the case of the barbecue food and cereals, the extract mixed with CH_3OH-H_2O (9 : 1, v/v, 150 mL) was heated at 80°C for 6 h. Then, the mixture was extracted by cyclohexane (100 mL × 2); the cyclohexane layer was washed by CH_3OH-H_2O (1 : 1, v/v, 100 mL × 3) and ddH_2O (200 mL × 2) successively. Next, the collected organic layer was condensed to about 40 mL and then washed with H_2SO_4 (60%, 100 mL) and ddH_2O (25 mL × 4) successively. The organic phase was dried by Na_2SO_4 and then condensed to about 2 mL.

All the extract was purified with a florisil column which was treated with cyclohexane/dichloromethane (1 : 1, v/v, 20 mL). Benzene (10 mL) was used as the leaching solution, and then the eluent was collected and condensed to near dryness. Afterwards, the residue was redissolved in acetonitrile (0.5 mL). The treated BaP samples above were detected by SA-ELISA after being diluted by 1000 times with PBS containing 5% DMSO as well as the HPLC method.

2.7. Establishment of the Immunoassay. The SA-ELISA determinations were performed in a 96-well microtitre plate. First, microplates were covered with 100 μL/tube of coating antigen at 4°C overnight. After washing, the unbound active sites were blocked with 200 μL/tube of blocking reagent. Then, the mixture was incubated for 1 h at 37°C. Next, 50 μL/tube of BaP standard and 50 μL/tube of bio-pAb-BaP dilution were added after additional washing and incubated for 1 h at 37°C. After another washing step, 100 μL/tube of SA-HRP dilution was added, and the mixture was incubated for 1 h at 37°C. After a final washing step, 100 μL/tube of TMB + H_2O_2 substrate solution was added. The enzymatic reaction was stopped by adding 50 μL terminating solution (2 mol L^{-1} H_2SO_4) after sufficient colour development. The absorbance of each well was immediately recorded in dual-wavelength mode (absorbance in 450 nm as test and 630 nm as reference).

Under optimal conditions, the standard curve was constructed by plotting average inhibition (%) values of six replicates against the logarithm of BaP concentrations. Limit of detection (LOD) was evaluated in terms of IC_{10} (concentration at which a compound inhibits a particular phenomenon by 10%). Inhibition was calculated as follows (A_{max} is the absorbance in absence of BaP; A_{min} is the absorbance of blank sample; and A_s is the absorbance of BaP standard solutions):

$$\text{Inhibition}(\%) = \frac{(A_{max} - A_{min}) - (A_s - A_{min})}{A_{max} - A_{min}} \times 100\%. \quad (1)$$

Table 1: Cross-reactivity of pAb-BaP with structural analogues.

Analogues	Structure	IC_{50} (ng mL^{-1})	Cross-reactivity (%)
Benzo(a)pyrene	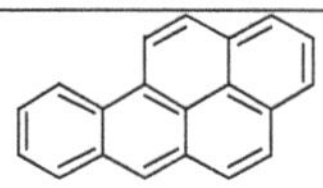	0.67	100
Benzo[ghi]perylene (BghiP)		30.45	2.2
Indeno[1,2,3-cd]pyrene	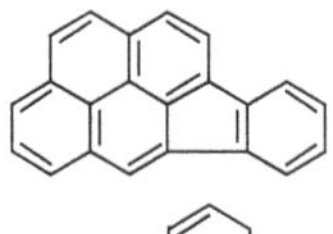	23.10	2.9
Perylene	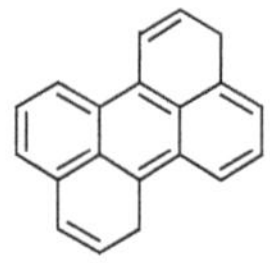	10.31	6.5
Benzo[k]fluoranthene	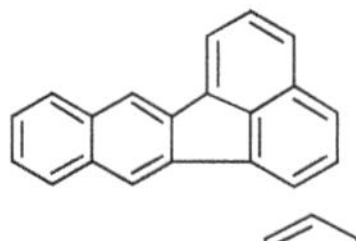	24.81	2.7
Dibenz[a,h]anthracene	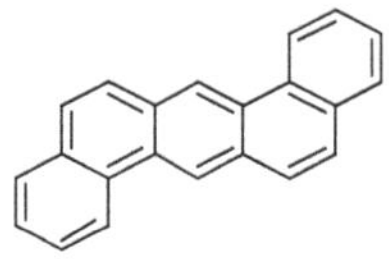	9.71	6.9
Benzo[b]fluoranthene	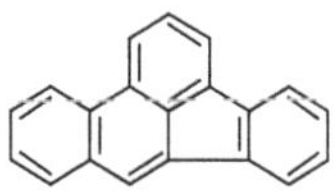	23.93	2.8
Chrysene	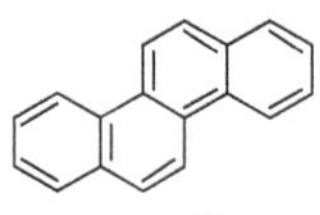	19.14	3.5
Fluoranthene	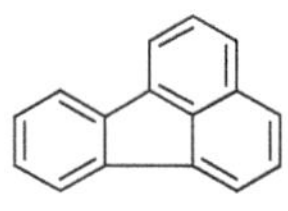	111.67	0.6
Benz[a]anthracene	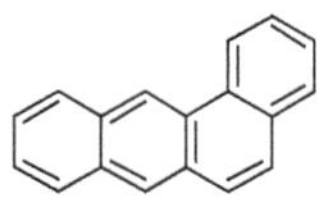	18.11	3.7
Pyrene	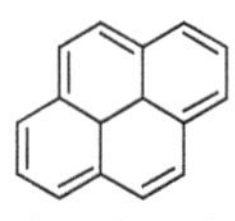	17.63	3.8
Anthracene	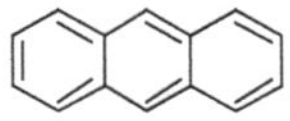	335.12	0.2
Phenanthrene	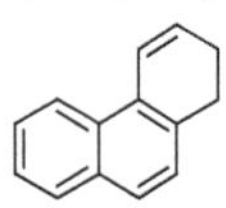	223.33	0.3
Naphthalene	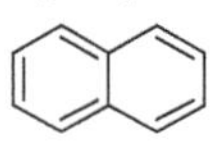	>3350	<0.02

3. Results and Discussion

3.1. Characterization of BaP Hapten and Antigens. The molecular weight of BaP hapten is 336.374. As shown in Figures 5, 6(a), and 6(b), $[M^+]$ ion peak was 337.16, Mass spectrometry was similar to the characterization of BaP = O. ^{1}HNMR (400 MHz $CDCl_3$): δ = 2.18 (t, 2H, 4′-H), 2.22 (s, 3H, 1′-H), 2.68 (t, 2H, 3′-H), 3.78 (t, 2H, 5′-H), 7.83–7.85 (m, 2H, 8-H, 9-H), 7.95 (d, 1H, 4-H), 7.97 (t, 1H, 2-H), 8.07 (d, 1H, 1-H), 8.21 (d, 1H, 12-H), 8.28 (d, 1H, 7-H), 8.34 (d, 1H, 5-H), 8.61–8.64 (m, 1H, 3-H), 9.07 (d, 1H, 11-H), 9.14–9.16 (m, 1H, 10-H) ppm. ^{13}CNMR (100 MHz $CDCl_3$): δ = 25.0 (C-4′), 27.4 (C-5′), 30.1 (C-1′), 43.1 (C-3′), 122.1 (C-11), 123.6 (C-10), 123.8 (C-12c), 124.4 (C-5), 124.6 (C-1), 125.0 (C-7), 125.6

Table 2: The optimization of coating antigen.

Coating antigen	A_{0max}	IC_{50} ($\mu g\,L^{-1}$)	A_{0max}/IC_{50}	Abs (negative group)
OVA-BaP	0.85	1.19	0.71	0.071
KLH-BaP	0.87	1.21	0.72	0.102
PLL-BaP	0.91	1.13	0.81	0.121

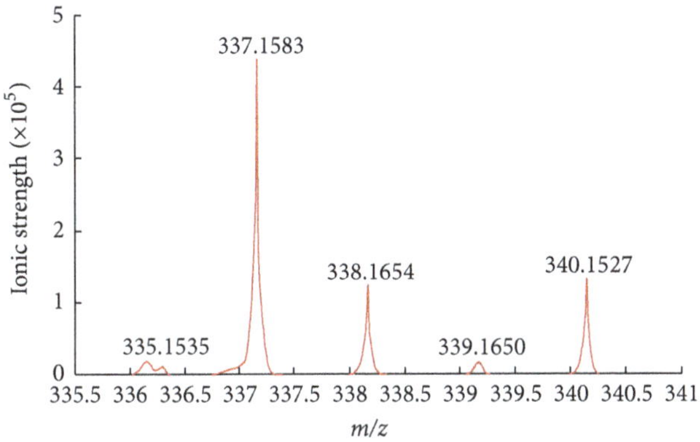

Figure 5: Mass spectrum of BaP hapten.

Table 3: The optimization of coating condition for OVA-BaP.

Coating condition	A_{0max}	IC_{50} ($\mu g\,L^{-1}$)	A_{0max}/IC_{50}
4°C overnight	1.07	1.73	0.62
25°C			
1 h	0.76	2.39	0.32
2 h	0.81	2.21	0.37
37°C			
1 h	0.92	2.26	0.41
2 h	1.05	1.78	0.59
43°C			
1 h	0.72	2.51	0.29
2 h	0.63	3.37	0.19

(C-3, C-9), 126.0 (C-2), 126.3 (C-8), 126.5 (C-10b), 126.9 (C-4, C-12b), 127.3 (C-5a), 128.1 (C-12), 128.3 (C-10a), 129.9 (C-6a), 131.2 (C-3a), 131.5 (C-12a), 132.6 (C-6), 208.8 (C-2′) ppm. According to the characteristic results above, the BaP hapten was synthesized successfully.

The prepared BaP-COOH-protein conjugates BSA-BaP and OVA-BaP were identified by the ultraviolet-visible (UV-vis) spectrophotometer, respectively. As shown in Figure 7, there were relationships and differences among the characteristic peaks of BaP-COOH, proteins, and conjugates. Moreover, new absorption peaks appeared at 361 nm (for BSA-BaP) and 373 nm (for OVA-BaP) in the UV-vis spectrum of BSA-BaP and OVA-BaP. The results revealed that the conjugates were successfully prepared. Measured by Coomassie Brilliant Blue assay, concentrations of BSA-BaP and OVA-BaP were 5.02 and 2.83 mg mL^{-1}, respectively.

3.2. Optimization of Reaction Conditions. Coating antigens prepared in different carrier proteins have different structures and molecular weights, which can lead to different abilities to react with the prepared antibody specifically. We selected KLH and PLL as carrier proteins and prepared another two types of coating antigens using the mixed anhydride method described above. Different coating antigens (OVA-BaP, KLH-BaP, and PLL-BaP) were compared. As shown in Table 2, there were no obvious differences between the IC_{50} values as well as A_{0max}/IC_{50}. Considering the background interferences, OVA-BaP, which obtained the minimum absorbance value (Abs) of the negative control group, was more appropriate for the immunoassay.

The interaction between proteins may affect the adsorption of proteins on the surface of polystyrene carrier at high temperature. Moreover, a long time of incubation has influences on the coating effect and the activity of coating antigens. Therefore, the appropriate incubation condition has a significant effect on the coating step. Several incubation conditions (4°C overnight, 25°C for 1 h and 2 h, 37°C for 1 h and 2 h, and 43°C for 1 h and 2 h) during the coating step were investigated. As shown in Table 3, IC_{50} value is the lowest and A_{0max}/IC_{50} is maximum in the condition of incubation at 4°C overnight. Therefore, the coating step was completed at 4°C overnight.

The optimal concentrations of bio-pAb-BaP and coating antigen were used to improve the sensitivity of the immunoassay. The number of immune complexes reduces rapidly if concentrations of the reagents are not appropriate. A chessboard assay was used to optimize the working concentrations of coating antigen and bio-pAb-BaP. The optimal reaction conditions are the lowest bio-pAb-BaP and coating antigen concentrations that resulted in the maximum absorbance (A_{0max}) of approximately 1.0. As shown in Table 4, the optimal coating antigen concentration was 5.66 $\mu g\,mL^{-1}$ and bio-pAb-BaP was at a dilution to 1:2000 (0.493 $\mu g\,mL^{-1}$).

Blocking is to eliminate the unoccupied sites on microplates; otherwise the unoccupied sites may interact with the components, such as bio-pAb-BaP and SA-HRP during the subsequent steps. The optimum blocking reagent should achieve the minimal background interference. Several blocking reagents (in PBS) were investigated. As shown in Figure 8(a), 1% gelatin, which achieved the minimal background interference (0.079), was used in the immunoassay.

Different incubation temperatures during the blocking step were compared. The optimum incubation temperature should cause the least background interferences, that is, the lowest absorbance value. As shown in Figure 8(b), the absorbance values obtained at 37°C and 43°C were obviously lower than others. Considering the effect on the activity of the coating antigen, 37°C was selected as incubation temperature during the blocking step.

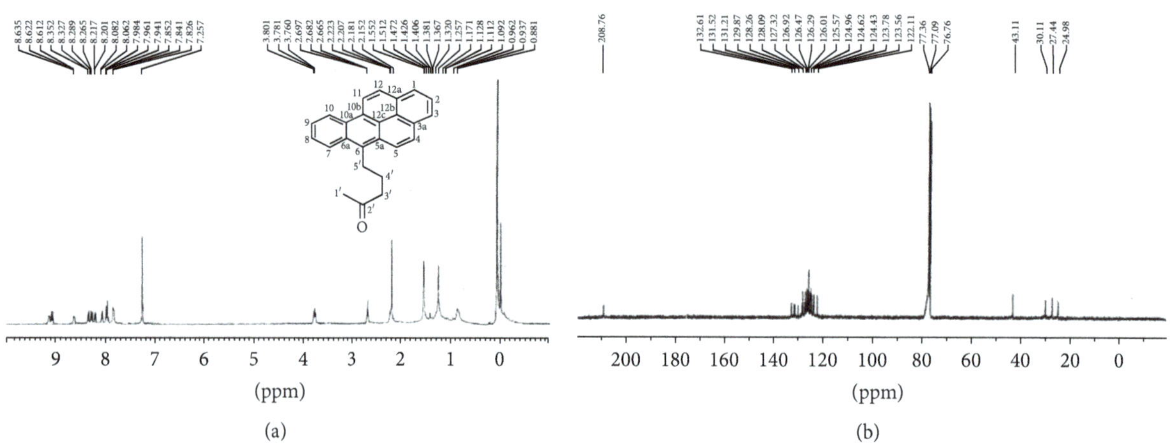

FIGURE 6: (a) The ^{1}H nuclear magnetic resonance (^{1}H-NMR) spectrum of BaP hapten. (b) The ^{13}C-NMR spectrum of BaP hapten.

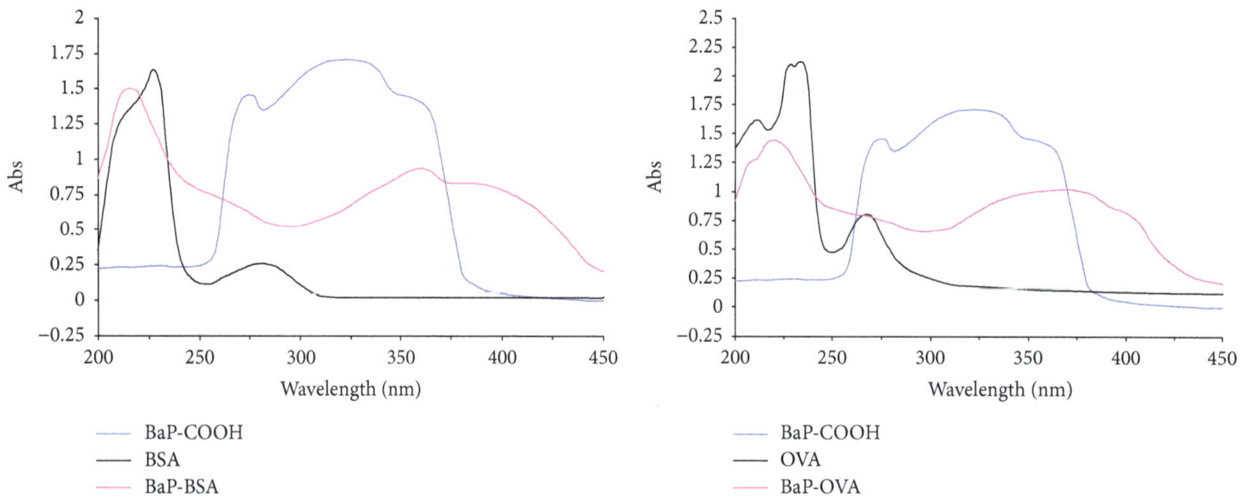

FIGURE 7: UV-vis spectrum of BaP-COOH, proteins, and conjugates. The proteins and conjugates were dissolved in PBS; the BaP-COOH was dissolved in DMSO.

As BaP is a type of hydrophobic organic compound, organic solvent needed to be added in the preparation of the BaP solution. Different concentrations of organic solvent may affect the adsorption capacity of the antibody. The effect of DMSO in different concentrations on the assay was investigated. The result is shown in Figure 8(c). PBS containing 5% DMSO, which induced the lowest IC_{50} value (0.22), was selected to improve the analyte solubility in the follow-up test.

Dilutions of SA-HRP (500, 1000, 1500, 2000, and 3000) were also investigated. The nonspecific reaction will increase at high SA-HRP concentrations; however the sensitivity of the immunoassay may be influenced at low SA-HRP concentrations. As shown in Figure 8(d), A_{0max}/IC_{50} was the maximum and the IC_{50} value was the lowest when the SA-HRP was diluted with PBST to 1 : 1000 ($IC_{50} = 0.49\,\mu g\,mL^{-1}$, $A_{0max} = 1.07$). Hence, the SA-HRP was at a dilution of 1 : 1000.

In the immune reaction of antigen and antibody stage, different incubation times (15, 30, 45, 60, 75, and 90 min) were investigated. As shown in Figure 8(e), the A_{0max} value increased with the increase of incubation time. The IC_{50} value (0.63) was lowest at 60 min and A_{0max}/IC_{50} was the maximum. So, an incubation time of 60 min was selected for the reaction system.

In the reaction system, the hydrogen bonding force of antigen and antibody may be influenced by pH value; different ionic strength also has an effect on the specificity of the immune reaction. As shown in Figures 8(f) and 8(g), A_{0max} decreased with the increase of salt concentrations and the lowest IC_{50} (1.81) was induced at an ionic strength of $0.1\,mol\,L^{-1}$. A_{0max} values decreased with the increase of pH and the appropriate IC_{50} (1.50) was induced at pH 7.40. So, the salt concentration was $0.1\,mol\,L^{-1}$ and the pH of buffer was 7.40 in the immunoassay.

3.3. Standard Curve and Linear Working Range for BaP. A series of concentrations of BaP standard solutions ($0.001\,ng\,mL^{-1}$, $0.005\,ng\,mL^{-1}$, $0.01\,ng\,mL^{-1}$, $0.05\,ng\,mL^{-1}$, $0.1\,ng\,mL^{-1}$, $0.5\,ng\,mL^{-1}$, $1\,ng\,mL^{-1}$, $5\,ng\,mL^{-1}$, $10\,ng\,mL^{-1}$, $50\,ng\,mL^{-1}$, $100\,ng\,mL^{-1}$, $500\,ng\,mL^{-1}$, and $1000\,ng\,mL^{-1}$)

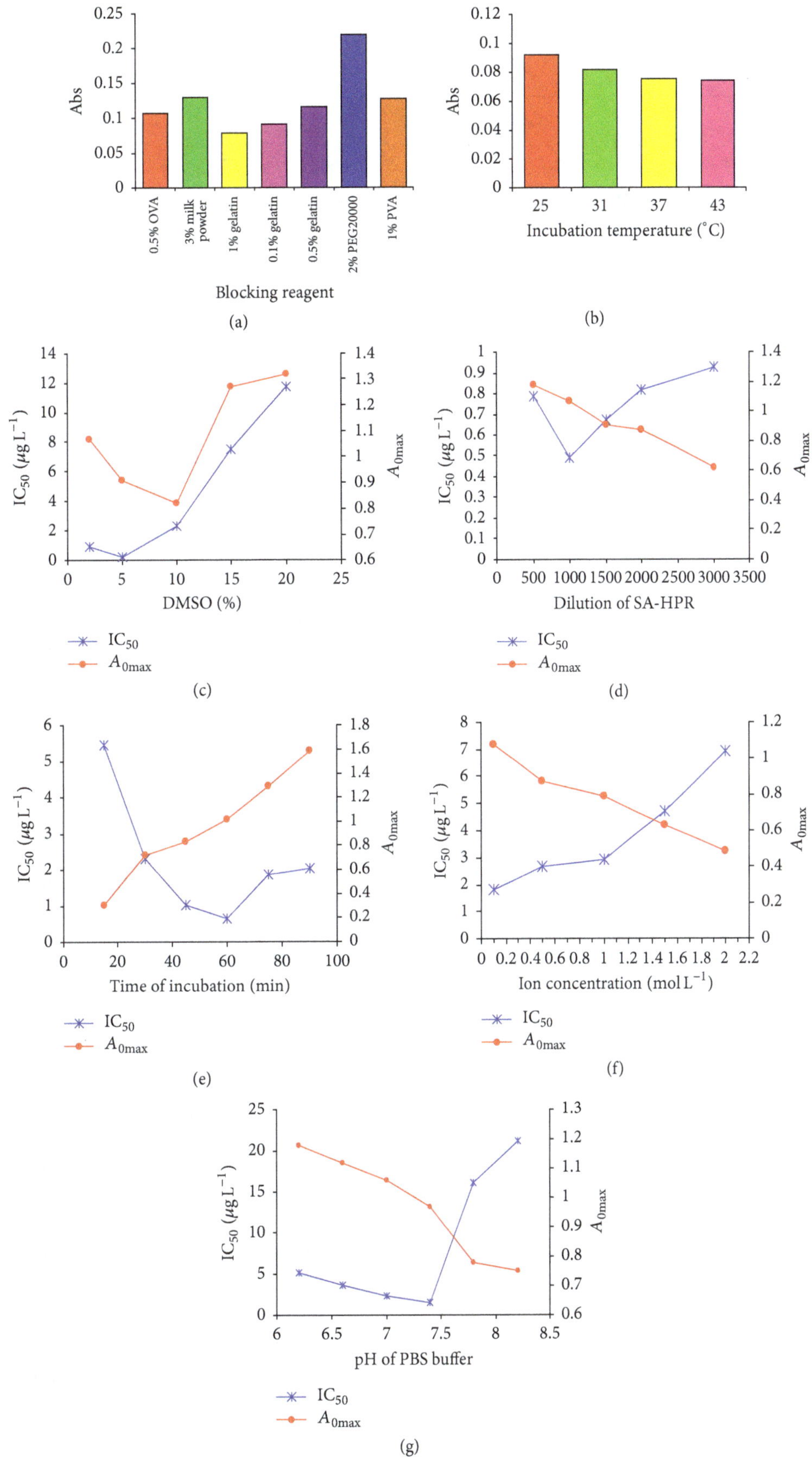

FIGURE 8: The optimal conditions of the immunoassay. (a) Blocking reagent. (b) Incubation temperature during blocking step. (c) DMSO concentration in PBS buffer. (d) SA-HRP concentration. (e) Incubation time. (f) Ion concentration of PBS. (g) pH of PBS.

Table 4: Optimal concentrations of OVA-BaP and bio-pAb-BaP.

Dilutions of bio-pAb-BaP[b]	OVA-BaP concentration (μg mL^{-1})[a]					
	28.3	14.15	5.66	2.83	1.42	0.71
100	2.855	2.166	2.011	1.972	2.244	2.612
500	1.997	1.657	1.721	1.851	1.879	2.446
1000	1.275	1.419	1.396	1.367	1.208	1.468
2000	0.865	1.036	1.039	0.910	0.820	0.907
4000	0.393	0.750	0.902	0.490	0.392	0.636
Blank	0.136	0.129	0.167	0.159	0.137	0.162

[a]The initial concentration of the prepared OVA-BaP was 2.83 mg mL^{-1}; the OVA-BaP was diluted with PBS buffer. [b]The initial concentration of the prepared bio-pAb-BaP was 0.985 mg mL^{-1}; the bio-pAb-BaP was diluted with CBS buffer.

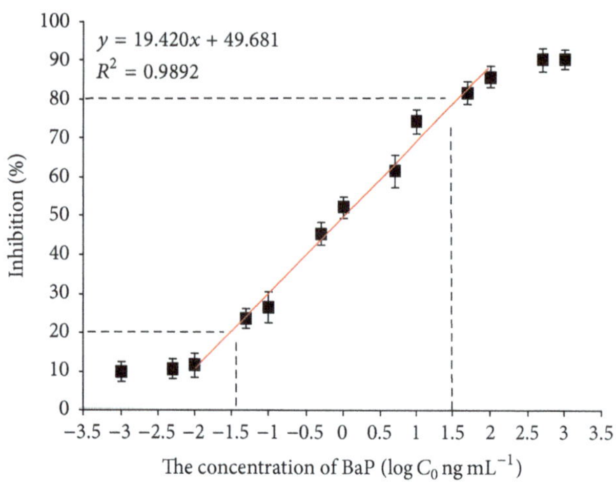

Figure 9: Standard curve of SA-ELISA.

were prepared by PBS with 5% DMSO, and the inhibition was determined under optimal conditions. As shown in Figure 9, linear regression equation is as follows: $y = 19.420 \log C_0 + 49.681$; $R^2 = 0.9892$. Linear working range, which was determined as concentration range that causes 20–80% colour inhibition, was 0.03–35.40 ng mL^{-1}. IC_{50}, which is an important criterion for evaluating sensitivity of the immunoassay, was 1.04 ng mL^{-1}. The limit of detection (IC_{10}) was 0.0094 ng mL^{-1}. In comparison with the sensitive indirect competitive enzyme-linked immunosorbent assay for detecting BaP [22], in which antibody-enzyme conjugates were used (LOD is 0.024 ng mL^{-1}), LOD of the SA-ELISA is obviously lower.

3.4. Repeatability of the Immunoassay. The stability of the proposed immunoassay was estimated by the coefficient of variation (CV). Different concentrations of BaP standard samples (0.05 ng mL^{-1}, 0.1 ng mL^{-1}, 0.5 ng mL^{-1}, 1 ng mL^{-1}, 5 ng mL^{-1}, 10 ng mL^{-1}, and 50 ng mL^{-1}) were run in the SA-ELISA method (Table 5). The intra-assay repeatability was evaluated after performing six replicates, and the interassay repeatability was evaluated over several weeks. The CV of intra-assay was <10% and the CV of interassay was <12%, indicating that repeatability and stability of the immunoassay were well.

3.5. Determination of BaP in Practical Samples. The proposed SA-ELISA method was used to detect BaP in several environmental and food samples. The results are shown in Table 6. As the combustion of fossil fuels and industrial production process contribute greatly to BaP emissions, BaP concentration of $PM_{2.5}$ samples collected at site 3 which is near a small industrial district was the highest. BaP concentration at site 2 was also significantly high, which was mainly caused by BaP emission of automobile exhaust and human daily lives. In addition, BaP concentration in the agricultural soil sample at site 4 was generally higher than that of site 5. This may be mainly caused by the influences of a small gas distribution plant near site 4.

These practical samples were also tested by HPLC to validate the immunoassay. As shown in Table 6, compared with HPLC analysis, the results detected by SA-ELISA were slightly higher. This may be caused by nonspecific adsorption of reagents used in the immunoassay and the cross-reactivity of pAb-BaP with other PAHs existed in the samples, which failed to be analyzed by HPLC method and could contribute to SA-ELISA-derived concentrations. However, the results of the immunoassay still have good consistency with those of HPLC. Statistical study showed that the correlation coefficient was 0.9887 (linear regression equation: $y = 0.9555x - 0.3447$), and p value from the paired sample t-test in comparison with the two methods was 0.0025. That means the difference was statistically significant at 0.05 levels. As the immunoassay was performed in a 96-well microtitre plate in a high throughput detecting manner, it took about 6 h to complete the analysis of 17 treated BaP samples, while HPLC spent about 76.5 h for the analysis (analysis for each sample needs about 15 min and the column needs to be washed after each detection (30 min); six reptiles were tested for each practical sample); it can be concluded that this immunoassay is rapid and effective.

3.6. Recovery Tests. The recovery of several practical samples spiked with BaP analyte was calculated to evaluate accuracy and precision of the immunoassay. Three practical samples (A_2, P_3, and spinach) were spiked with BaP standard in different concentrations ranging from 1–50 ng g^{-1} before extraction. The spiked samples were treated as described above and then tested by the SA-ELISA method. As shown in Table 7, the recovery of SA-ELISA was 91.12–109.23% and the

TABLE 5: The variability of intra-assay and interassay for BaP.

Object	BaP concentration (ng mL^{-1})						
	0.05	0.1	0.5	1	5	10	50
Intra-assay CV (%)	9.69	7.71	6.27	5.99	4.81	4.06	3.52
Interassay CV (%)	11.91	10.96	10.61	9.69	9.02	7.65	6.79

TABLE 6: Determination of BaP concentration by SA-ELISA and HPLC.

Samples (n = 17)	BaP concentration (ng g^{-1}, mean ± SD); n = 6	
	SA-ELISA	HPLC
$PM_{2.5}$ particles (n = 3)		
P_1	5.91 ± 0.106	5.67 ± 0.046
P_2	12.11 ± 0.129	11.96 ± 0.053
P_3	31.03 ± 0.201	30.09 ± 0.101
Agricultural soil (n = 2)		
A_1	7.99 ± 0.142	6.02 ± 0.061
A_2	12.98 ± 0.119	10.11 ± 0.049
Vegetables (n = 5)		
Ipomoea aquatica	0.75 ± 0.061	0.69 ± 0.018
Brassica alboglabra Bailey	1.01 ± 0.087	0.95 ± 0.035
Radish	0.95 ± 0.079	0.87 ± 0.034
Spinach	1.62 ± 0.077	1.52 ± 0.020
Chinese cabbage	1.79 ± 0.095	1.51 ± 0.031
Cereals (n = 3)		
Rice	2.81 ± 0.052	2.27 ± 0.015
Millet	3.92 ± 0.107	3.12 ± 0.038
Corn	2.91 ± 0.091	2.51 ± 0.026
Barbecue food (n = 4)		
Roast meat	6.21 ± 0.122	6.01 ± 0.051
Roast fish	7.97 ± 0.136	6.21 ± 0.048
Roast bean curd	5.97 ± 0.127	5.72 ± 0.053
Roasted duck	6.62 ± 0.112	6.45 ± 0.040

CV was 4.63–8.79%. It can be concluded the immunoassay presents good accuracy.

3.7. Comparison with the Correlated Immunoassay. In the highly sensitive BA-IPCR method for detecting BaP (linear range is 5 pg L^{-1}–10 ng L^{-1}; LOD is 2.85 pg L^{-1}) [21], the analysis is based on the PCR technique and the fluorescence labeled in DNA is used as detection signal. The BA-IPCR assay requires an additional step to pretreat PCR tubes to improve the adsorption capacity while there is no need for this operation in the SA-ELISA method. Moreover, after the similar coating and blocking processes, it only needs to add bio-pAb-BaP for immune reaction and to add SA-HRP for chromogenic reaction prior to the final instrumental detection in the SA-ELISA method; however, the BA-IPCR needs step-by-step use of bio-pAb-BaP, streptavidin, biotinylated DNA (bio-DNA), and then PCR mixture, which is much more complicated. With the 96-well microtitre plate, the analysis of nearly 100 BaP samples can be achieved within 6 h while the BA-IPCR spends about 12 h (including the pretreatment of PCR tubes). Therefore, the proposed SA-ELISA is more rapid and convenient. Although the BA-IPCR assay is more sensitive, the actual linear working range is limited due to its trace levels, which may lead to a certain limitation for the detection of some practical samples. In addition, the price of bio-DNA used for PCR analysis is very expensive; this increases the economic cost greatly. Therefore, we consider the SA-ELISA method has a better applicability for detecting BaP samples.

4. Conclusions

In this study, a specific BaP hapten was synthesized, and diverse BaP antigens and relative polyclonal antibody were prepared; then, a streptavidin-horseradish peroxidase-based enzyme-linked immunosorbent assay was developed and used for the sensitive and high throughput monitoring of BaP in environmental and food samples. Under optimal conditions, this proposed method had a low limit of 0.0094 ng mL^{-1}. Linear working range was from 0.03 ng mL^{-1} to 35.40 ng mL^{-1}. The recovery of spiked samples ranged from 91.12% to 109.23%. IC_{50} value of the assay

TABLE 7: Determination of BaP in spiked samples.

Samples	BaP concentration ($ng\,g^{-1}$)	Spiked level ($ng\,g^{-1}$)	Average recovery (%) ± CV (%) ($n = 6$)
Spinach	1.62	1	105.61 ± 5.49
		2	102.35 ± 7.61
		5	91.12 ± 4.63
A_2	12.98	5	97.03 ± 6.22
		10	103.11 ± 7.09
		20	95.77 ± 5.26
P_3	31.03	10	99.09 ± 8.79
		20	109.23 ± 7.61
		50	92.71 ± 8.26

was $1.04\,ng\,mL^{-1}$. The SA-ELISA could selectively detect BaP with negligible CR values below 7% against several structural BaP analogues. Meanwhile, the results of BaP detection for several practical samples were consistent with those obtained by HPLC, indicating the proposed method was accurate and reliable. Moreover, due to the application of a 96-well microtitre plate, this immunoassay had potential for the batch-testing detection of BaP samples. Therefore, the established SA-ELISA is a useful option for the batch detection of BaP in environmental and food samples with specificity and sensitivity.

Conflicts of Interest

The authors have no conflicts of interest regarding the publication of this paper.

Acknowledgments

The research was supported by the National Natural Science Foundation of China (Projects nos. 21177082 and 21477074), the Science and Technology Commission of Shanghai Municipality (Key Project for Fundamental Research no. of Shanghai Municipality in China (Key Project for Science 09JC1407600), and the Science and Technology Commission and Technology Research no. 09231202805).

References

[1] A. O. Barakat, A. Mostafa, T. L. Wade, S. T. Sweet, and N. B. El Sayed, "Spatial distribution and temporal trends of polycyclic aromatic hydrocarbons (PAHs) in sediments from Lake Maryut, Alexandria, Egypt," *Water, Air, & Soil Pollution*, vol. 218, no. 1-4, pp. 63–80, 2011.

[2] M. C. Liao, P. C. Chio, Y. W. Chen et al., "Lung cancer risk in relation to traffic-related nano/ultrafine particle-bound PAHs exposure: a preliminary probabilistic assessment," *Journal of Hazardous Materials*, vol. 190, no. 1-3, pp. 150–158, 2011.

[3] H. Nakata, K. Uehara, Y. Goto et al., "Polycyclic aromatic hydrocarbons in oysters and sediments from the Yatsushiro Sea, Japan: Comparison of potential risks among PAHs, dioxins and dioxin-like compounds in benthic organisms," *Ecotoxicology and Environmental Safety*, vol. 99, pp. 61–68, 2014.

[4] P. Mottier, V. Parisod, and R. J. Turesky, "Quantitative determination of polycyclic aromatic hydrocarbons in barbecued meat sausages by gas chromatography coupled to mass spectrometry," *Journal of Agricultural and Food Chemistry*, vol. 48, no. 4, pp. 1160–1166, 2000.

[5] N. Kazerouni, R. Sinha, C.-H. Hsu, A. Greenberg, and N. Rothman, "Analysis of 200 food items for benzo[a]pyrene and estimation of its intake in an epidemiologic study," *Food and Chemical Toxicology*, vol. 39, no. 5, pp. 423–436, 2001.

[6] E. Manoli, A. Kouras, and C. Samara, "Profile analysis of ambient and source emitted particle-bound polycyclic aromatic hydrocarbons from three sites in northern Greece," *Chemosphere*, vol. 56, no. 9, pp. 867–878, 2004.

[7] R. V. Galiulin and R. A. Galiulina, "Geoecology of benzo[a]pyrene as a coal pyrolysis product," *Solid Fuel Chemistry*, vol. 44, no. 5, pp. 364-365, 2010.

[8] S. S. Xu, W. X. Liu, and S. Tao, "Emission of polycyclic aromatic hydrocarbons in China," *Environment Science & Technology*, vol. 40, no. 3, pp. 702–708, 2006.

[9] H.-W. Chen, "Determination of polycyclic aromatic hydrocarbons in water by solid-phase microextraction and liquid chromatography," *Analytical Sciences*, vol. 20, no. 10, pp. 1383–1388, 2004.

[10] L. Zhu, B. Chen, J. Wang, and H. Shen, "Pollution survey of polycyclic aromatic hydrocarbons in surface water of Hangzhou, China," *Chemosphere*, vol. 56, no. 11, pp. 1085–1095, 2004.

[11] K. Ravindra, E. Wauters, and R. Van Grieken, "Variation in particulate PAHs levels and their relation with the transboundary movement of the air masses," *Science of the Total Environment*, vol. 396, no. 2-3, pp. 100–110, 2008.

[12] P. Anandakumar, S. Kamaraj, S. Jagan et al., "Capsaicin inhibits benzo(a)pyrene-induced lung carcinogenesis in an in vivo mouse model," *Inflammation Research*, vol. 61, no. 11, pp. 1169–1175, 2012.

[13] C. E. Boström, P. Gerde, A. Hanberg et al., "Cancer risk assessment, indicators, and guidelines for polycyclic aromatic hydrocarbons in the ambient air," *Environmental Health Perspectives*, vol. 110, supplement 3, pp. 451–488, 2002.

[14] D. M. J. Curfs, E. Lutgens, M. J. J. Gijbels, M. M. Kockx, M. J. A. P. Daemen, and F. J. Van Schooten, "Chronic Exposure to the Carcinogenic Compound Benzo[a]Pyrene Induces Larger

and Phenotypically Different Atherosclerotic Plaques in ApoE-Knockout Mice," *The American Journal of Pathology*, vol. 164, no. 1, pp. 101–108, 2004.

[15] K.-C. Lo, C.-Y. Sun, Y.-J. Ding et al., "Nephrotoxicity assessments of benzo(a)pyrene during zebrafish embryogenesis," *Research on Chemical Intermediates*, vol. 40, no. 6, pp. 2177–2185, 2014.

[16] H. Alomirah, S. Al-Zenki, S. Al-Hooti et al., "Concentrations and dietary exposure to polycyclic aromatic hydrocarbons (PAHs) from grilled and smoked foods," *Food Control*, vol. 22, no. 12, pp. 2028–2035, 2011.

[17] J. P. Butler, G. B. Post, P. J. Lioy, J. M. Waldman, and A. Greenberg, "Assessment of carcinogenic risk from personal exposure to benzo[a]pyrene in the total human environmental exposure study (THEES)," *Journal of The Air & Waste Management Association*, vol. 43, no. 7, pp. 970–977, 1993.

[18] Y. Zhang, X. Gao, A. Gao, and M. Fan, "A Biotin-Streptavidin Amplified Enzyme-Linked Immunosorbent Assay with Improved Sensitivity for Rapid Detection of Ractopamine in muscular tissue: Development and Nonspecific Adsorption Reduction," *Food Analytical Methods*, vol. 5, no. 5, pp. 1214–1220, 2012.

[19] M.-C. Zhang, Q.-E. Wang, and H.-S. Zhuang, "Determination of dibutyl o-phthalate by antigen-coated competitive fluorescence immunoassay," *Analytical Letters*, vol. 40, no. 1, pp. 127–137, 2007.

[20] J. Guo, K. Luo, D. Chen, X. Tan, and Z. Song, "A rapid and sensitive method for the determination of dibutyl phthalate in wine by flow-injection chemiluminescence analysis," *Journal of Food Composition and Analysis*, vol. 31, no. 2, pp. 226–231, 2013.

[21] Z. Ma and H. Zhuang, "A Highly Sensitive Real-time Immuno-PCR Assay for Detecting Benzo[a]pyrene in Food Samples by Application of Biotin-Streptavidin System," *Food Analytical Methods*, pp. 1–11, 2017.

[22] D. Matschulat, A. Deng, R. Niessner, and D. Knopp, "Development of a highly sensitive monoclonal antibody based ELISA for detection of benzo[a]pyrene in potable water," *Analyst*, vol. 130, no. 7, pp. 1078–1086, 2005.

9

Quantum Chemical Investigation on the Antioxidant Activity of Neutral and Anionic Forms of Juglone: Metal Chelation and Its Effect on Radical Scavenging Activity

Aymard Didier Fouegue Tamafo,[1] Julius Numbonui Ghogomu,[1] Nyiang Kennet Nkungli,[1] Désiré Bikélé Mama,[2] and Elie Younang[3]

[1] *Department of Chemistry, Faculty of Science, University of Dschang, P.O. Box 67, Dschang, Cameroon*
[2] *Department of Chemistry, Faculty of Science, University of Douala, P.O. Box 24157, Douala, Cameroon*
[3] *Department of Inorganic Chemistry, Faculty of Science, University of Yaoundé I, P.O. Box 812, Yaoundé, Cameroon*

Correspondence should be addressed to Julius Numbonui Ghogomu; ghogsjuju@hotmail.com

Academic Editor: Zhongfang Chen

The chelation ability of divalent Mg, Ca, Fe, Co, Ni, Cu, Zn, and monovalent Cu ions by neutral and anionic forms of juglone has been investigated at DFT/B3LYP/6-31+G(d,p) level of theory in gas and aqueous phases. It is noteworthy that only the 1:1 stoichiometry was considered herein. The effects of these metals on the radical scavenging activity of neutral juglone were evaluated via the usual descriptors of hydrogen atom transfer. According to our results, metal chelation by the two forms of juglone was spontaneous and exothermic in both media. Based on the binding energies, Cu(II) ion showed the highest affinity for the ligands. QTAIM analyses identified the metal-ligand bonds as intermediate type interactions in all the chelates, except those of Ca and Mg. It was also found that the chelates were better radical scavengers than the ligands. In the gas phase, the scavenging activity of the compounds was found to be governed by direct hydrogen atom transfer, the Co(II) chelate being the most reactive. In the aqueous phase also, the sequential proton loss electron transfer was preferred by all the molecules, while the Cu(II) chelates were the most reactive.

1. Introduction

In living organisms, transition metal ions (M^{n+}) in their lower oxidation states ($n = 1$ or 2) can catalyze oxidative reactions by reducing their activation energy. By doing so, this may lead to the formation of very dangerous reactive oxygen species on the one hand, most common of which are the hydroxyl radicals ($HO^{\bullet}$), reactive nitrogen species [1–5], and oxidized metal ions $M^{(n+1)+}$ (1), via Fenton reactions on the other [2, 6]. Direct consequences of this are diseases such as cancer, cataract, Alzheimer and Parkinson diseases, accelerated ageing, arthritis, and compromised immune systems:

$$M^{n+} + HOOH \longrightarrow M^{(n+1)+} + HO^{-} + HO^{\bullet} \qquad (1)$$

In the presence of the superoxide radical ($O_2^{-\bullet}$) or some other antioxidant molecules like ascorbic acid (AH_2) or polyphenols (PhOH) ((2) and (3)), the oxidized state of the metal ions $M^{(n+1)+}$ is reduced to M^{n+} [3], which in turn undergoes another Fenton reaction.

$$M^{(n+1)+} + 2AH_2 \longrightarrow M^{n+} + 2AH^{\bullet} + 2H^{+} \qquad (2)$$

$$M^{(n+1)+} + PhOH \longrightarrow M^{n+} + PhO^{\bullet} + H^{+} \qquad (3)$$

Metal-chelating compounds encompassing antioxidants that are capable of altering their redox potentials, that is, capable of removing metals from their surrounding environments, are very important in foods processing, beverages, cosmetics, metal corrosion, pharmaceutical formulations, and living organisms [3, 7–11]. From the foregoing, a high affinity for the targeted metal ions by the antioxidant, which results in their greater capability of scavenging the free active metal ions, is one of the key features of a good chelating agent [12]. Indeed, natural polyphenols can chelate metal ions in order to prevent the generation of reactive hydroxyl radicals [13–15]. Typically, these polyphenols bind to transition metal ions via their hydroxyl, carbonyl, and carboxyl groups. This binding process occurs through an acid-base reaction in which the metal ion acts as the Lewis acid and the polyphenol functions as the Lewis base. The above-mentioned solutions toward mitigating the excesses of transition metal ions in living organisms have been subject to many studies [9–12].

In a previous paper, we evaluated the Fe^{2+} ion chelating ability of neutral juglone (5-hydroxy-1,4-naphthoquinone) and two of its derivatives by means of the density functional theory (DFT) [16]. In this paper, the effect of the Fe(II) ion chelation on the antioxidant activity of juglone and its derivatives was evaluated. However, apart from Fe^{2+} ion, food and cosmetic ingredients, lipids, polymer materials, and living organisms also contain many other transition as well as alkali-earth metal ions. Unfortunately, some of these transition metal ions can catalyze the formation of free radicals. Against this backdrop, the present article aims at identifying the metal ions in living organisms having the highest affinity toward neutral and deprotonated juglone. Moreover, the effects of these metals on the radical scavenging activity of juglone in gas and aqueous media have been investigated herein. To attain this objective, the ions Fe^{2+}, Co^{2+}, Ni^{2+}, Cu^{2+}, Zn^{2+}, Ca^{2+}, and Mg^{2+} as well as Cu^{+} have been chosen because they are of paramount importance in bioinorganic chemistry and also for their abundance in enzymes. Additionally, these ions are highly involved in the strengthening of living organisms. To the best of our knowledge, the present work is the first attempt of utilizing quantum chemical tools in studying the metal selectivity of juglone toward the above-mentioned metals, as well as the antiradical properties of the resulting chelates.

To achieve these objectives, the thermodynamic parameters (reaction enthalpies and free energies of formation) as well as interaction energies have been employed to study the stability of the chelates. Bader's Quantum Theory of Atom in Molecules (QTAIM) and the Natural Bond Analysis (NBO) were used to study the metal-ligand interactions. The usual thermodynamic descriptors of antioxidant activity (AOA) were used to evaluate the effect of the metal ions on the radical scavenging activity of neutral juglone. These descriptors are [17] as follows:

(i) the bond dissociation enthalpy (BDE) characterizing the direct hydrogen atom transfer from the antioxidant (MJ-OH) to a free radical as

$$\text{MJ-OH} \longrightarrow \text{MJ-O}^{\bullet} + \text{H}^{\bullet} \quad (4)$$

(ii) the ionization potential (IP) and proton dissociation enthalpy (PDE), describing the electron transfer-proton transfer mechanism (ET-PT) as

$$\text{MJ-OH} \longrightarrow \text{MJ-OH}^{\bullet+} + e^{-} \longrightarrow \text{MJ-O}^{\bullet} + \text{H}^{\bullet} \quad (5)$$

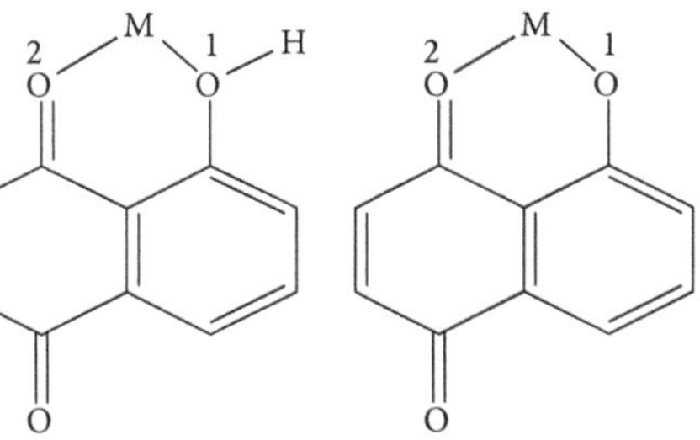
$M = Mg^{2+}, Ca^{2+}, Fe^{2+}, Co^{2+}, Ni^{2+}, Cu^{2+}, Cu^{+}$ & Zn^{2+}

FIGURE 1: Structures of the studied chelates.

(iii) the proton affinity and the electron transfer enthalpy (PA and ETE, (6)) describing the sequential proton transfer electron transfer (SPT-ET) mechanism:

$$\text{MJ-OH} \longrightarrow \text{MJ-O}^{-} + \text{H}^{+} \longrightarrow \text{MJ-O}^{\bullet} + \text{H}^{\bullet} \quad (6)$$

2. Theoretical Background and Computational Details

All electronic structure calculations have been performed at DFT/B3LYP/6-31+G(d,p) level [18–24] in the gas and the aqueous phases, using the Gaussian 09 package [25]. No symmetry constraints were used during geometry optimizations. With the aim of determining the nature of all the stationary points, frequency calculations were performed on all the optimized geometries at the same level of theory. The DFT/B3LYP method was used because it is frequently employed for the studies on metalation of natural antioxidant [9–11, 26]. The calculations on closed shell systems were performed by the means of the Restricted Kohn-Sham formalism whereas the Unrestricted Kohn-Sham approach was used for open shell systems [23, 27]. The IEF-PCM (Integral Equation Formulation-Polarizable Continuum Model) solvation model [28, 29] has been used in this work to predict the effect of water on the studied properties.

The enthalpies and free energies of formation for the different chelates (Figure 1) as well as their interaction energies were calculated via (7). In this equation, Q stands for the enthalpy, the free energy or the thermal energy of the chelates, the ligands, and the central metal ions. Here, the ligands are both the neutral and deprotonated forms of juglone.

$$\Delta Q = Q_{L-M} - Q_{L} - Q_{M^{n+}} \quad (7)$$

The Natural Population Analysis (NPA) charge [30] on the central metal ions has been calculated and used to evaluate the degree of reduction of these metals after their chelation by the ligands. The nature of all metal-ligand bonds was studied via the Quantum Theory of Atom in Molecules (QTAIM) proposed by Bader [31], as implemented in Multiwfn [32].

The effects of all eight metal ions studied on the antioxidant activity of juglone have been investigated using the

Table 1: Gas phase energies (Hartree) of the Fe(II), Co(II), and Ni(II) chelates related to their multiplicities calculated at B3LYP/6-31+G(d,p) level of theory.

	Juglone		Deprotonated juglone	
	Low multiplicity	High multiplicity	Low multiplicity	High multiplicity
Fe(II)	−1873.11248	−1873.19968	−1872.96805	−1873.04352
Co(II)	−1992.19774	−1992.21763	−1992.17881	−1992.06371
Ni(II)	−2117.71185	−2117.74145	−2117.57340	−2117.59036

Table 2: Gas and water phases values of the O-metal and O-H bond lengths (angstrom) of the chelates obtained from the optimized geometries at DFT/B3LYP/6-31+(d,p) level.

	Juglone						Deprotonated juglone			
	Gas			Water			Gas		Water	
	O_1-M	O_2-M	O-H	O_1-M	O_2-M	O-H	O_1-M	O_2-M	O_1-M	O_2-M
J-Mg	1.967	1.854	0.975	2.101	1.976	0.972	1.839	1.874	1.951	1.985
J-Ca	2.386	2.180	0.973	2.664	2.431	0.971	2.152	2.205	2.331	2.394
J-Fe	1.973	1.867	0.977	2.136	1.952	0.973	1.799	1.916	1.901	1.971
J-Co	1.934	1.891	0.979	2.032	1.936	0.973	1.790	1.912	1.868	1.976
J-Ni	1.932	1.845	0.979	2.013	1.909	0.973	1.781	1.873	1.854	1.915
J-Cu	2.080	1.910	0.974	1.936	1.834	0.976	1.925	1.950	1.816	1.846
J-Cu(I)	2.087	1.918	0.970	2.123	1.942	0.971	1.888	1.914	1.921	1.939
J-Zn	1.926	1.832	0.976	2.106	1.962	0.972	1.812	1.847	1.905	1.945

following equations, which are based on the hydrogen atom transfer mechanisms ((4)–(6)):

$$\mathrm{BDE} = \mathrm{H}_{\mathrm{MJ\text{-}O}^\bullet} + \mathrm{H}_{\mathrm{H}^\bullet} - \mathrm{H}_{\mathrm{MJ\text{-}OH}} \quad (8)$$

$$\mathrm{IP} = \mathrm{H}_{\mathrm{MJ\text{-}OH}^{+\bullet}} + \mathrm{H}_{e^-} - \mathrm{H}_{\mathrm{MJ\text{-}OH}} \quad (9)$$

$$\mathrm{PDE} = \mathrm{H}_{\mathrm{MJ\text{-}O}^\bullet} + \mathrm{H}_{\mathrm{H}^+} - \mathrm{H}_{\mathrm{MJ\text{-}OH}^{+\bullet}} \quad (10)$$

$$\mathrm{PA} = \mathrm{H}_{\mathrm{MJ\text{-}O}^-} + \mathrm{H}_{\mathrm{H}^+} - \mathrm{H}_{\mathrm{MJ\text{-}OH}} \quad (11)$$

$$\mathrm{ETE} = \mathrm{H}_{\mathrm{MJ\text{-}O}^\bullet} + \mathrm{H}_{e^-} - \mathrm{H}_{\mathrm{MJ\text{-}O}^-} \quad (12)$$

In these equations, H stands for the enthalpy of formation of the species calculated at 298.15 K and 1 atm. Aimed at studying the spontaneity of the AOA mechanisms, the free energies of these mechanisms have also been calculated by replacing H with G, the free energy of formation.

3. Results and Discussion

3.1. Effect of Multiplicity. Since the multiplicities of the chelates and complexes are dependent on the strength of the ligands, the first section of the results is devoted to the study of the energies of the highest and lowest multiplicity states of iron, cobalt, and nickel chelates. These states are singlet and quintet, singlet and quartet, and singlet and triplet, respectively, for the iron, cobalt, and nickel chelates. This investigation was carried out in gas phase for juglone and deprotonated juglone chelates, and the states of multiplicity adopted have been used in the aqueous phase. The energies of the aforementioned multiplicity states are presented in Table 1. In the case of the Mg(II), Ca(II), Cu(I), and Cu(II) as well as Zn(II) chelates, the effect of multiplicity has not been investigated because only one multiplicity state (doublet for Cu(II) and singlet for the rest of chelates) is adopted.

It is clear from Table 1 that, for all the chelates, the highest multiplicity state is the most stable (that with the lowest energy). Therefore, the quintet, quartet, and triplet states of Fe (II), Co(II), and Ni(II) chelates will be adopted throughout the rest of the work. The $\langle S^2 \rangle$ values of the high multiplicity state for the Fe(II), Co(II), and Ni(II) chelates of juglone are, respectively, 6.00, 3.75, and 2.00. For the deprotonated juglone compounds, these $\langle S^2 \rangle$ values are 6.00, 3.75, and 2.00, respectively, for the Fe(II), Co(II), and Ni(II) chelates. Interestingly, all of these values are equal to the $\langle S^2 \rangle$ values of the pure quintet, quartet, and triplet wave functions, which are 6.00, 3.75, and 2.00, respectively.

3.2. Geometric Parameters. The gas phase optimized geometries of all the chelates are presented in Figure 2. The values of the metal-ligand bond lengths as well as those of the O-H bonds are reported in Table 2. The variation of these metal-ligand bond lengths has been plotted and presented in Figure 3. The labeling has been described in Figure 1.

It can be seen from Figure 3 that, in both gas and aqueous phases, the lengths of the M-O_1 bonds are longer than those of the M-O_2 bonds in juglone chelates. It is clearly evidenced from the figure that, apart from the Cu(II) chelate, the solvent environment increases the two M-O bond lengths, which is apparently responsible for the interaction energy increment ongoing from the gas to the solvent phase. A similar observation is made in the case of the chelates of deprotonated juglone as depicted in Figure 3. However, in

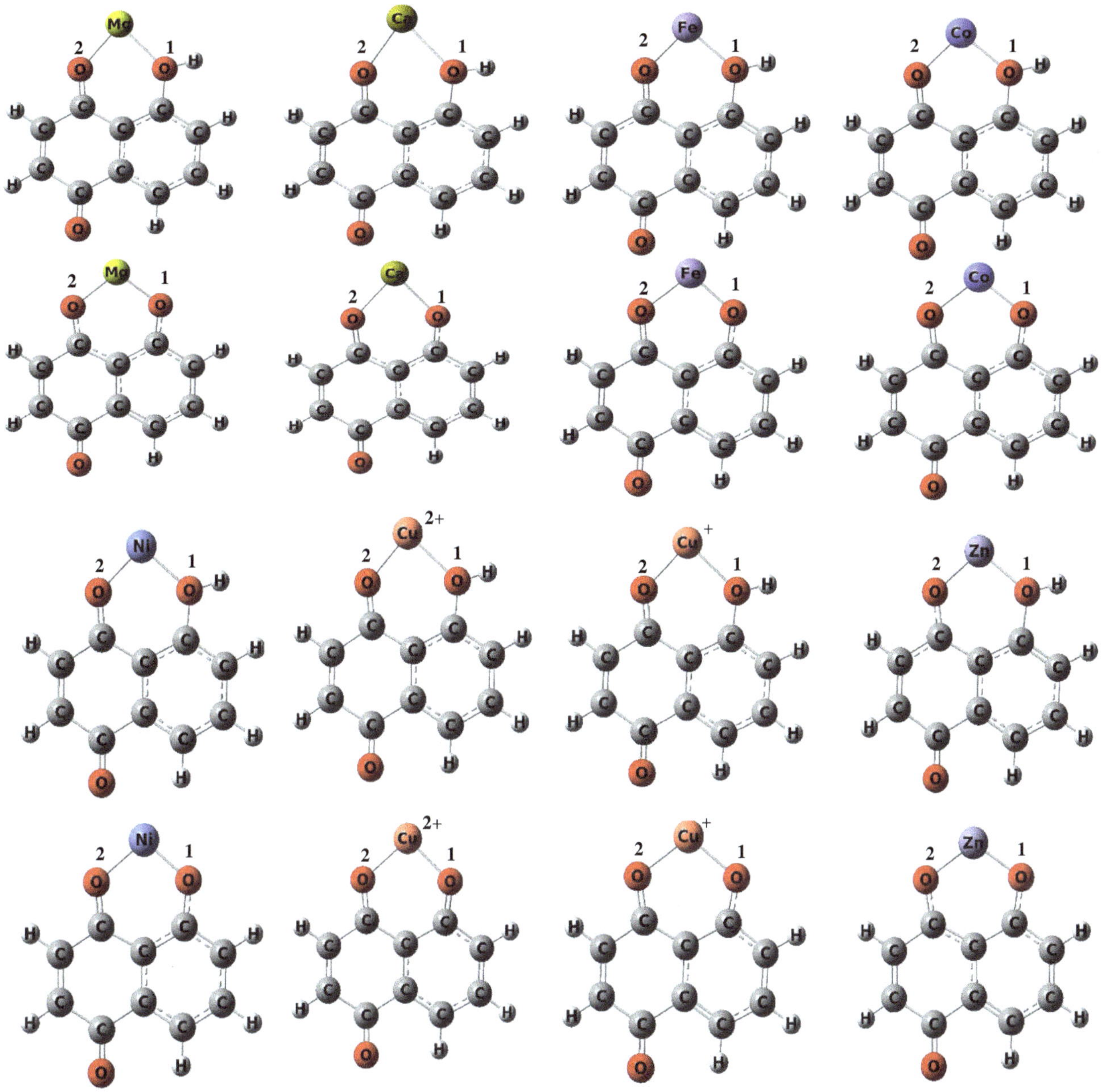

FIGURE 2: Gas phase optimized geometries of the chelates obtained at B3LYP/6-31+G(d,p) level of theory.

these compounds the M-O_1 bonds are shorter than the M-O_2 bonds owing to the negative charge carried by the O_1 atom, which increases its basicity relative to that of O_2. Because of this negative charge, the M-O bond lengths are shorter in the chelates of anionic juglone than in those of neutral juglone.

The lengths of the O-H bond in the chelates of the neutral form of juglone are presented in Table 2. In gas phase, the length of this bond lies within the range 0.970 to 0.979 Å, with the Co and Ni chelates having the highest values. Apart from the two Cu chelates, the presence of water leads to slight discrepancies in the O-H bond lengths. In this solvent, the length of this bond lies within 0.971 and 0.976 Å, with the Cu(II) chelate having the highest values. Based on these results, the best antioxidants in gas phase are certainly the Ni(II) and Co(II) chelates whereas the best antioxidant in aqueous phase is the Cu(II) chelate.

3.3. Chelation Ability. The chelating ability of both neutral and deprotonated juglone toward the Mg(II), Ca(II), Fe(II), Co(II), Ni(II), Cu(II), Cu(I), and Zn(II) ions has been evaluated through the enthalpies (ΔH) and free energies (ΔG) of formation of the chelates, as well as the complexation (binding) energies (ΔE) as summarized in (7). The values of these thermodynamic parameters calculated at 298.15 K and 1 atm are reported in Table 3. It is obvious from the values in Table 3 that the formation of all the juglone chelates in gas phase is exothermic and exergonic, meaning that these reactions are thermodynamically feasible. The formation of

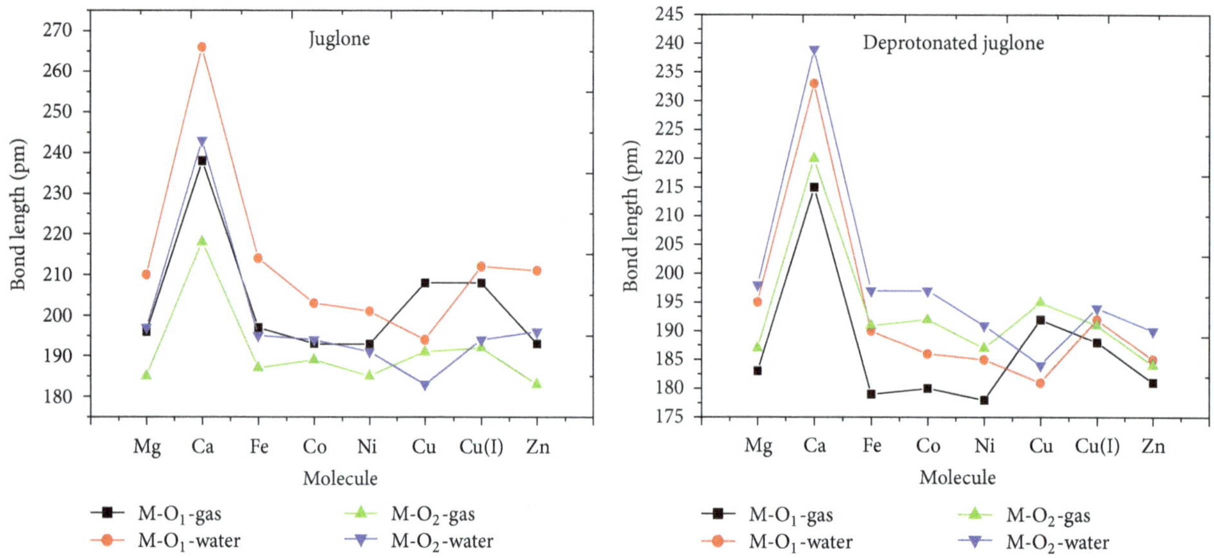

FIGURE 3: Variation of the metal-oxygen bond lengths (pm) of the studied chelates obtained from the optimized geometries at DFT/B3LYP/6-31+(d,p) level.

TABLE 3: Thermodynamic parameters (kJ/mol), atomic charge of the central metal(e) (ligand-metal charge transfer), and the correlation coefficient (R^2) all obtained at B3LYP/6-31+G(d,p) level.

	Juglone					Deprotonated juglone				
	M(e)	ΔH	ΔG	ΔE	R^2	M(e)	ΔH	ΔG	ΔE	R^2
Gas										
Mg(II)	1.84(0.16)	−674.1	−640.4	−672.2		1.76(0.24)	−1623.6	−1582.4	−1618.5	
Ca(II)	1.93(0.07)	−416.1	−384.1	−415.1		1.88(0.12)	−1275.3	−1237.6	−1227.2	
Fe(II)	1.58(0.42)	−769.6	−735.7	−767.5		1.43(0.57)	−1776.1	−1735.3	−1773.6	
Co(II)	1.51(0.49)	−838.9	−805.9	−837.9	0.746	1.28(0.72)	−1851.3	−1811.9	−1853.8	0.639
Ni(II)	1.31(0.69)	−945.1	−915.7	−945.6		1.17(0.83)	−1966.5	−1926.7	−1963.0	
Cu(II)	1.08(0.92)	−1027.0	−996.0	−1027.0		0.97(1.03)	−2038.6	−2001.4	−2036.9	
Cu(I)	0.89(0.11)	−244.8	−214.1	−244.7		0.81(0.19)	−772.7	−733.5	−769.0	
Zn(II)	1.66(0.34)	−841.5	−808.3	−841.6		1.55(0.45)	−1841.0	−1801.1	−1838.7	
Water										
Mg(II)	1.93(0.07)	−48.6	−14.1	−46.1		1.89(0.11)	−209.1	−168.2	−205.2	
Ca(II)	1.98(0.02)	9.4	39.3	10.6		1.96(0.04)	−103.2	−62.4	−99.4	
Fe(II)	1.85(0.15)	−51.1	−15.0	−49.5		1.74(0.26)	−239.9	−197.3	−235.5	
Co(II)	1.80(0.20)	−94.7	−58.9	−92.2	0.853	1.66(0.34)	−262.2	−221.1	−259.8	0.871
Ni(II)	1.76(0.24)	−117.9	−81.7	−115.3		1.58(0.42)	−321.6	−280.7	−317.2	
Cu(II)	1.49(0.51)	−368.7	−333.4	−366.4		1.37(0.63)	−630.8	−590.2	−627.7	
Cu(I)	0.98(0.02)	−92.2	−567.4	−90.0		0.86(0.14)	−222.6	−185.2	−219.5	
Zn(II)	1.88(0.12)	−63.1	−27.7	−62.8		1.77(0.23)	−242.4	−203.8	−239.5	

the Cu(II) chelate is the most favorable, since it has the lowest free energy of formation (−996 kJ/mol) in gas phase. The negative values of the binding (interaction) energies of all the juglone chelates are indicative of high stability of these molecules in the gas phase. The gas phase binding energies of the juglone chelates decrease in the order: Cu(I) > Ca(II) > Mg(II) > Fe(II) > Co(II) > Zn(II) > Ni(II) > Cu(II). This ranking is in good agreement with the relative affinity of the metal ion to nonmetal atoms or anions proposed by Williams and Fraústo [33] for the first row transition elements. It is evidenced from Table 3 that water as solvent greatly increases the values of the said thermodynamic parameters. Indeed, in water the formation of the Ca(II) chelate is even endothermic and endergonic. Besides the Ca(II) chelate, the interaction energies of all the other chelates remain negative in aqueous phase, although they greatly increase in magnitude. The least

ΔE difference between the aqueous and gas phases is found to be 154 kJ/mol for Cu(I), whereas the highest (830 kJ/mol) is observed for the Ni(II) chelate. Like in gas phase, the Cu(II) remains the metal ion with the highest affinity toward juglone (i.e., the chelate with the lowest binding energy).

It is also clear from Table 3 that the formation of all the chelates of deprotonated juglone is thermodynamically favorable, since it is exothermic and spontaneous in gas and aqueous phases. Furthermore, these chelates are all stable since their binding energies are negative in both media. In each of these study media, the values of ΔH, ΔG, and ΔE are more negative relative to those of juglone chelates. This result agrees with those obtained by Leopoldini and coworkers [11] in the study of iron chelation by Quercetin. In both media, Cu(II) is still found to be the metal with the highest affinity toward deprotonated juglone, while Cu(I) and Ca(II) also remain the metals with the lowest affinity to the ligand in the gas and the aqueous phases, respectively. As is the case with juglone chelates, the presence of water as solvent also leads to an increase in the ΔH, ΔG, and ΔE values of deprotonated juglone chelates.

The NPA charges on the central metal ions of the different chelates presented in Table 3 show that the charges on the metals are lower than their formal charges, which are +2 or +1. This implies that ligand-metal charge transfer is effective. From the difference between the formal and actual charges on the central metal ions (presented in brackets in Table 2), it has been found that the highest ligand-metal charge transfer is obtained for Cu(II) in both gas and aqueous phases. This result is in good agreement with the high stability observed with the Cu(II) chelates. In addition, the charges carried by the metals in deprotonated juglone chelates are lower than those on the metals in juglone chelates, which is also in good agreement with their binding energy values. Moreover, ligand-metal charge transfer has been found to decrease from gas to aqueous phase, in conformity with the increment in the values of ΔH, ΔG, and ΔE observed in water. In order to determine the relationship between the ligand-metal charge transfer and the interaction energies, a linear plot between the two parameters has been done. The values of the correlation coefficient (R^2) reported in Table 3 are all higher than 0.6. Better correlation coefficients are obtained in water, wherein R^2 values of 0.853 and 0.871 are calculated for the chelates of juglone and deprotonated juglone, respectively. From the foregoing results, it can be concluded that both the neutral and anionic forms of juglone can be efficiently used in the fight against metal ions toxicity and the adverse effects of metal overload in foods, polymers, and living organisms.

3.4. QTAIM Analysis of the M-O Bond Characters. Among the several techniques used in studying bonding interactions in molecules, those based on electron density analysis are the most reliable because electron density ($\rho(\mathbf{r})$) is an observable. Bader's atoms in molecules method is one of the most used for this purpose, since it is based on $\rho(\mathbf{r})$ and its Laplacian ($\nabla^2\rho(\mathbf{r})$) [31, 34]. This method also known as Bader's topological analysis, specifically studies $\rho(\mathbf{r})$ and $\nabla^2\rho(\mathbf{r})$ at the bond critical points (BCP), a BCP being a region between a pair of nuclei, where $\nabla\rho(\mathbf{r}) = 0$. Generally, the character of a chemical interaction is determined based on the following criteria at the BCPs [35–37]:

(i) high values of $\rho(\mathbf{r})$, $\nabla^2\rho(\mathbf{r}) < 0$, and $[-g(\mathbf{r})/v(\mathbf{r})] < 1$ are associated with covalent bonds (shared interactions);

(ii) low values of $\rho(\mathbf{r})$ (generally less than 0.1 a.u.), $\nabla^2\rho(\mathbf{r}) > 0$, and $0.5 < [-g(\mathbf{r})/v(\mathbf{r})] < 1$ signify intermediate type interactions such as polar covalent bonds, strong hydrogen bonds, and coordination bonds;

(iii) low values of $\rho(\mathbf{r})$ (typically less than 0.1 a.u.), $\nabla^2\rho(\mathbf{r}) > 0$, and $[-g(\mathbf{r})/v(\mathbf{r})] > 1$ are associated with closed shell interactions (weak interactions) like ionic, weak hydrogen and van der Waals interactions.

Here, $g(\mathbf{r})$ is the kinetic energy density at the BCP (always positive), while $v(\mathbf{r})$ is the potential energy density at the BCP (always negative). The above-mentioned criteria were exploited in this research endeavor to evaluate the characters of the M-O bonds. For all the chelates, the Poincaré-Hopf relation was fulfilled [31]. Our calculated values of these parameters are reported in Table 4.

From Table 4, it can be seen that the M-O bonds in the Mg(II) and Ca(II) chelates are characterized by low values of $\rho(\mathbf{r})$, $\nabla^2\rho(\mathbf{r}) > 0$, and $[-g(\mathbf{r})/v(\mathbf{r})] > 1$; hence, they are weak interactions. In the case of the M-O bonds in the rest of the chelates, the values of $\rho(\mathbf{r})$ are low, $\nabla^2\rho(\mathbf{r}) > 0$, and $0.5 < [-g(\mathbf{r})/v(\mathbf{r})] < 1$, meaning that they are intermediate type interactions. These observations are similar in both gas and aqueous phases, although the values of the parameters change slightly.

For the juglone chelates, the $\rho(\mathbf{r})$ values of the O_2-M bonds are found to be higher than those of the O_1-M bonds, supporting the fact that the former bonds are shorter than the latter. The reverse trend is observed for the chelates of deprotonated juglone because the O_1 atom carries a negative charge, which increases electron density around the O_1-M bonds. As evidenced from Table 3, the electron density at the BCPs of the M-O bonds reduces ongoing from gas to aqueous phase, matching with the increase in their bond lengths from gas to aqueous phase as previously observed in Section 3.3. From the values of $\rho(\mathbf{r})$, $\nabla^2\rho(\mathbf{r})$, and $[-g(\mathbf{r})/v(\mathbf{r})]$, it is clear that the O-H bonds are covalent in nature. The electron densities on all the O-H bonds are nearly the same and lie between 0.344 and 0.357 in gas phase and between 0.348 and 0.354 in water. This serves as a plausible explanation for the small discrepancies observed in the O-H bond lengths of the compounds (Section 3.3). In this light, the slight reductions in the lengths of these bonds in water are attributable to the slight changes in their $\rho(\mathbf{r})$ values.

3.5. Radical Scavenging Activity. Motivated by the stability of the juglone chelates as earlier revealed by studies carried out on thermodynamic parameters and binding energies, we have studied the different mechanisms of hydrogen atom transfer by antioxidants in order to investigate the effect of the metal ions studied on the radical scavenging activity of juglone. The values of the parameters describing these mechanisms, calculated using (8)–(12), are reported in Table 5. The gas

TABLE 4: Topological characteristics at the BCPs of the O-M and O-H bonds of the chelates.

	Juglone						Deprotonated juglone			
		Gas			Water		Gas		Water	
	O_1-H	O_1-M	O_2-M	O_1-H	O_1-M	O_2-M	O_1-M	O_2-M	O_1-M	O_2-M
Mg(II)										
$\rho(\mathbf{r})$	0.348	0.046	0.064	0.351	0.031	0.044	0.068	0.062	0.049	0.044
$\nabla^2\rho(\mathbf{r})$	−2.124	0.369	0.571	−2.156	0.224	0.357	0.611	0.538	0.398	0.350
$-g(\mathbf{r})/v(\mathbf{r})$	0.093	1.226	1.184	0.092	1.240	1.248	1.165	1.182	1.220	1.232
Ca(II)										
$\rho(\mathbf{r})$	0.354	0.032	0.051	0.356	0.015	0.027	0.056	0.050	0.035	0.030
$\nabla^2\rho(\mathbf{r})$	−2.096	0.171	0.324	−2.131	0.077	0.147	0.356	0.302	0.202	0.164
$-g(\mathbf{r})/v(\mathbf{r})$	0.101	1.147	1.120	0.099	1.192	1.198	1.086	1.107	1.146	1.167
Fe(II)										
$\rho(\mathbf{r})$	0.344	0.083	0.111	0.350	0.053	0.086	0.137	0.099	0.102	0.083
$\nabla^2\rho(\mathbf{r})$	−2.112	0.452	0.684	−2.162	0.256	0.497	0.799	0.594	0.582	0.489
$-g(\mathbf{r})/v(\mathbf{r})$	0.091	0.968	0.950	0.091	0.953	0.974	0.908	0.957	0.951	0.973
Co(II)										
$\rho(\mathbf{r})$	0.344	0.087	0.101	0.348	0.069	0.089	0.134	0.096	0.109	0.080
$\nabla^2\rho(\mathbf{r})$	−2.124	0.551	0.635	−2.174	0.345	0.508	0.870	0.606	0.642	0.485
$-g(\mathbf{r})/v(\mathbf{r})$	0.089	0.959	0.940	0.089	0.913	0.934	0.916	0.942	0.925	0.945
Ni(II)										
$\rho(\mathbf{r})$	0.344	0.090	0.113	0.348	0.075	0.094	0.136	0.106	0.115	0.940
$\nabla^2\rho(\mathbf{r})$	−2.127	0.473	0.651	−2.173	0.322	0.527	0.824	0.607	0.592	0.519
$-g(\mathbf{r})/v(\mathbf{r})$	0.088	0.875	0.891	0.088	0.842	0.897	0.879	0.892	0.862	0.892
Cu(II)										
$\rho(\mathbf{r})$	0.349	0.063	0.096	0.343	0.089	0.119	0.095	0.089	0.126	0.116
$\nabla^2\rho(\mathbf{r})$	−2.134	0.269	0.450	−2.176	0.405	0.550	0.429	0.399	0.585	0.546
$-g(\mathbf{r})/v(\mathbf{r})$	0.093	0.831	0.817	0.085	0.824	0.817	0.809	0.811	0.812	0.820
Cu(I)										
$\rho(\mathbf{r})$	0.357	0.062	0.095	0.354	0.057	0.088	0.105	0.098	0.096	0.092
$\nabla^2\rho(\mathbf{r})$	−2.126	0.274	0.455	−2.155	0.253	0.421	0.506	0.465	0.451	0.429
$-g(\mathbf{r})/v(\mathbf{r})$	0.100	0.839	0.821	0.095	0.853	0.821	0.816	0.814	0.813	0.814
Zn(II)										
$\rho(\mathbf{r})$	0.343	0.094	0.122	0.350	0.058	0.085	0.130	0.119	0.101	0.091
$\nabla^2\rho(\mathbf{r})$	−2.132	0.387	0.517	−2.167	0.247	0.343	0.564	0.501	0.411	0.366
$-g(\mathbf{r})/v(\mathbf{r})$	0.089	0.787	0.774	0.137	0.854	0.791	0.774	0.772	0.777	0.783

phase enthalpy and free energy of electron employed here are 3.145 kJ/mol and −3.42 kJ/mol, respectively [38]. In water, the values −105 kJ/mol [39] and −156.8 kJ/mol [40] for the enthalpy and the free energy of electron, respectively, were used. On the other hand, the enthalpies of formation for the proton in gas phase and water are 6.197 kJ/mol [38] and −1022 kJ/mol [39], respectively. The values of the proton hydration free energies in these media are, respectively, −26.26 kJ/mol [38] and −1104.5 kJ/mol [40].

3.5.1. Direct Hydrogen Atom Transfer Mechanism. The BDE and BDFE values presented in Table 4 are all positive, indicating that this mechanism is endothermic and nonfavorable at 298.15 K and 1 atm. To facilitate the analysis of BDE and BDFE, their values have been plotted in the graphs shown in Figure 4. It is clear from this figure that the BDE and BDFE values of the chelates are lower than that of juglone in the two media investigated. This is a clear indication of the fact that the presence of these metal ions increases the AOA of juglone, which is in good agreement with the findings in the literature [9, 26]. In gas phase, the Co(II) chelate has been found to possess the lowest BDE value (262 kJ/mol), which corroborates with its highest O-H bond length observed in the said phase (Section 3.3). The BDE values of the chelates studied are found to increase in the order: Co < Fe < Zn < Mg = Ni < Ca < Cu(II) < Cu(I). This ranking shows that the copper chelates are the least reactive in gas phase. This low AOA may be attributed to the observed low electron spin density distribution in the radicals of these copper chelates, as shown in Figure 5. This figure shows a huge distribution of electron spin density in the two most reactive chelates (Co and Fe chelates), which stabilizes the radical.

TABLE 5: Gas and aqueous phases values of the antioxidant activity descriptors of the chelates studied, calculated at DFT/B3LYP/6-31+G(d,p) level.

Media	Mol.	BDE (kJ/mol)	BDFE (kJ/mol)	IP (eV)	IPFE (eV)	PDE (kJ/mol)	PDFE (kJ/mol)	PA (kJ/mol)	PAFE (kJ/mol)	ETE (kJ/mol)	ETFE (kJ/mol)
Gas	J	417.1	377.3	8.64	8.54	899.2	865.2	1422.0	1386.1	312.7	302.2
	Mg(II)	290.0	256.8	16.44	16.34	20.7	−9.1	473.5	444.4	1133.1	1122.8
	Ca(II)	297.0	263.7	15.30	15.13	137.4	113.9	562.8	533.9	1050.8	1040.3
	Fe(II)	274.3	242.4	16.27	16.19	20.6	−8.7	414.5	386.6	1176.5	1167.3
	Co(II)	262.8	231.7	16.23	16.13	12.8	−14.1	408.6	380.4	1169.8	1161.8
	Ni(II)	290.4	261.3	16.77	16.72	−11.0	−40.8	400.7	375.4	1206.4	1197.3
	Cu(II)	299.7	268.3	16.48	16.35	25.6	2.7	411.4	381.1	1203.9	1198.7
	Cu(I)	324.6	291.4	12.74	12.66	411.4	381.0	894.1	867.4	746.1	735.5
	Zn(II)	286.2	253.6	16.49	16.39	11.6	−17.6	422.5	393.6	1180.3	1170.5
Water	J	400.4	360.2	5.57	4.99	41.6	−42.6	192.2	109	386.2	330.0
	Mg(II)	346.4	314.7	6.38	5.82	−89.2	−168.6	33.4	−45.1	493.3	438.4
	Ca(II)	355.5	319.2	5.82	5.28	−27.2	−110.7	82.3	7.3	452.5	391.5
	Fe(II)	324.8	291.8	6.42	5.85	−115.2	−193.8	7.1	−73.3	497.0	443.7
	Co(II)	346.9	313.8	6.68	6.10	−118.6	−195.7	26.4	−54.2	499.8	446.7
	Ni(II)	363.1	328.8	6.86	6.33	−118.7	−203.2	−8.8	−90.0	552.1	497.4
	Cu(II)	312.1	276.2	6.92	6.41	−174.9	−262.6	−67.3	−147.8	559.7	503.6
	Cu(I)	315.8	279.5	5.83	5.25	−67.3	−147.8	65.5	−20.5	429.6	379.4
	Zn(II)	343.8	306.5	6.39	5.80	−93.5	−174.8	15.6	−66.3	507.5	451.5

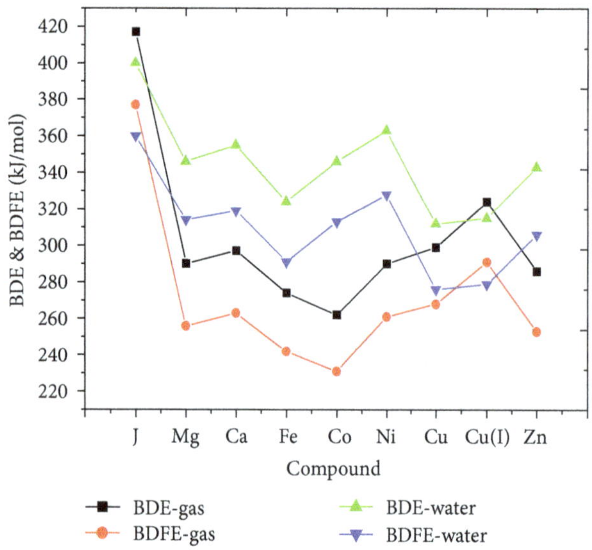

FIGURE 4: Superposition of BDE and BDFE of the studied chelates in gas and aqueous phases.

Apart from the Cu(I) chelate, the BDE values of the other chelates increase in water as shown in Figure 3. This result agrees with the slight decrease in the O-H bond length (and the increase for the Cu chelates) previously observed. As a result of the O-H bond length variations ongoing from gas phase to water, the Cu(II) chelate has the lowest BDE value (312 kJ/mol) in water and is thus the best antioxidant among those studied. The BDE values in water are in the order: Cu(II) < Cu(I) < Fe < Zn < Mg = Co < Ca < Ni. It is clear from Figure 3 that the BDE and BDFE values of the chelates follow the same trend in both gas and aqueous phases.

3.5.2. Electron Transfer-Proton Transfer Mechanism. IP is the first and the determining step of this mechanism. The IP and IPFE values of the chelates reported in Table 4 are plotted in Figure 6. As depicted in this figure, IP and IPFE like BDE and BDFE follow the same trend in both gas and aqueous phases. However, the differences between IPs and IPFEs are very small compared to those between BDEs and BDFEs in a given milieu. It can also be seen from Table 5 that the IP values are higher than BDEs in gas and aqueous phases. It is evident from Figure 4 that, in gas phase, the presence of the metal ions leads to a great increment in the IP value of juglone. This result agrees with those obtained by other authors in similar studies [9, 26]. In gas phase, the Cu(I) chelate has the lowest IP value (12.74 eV). Although the IPs and IPFEs of all molecules greatly decrease in presence of water, the IP values of the chelates are close to that of juglone in the solvent phase, whereas in gas phase the IP values of the chelates are significantly higher than that of juglone. In water phase, the Cu(I) chelate still has the lowest IP value (5.83 eV).

Also presented in Table 4 are the PDE and PDFE values of the chelates. The PDE values are found to be much lower than that of juglone, with that of the Ni chelate being even negative (−11 kJ/mol). From this result, it is clear that the cationic radicals obtained from the chelates are very reactive. The PDFE values show that the proton transfer reaction by the cationic radicals of the Mg, Fe, Co, Ni, and Zn chelates is exergonic in gas phase. A close inspection of Table 4 also shows that, in water, the PDE and PDFE values of all the chelate are negative, meaning that the reaction is spontaneous under standard conditions. This is certainly due to the low values of the enthalpy and free energy of proton in water.

3.5.3. Sequential Proton Loss Electron Transfer Mechanism. The first step of this mechanism corresponds to a proton transfer by the antioxidant and is described by PA. Low PA value indicates a high acidic hydroxyl group and thus good AOA. The PA and PAFE values are presented in Table 4. For better comprehension of their analysis, the PA and PAFE values have been plotted and their trends are shown in Figure 7.

It is clear from Figure 6 that the PA values of the chelates are lower relative to that of juglone in gas and aqueous phases. This signifies that metal chelation increases the acidity of the OH group. This result is in good agreement with those obtained by Holtomo and coworkers [9]. In gas phase, the PA values of the chelates increase in the order: Ni(II) < Co(II) < Cu(II) < Fe(II) < Zn(II) < Mg(II) < Ca(II) < Cu(I). According to this ranking, the highest acidity of the O-H group of juglone in gas phase is obtained in the presence of the Ni(II) ion. In aqueous solution, the PA and PAFE values are found to decrease due to the low values of the enthalpy and free energy of proton in water. In this solvent, the PA values of the Ni(II) and Cu(II) chelates are even negative, indicating that the proton transfer reactions by these chelates are exothermic. It can be seen from Table 5 that, apart from the Ca (II) chelate, the PAFE values of other chelates are negative in water, showing that the reaction is spontaneous. In water, the PA values of the chelates are found to increase in the order: Cu(II) < Ni(II) < Fe(II) < Zn(II) < Co(II) < Mg(II) < Cu(I) < Ca(II). Based on this classification, it can be concluded that the Cu(II) chelate is the most acidic. From Figure 7, it can be seen that PA and PAFE values follow the same trend in the two study milieus. From Table 5, it can also be observed that the PA and PAFE values are always higher than the PDE and PDFE values, buttressing the higher reactivity of the cationic radicals.

The ETE and ETFE values describing the electron transfer reaction by the anions are presented in Table 4. These values revealed that, in gas phase and water, ETEs and ETFEs are lower than IPs and IPFEs. The ETE and ETFE values can be used to describe the AOA of the chelate formed from deprotonated juglone, since these molecules no longer contain any hydroxyl group. Indeed, by comparing the ETE values of the said chelates and the IP value of juglone in gas phase, it can be concluded that the only chelate which could be considered to be a better reducing agent than juglone is that of Cu(I). In water, however, the ETE values of the chelates are lower than the IP value of juglone except for the Cu(II) chelate. This means that nearly all of the deprotonated juglone chelates are better reducing agents and as such are better antioxidants than juglone in water. The ETEs and ETFEs values follow the same trend in both media, thus leading to the same conclusions.

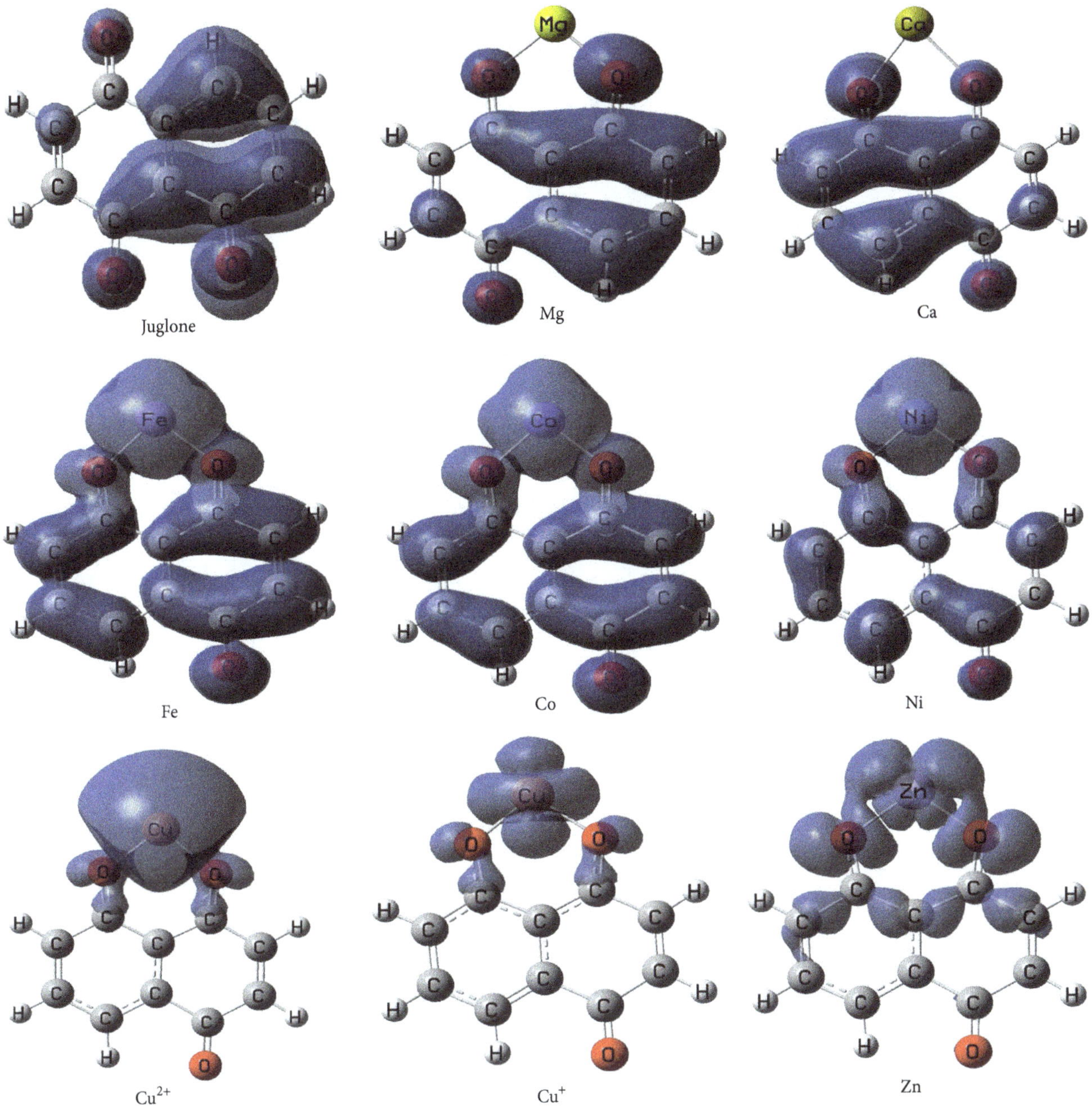

FIGURE 5: Spin density distribution for the radical formed from the chelates after the hydrogen atom abstraction, obtained at isovalue = 0.04.

3.5.4. Thermodynamic Preferred Mechanism. To determine the thermodynamically preferred mechanism of the juglone chelates, the free energies of the first steps of the hydrogen transfer mechanisms have been plotted in the graphs shown in Figure 8.

The following ranking of the free energies for the first step reactions of the chelates in gas phase, BDFE < PAFE < IPFE, is obvious from Figure 8. According to this classification, the direct HAT is the preferred mechanism in the gas phase for neutral juglone and its chelates. Figure 8 also shows that, in gas phase, the chelates are more reactive than the ligand, the Co chelate being the most reactive. From the graph in Figure 8, the following ranking of free energies can be made in water as solvent for all the compounds: PAFE < BDFE < IPFE. This shows that the sequential proton loss electron transfer is the preferred mechanism in water. This result is in good agreement with those found in the literature with the polar solvents [41–48]. In water, the Cu(II) chelate is found to be the most stable and the most reactive.

4. Conclusion

In this paper, the antioxidant activity of neutral and deprotonated juglone has been investigated in the gas and the aqueous phases at DFT/B3LYP/6-31+G(d,p) level of theory. In this regard, the chelating ability of these ligands toward Mg^{2+},

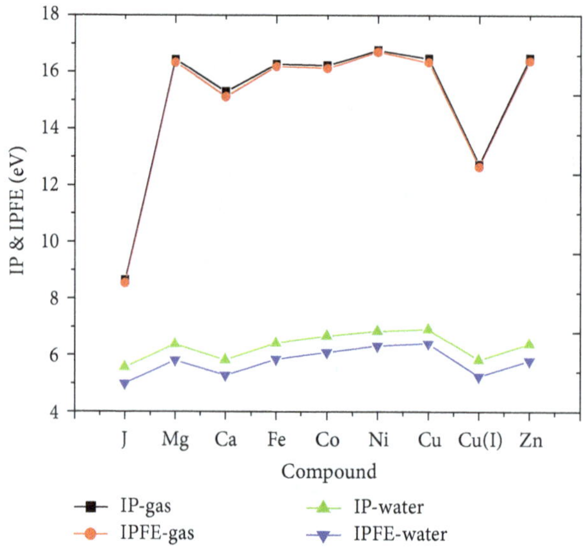

FIGURE 6: Superposition of IP and IPFE of the studied chelates in gas and aqueous phases.

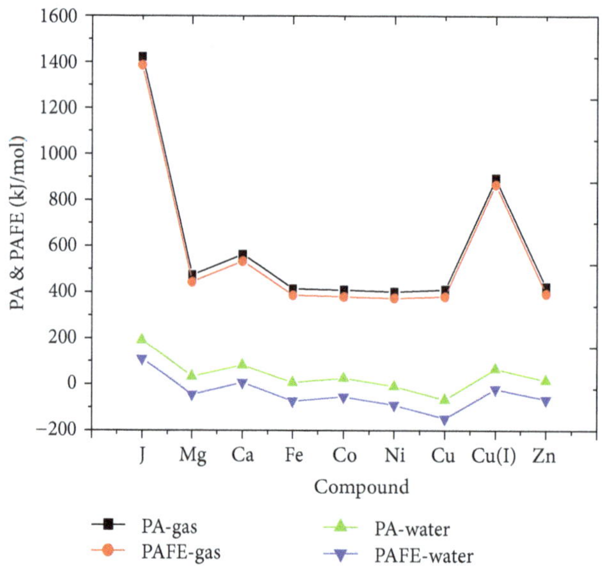

FIGURE 7: Superposed values of PA and PAFE of the chelates studied in the gas and the aqueous phases.

Ca^{2+}, Fe^{2+}, Co^{2+}, Ni^{2+}, Cu^{2+}, Cu^{+}, and Zn^{2+} ions usually involved in the generation of free radicals in living organisms was studied. Furthermore, the effect of these metals on the radical scavenging activity of neutral juglone was investigated by means of the usual hydrogen atom transfer mechanisms. The analysis of the thermodynamic parameters (ΔH and ΔG) and the interaction energies calculated herein showed that the neutral and anionic forms of juglone are capable of chelating these metal ions in gas and aqueous phases. For both ligands, the highest metal ion affinity has been exhibited toward the Cu(II) ions. This trend has been confirmed by the fact that the highest reduction effect due to chelation is observed for the Cu(II) chelate. Bader's topological analyses revealed that the metal-ligand bonds in the Mg and Ca chelates are weak interactions, while those in the other chelates are intermediate type interactions, in both gas and aqueous phases. Analyses of the BDE, IP, and PA along with their free energies showed that direct HAT transfer is the preferred AOA mechanism of the neutral juglone chelates in gas phase, with the Co(II) chelate being the most reactive. In water, sequential proton transfer electron transfer has been found to be the most preferred AOA mechanism, for which the Cu(II) chelate is found to be the most reactive. However, to gain further insight into the antioxidant mechanisms of the molecules currently studied, we intend to perform kinetic studies on the molecules.

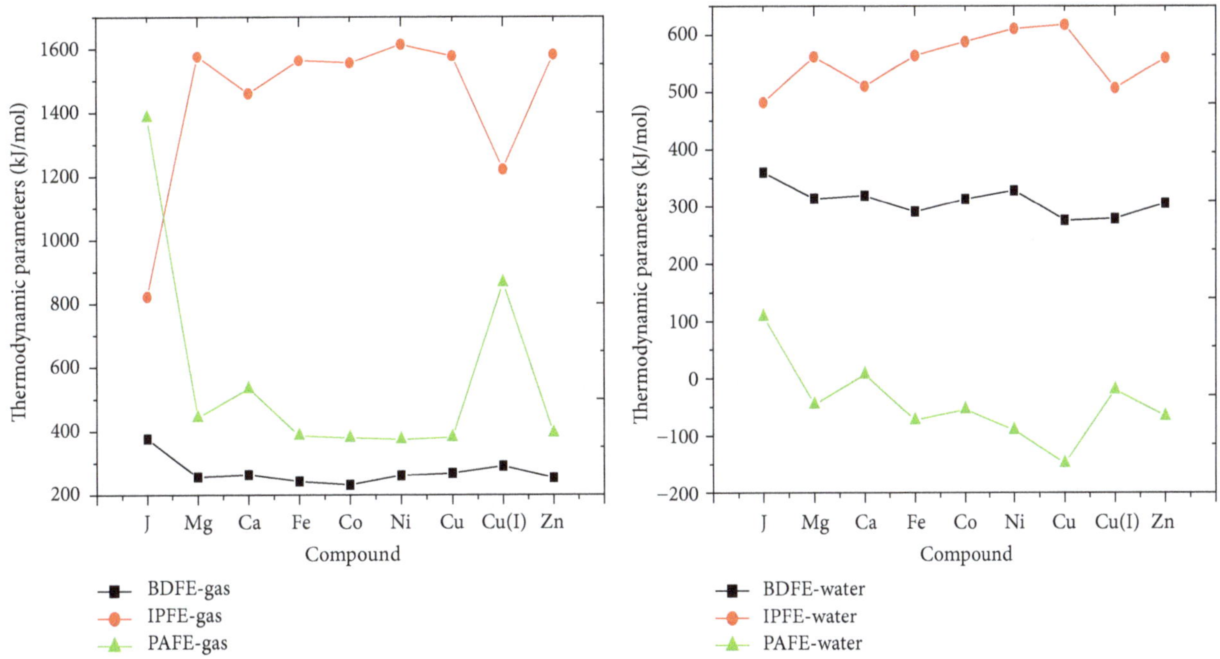

FIGURE 8: Superposed values of BDFE, IPFE, and PAFE of the chelates of neutral juglone in gas and aqueous phases.

Conflicts of Interest

The authors declare that there are no conflicts of interest regarding the publication of this paper.

Acknowledgments

The work has been supported by the annual research grant of the Laboratory of Noxious Chemistry and Environmental Engineering of the University of Dschang, Cameroon. It also benefited from the financial assistance of the Ministry of External Affairs of India and FICCI (Federation of Indian Chambers of Commerce and Industry) through a CV Raman International Fellowship award (Grant no. 101F102) for African Researchers that was realized at IIT Kanpur, Kanpur, India.

References

[1] S. S. Gill and N. Tuteja, "Reactive oxygen species and antioxidant machinery in abiotic stress tolerance in crop plants," *Plant Physiology and Biochemistry*, vol. 48, no. 12, pp. 909–930, 2010.

[2] V. Chobot and F. Hadacek, "Iron and its complexation by phenolic cellular metabolites: from oxidative stress to chemical weapons," *Plant Signaling and Behavior*, vol. 5, no. 1, pp. 4–8, 2010.

[3] E. A. Decker, J. E. Ryan, and D. J. Mc Clements, *Oxidation in Foods And Beverages and Antioxidant Applications*, Wood Head Publishing Limited, New Dehli, India, 2010.

[4] M. Valko, D. Leibfritz, J. Moncol, M. T. D. Cronin, M. Mazur, and J. Telser, "Free radicals and antioxidants in normal physiological functions and human disease," *International Journal of Biochemistry and Cell Biology*, vol. 39, no. 1, pp. 44–84, 2007.

[5] K. M. Lincoln, T. E. Richardson, L. Rutter, P. Gonzalez, J. W. Simpkins, and K. N. Green, "An N-heterocyclic amine chelate capable of antioxidant capacity and amyloid disaggregation," *ACS Chemical Neuroscience*, vol. 3, no. 11, pp. 919–927, 2012.

[6] M. Leopoldini, N. Russo, and M. Toscano, "The molecular basis of working mechanism of natural polyphenolic antioxidants," *Food Chemistry*, vol. 125, no. 2, pp. 288–306, 2011.

[7] I. Gülçin, Z. Huyut, M. Elmastaş, and H. Y. Aboul-Enein, "Radical scavenging and antioxidant activity of tannic acid," *Arabian Journal of Chemistry*, vol. 3, no. 1, pp. 43–53, 2010.

[8] Y. Qiu, Z. Wang, A. C. E. Owens et al., "Antioxidant chemistry of graphene-based materials and its role in oxidation protection technology," *Nanoscale*, vol. 6, no. 20, pp. 11744–11755, 2014.

[9] O. Holtomo, M. Nsangou, J. J. Fifen, Z. Dhaouadi, and O. Motapon, "Antioxidative potency and UV-Vis spectra features of the compounds resulting from the chelation of Fe^{2+} by caffeic acid phenethyl ester and two of its derivatives," *Computational and Theoretical Chemistry*, vol. 1067, pp. 135–147, 2015.

[10] M. M. Kabanda, V. T. Tran, K. M. Seema et al., "Conformational, electronic and antioxidant properties of lucidone, linderone and methyllinderone: DFT, QTAIM and NBO studies," *Molecular Physics*, vol. 113, no. 7, pp. 683–697, 2015.

[11] M. Leopoldini, N. Russo, S. Chiodo, and M. Toscano, "Iron chelation by the powerful antioxidant flavonoid quercetin," *Journal of Agricultural and Food Chemistry*, vol. 54, no. 17, pp. 6343–6351, 2006.

[12] T. Marino, M. Pavelka, M. Toscano, and N. Russo, "Structural and binding properties of metal ion chelators relevant to Alzheimer's disease. A theoretical investigation," *International Journal of Quantum Chemistry*, vol. 112, no. 9, pp. 2109–2114, 2012.

[13] P. Mladěnka, K. Macáková, T. Filipský et al., "In vitro analysis of iron chelating activity of flavonoids," *Journal of Inorganic Biochemistry*, vol. 105, no. 5, pp. 693–701, 2011.

[14] N. R. Perron and J. L. Brumaghim, "A review of the antioxidant mechanisms of polyphenol compounds related to iron binding," *Cell Biochemistry and Biophysics*, vol. 53, no. 2, pp. 75–100, 2009.

[15] M. Leopoldini, T. Marino, N. Russo, and M. Toscano, "Antioxidant properties of phenolic compounds: H-atom versus electron transfer mechanism," *The Journal of Physical Chemistry A*, vol. 108, no. 22, pp. 4916–4922, 2004.

[16] F. A. D. Tamafo, J. N. Ghogomu, M. D. Bikélé, N. K. Nkungli, and E. Younang, "Structural and antioxidant properties of compounds obtained from Fe^{2+} chelation by juglone and two of its derivatives: DFT, QTAIM, and NBO studies," *Bioinorganic Chemistry and Applications*, vol. 2016, Article ID 8636409, 13 pages, 2016.

[17] A. Galano, G. Mazzone, R. Alvarez-Diduk, T. Marino, J. R. Alvarez-Idaboy, and N. Russo, "The food antioxidants: chemical insights at the molecular level," *Annual Review of Food Science and Technology*, vol. 7, no. 1, pp. 335–352, 2016.

[18] R. Ditchfield, W. J. Hehre, and J. A. Pople, "Self-consistent molecular-orbital methods. IX. An extended gaussian-type basis for molecular-orbital studies of organic molecules," *The Journal of Chemical Physics*, vol. 54, no. 2, pp. 720–723, 1971.

[19] W. J. Hehre, K. Ditchfield, and J. A. Pople, "Self-consistent molecular orbital methods. XII. Further extensions of gaussian-type basis sets for use in molecular orbital studies of organic molecules," *The Journal of Chemical Physics*, vol. 56, no. 5, pp. 2257–2261, 1972.

[20] A. D. Becke, "Density-functional thermochemistry. III. The role of exact exchange," *The Journal of Chemical Physics*, vol. 98, no. 7, pp. 5648–5652, 1993.

[21] C. Lee, W. Yang, and R. G. Parr, "Development of the Colle-Salvetti correlation-energy formula into a functional of the electron density," *Physical Review B*, vol. 37, no. 2, pp. 785–789, 1988.

[22] E. Lewars, *Computational chemistry: Introduction to The Theory And Applications of Molecular and Quantum Mechanics*, Klumer Academy Publishers, New York, NY, USA, 2003.

[23] F. Jensen, *Introduction to Computational Chemistry*, John Wiley & Sons, New York, NY, USA, 1999.

[24] K. I. Ramachandran, G. Deepa, and K. Namboori, *Computational Chemistryand Molecular Modeling : Principles and Applications*, Springer, Berlin, Germany, 2008.

[25] M. J. Frisch, G. W. Trucks, H. B. Schlegel et al., *Gaussian 09, Revision A.02*, Gaussian, Inc, Wallingford, UK, 2009.

[26] J. J. Fifen, M. Nsangou, Z. Dhaouadi, O. Motapon, and S. Lahmar, "Single or double hydrogen atom transfer in the reaction of metal—associated phenolic acids with •OH radical: DFT study," *Journal of Molecular Structure: THEOCHEM*, vol. 901, no. 1-3, pp. 49–55, 2009.

[27] D. C. Young, *Computational Chemistry: A Practical Guide for ApplyingTechniques to Real-World Problems*, John Wiley and Sons Inc, New York, NY, USA, 2001.

[28] P. W. Atkins and R. S. Friedman, *Molecular Quantum Mechanics*, Oxford university press, Oxford, UK, 3rd edition, 1996.

[29] I. N. Levine, *Quantum Chemistry*, Printice Hall Inc, Upper Saddle River, NJ, USA, 5th edition, 2000.

[30] E. D. Glendening, R. L. Clark, and F. Weinhold, "Natural bond orbital methods," *WIREs Compututational Molecular Sciences*, vol. 2, pp. 1–42, 2012.

[31] R. F. W. Bader, *Atoms in Molecules: A Quantum Theory*, Oxford University Press, Oxford, UK, 1990.

[32] T. Lu and F. Chen, "Calculation of molecular orbital composition," *Acta Chimica Sinica*, vol. 69, no. 20, pp. 2393–2406, 2011.

[33] R. J. P. Williams and S. J. J. R. Fraústo, *The Chemistry of Evolution*, Elsevier, Amsterdam, The Netherland, 2006.

[34] P. L. A. Popelier, "On the full topology of the Laplacian of the electron density," *Coordination Chemistry Reviews*, vol. 197, no. 1, pp. 169–189, 2000.

[35] G. V. Baryshnikov, B. F. Minaev, V. A. Minaeva, A. T. Podgornaya, and H. Ågren, "Application of Bader's atoms in molecules theory to the description of coordination bondsin the complex compounds of Ca^{2+} and Mg^{2+} with methylidene rhodanine and its anion," *Russian Journal of General Chemistry*, vol. 82, no. 7, pp. 1254–1262, 2012.

[36] I. L. Kirby, M. Brightwell, M. B. Pitak, C. Wilson, S. J. Coles, and P. A. Gale, "Systematic experimental charge density analysis of anion receptor complexes," *Physical Chemistry Chemical Physics*, vol. 16, no. 22, pp. 10943–10958, 2014.

[37] N. Trendafilova, G. Bauer, and T. Mihaylov, "DFT and AIM studies of intramolecular hydrogen bonds in dicoumarols," *Chemical Physics*, vol. 302, no. 1–3, pp. 95–104, 2004.

[38] J. E. Bartmess, "Thermodynamics of the electron and the proton," *The Journal of Physical Chemistry*, vol. 98, no. 25, pp. 6420–6424, 1994.

[39] J. Rimarcik, V. Lukeš, E. Klein, and M. Ilčin, "Study of the solvent effect on the enthalpies of homolytic and heterolytic nh bond cleavage in *p*-phenylenediamine and tetracyano-*p*-phenylenediamine," *Journal of Molecular Structure*, vol. 952, no. 1–3, pp. 25–30, 2010.

[40] J. J. Fifen, Z. Dhaouadi, M. Nsangou, O. Holtomo, and N. Jaidane, "Proton-coupled electron transfer in the reaction of 3,4-dihydroxyphenylpyruvic acid with reactive species in various media," *International Journal of Chemical Physics*, vol. 2015, Article ID 835707, 13 pages, 2015.

[41] Z. Marković, D. Milenković, J. Đorović et al., "Free radical scavenging activity of morin 2′-O phenoxide anion," *Food Chemistry*, vol. 135, no. 3, pp. 2070–2077, 2012.

[42] M. Szeląg, A. Urbaniak, and H. A. R. Bluyssen, "A theoretical antioxidant pharmacophore for natural hydroxycinnamic acids," *Open Chemistry*, vol. 13, no. 1, pp. 17–31, 2015.

[43] Y. Chen, H. Xiao, J. Zheng, and G. Liang, "Structure-thermodynamics-antioxidant activity relationships of selected natural phenolic acids and derivatives: an experimental and theoretical evaluation," *PLoS ONE*, vol. 10, no. 3, Article ID e0121276, 2015.

[44] M. Najafi, "On the antioxidant activity of the *Ortho* and *Meta* substituted Daidzein derivatives in the gas phase and solvent environment," *Journal of the Mexican Chemical Society*, vol. 58, no. 1, pp. 36–45, 2014.

[45] N. M. Thong, D. T. Quang, N. H. T. Bui, D. Q. Dao, and P. C. Nam, "Antioxidant properties of xanthones extracted from the pericarp of *Garcinia mangostana* (Mangosteen): a theoretical study," *Chemical Physics Letters*, vol. 625, pp. 30–35, 2015.

[46] Z. Marković, J. Dorović, Dimitrić M. J. M., R. Biocanin, and D. Amic, "Comparative density functional study of antioxidative

activity of the hydroxybenzoic acids and their anions," *Turkish Journal of Chemistry*, vol. 40, no. 3, pp. 499–509, 2016.

[47] G. Mazzone, N. Malaj, A. Galano, N. Russo, and M. Toscano, "Antioxidant properties of several coumarin-chalcone hybrids from theoretical insights," *Royal Society of Chemistry Advances*, vol. 5, no. 1, pp. 565–575, 2015.

[48] G. Mazzone, N. Russo, and M. Toscano, "Antioxidant properties comparative study of natural hydroxycinnamic acids and structurally modified derivatives: computational insights," *Computational and Theoretical Chemistry*, vol. 1077, pp. 39–47, 2016.

Preparation and Performance of an Oil-Soluble Polyethylene Wax Particles Temporary Plugging Agent

Cuixia Li, Xiaoping Qin, Liangchuan Li, Fei Yan, Jinzhong Wang, and Yixin Zhang

Drilling and Production Technology Research Institute, PetroChina Jidong Oilfield Company, Tangshan 063000, China

Correspondence should be addressed to Xiaoping Qin; 948801727@qq.com

Academic Editor: Carola Esposito Corcione

In order to solve the problem of low strength, low permeability recovery rate, and narrow suitable temperature range of the oil-soluble temporary plugging agent, the idea of preparing the oil-soluble polyethylene wax particles temporary plugging agent (OPPTA) using polyethylene wax as the main raw material was proposed. Emulsifier and stabilizer were selected, and the dosage of emulsifier, stabilizer, and stirring speed was optimized. The method for preparing OPPTA was formed. The temporary plugging performance of OPPTA for the natural core was studied. The plugging rate and the recovery rate of core permeability after plugging removal of OPPTA were 99.9% and 90.1%, respectively. The temporary plugging performance of OPPTA was better than that of a conventional oil soluble temporary plugging agent W-11.

1. Introduction

When a workover operation is performed in a reservoir with fractures or dominant percolation channels, in order to prevent the workover fluid entering the reservoir to form a blockage, the temporary plugging agent is used to temporarily block the fractures or the dominant percolation channels [1–3]. After the completion of the workover, the temporary plugging agent loses the plugging effect, and the permeability of the fractures and the dominant percolation channels can be restored, so as to realize the protection of the reservoir [4–12].

At present, the temporary plugging technology mainly includes the selective temporary plugging, crude oil thickening temporary plugging, shielding temporary plugging, and so on. The oil-soluble temporary plugging agent is a selective temporary plugging because it can be dissolved in oil and taken out of the reservoir during plugging removal, the permeability of the reservoir is recovered, and the damage to the reservoir is small. Therefore, the oil-soluble temporary plugging agent has a good application prospect. The oil-soluble resin temporary plugging agent is mainly made of petroleum resin products, such as oil-soluble phenolic resin, rosin and rosin modifier, and polyurethane, with higher melting point. This kind of material is insoluble in water, but it has good solubility in oil, so it is less harmful to the reservoir. However, the price of this kind of material is higher, and it is difficult to be popularized and applied on a large scale. Because of the brittleness and softening point of this kind of material, the temporary plugging effect for the oil wells with high temperature, fracture, or dominant percolation channel is usually not ideal, and the plugging strength needs to be further strengthened. In addition, the melting point of the oil-soluble temporary plugging agent is generally fixed at present. So its application range is small [13–15].

Polyethylene wax (PEW), also known as high molecular wax, has the following characteristics: (1) high melting point, (2) miscible with crude oil at high temperature, (3) controllable particle size, (4) high hardness, and (5) stable chemical properties. Because the melting point of polyethylene wax varies with the molecular weight, the temporary plugging can be carried out at different reservoir temperatures by controlling the molecular weight of polyethylene wax [16–23]. However, the existing production technology can not produce the oil-soluble polyethylene wax particles temporary plugging agent (OPPTA) which can meet the temporary plugging demand of the reservoir with fractures or dominant percolation channels.

2. Experimental

2.1. Reagents and Instruments. Polyethylene wax (PEW-95, PEW-110, PEW-125, PEW-140, and PEW-160; median particle size 2 mm) was purchased in Yangzhou New Material Co., Ltd. The melting points of PEW-95, PEW-110, PEW-125, PEW-140, and PEW-160 were 95°C, 110°C, 125°C, 140°C, and 160°C, respectively. Span-60 (S-60), Tween-20 (T-20), sodium hexadecyl sulphate (AS-1), sodium dodecyl sulphate (SDS), two methyl silicone oil, partially hydrolyzed polyacrylamide (HPAM, molecular weight 25 million, hydrolysis degree 20%), and guar gum (TR) were purchased from the Sichuan Kelong Chemical Reagent Factory, and all the reagents were analytically pure and were not purified before use. Hydrochloric acid of mass concentration of 10% and hydrofluoric acid of mass concentration of 10% were purchased from Sichuan Kelong Chemical Reagent Factory. The oil-soluble temporary plugging agent W-11 and stabilizer A-1 were all self-made in the laboratory.

The ion compositions of the formation water and injection water are shown in Table 1.

The parameters of natural cores are shown in Table 2.

The natural cores with the artificial fracture are shown in Figure 1.

The main instruments were the particle size analyzer, multifunction core displacement test device, electronic balance, microscope, and German IKA top mechanical stirrer.

2.2. Performance Evaluation of Polyethylene Wax

2.2.1. Oil Solubility of Polyethylene Wax. 100 g polyethylene wax PEW-95 and 300 ml two methyl silicone oil were added to the beaker with a capacity of 1000 ml. The quality of undissolved PEW-95 at 75°C, 85°C, and 95°C was measured. The oil soluble rate of polyethylene wax was calculated by the following formula:

$$y = \frac{m_1 - m_2}{m_1} \times 100, \quad (1)$$

where y is the oil soluble rate of polyethylene wax (%), m_1 is the quality of the added polyethylene wax (g), and m_2 is the quality of undissolved polyethylene wax (g). According to the above methods, the oil-soluble rate of polyethylene wax PEW-110, PEW-125, PEW-140, and PEW-160 was determined at 10°C below the melting point, 20°C below the melting point, and melting point, respectively.

2.2.2. Acid Solubility of Polyethylene Wax. 100 g PEW-95 and 300 ml hydrochloric acid with a mass concentration of 10% (or the mixed solution of hydrochloric acid and hydrofluoric acid, hydrochloric acid with a mass concentration of 10%, and hydrofluoric acid with a mass concentration of 3%) were added to a beaker, and then, the mixture was stirred at room temperature for 24 h. The quality of insoluble polyethylene wax was determined. The acid solving rate of polyethylene wax was calculated by the following formula:

Table 1: Ion compositions of injected water and formation water.

Ion	$Na^+ + K^+$	Ca^{2+}	Mg^{2+}	Cl^-	SO_4^{2-}	HCO_3^-
Injected water ion content (%)	0.0616	0.0058	0.0032	0.0420	0.0103	0.1117
Formation water ion content (%)	0.2235	0.0081	0.0020	0.2798	0.0069	0.1380

$$Y = \frac{M_1 - M_2}{M_1} \times 100, \quad (2)$$

where Y is the acid solving rate of polyethylene wax (%), M_1 is the quality of polyethylene wax added (g), and M_2 is the quality of insoluble polyethylene wax (g).

2.3. Preparing Method of OPPTA. OPPTA was prepared through the following method:

(1) A certain amount of the emulsifier and 38 g polyethylene wax (PEW-95) were added to a flask and heated until polyethylene wax was completely melted.

(2) A certain amount of the stabilizer, 0.6 g dispersants T-20, and 58.8 g injected water were added to a beaker and stirred until complete dissolving, and then the mixture solution was heated to the temperature of the polyethylene wax melting point.

(3) At a certain stirring speed, 10 g mixture solution was dripped into the melted polyethylene wax, and the stirring time was 2 minutes. Then, the rest of the mixture solution was added to the melted polyethylene wax, and the stirring time was 20 minutes. The products were cooled at room temperature.

2.4. Optimization of the Emulsifier and Its Dosage. The effects of different emulsifiers (S-60, T-20, AS-1, and SDS) on the particle size and stability of OPPTA were studied. The dosage of the emulsifier was 2 g. The stirring speed was 800 r/min. OPPTA was prepared in accordance with the above preparation method, and no stabilizer was used.

After the emulsifier was selected, the influence of emulsifier dosage on the particle size and stability of OPPTA was studied by changing the dosage of the emulsifier.

2.5. Optimization of Stirring Speed. The influence of stirring speed on the particle size and stability of OPPTA was studied. Emulsifier was 2 g S-60. At different stirring speeds, OPPTA was prepared in accordance with the above preparation method, and no stabilizer was used.

2.6. Optimization of the Stabilizer and Its Dosage. The influence of the stabilizers (HPAM, TR, and A-1) on the particle size and stability of OPPTA was studied. Emulsifier

Table 2: The parameters of natural cores with the artificial fracture.

Core number	Length (cm)	Diameter (cm)	Porosity (%)	Pore volume, PV (ml)	Permeability (mD)	Temporary plugging agent
NP-7	5.76	2.51	18.1	5.2	421.3	OPPTA
NP-8	5.41	2.52	18.7	5.0	415.7	W-11

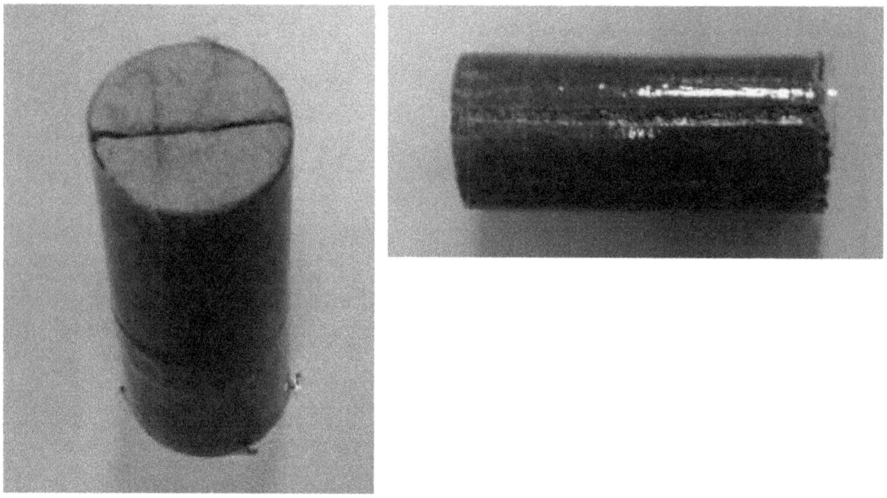

Figure 1: The natural cores with the artificial fracture (NP-7).

was 2 g S-60, and the stirring speed was 900 r/min. The dosage of the stabilizer was 0.5 g. OPPTA was prepared in accordance with the above preparation method.

After the stabilizer was selected, the influence of stabilizer dosage on the particle size and stability of OPPTA was studied by changing the dosage of the stabilizer.

2.7. Temporary Plugging Ability of OPPTA. The natural core with artificial fracture was used to study temporary plugging performance of OPPTA. The experimental procedures for determining the temporary plugging ability of OPPTA on the core were as follows [24, 25]:

(1) 100 g OPPTA was prepared using PEW-95 according to the above preparation method. The emulsifier was 2 g S-60, the stirring speed was 900r/min, and the stabilizer was 0.6 g A-1.

(2) At 65°C, the formation water was injected into the natural core at the rate of 0.2 ml/min until the injection pressure was basically stable; OPPTA was injected into the core positively at the rate of 0.2 ml/min, and the volume of OPPTA was 6 pore volume (PV); and the injected water was injected positively into the core until the injection pressure was essentially smooth.

(3) At 65°C, the formation water was injected into the core in the reverse direction at the rate of 0.2 ml/min until the injection pressure was basically stable; the kerosene was injected in the reverse direction into the core at the rate of 0.2 ml/min until the injection pressure was basically stable; and the formation water was injected in the reverse direction into the core at the rate of 0.2 ml/min until the injection pressure was basically stable.

The permeability of the core could be calculated by the Darcy formula in different experimental process. The plugging rate of OPPTA in the core was calculated by Formula (3), and the permeability recovery rate after plugging was calculated by Formula (4):

$$D = \frac{K_b - K_a}{K_b} \times 100, \quad (3)$$

$$Q = \frac{K_r}{K_b} \times 100, \quad (4)$$

where D is the plugging rate (%), Q is the permeability recovery rate (%), K_b is the core permeability before OPPTA injection (mD), K_a is the core permeability after OPPTA injection (mD), and K_r is the core permeability after plugging removal (mD).

3. Results and Discussion

3.1. Oil Solubility of Polyethylene Wax. The oil solubility of polyethylene wax with different melting points at different temperatures had been studied. The results are shown in Table 3. Polyethylene wax had good oil solubility. When the temperature was lower than its melting point 10 or 20 degrees centigrade, the oil solubility could reach above 97%. The melting point of polyethylene wax was related to molecular weight. Polyethylene wax with different melting points could be obtained by controlling the molecular weight of polyethylene wax during its production process. Thus, temporary plugging and plugging removal could be realized

Table 3: The results of oil solubility of polyethylene wax.

PEW	PEW-95			PEW-110			PEW-125			PEW-140			PEW-160		
Temperature (°C)	75	85	95	90	100	110	105	115	125	120	130	140	140	150	160
Undissolved mass (g)	1.52	0.37	0.01	0.98	0.35	0.03	1.31	0.46	0.01	1.72	0.62	0.01	1.64	0.22	0.01
Oil soluble rate (%)	97.97	99.56	99.99	98.91	99.65	99.97	98.75	99.60	99.99	98.57	99.52	99.99	98.83	99.85	99.99

for different temperature reservoirs. Good oil solubility was the key performance of polyethylene wax in the preparation of the temporary plugging agent.

3.2. Acid Solubility of Polyethylene Wax. The solubility of polyethylene wax in different acid solutions was studied. The results are shown in Table 4. The acid dissolution rate of polyethylene wax in different acid solutions was less than 1%. Hydrochloric acid and soil acid had less acid corrosion on polyethylene wax. The polyethylene wax had no reaction with the acid solution, indicating that polyethylene wax could be used for temporary plugging protection of the reservoir, when the acid workover fluid was used. This phenomenon could be explained by the very stable chemical properties of polyethylene wax at room temperature and atmospheric pressure.

Table 4: The results of acid solubility of polyethylene wax.

PEW	Acid	Undissolved mass (g)	Acid dissolution rate (%)
PEW-95	/	100	/
	10% HCl	99.3	0.70
	10% HCl + 3% HF	99.1	0.91
PEW-110	/	100	/
	10% HCl	99.5	0.50
	10% HCl + 3% HF	99.2	0.81
PEW-125	/	100	/
	10% HCl	99.4	0.60
	10% HCl + 3% HF	99.1	0.91
PEW-140	/	100	/
	10% HCl	99.7	0.30
	10% HCl + 3% HF	99.1	0.91
PEW-160	/	100	/
	10% HCl	99.4	0.60
	10% HCl + 3% HF	99.2	0.81

3.3. Optimization of the Emulsifier and Its Dosage. The influence of the emulsifier on the particle size of OPPTA was shown in Table 5. The emulsifying effect of S-60 was the best in the four kinds of emulsifiers. The OPPTA had a particle size of 51.5 μm and a stable time of 28 hours.

The influence of emulsifier S-60 dosage on the particle size and stability time of OPPTA was shown in Figure 2. With the increase of emulsifier S-60 dosage, the median diameter of OPPTA gradually was decreased and the stability time was prolonged. When the emulsifier S-60 dosage was increased to 2%, the particle size of the temporary plugging agent was 51.5 μm, and the stability time was 28 hours. Subsequently, increasing the amount of S-60 had little influence on the particle size and stability time of the temporary plugging agent.

Table 5: Influence of the emulsifier on particle size and stability time of the temporary plugging agent.

Emulsifiers	S-60	T-20	AS-1	SDS
Median particle size (μm)	51.5	56.9	79.2	97.3
Stable time (h)	28	24	20	12

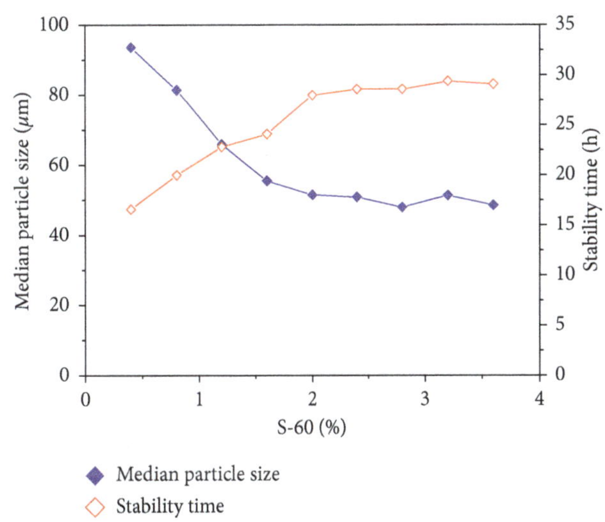

Figure 2: Influence of emulsifier S-60 dosage on particle size and stability time of the temporary plugging agent.

3.4. Optimization of Stirring Speed. The influence of stirring speed on the particle size of OPPTA was shown in Figure 3. With the increase of stirring speed, the dispersion effect of polyethylene wax was better, and the particle size of OPPTA became smaller. When the stirring speed was 900 r/min, the corresponding particle size was 50.7 μm, and the stability time was 28.8 h.

3.5. Optimization of the Stabilizer and Its Dosage. The influence of stabilizers on the particle size of OPPTA was shown in Table 6. The stability effect of A-1 was the best in the three kinds of stabilizers. The particle size of OPPTA prepared by A-1 was 50.2 μm, and the stability time was 36 hours.

The influence of stabilizer A-1 dosage on the particle size and stability time of OPPTA was shown in Figure 4. The effect of stabilizer A-1 dosage on the median diameter of OPPTA was small, and the median particle size was kept at about 50 μm. When the amount of stabilizer A-1 was

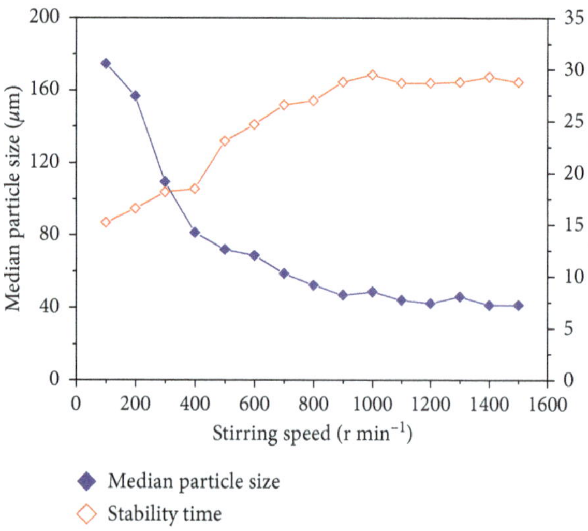

FIGURE 3: Effect of stirring speed on particle size and stability time of the temporary plugging agent.

TABLE 6: Influence of the stabilizer on particle size and stability time of the temporary plugging agent.

Stabilizers	HPAM	TR	A-1
Median particle size (μm)	52.2	51.4	50.2
Stability time (h)	32	33	36

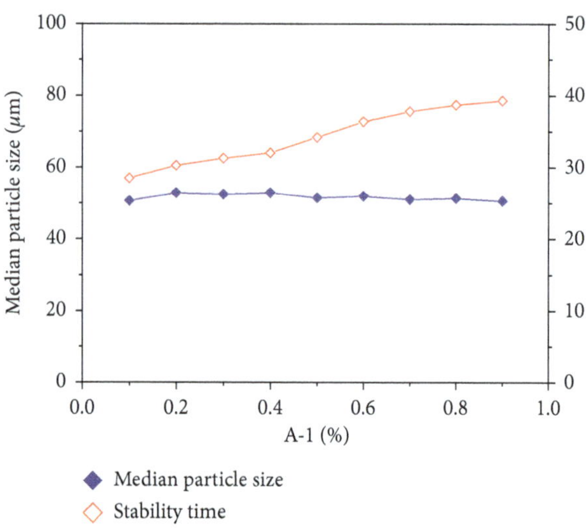

FIGURE 4: Influence of stabilizer A-1 dosage on particle size and stability time of the temporary plugging agent.

increased to 0.6%, the stability time of the temporary plugging agent reached 36 hours, and it could meet the requirements of conventional workover operation. Considering the cost and injection of the temporary plugging agent, the optimum dosage of stabilizer A-1 was 0.6%.

3.6. Preparing Method of OPPTA. 2 g S-60 and 38 g polyethylene wax were added to a flask and heated until polyethylene wax was completely melted. 0.6 g A-1, 0.6 g dispersants T-20, and 58.8 g injected water were added to

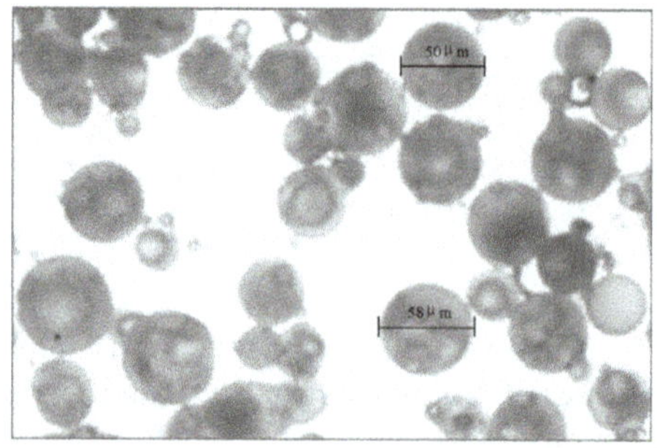

FIGURE 5: The micromorphology of OPPTA at room temperature.

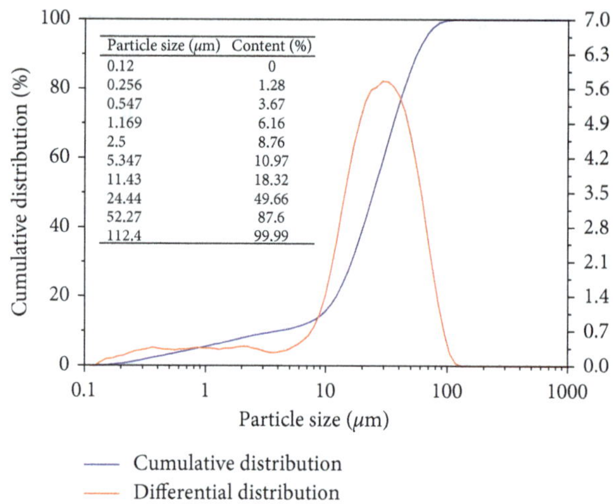

FIGURE 6: The particle size distribution diagram of OPPTA at room temperature.

a beaker and stirred until dissolved completely, and then the mixture solution was heated to the temperature of the polyethylene wax melting point. At the stirring speed of 900 r/min, 10 g mixture solution was dripped into the melted polyethylene wax, and the stirring time was 2 minutes. Then, the rest of the mixture solution was added to the melted polyethylene wax, and the stirring time was 20 minutes. The products were cooled at room temperature.

The micromorphology of OPPTA was observed under a microscope at room temperature, as shown in Figure 5. The particle size distribution diagram of OPPTA was obtained using the laser particle size analyzer at room temperature, as shown in Figure 6. The experimental results showed that OPPTA was polyethylene wax microspheres with good sphericity. Microspheres with a diameter of 11.43 to 52.27 μm accounted for 69.28%.

3.7. Temporary Plugging Effect of OPPTA on Core. As shown in Figure 7, when the permeability of the core was measured by water injection, the oil-soluble polyethylene wax particles temporary plugging agent was injected. The injection pressure continued to rise, and the injection pressure increased to 1.1 MPa when the 6PV OPPTA was injected. Then, the positive injection was carried out, and the pressure remained at about 0.49 MPa. The core corresponding permeability was 0.6 mD, and the plugging rate was 99.9%. In the process of

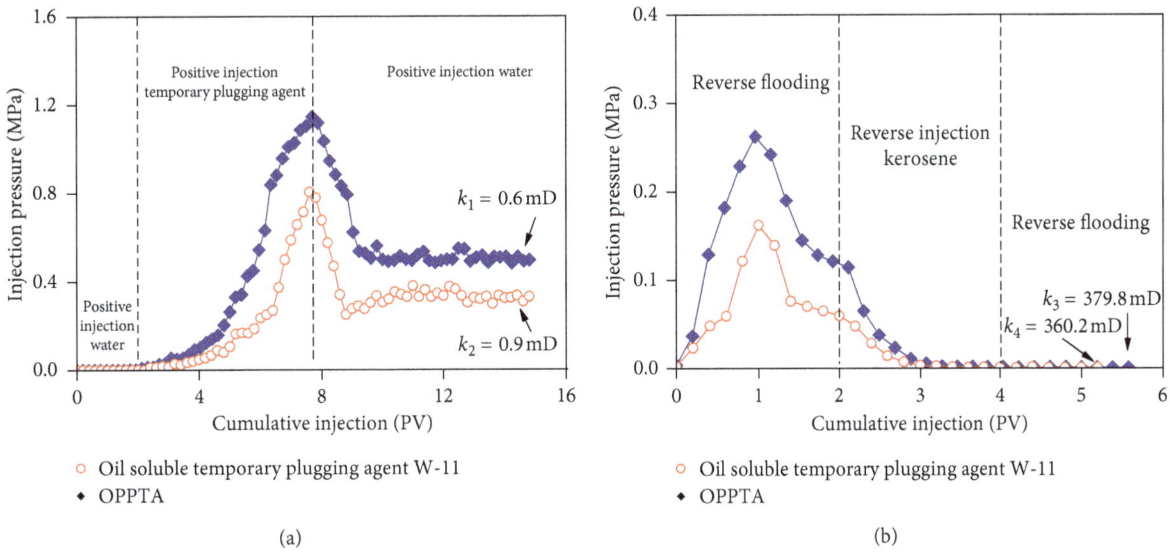

FIGURE 7: Temporary plugging effect of the temporary plugging agent on the crack: (a) sealing process; (b) plugging removal process.

plugging removal, reverse water injection was first carried out, and the injection pressure was kept at about 0.12 MPa; using kerosene to remove plugging, the injection pressure reduced to about 0.0015 MPa. In the reverse flooding process, the injection pressure basically decreased to the pressure level of the forward water flooding. The recovery rate of core permeability of OPPTA reached 90.1% after plugging removal. Under the same conditions, the plugging rate of the temporary plugging agent W-11 was 99.8%, and the recovery rate of core permeability after plugging removal was 86.6%. The experimental results showed that the plugging rate and the recovery rate of core permeability of OPPTA were higher than that of W-11 under the same experimental conditions. These experimental data confirmed that the OPPTA had a better temporary plugging effect.

4. Conclusion

The findings of the study are as follows:

(1) A method for preparing OPPTA using polyethylene wax as the main raw material was put forward, which solved the problems of low plugging strength of the existing temporary plugging agent and low recovery rate of core permeability after plugging removal.

(2) Compared with the conventional oil-soluble temporary plugging agent W-11, the plugging rate and the recovery rate of core permeability after plugging removal of OPPTA were higher, which indicated that it had a better temporary plugging effect.

(3) By changing the preparation conditions, a series of OPPTA with different particle sizes could be prepared to meet the temporary plugging requirements of different reservoirs.

Conflicts of Interest

The authors declare that there are no conflicts of interest.

Acknowledgments

The authors are grateful to the Experimental Center of Drilling Technology Research Institute for assistance in temporary plugging ability evaluation experiment of OPPTA.

References

[1] Y. Kang, L. You, X. Xu et al., "Prevention of formation damage induced by mud lost in deep fractured tight gas reservoir in western Sichuan basin," *Journal of Canadian Petroleum Technology*, vol. 51, no. 1, pp. 46–51, 2012.

[2] Y. Tang, T. Yildiz, E. Ozkan et al., "Effects of formation damage and high velocity flow on the productivity of perforated horizontal wells," *SPE Reservoir Evaluation and Engineering*, vol. 8, no. 4, pp. 315–324, 2005.

[3] X. Tang, L. Yang, Y. Liu et al., "A new in-depth fluid diverting agent of inorganic gel coating," *Petroleum Exploration and Development Online*, vol. 39, no. 1, pp. 82–87, 2012.

[4] A. Moradi, "A review of thermally stable gels for fluid diversion in petroleum production," *Journal of Petroleum Science and Engineering*, vol. 26, no. 1-4, pp. 1-10, 2000.

[5] A. C. Soares and F. H. E. A. Ferreira Jr, "An experimental study for mechanical formation damage," *SPE Reservoir Evaluation and Engineering*, vol. 5, no. 6, pp. 480–487, 2002.

[6] A. Lohne, L. Han, C. V. D. Zwaag et al., "Formation-damage and well-productivity simulation," *SPE Journal*, vol. 15, no. 3, pp. 751–769, 2010.

[7] L. Zhao, Y. Pei, G. Du et al., "Thermo-responsive temporary plugging agent based on multiphase transitional supramolecular gel," *Petroleum Chemistry*, vol. 58, no. 1, pp. 94–101, 2018.

[8] A. S. Markowski, "A review of layer of protection analysis techniques for oil and gas industry," *International Journal of Oil Gas and Coal Technology*, vol. 5, no. 1, pp. 66–79, 2012.

[9] X. Wu, H. Pu, K. Zhu et al., "Formation damage mechanisms and protection technology for Nanpu nearshore tight gas reservoir," *Journal of Petroleum Science and Engineering*, vol. 158, pp. 509–515, 2017.
[10] M. Lei, W. Huang, N. Li et al., "The damage mechanism of oil-based drilling fluid for tight sandstone gas reservoir and its optimization," *Journal of Petroleum Science and Engineering*, vol. 158, pp. 616–625, 2017.
[11] H. Mao, Z. Qiu, Z. Shen et al., "Hydrophobic associated polymer based silica nanoparticles composite with core–shell structure as a filtrate reducer for drilling fluid at ultra-high temperature," *Journal of Petroleum Science and Engineering*, vol. 129, pp. 1–14, 2015.
[12] H. J. Khan, M. S. Mirabolghasemi, H. Yang et al., "Study of formation damage caused by retention of bi-dispersed particles using combined pore-scale simulations and particle flooding experiments," *Journal of Petroleum Science and Engineering*, vol. 158, pp. 293–308, 2017.
[13] K. S. Pedersen and H. P. Rønningsen, "Influence of wax inhibitors on wax appearance temperature, pour point, and viscosity of waxy crude oils," *Energy and Fuels*, vol. 17, no. 2, pp. 321–328, 2003.
[14] H. Li and J. Gong, *The effect of pressure on wax disappearance temperature and wax appearance temperature of water cut crude oil*, Mccarthy, 2010.
[15] D. Liu, X. Shi, X. Zhong et al., "Properties and plugging behaviors of smectite-superfine cement dispersion using as water shutoff in heavy oil reservoir," *Applied Clay Science*, vol. 147, pp. 160–167, 2017.
[16] M. J. Hato and A. S. Luyt, "Thermal fractionation and properties of different polyethylene/wax blends," *Journal of Applied Polymer Science*, vol. 104, no. 4, pp. 2225–2236, 2007.
[17] X. Xu, H. Xiao, Y. Guan et al., "Permanent antistatic polypropylene based on polyethylene wax/polypropylene wax-grafting sodium acrylate," *Journal of Applied Polymer Science*, vol. 126, no. 1, pp. 83–90, 2012.
[18] A. S. Luyt and M. J. Hato, "Thermal and mechanical properties of linear low-density polyethylene/low-density polyethylene/wax ternary blends," *Journal of Applied Polymer Science*, vol. 96, no. 5, pp. 1748–1755, 2005.
[19] A. Liskova and H. Berghmans, "Phase-separation phenomena in the polymerization of styrene in the presence of polyethylene wax," *Journal of Applied Polymer Science*, vol. 91, no. 4, pp. 2234–2243, 2010.
[20] J. Taheri-Shakib, M. Rajabi-Kochi, E. Kazemzadeh et al., "A comprehensive study of the impact of wax compositions on the wax appearance temperature (WAT) of some Iranian crude oils: an experimental investigation," *Journal of Petroleum Science and Engineering*, vol. 165, pp. 67–80, 2018.
[21] J. Taheri-Shakib, A. Shekarifard, and H. Naderi, "Characterization of the wax precipitation in Iranian crude oil based on Wax Appearance Temperature (WAT): Part 1. The influence of electromagnetic waves," *Journal of Petroleum Science and Engineering*, vol. 161, pp. 530–540, 2018.
[22] J. C. Via, C. L. Braue, and L. T. Taylor, "Supercritical fluid fractionation of a low molecular weight, high-density polyethylene wax using carbon dioxide, propane, and propane-modified carbon dioxide," *Analytical Chemistry*, vol. 66, no. 5, pp. 603–609, 1994.
[23] I. Krupa and A. S. Luyt, "Mechanical properties of uncrosslinked and crosslinked linear low-density polyethylene/wax blends," *Journal of Applied Polymer Science*, vol. 81, no. 4, pp. 973–980, 2001.
[24] N. Lai, D. Wan, Z. Ye et al., "A water-soluble acrylamide hydrophobically associating polymer: synthesis, characterization, and properties as EOR chemical," *Journal of Applied Polymer Science*, vol. 129, no. 4, pp. 1888–1896, 2013.
[25] N. Lai, X. Qin, Z. Ye et al., "The study on permeability reduction performance of a hyperbranched polymer in high permeability porous medium," *Journal of Petroleum Science and Engineering*, vol. 112, no. 3, pp. 198–205, 2013.

11

Mechanism of Chlorination Process: From Propanoic Acid to α-Chloropropanoic Acid and Byproducts Using Propanoic Anhydride as Catalyst

Da Xue,[1] Yuna Li,[1] Jisheng Liu,[1] Chuan Gao,[2] Jianwei Xue,[1] Fuxiang Li,[1] and Zhiping Lv[1]

[1]*College of Chemistry and Chemical Engineering, Taiyuan University of Technology, Taiyuan 030024, China*
[2]*Faculty of Arts and Social Sciences, The University of Sydney, Sydney, NSW, Australia*

Correspondence should be addressed to Jianwei Xue; xuejianwei@tyut.edu.cn

Academic Editor: Gulaim A. Seisenbaeva

This article reports on findings regarding the mechanism of chlorination process. In this experiment, propanoic acid was chlorinated to α-chloropropanoic acid in a lab-scale glass tube reactor operating at 130°C. Propanoic anhydride and concentrated sulfuric acid were, respectively, used as the catalyst and the promoter. This experiment adopted the DFT method to calculate the activation energy of routes for the synthesis α-chloropropanoic acid, β-chloropropanoic acid, α,α-dichloropropanoic acid, and α,β-dichloropropanoic acid. The results showed that the main route of α-chloropropanoic acid was formed through an ionic mechanism when propanoic anhydride was used as the catalytic agent. Activation energy of 1-propen-1-ol,1-chloro, which was formed from 1-prop-anol,1-chloro-, was the highest in the process of ionic mechanism. In addition, α,α-dichloropropanoic acid was formed via a consecutive ionic chlorination path from α-chloropropanoic acid. β-Chloropropanoic acid was produced from propanoic acid through a chlorination radical mechanism. α,β-Dichloropropanoic acid was formed via a consecutive radical chlorination path.

1. Introduction

α-Chloropropanoic acid, as a kind of active organic chemical, has been widely used to produce pesticides, dyes, chemicals of agriculture and forestry, intermediates, and so forth. It plays an important role in the fine chemical industry. Due to the existence of a chlorine atom in the α-chloropropanoic acid's molecular structure, its acidic density and reactivity are stronger than those of propanoic acid, which is more active in chemical reactions [1–4].

Sioli et al. [5] proposed the acid-catalyzed enolization of ionic mechanism for the first time to explore the synthesis of α-chloropropanoic acid. Paatero et al. [6] found that acid catalysis was a process of autocatalysis when using chlorosulfonic acid as an enolizing agent. Salmi et al. [7, 8] and Mäki-Arvela et al. [9] identified an autocatalysis effect. They found that the monochloropropanoic acid created in the reaction was much stronger than the propanoic acid [10], which led to an increasing rate of acid catalysis and autocatalysis. The autocatalytic kinetic was explained by a reaction involving the acid-catalyzed enolization of the key intermediate, propanoyl chloride, as a rate-determining step.

Ogata et al. [11] and Ham et al. [12] found that the generation of β-chloropropanoic acid was by radical chlorination in the α-chloropropanoic acid synthesis process. Paatero et al. [6] identified that the formation route of α,α-dichloropropanoic acid was the same as α-chloropropanoic acid formation, the parallel acid-catalyzed enolization mechanism. Mäki-Arvela et al. [10] also found that α,α-dichloropropanoic acid generation has two routes: the ionic mechanism and the radical mechanism. There exists a competition between these two mechanisms. α,β-Dichloropropanoic acid was formed through further chlorination of α-chloropropanoic acid.

However, the reaction mechanisms of byproducts, for instance, the α,α-dichloropropanoic acid, have received different opinions. Thus, this article aims to report on findings on the reaction mechanism of chlorination products through experimenting with propanoic acid chlorinated by using the density function method with $Dmol^3$ program of Materials

Studio 5.5. The results could contribute to production optimization of α-chloropropanoic acid.

2. Experimental and Calculation Section

2.1. Experimental. Propanoic acid chlorination was carried out in a homemade glass tube equipped with a magnetic stirring apparatus and an oil bath heater with a temperature control. Chlorine gas was dispersed into the reaction mixture which was metered by a rotameter. A reflux condenser equipped with low temperature cooling circulating pump was placed on the top of the reactor. The gaseous compounds passed through water, sodium bicarbonate solution, and concentrated sodium hydroxide solution before being released to the air.

The propanoic acid to be chlorinated was placed in the reaction vessel. A slightly excess chlorine feed was introduced, and the liquid phased was heated to the desired reaction temperature. A certain amount of propanoic anhydride was added. Samples of the liquid phase were analyzed with gas chromatograph after esterification [10].

2.2. Calculation Methods. Density functional theory method quantum chemistry calculation was adopted and the calculations were performed by using the $Dmol^3$ program mounted on Materials Studio 5.5 package. All the reactants and products in the chlorination reaction were optimized and their stable structures were obtained at the same time. All transition states of the reactions were trialed and searched carefully by using an LST/QST method, getting the primitive activation energy and the heat of reaction. The nonlocal exchange and correlation energies were calculated by the PW91 function of the generalized gradient approximation (GGA) [13] and a double numerical plus polarization (DNP) basis set in the level of All Electron. The convergence criteria included threshold values of 1×10^{-5} Hartree, 0.02 Hartree per nm, and 0.0005 nm for energy, force, and displacement convergence, respectively, while the self-consistent field (SCF) density convergence threshold value was specified as 1×10^{-6} Hartree. Fermi smearing of 0.005 Hartree was employed to improve the computational performance, and the solvent effect was also considered during the calculations. The geometries of all stationary points were optimized at this level.

3. Results and Discussion

3.1. The Experimental Results. The experiment first investigated the effect of propanoic acid catalytic chlorine (see Figure 1) in the condition of 15 ml propanoic acid, 5 ml propanoic anhydride, 130°C reaction temperature, and 40 ml/min chlorine flow rate with micro excess. The generation of α-chloropropanoic acid was little in the first 30 min and increased rapidly during the period from 30 min to 150 min. Its generation did not increase obviously thereafter. The generation of β-chloropropanoic acid increased steadily in the first 150 min but increased slowly from 150 min to 360 min. The generation of α,α-dichloropropanoic acid increased quite slowly before 250 min and increased rapidly in the later reaction.

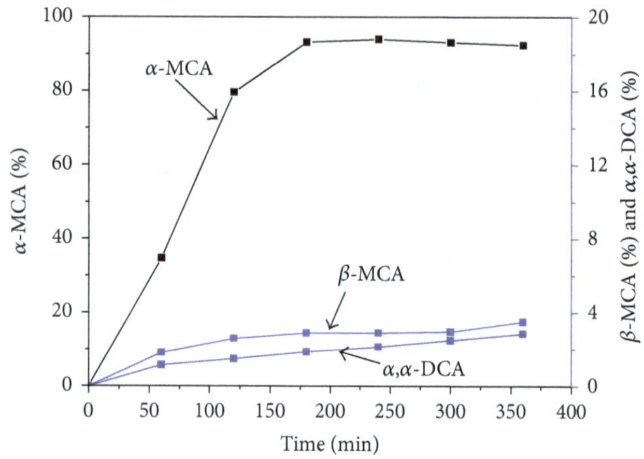

Figure 1: The reaction products change over time at 130°C.

In order to verify the catalytic principle, 0.4 ml concentrated sulfuric acid was added to propanoic anhydride catalytic chlorination to observe the effect of the chlorination reaction, compared with the control experiment. Chloride products were analyzed by gas chromatography (Table 1). It was found that the generation of α-chloropropanoic acid increased obviously, while the generation of both β-chloropropanoic acid and α,α-dichloropropanoic acid reduced after concentrated sulfuric acid was added. Also, the generation of α-chloropropanoic acid (Table 1, Entry α-MCA (%)) increased rapidly in the first 90 min, increased slowly from 90 min to 150 min, and barely increased from 150 min to 180 min. The generation of β-chloropropanoic acid (Table 1, Entry β-MCA (%)) shrank with time. The generation of α,α-dichloropropanoic acid (Table 1, Entry α,α-DCA (%)) increased in the first 30 min but reduced from 30 min to 180 min.

3.2. Synthesis Mechanism of α-Chloropropanoic Acid. In order to investigate the formation route of α-chloropropanoic acid, two possible reaction paths were constructed: the radical chlorination mechanism and ionic chlorination mechanism (Scheme 1). Path 1 is the formation route of α-chloropropanoic acid via radical chlorination mechanism (Scheme 1; see (1)). Path 2 is the formation route of α-chloropropanoic acid via ionic chlorination mechanism (Scheme 1; see (2)–(6)).

The configurations of reactants and products were built and the geometry structure was optimized. Transition states were investigated and calculated on the basis of reactants and products. Each reaction transition state was derived from the potential energy surface at the highest point of the potential energy corresponding to the configuration. Frequency analysis showed that these configurations were the only virtual frequency. Energy of reactants, products, and transition states to relevant primitives and the corresponding virtual frequency of the transition state are presented (Table 2). At the same time, energy potential diagrams of the

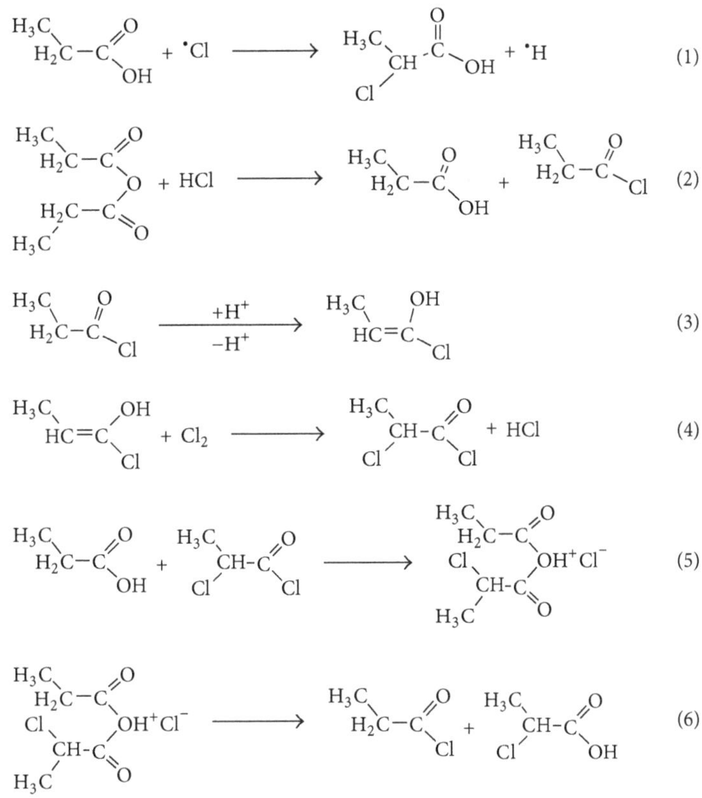

Scheme 1: Paths of the generated α-chloropropanoic acid.

Table 1: Different chloride products change over time without concentrated sulfuric acid and joining concentrated sulfuric acid.

Catalyst	T (min)	Propanoic acid (%)	α-MCA (%)	β-MCA (%)	α,α-DCA (%)
20 mol% propanoic anhydride	30	80.05	18.23	1.24	0.47
	60	62.56	34.63	1.83	0.98
	90	39.74	56.73	2.11	1.42
	120	11.13	84.84	2.47	1.56
	150	5.80	89.79	2.63	1.78
	180	0.51	94.57	2.91	2.01
20 mol% propanoic acid anhydride + concentrated sulfuric acid	30	46.52	52.18	0.59	0.71
	60	30.66	67.71	0.82	0.81
	90	13.16	85.04	0.93	0.87
	120	5.8	92.18	1.04	0.98
	150	3.23	94.53	1.18	1.06
	180	0.41	97.22	1.24	1.13

primitive reactions with corresponding reactants, transition states, and products of configuration are shown in Figure 2.

The formation of α-chloropropanoic acid via ionic chlorination mechanism is as follows. First, propanoic anhydride conducted initiation (Scheme 1; see (2)) and formed propanoyl chloride. Its reaction activation energy was 12.54 kJ/mol. Propanoyl chloride then conducted enolization (Scheme 1; see (3)) and formed 1-propen-1-ol,1-chloro, and the corresponding primitive activation energy was 125.73 kJ/mol. Since then, the double bond in 1-propen-1-ol, 1-chloro was apt to react with chlorine (Scheme 1; see (4)) and formed α-chloropropanoyl chloride, and the corresponding primitive activation energy was 20.11 kJ/mol. Finally, the OH and Cl exchange reaction generated α-chloropropanoic acid (Scheme 1; see (5) and (6)) and its activation energy was 85.61 kJ/mol and 90.48 kJ/mol. Therefore, reaction (2) was the catalyst initiation reaction among whole reactions (2)–(6) and propanoic acid anhydride formed a real catalyst, propanoyl chloride. The whole reaction process became an acid catalytic reaction once propanoyl chloride was formed.

TABLE 2: Possible elementary reactions involved in α-chloropropanoic acid and byproducts synthesis together with the activation energies (Ea) and reaction energies (ΔH).

Step	Chemical reaction	Frequency (cm^{-1})	Ea ($kJ\cdot mol^{-1}$)	ΔH ($kJ\cdot mol^{-1}$)
(1)	$C_2H_5COOH + {}^{\bullet}Cl \longrightarrow H^+ + CH_3CHClCOOH$	−237.61	158.37	131.50
(2)	$(C_2H_5CO)_2O + HCl \longrightarrow C_2H_5COOH + C_2H_5COCl$	−118.79	12.54	−9.68
(3)	$C_2H_5COCl + H^+ \longrightarrow H^+ + CH_3CHCClOH$	−117.24	125.73	117.19
(4)	$CH_3C_2HClOH + Cl_2 \longrightarrow HCl + C_2H_4ClCOCl$	−48.63	20.11	−170.74
(5)	$C_2H_4ClCOCl + C_2H_5COOH \longrightarrow C_6H_{10}O_3Cl_2$	−161.81	85.61	19.05
(6)	$C_6H_{10}O_3Cl_2 \longrightarrow C_2H_5COCl + CH_3CHClCOOH$	−138.82	90.48	−116.71
(7)	$C_2H_5COOH + {}^{\bullet}Cl \longrightarrow H^+ + CH_2ClCH_2COOH$	−343.25	217.04	146.69
(8)	$CH_3C_2HClOH + Cl_2 \longrightarrow HCl + CH_2ClCH_2COCl$	−294.33	110.27	−187.88
(9)	$CH_2ClCH_2COCl + C_2H_5COOH \longrightarrow C_6H_{10}O_3Cl_2$	−121.03	188.14	106.79
(10)	$C_6H_{10}O_3Cl_2 \longrightarrow C_2H_5COCl + CH_2ClCH_2COOH$	−208.36	34.54	−122.72
(11)	$CH_3CHClCOOH + {}^{\bullet}Cl \longrightarrow H^+ + CH_3CCl_2COOH$	−758.10	162.64	134.13
(12)	$CH_3CHClCOOH + {}^{\bullet}Cl \longrightarrow H^+ + CH_2ClCHClCOOH$	−486.96	180.08	138.43
(13)	$CH_2ClCH_2COOH + {}^{\bullet}Cl \longrightarrow H^+ + CH_2ClCHClCOOH$	−887.32	178.87	153.22
(14)	$CH_3CHClCOCl + H^+ \longrightarrow H^+ + CH_3CClCClOH$	−1103.93	130.81	117.22
(15)	$CH_3CClCClOH + Cl_2 \longrightarrow HCl + CH_3CCl_2COCl$	−84.36	21.99	−129.19
(16)	$CH_3CCl_2COCl + C_2H_5COOH \longrightarrow C_6H_9O_3Cl_3$	−69.22	82.61	−168.57
(17)	$C_6H_9O_3Cl_3 \longrightarrow CH_3CCl_2COOH + C_2H_5COCl$	−365.41	109.73	−95.67

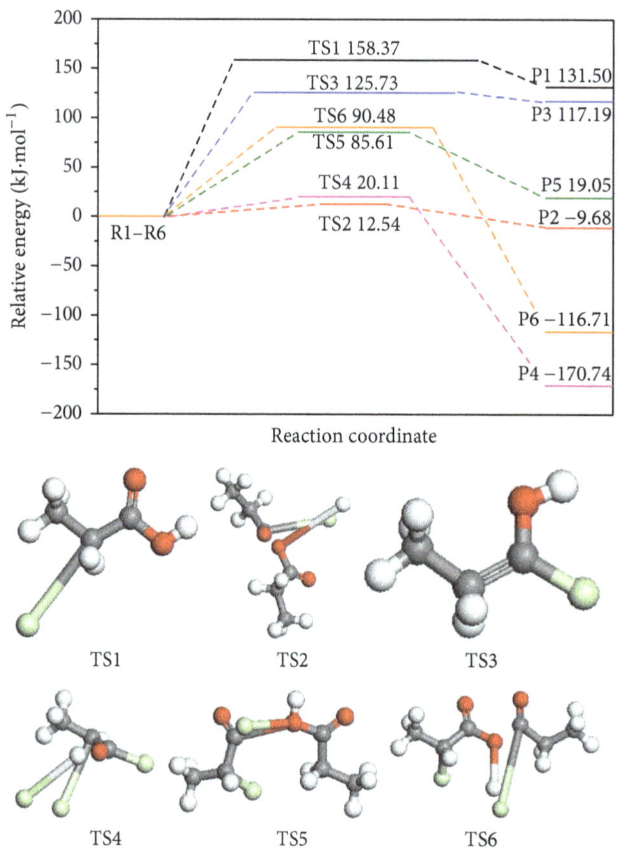

FIGURE 2: Potential energy diagram of α-chloropropanoic acid synthesis together with the initial states, transition states, and final states.

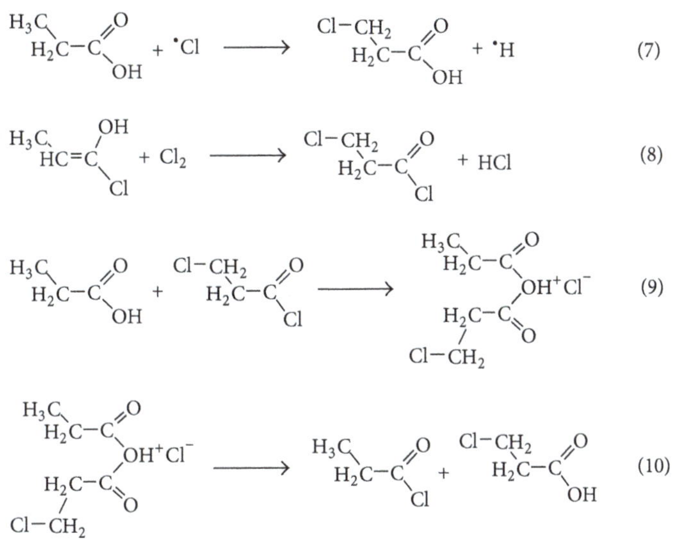

SCHEME 2: Radical chlorination mechanism of byproducts.

More importantly, the acidity of α-chloropropanoic acid was stronger than that of propanoic acid, so the acid catalytic reaction was accelerated.

The activation energy 125.73 kJ/mol (Scheme 1; see (3)) was the highest as the rate-determining step after propanoyl chloride was formed. It is consistent with the results of Paatero. The results also showed that the formation rate of α-chloropropanoic acid was slow at the beginning and was rapid in the middle of the reaction. The reason for it could be that catalyst initiation reaction was slow, and the activation energy of the whole process would reduce once propanoyl chloride was formed. Meanwhile, the rate of acid catalysis and autocatalysis (Scheme 1; see (3)) was accelerated due to the fact that the acidity of α-chloropropanoic acid was stronger than that of propanoic acid. Thus, the generation speed of α-chloropropanoic acid was accelerated. In addition, the generation of α-chloropropanoic acid increased significantly when concentrated sulfuric acid was added. The added H^+ in the reaction system promoted the progress of acid catalysis and autocatalysis (Scheme 1; see (3)). In turn, it led to the result that the reaction was conducted toward the generation of α-chloropropanoic acid. Therefore, it can be drawn from the theoretical calculation and experimental results that the whole process became an acid catalytic reaction when propanoyl chloride was formed. The generation of α-chloropropanoic acid could be accelerated as long as propanoyl chloride acid-catalyzed enolization (Scheme 1; see (3)) was expedited. It could also inhibit the generation of β-chloropropanoic acid and α,α-dichloropropanoic acid.

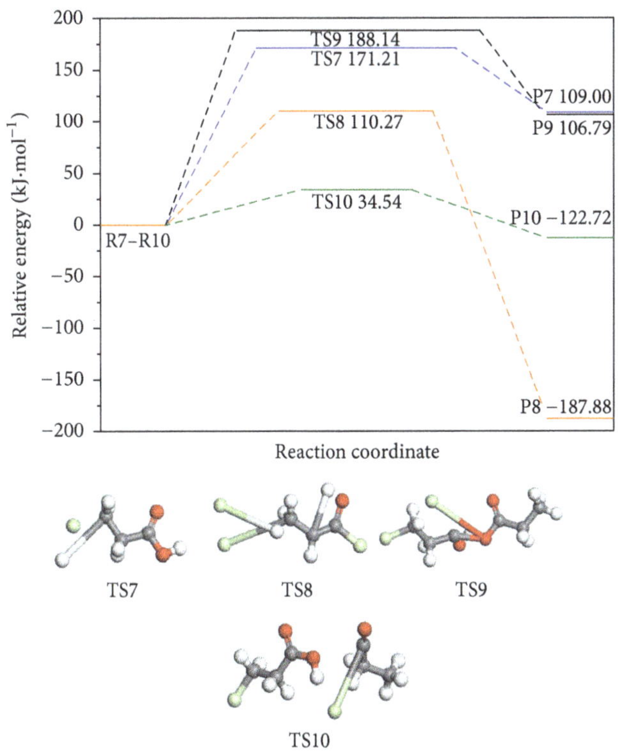

FIGURE 3: Potential energy diagram of the radical chlorination reactions related of byproducts together with the initial states, transition states, and final states.

3.3. Synthesis Ionic Chlorination Mechanism of β-Chloropropanoic Acid. In order to investigate the formation route of β-chloropropanoic acid, two possible reaction paths were constructed: the radical chlorination mechanism and ionic chlorination mechanism (Scheme 2, Figure 3). Path 1 is the formation route of β-chloropropanoic acid via radical chlorination mechanism (Scheme 2; see (7)). Path 2 is the formation route of β-chloropropanoic acid via ionic chlorination mechanism (Scheme 2; see (8)–(10)).

The formation of β-chloropropanoic acid via ionic chlorination mechanism is as follows. First, propanoyl chloride conducted enolization (Scheme 1; see (3)) and formed 1-propen-1-ol,1-chloro, and the corresponding primitive activation energy was 125.73 kJ/mol. Then, the double bond

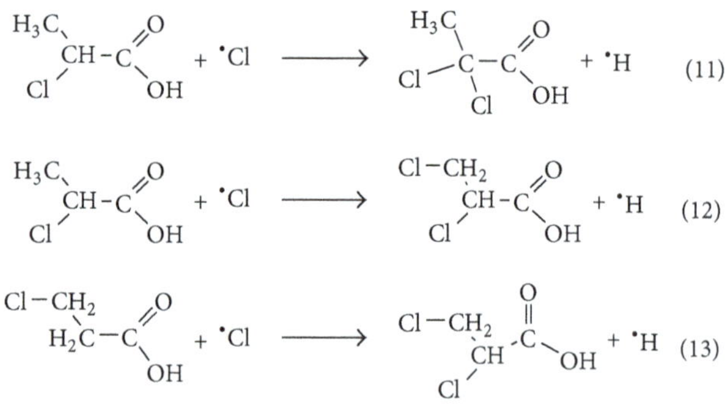

SCHEME 3: The radical chlorination mechanism of byproducts.

in 1-propen-1-ol, 1-chloro was apt to react with chlorine (Scheme 2; see (8)) and formed β-chloropropanoyl chloride, and the corresponding primitive activation energy was 110.27 kJ/mol. Finally, the OH and Cl exchange reaction generated β-chloropropanoic acid (Scheme 2; see (9) and (10)) and its activation energy was 188.14 kJ/mol and 34.54 kJ/mol. Compared to these two kinds of paths which generated β-chloropropanoic acid and α-chloropropanoic acid, the activation energy of the formed β-chloropropanoic acid (Scheme 2; see (9)) was higher than that of the formed α-chloropropanoic acid (Scheme 1; see (6)). Therefore, it is difficult to form β-chloropropanoic acid by ionic chlorination.

3.4. Radical Chlorination Mechanism of Byproducts. The radical chlorination mechanism consists of successive chlorination steps. Initially, the α-chloropropanoic acid and β-chloropropanoic acid were formed and then further chlorinated to α,α-dichloropropanoic acid and α,β-dichloropropanoic acid (Scheme 3). The activation energies (Ea) and reaction energies (ΔH) calculation results are shown in Table 1. The energy potential diagram of the primitive reaction and the corresponding reactants, transition states, and products configuration is presented in Figure 4.

The activation energy of radical chlorination reaction which generated β-chloropropanoic acid (Scheme 2; see (7)) and α,α-dichloropropanoic acid (Scheme 3; see (11)) was 171.21 kJ/mol and 162.64 kJ/mol, respectively. They were lower than the activation energy of generated α,β-dichloropropanoic acid, which were 180.08 kJ/mol and 178.87 kJ/mol (Scheme 3; see (12)) and 187.87 kJ/mol (Scheme 3; see (13)). Therefore, β-chloropropanoic acid and α,α-dichloropropanoic acid are the main byproducts.

The activation energy of generated β-chloropropanoic acid (Scheme 2; see (7)) was 171.21 kJ/mol, higher than 125.73 kJ/mol (Scheme 1; see (3)), which was the rate-determining step of generating α-chloropropanoic acid. The amount of generated β-chloropropanoic acid changed with time which was different from α-chloropropanoic acid in this experiment. Thus, the generation of α-chloropropanoic acid was mainly through the acid-catalyzed enolization route, while β-chloropropanoic acid was mainly through the radical chlorination route.

Compared to these two kinds of paths which generated α, β-dichloropropanoic acid, the activation energy of

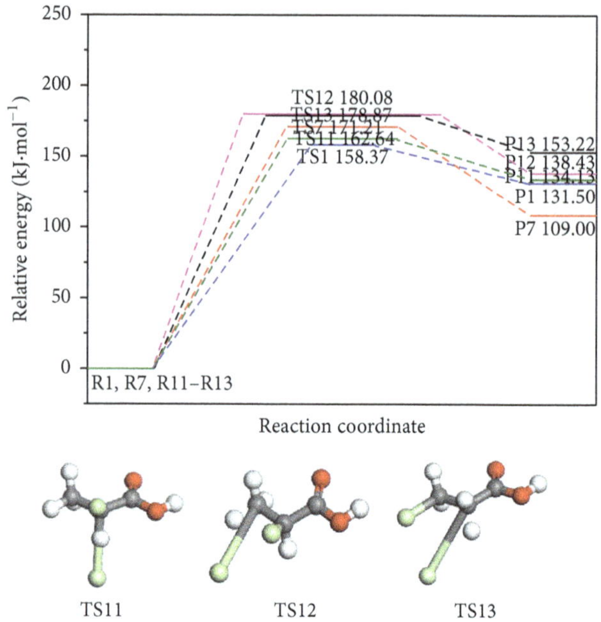

FIGURE 4: Potential energy diagram of the radical chlorination mechanism of byproducts together with the initial states, transition states, and final states.

the formed β-chloropropanoic acid (Scheme 2; see (7)) was higher than that of the formed α-chloropropanoic acid (Scheme 1; see (3)). Therefore, the formation of α,β-dichloropropanoic acid could be inhibited through inhibiting the formation of β-chloropropanoic. But the quantity of α,β-dichloropropanoic acid is hardly detected in the experiment.

3.5. Ionic Chlorination Mechanism of α,α-Dichloropropanoic Acid. In order to explore whether the formation between α,α-dichloropropanoic acid and α-chloropropanoic acid was a consecutive reaction, the paths in ionic chlorination mechanism were designed (Scheme 4). The path (Scheme 4; see (14)), the intermediate in the process generating α-chloropropanoyl chloride, can also generate 1-propenol, 1,2-dichloro. The consecutive chlorination of 1-propenol, 1,2-dichloro with chlorine generated α,α-dichlorine propanoyl chloride (Scheme 4; see (15)). α,α-Dichlorine propanoyl chloride with propanoic acid generated α,α-dichloropropanoic acid (Scheme 4; see (16) and (17)). The OH-Cl exchange reaction generated the final product. At the same time, energy potential diagram of the primitive reaction, the corresponding reactants, transition states, and products of configuration is shown in Figure 5.

It was observed that activation energy 130.81 kJ/mol (Scheme 4; see (14)) was higher than 109.73 kJ/mol (Scheme 4; see (15)) which was the rate-determining step in ionic chlorination mechanism. α,α-Dichloropropanoic acid and α-chloropropanoic acid were generated at the same time in the process of ionic chlorination mechanism. However, further comparison showed that radical chlorination activation energy 162.64 kJ/mol (Scheme 3; see (11)) was higher than 130.81 kJ/mol (Scheme 4; see (14)). This implies that radical

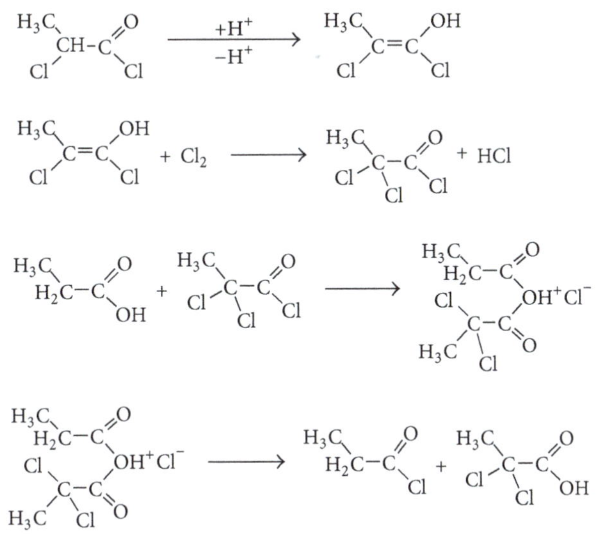

(14)

(15)

(16)

(17)

SCHEME 4: The ionic chlorination mechanism of α,α-dichloropropanoic acid.

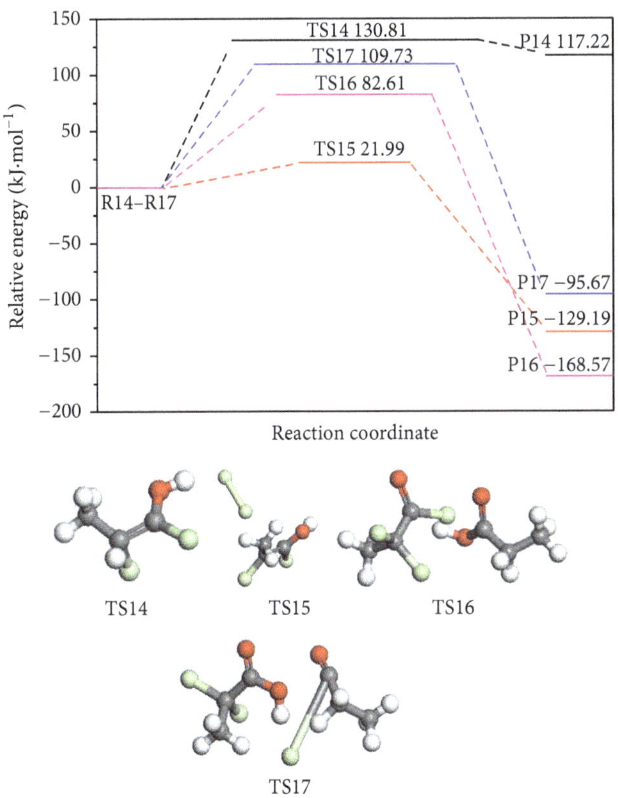

FIGURE 5: Potential energy diagram of ionic chlorination of α,α-dichloropropanoic acid together with the initial states, transition states, and final states.

chlorination was not easier to occur than ionic chlorination, which means that ionic chlorination was the main reaction path. Although higher temperature was the main cause of generation of byproducts [10], it was also observed that the generation of α,α-dichloropropanoic acid was less before concentrated sulfuric acid was added. This implies that the increase of H^+ does accelerate the acid-catalyzed path where the product was α,α-dichloropropanoic acid. It proved that the generation path of α,α-dichloropropanoic acid was ionic chlorination mechanism which also was the cause of the phenomenon of the generation of byproducts. The α,α-dichloropropanoic acid increased rapidly under the reaction conditions with higher temperature or excessive chlorine in intermittent industrial manufacture.

4. Conclusions

Based on the investigation of the mechanism of chlorination from propanoic acid to α-chloropropanoic acid, the results could be summarized as follows:

(1) The activation energy of the ionic mechanism for α-chloropropanoic acid formation was lower than that of the radical mechanism. The ionic mechanism involved four reaction routes and was initiated by the acid-catalyzed enolization of propanoyl chloride. Additionally, the activation energy of the rate-determining step was the process of 1-propanol,1-chloro formed 1-propen-1-ol,1-chloro.

(2) The formation route of β-chloropropanoic acid was mainly through propanoic acid radical chlorination mechanism. α,β-Dichloropropanoic acid is formed via a consecutive radical chlorination mechanism.

(3) The ionic mechanism of α,α-dichloropropanoic acid showed that the ionic chlorination was the main reaction path, and it was not easier for radical chlorination to occur compared with ionic chlorination for generation of α,α-dichloropropanoic acid. Hence, the formation route of α,α-dichloropropanoic acid and α-chloropropanoic acid was consecutive reaction. The formation of α,α-dichloropropanoic acid was mainly by α-chloropropanoic acid consecutive ionic chlorination mechanism which indicated that α-chloropropanoic acid is the key intermediate for α,α-dichloropropanoic acid formation process.

Conflicts of Interest

The authors declare that there are no conflicts of interest regarding the publication of this paper.

Acknowledgments

All calculations in this paper were accomplished in the Key Laboratory of Coal Science and Technology, Ministry of Education and Shanxi Province. The authors are indebted to Professor Wang Baojun and Professor Zhang Riguang.

References

[1] Y. Ogata and S. Watnabe, "Kinetics of the chlorosulfonic acid promoted a iodination of propionic acid," *Journal of Organic Chemistry*, vol. 44, pp. 2768–2770, 1979.

[2] W. G. Pan and H. J. Zhang, "Study on the Synthesis of L-2-Chloropropanoic Acid," *Chemical World (China)*, vol. 50, pp. 40–42, 2010.

[3] T. Salmi, P. Mäki-Arvela, E. Paatero, and R. Byggningsbacka, "Acid catalytic effects in the chlorination of propanoic acid," *Journal of Chemical Technology and Biotechnology*, vol. 75, pp. 89–97, 2000.

[4] N. V. Tsirul'nikova, N. Podmareva, E. S. Dernovaya, S. K. Belus, and T. S. Fetisova, "Method of producing ethylenediamine-N, N'-di-α-propionic acid," *Pat. RU2554095*, 2016.

[5] G. Sioli, P. M. Spaziante, and L. Giuffre, "Make MCPA in two stages," *Hydrocarbon Process*, vol. 58, pp. 111-112, 1979.

[6] E. Paatero, T. Salmi, and K. Fagerstolt, "Selective synthesis of α-chlorocarboxylic acids," *Industrial and Engineering Chemistry Research*, vol. 31, no. 11, pp. 2425–2437, 1992.

[7] T. Salmi, E. Paatero, and K. Fagerstolt, "Kinetic model for the synthesis of α-chlorocarboxylic acids," *Chemical Engineering Science*, vol. 48, no. 4, pp. 735–751, 1993.

[8] T. Salmi, E. Paatero, and K. Fagerstolt, "Optimal degree of backmixing in autocatalytic reactions a case study: chlorination of dodecanoic acid," *Chemical Engineering Research and Design*, vol. 71, no. 5, pp. 531–542, 1993.

[9] P. Mäki-Arvela, T. Salmi, E. Paatero, and R. Sjöholm, "Selective synthesis of α-chloropropanoic acid," *Industrial and Engineering Chemistry Research*, vol. 34, no. 6, pp. 1976–1993, 1995.

[10] P. Mäki-Arvela, T. Salmi, and E. Paatero, "Kinetics of the chlorination of propanoic acid in the presence of chlorosulphonic acid," *Chemical Engineering Science*, vol. 50, no. 14, pp. 2275–2288, 1995.

[11] Y. Ogata, T. Harada, K. Matsuyama, and I. Kazuo, "α-Chlorination of aliphatic acids by molecular chlorine," *Journal of Organic Chemistry*, vol. 40, no. 20, pp. 2960–2962, 1975.

[12] J. S. Ham, M. K. Davis, and J. S. Song, "The radical trapping effect of chlorine has been studied by means of electron paramagnetic resonance," *Journal of Polymer Science, Polymer Physics Edition*, vol. 11, 217 pages, 1973.

[13] C. Xia, L. Ju, Y. Zhao et al., "Heterogeneous oxidation of cyclohexanone catalyzed by TS-1: Combined experimental and DFT studies," *Cuihua Xuebao/Chinese Journal of Catalysis*, vol. 36, no. 6, pp. 845–854, 2015.

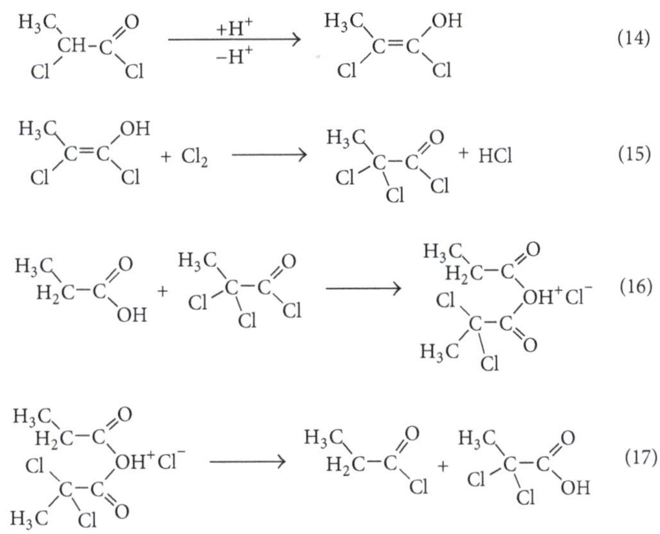

SCHEME 4: The ionic chlorination mechanism of α,α-dichloropropanoic acid.

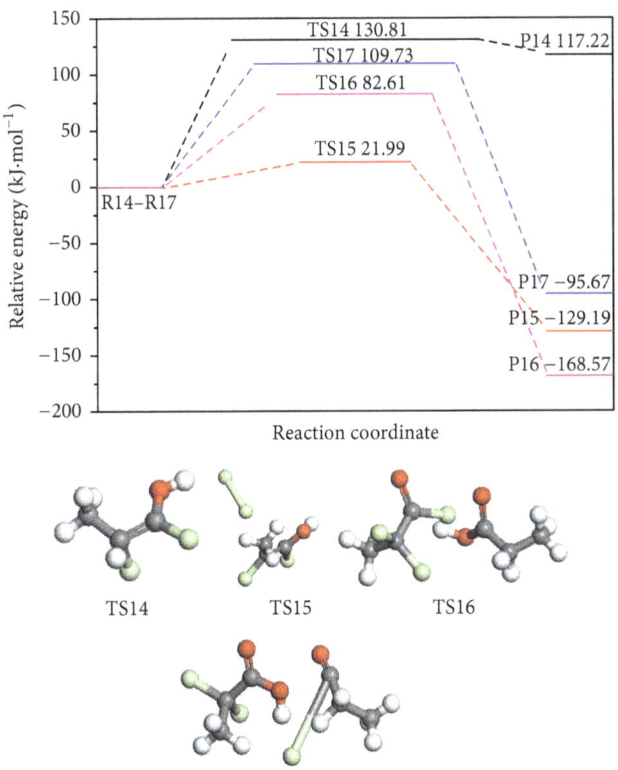

FIGURE 5: Potential energy diagram of ionic chlorination of α,α-dichloropropanoic acid together with the initial states, transition states, and final states.

chlorination was not easier to occur than ionic chlorination, which means that ionic chlorination was the main reaction path. Although higher temperature was the main cause of generation of byproducts [10], it was also observed that the generation of α,α-dichloropropanoic acid was less before concentrated sulfuric acid was added. This implies that the increase of H^+ does accelerate the acid-catalyzed path where the product was α,α-dichloropropanoic acid. It proved that the generation path of α,α-dichloropropanoic acid was ionic chlorination mechanism which also was the cause of the phenomenon of the generation of byproducts. The α,α-dichloropropanoic acid increased rapidly under the reaction conditions with higher temperature or excessive chlorine in intermittent industrial manufacture.

4. Conclusions

Based on the investigation of the mechanism of chlorination from propanoic acid to α-chloropropanoic acid, the results could be summarized as follows:

(1) The activation energy of the ionic mechanism for α-chloropropanoic acid formation was lower than that of the radical mechanism. The ionic mechanism involved four reaction routes and was initiated by the acid-catalyzed enolization of propanoyl chloride. Additionally, the activation energy of the rate-determining step was the process of 1-propanol,1-chloro formed 1-propen-1-ol,1-chloro.

(2) The formation route of β-chloropropanoic acid was mainly through propanoic acid radical chlorination mechanism. α,β-Dichloropropanoic acid is formed via a consecutive radical chlorination mechanism.

(3) The ionic mechanism of α,α-dichloropropanoic acid showed that the ionic chlorination was the main reaction path, and it was not easier for radical chlorination to occur compared with ionic chlorination for generation of α,α-dichloropropanoic acid. Hence, the formation route of α,α-dichloropropanoic acid and α-chloropropanoic acid was consecutive reaction. The formation of α,α-dichloropropanoic acid was mainly by α-chloropropanoic acid consecutive ionic chlorination mechanism which indicated that α-chloropropanoic acid is the key intermediate for α,α-dichloropropanoic acid formation process.

Conflicts of Interest

The authors declare that there are no conflicts of interest regarding the publication of this paper.

Acknowledgments

All calculations in this paper were accomplished in the Key Laboratory of Coal Science and Technology, Ministry of Education and Shanxi Province. The authors are indebted to Professor Wang Baojun and Professor Zhang Riguang.

References

[1] Y. Ogata and S. Watnabe, "Kinetics of the chlorosulfonic acid promoted a iodination of propionic acid," *Journal of Organic Chemistry*, vol. 44, pp. 2768–2770, 1979.

[2] W. G. Pan and H. J. Zhang, "Study on the Synthesis of L-2-Chloropropanoic Acid," *Chemical World (China)*, vol. 50, pp. 40–42, 2010.

[3] T. Salmi, P. Mäki-Arvela, E. Paatero, and R. Byggningsbacka, "Acid catalytic effects in the chlorination of propanoic acid," *Journal of Chemical Technology and Biotechnology*, vol. 75, pp. 89–97, 2000.

[4] N. V. Tsirul'nikova, N. Podmareva, E. S. Dernovaya, S. K. Belus, and T. S. Fetisova, "Method of producing ethylenediamine-N, N'-di-α-propionic acid," *Pat. RU2554095*, 2016.

[5] G. Sioli, P. M. Spaziante, and L. Giuffre, "Make MCPA in two stages," *Hydrocarbon Process*, vol. 58, pp. 111-112, 1979.

[6] E. Paatero, T. Salmi, and K. Fagerstolt, "Selective synthesis of α-chlorocarboxylic acids," *Industrial and Engineering Chemistry Research*, vol. 31, no. 11, pp. 2425–2437, 1992.

[7] T. Salmi, E. Paatero, and K. Fagerstolt, "Kinetic model for the synthesis of α-chlorocarboxylic acids," *Chemical Engineering Science*, vol. 48, no. 4, pp. 735–751, 1993.

[8] T. Salmi, E. Paatero, and K. Fagerstolt, "Optimal degree of backmixing in autocatalytic reactions a case study: chlorination of dodecanoic acid," *Chemical Engineering Research and Design*, vol. 71, no. 5, pp. 531–542, 1993.

[9] P. Mäki-Arvela, T. Salmi, E. Paatero, and R. Sjöholm, "Selective synthesis of α-chloropropanoic acid," *Industrial and Engineering Chemistry Research*, vol. 34, no. 6, pp. 1976–1993, 1995.

[10] P. Mäki-Arvela, T. Salmi, and E. Paatero, "Kinetics of the chlorination of propanoic acid in the presence of chlorosulphonic acid," *Chemical Engineering Science*, vol. 50, no. 14, pp. 2275–2288, 1995.

[11] Y. Ogata, T. Harada, K. Matsuyama, and I. Kazuo, "α-Chlorination of aliphatic acids by molecular chlorine," *Journal of Organic Chemistry*, vol. 40, no. 20, pp. 2960–2962, 1975.

[12] J. S. Ham, M. K. Davis, and J. S. Song, "The radical trapping effect of chlorine has been studied by means of electron paramagnetic resonance," *Journal of Polymer Science, Polymer Physics Edition*, vol. 11, 217 pages, 1973.

[13] C. Xia, L. Ju, Y. Zhao et al., "Heterogeneous oxidation of cyclohexanone catalyzed by TS-1: Combined experimental and DFT studies," *Cuihua Xuebao/Chinese Journal of Catalysis*, vol. 36, no. 6, pp. 845–854, 2015.

12

Hantzsch Reaction Starting Directly from Alcohols through a Tandem Oxidation Process

Xiaobing Liu[1] **and Bin Liu**[2]

[1]*School of Chemistry and Chemical Engineering, Jinggangshan University, Ji'an 343009, China*
[2]*School of Basic Medicine and Pharmacy, Jinggangshan University, Ji'an 343009, China*

Correspondence should be addressed to Xiaobing Liu; xiaobingliu928@126.com

Academic Editor: Pedro M. Mancini

A Brønsted acidic ionic liquid, 3-(N,N-dimethyldodecylammonium) propanesulfonic acid hydrogen sulphate ([DDPA][HSO_4]), has been successfully applied to catalyze sequential oxidation of aromatic alcohols with $NaNO_3$ followed by their condensation with dicarbonyl compound and ammonium acetate. The corresponding pyridine analogues of Hantzsch 1,4-dihydropyridines could be obtained as a major product with high yields by the multicomponent reaction. The present work utilizing alcohols instead of aldehyde in Hantzsch reaction is a valid and green alternative to the classical synthesis of the corresponding pyridine analogues of Hantzsch 1,4-dihydropyridines.

1. Introduction

1,4-Dihydropyridines and their derivatives are an important class of heterocycles in biologically active and naturally occurring molecules [1–3]. Many methods for the preparation of 1,4-dihydropyridines have been reported and most of them are mainly confined to the modification and optimization of the Hantzsch multicomponent synthesis between an aldehyde, a 1,3-dicarbonyl compound, and a source of ammonia [4–6]. At the same time, the oxidative aromatization of 1,4-dihydropyridines has also been extensively studied and numerous oxidizing reagents, namely, $CuBr_2$ [7], SiO_2/P_2O_5-SeO_2 [8], $H_2O_2/Co(OAc)_2$ [9], and $Na_2S_2O_4$/TBHP [10], have been used for the aromatization of 1,4-dihydropyridines. Some reports have observed that the direct oxidative procedure was very efficient for the preparation of Hantzsch pyridines under microwave-assisted condition or solvent-free condition [11–13]. However, the search for the new readily available and green methods is still being actively pursued.

The aldehydes in Hantzsch reaction are more volatile, toxic, or unstable than their corresponding alcohols. Recently, tandem oxidative processes (TOP) [14–16] in which the alcohol is oxidized to the aldehyde in situ have developed into an innovative addition to synthetic organic chemistry. Garima et al. [17] reported an efficient oxidative access to 3,4-dihydropyrimidin-2-(1H)-ones directly from aromatic alcohols. In the literature, a Brønsted acidic ionic liquid has been successfully applied to catalyze sequential oxidation of aromatic alcohols with $NaNO_3$ followed by their condensation with urea and dicarbonyl compounds.

Ionic liquids have attracted extensive interest and played a dual solvent-catalyst role in organic synthesis due to their favourable properties [18]. Brønsted acidic ionic liquids combining the advantageous characteristics of solid acids and mineral acids are designed to replace traditional mineral liquid acids in catalytic action [19]. In continuation of our interest in exploring green catalysis of ionic liquids [20, 21], we herein report a Brønsted acidic ionic liquid 3-(N,N-dimethyldodecylammonium) propanesulfonic acid hydrogen sulphate ([DDPA][HSO_4]) catalyzed sequential oxidation of aromatic alcohols with $NaNO_3$ followed by their condensation with dicarbonyl compound and ammonium acetate. We have found that the corresponding pyridine analogues of 1,4-dihydropyridines could be obtained as a major product by the multicomponent reaction, whereas only trace amounts of 1,4-dihydropyridines were detected (Scheme 1).

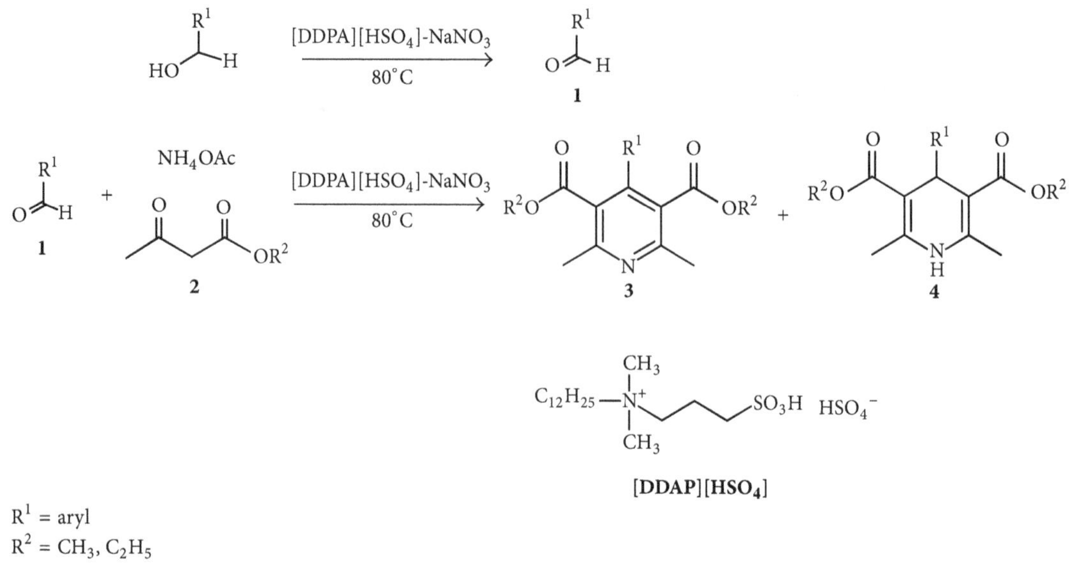

SCHEME 1: Synthesis of the corresponding pyridine analogues of 1,4-dihydropyridines directly from aromatic alcohols.

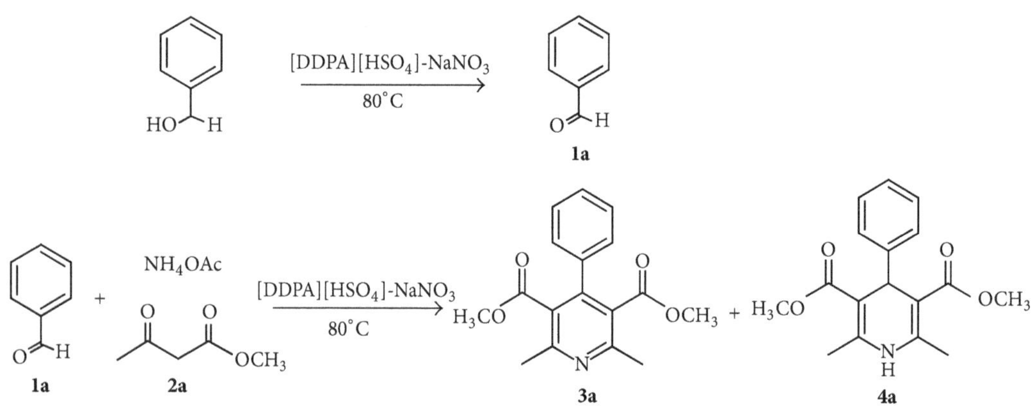

SCHEME 2: Model Hantzsch reaction directly from benzyl alcohol in [DDPA][HSO_4].

2. Materials and Methods

2.1. General Reagents and Instrumentation. All starting chemicals (AR grade) were purchased from commercial suppliers and used without further purification. Melting points were determined on a Thomas-Hoover capillary apparatus and were uncorrected. The IR spectra were recorded with a Bomem Michelson model 102 FTIR. ^{1}H NMR spectra were recorded on Bruker DRX (500 MHz) spectrometer. Gas chromatography (GC) was recorded on a HP 6890 Plus GC instrument. Elemental analyses were performed on a Yanagimoto MT3CHN recorder. The ionic liquid [DDPA][HSO_4] was synthesized according to the literature procedures [22].

2.2. Experimental Procedures. A mixture of an aromatic alcohol (3 mmol), sodium nitrate (6 mmol), and [DDPA][HSO_4] (1.2 mmol) was heated with stirring at 80°C for 5–30 min in a round-bottomed flask. After oxidation of the aromatic alcohol (monitored by TLC), dicarbonyl compound (3 mmol) and ammonium acetate (3 mmol) were added. The mixture was heated with stirring at 80°C for 1–3 h. The reaction was monitored by TLC after an interval of 15 min. After completion, the reaction mixture was cooled to rt and extracted with ethyl acetate. The organic layer was separated and washed with brine solution and then dried over anhydrous Na_2SO_4. Solvent was removed in vacuo and the crude product was purified by chromatography over silica gel. The aqueous layer (containing the ionic liquid) could be reused directly in the next run without further purification.

3. Results and Discussion

In the initial experiments, the multicomponent reaction of benzyl alcohol, ethyl acetoacetate, and ammonium acetate was selected as a model reaction (Scheme 2). We studied the tandem oxidative cyclocondensation of benzyl alcohol, ethyl acetoacetate, and ammonium acetate in the presence of [DDPA][HSO_4] with $NaNO_3$ at 80°C, but no reaction took place even after stirring the reaction mixture for 12 h. However, we observed the formation of 1,4-dihydropyridine

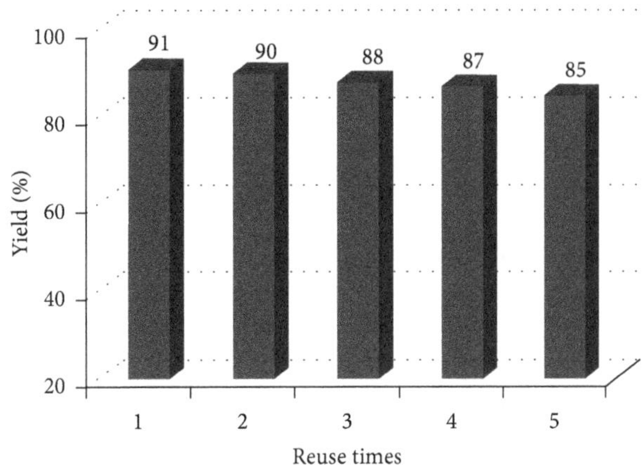

FIGURE 1: Reusability of the ionic liquid [DDPA][HSO_4].

TABLE 1: Model of Hantzsch reaction directly from benzyl alcohol in different ionic liquids[a].

Entry	Ionic liquid	Ratio[b]	Yield (%)[c]
1	[DDPA][HSO_4]	**3a** : **4a** = 93 : 7	91
2	[Hmim][H_2PO_4]	**3a** : **4a** = 86 : 14	37
3	[Hmim]NO_3	—	—
4	[bmin]BF_4	—	—

[a]Benzyl alcohol (3 mmol), ethyl acetoacetate (3 mmol), ammonium acetate (3 mmol), ionic liquid (1.2 mmol), and $NaNO_3$ (6 mmol), 1 h. [b]Yields and ratios were based on GC analysis. [c]Yields referred to the desired product **3a**.

and its corresponding pyridine when ethyl acetoacetate and ammonium acetate were added to the reaction after the oxidation of the alcohol to the aldehyde was complete.

Subsequently, different ionic liquids were screened for the same model reaction. The results are summarized in Table 1. It was shown that the reaction was unsuccessful in [Hmim]NO_3 and [bmin]BF_4 because an acidic hydrogen is absent in the two ionic liquids. The reaction could proceed effectively in [DDPA][HSO_4] but relatively low yields of product in [Hmim][H_2PO_4]. This is probably due to the lower Brønsted acidity associated with [H_2PO_4]. Thus, [DDPA][HSO_4] plays a dual role as an acid catalyst and as a solvent for both the oxidation of alcohol and the subsequent condensation. The product 4a could be converted to product 3a in the exposure of air and the $NaNO_3$-[DDPA][HSO_4] system could speed up the process.

To explore the stability of the ionic liquid after reaction, the recycling performance of [DDPA][HSO_4] in the same model reaction was investigated. After the reaction, the products were separated from the catalytic system by extraction with ethyl acetate. The aqueous layer (containing the catalyst) was reused in the next run without further purification. As shown in Figure 1, the catalyst could be reused at least five times without significant decrease in catalytic activity.

With these results in hand, we decided to explore the scope of this method. A variety of substituted aromatic alcohols was extended in the procedure in the presence of [DDPA][HSO_4] and the obtained results are summarized in Table 2. It can be seen that the tandem oxidation of aromatic alcohols followed by their condensation with dicarbonyl compound and ammonium acetate in the presence of [DDPA][HSO_4] proceeded smoothly to give the corresponding products in good yields. Aromatic alcohols with a strong electron-withdrawing group required a longer time for the reaction (Table 2, entries 8 and 9), whereas aromatic alcohols with electron-donating substitutes gave excellent yields (Table 1, entries 2, 4, 5, and 6) in a shorter reaction time. An important feature of the present protocol is the ability to tolerate variation in all components simultaneously.

A plausible mechanism for the sequential oxidation of aromatic alcohols with $NaNO_3$ followed by their condensation with dicarbonyl compound and ammonium acetate is depicted in Scheme 3. In acidic ionic liquid [DDPA][HSO_4], aromatic alcohols were oxidized to aromatic aldehydes with $NaNO_3$. After the oxidation completion, a multicomponent reaction of the aromatic aldehydes, dicarbonyl compound, and ammonium acetate was conducted in [DDPA][HSO_4] and then the Hantzsch 1,4-dihydropyridines were formed. Subsequently, the Hantzsch 1,4-dihydropyridines could be further oxidized to the desired corresponding pyridine analogues with $NaNO_3$-[DDPA][HSO_4] in air.

4. Conclusions

In conclusion, we have described an efficient protocol for Hantzsch reaction starting from alcohols using $NaNO_3$-[DDPA][HSO_4] system. The protocol involves [DDPA][HSO_4] catalyzed sequential oxidation of aromatic alcohols with $NaNO_3$ followed by their condensation with dicarbonyl compound and ammonium acetate. The corresponding pyridine analogues of Hantzsch 1,4-dihydropyridines could be obtained as a major product with high yields by the multicomponent reaction. The present work utilizing alcohols instead of aldehyde in Hantzsch reaction is a valid and green alternative to the classical synthesis of the corresponding pyridine analogues of Hantzsch 1,4-dihydropyridines.

TABLE 2: Hantzsch reaction of different substrates catalyzed by $[DDPA][HSO_4]^a$.

Entry	R^1	R^2	Time (h)	Yield (%)[b]	Mp °C (lit. Mp °C)[c]
1	C_6H_5	CH_3	1	91	118 (118) [7]
2	4-MeO-C_6H_4	CH_3	1	93	91 (92) [7]
3	4-Cl-C_6H_4	CH_3	2	85	118 (120) [7]
4	C_6H_5	CH_3CH_2	1	90	60-61 (63-64) [23]
5	2-MeO-C_6H_4	CH_3CH_2	1	93	58 (55–57) [24]
6	4-MeO-C_6H_4	CH_3CH_2	1	91	52 (51-52) [7]
7	4-Cl-C_6H_4	CH_3CH_2	2	83	66 (65–67) [23]
8	3-NO_2-C_6H_4	CH_3CH_2	3	76	60 (60) [7]
9	4-NO_2-C_6H_4	CH_3CH_2	3	72	115 (114-115) [23]

[a]Benzyl alcohol (3 mmol), ethyl acetoacetate (3 mmol), ammonium acetate (3 mmol), ionic liquid (1.2 mmol), and $NaNO_3$ (6 mmol). [b]Yields were based on GC analysis and referred to the desired product **3**. [c]Mp of product **3**.

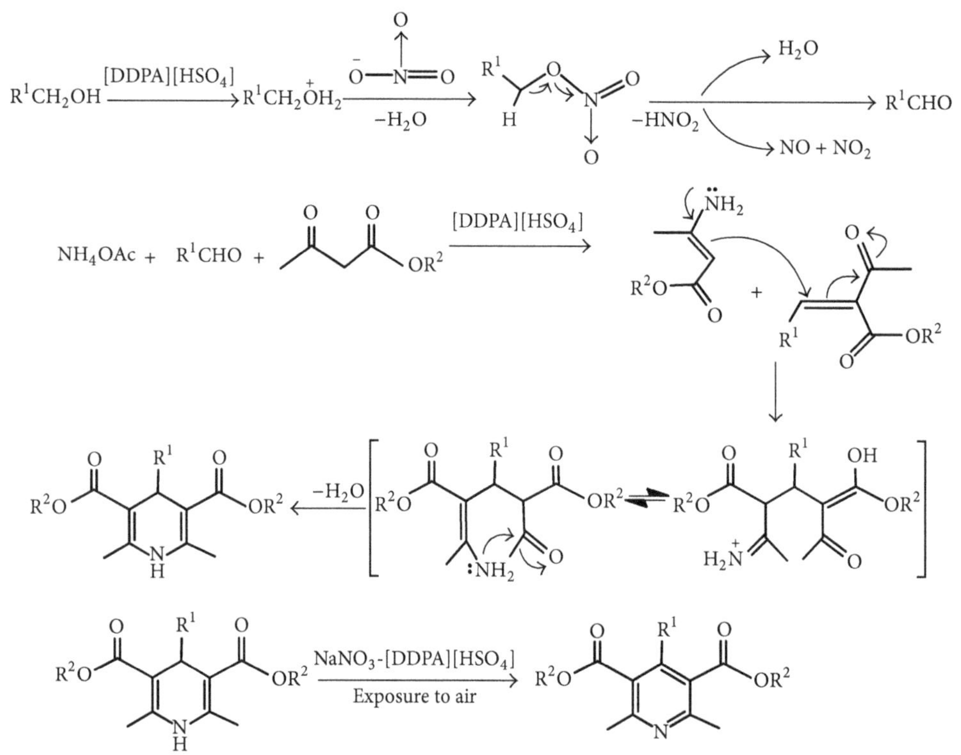

SCHEME 3: Plausible mechanism for Hantzsch reaction directly from alcohol in $[DDPA][HSO_4]$.

Conflicts of Interest

All authors declare no conflicts of interest.

Acknowledgments

This research was supported by National Natural Science Foundation of China (Grant no. 31360228) and Doctoral Foundation of Jinggangshan University (Grant no. JZB1322).

References

[1] F. Lentz, M. Hemmer, N. Reiling, and A. Hilgeroth, "Discovery of novel N-phenyl 1,4-dihydropyridines with a dual mode of antimycobacterial activity," *Bioorganic and Medicinal Chemistry Letters*, vol. 26, no. 24, pp. 5896–5898, 2016.

[2] S.-J. Choi, J.-H. Cho, I. Im et al., "Design and synthesis of 1,4-dihydropyridine derivatives as BACE-1 inhibitors," *European Journal of Medicinal Chemistry*, vol. 45, no. 6, pp. 2578–2590, 2010.

[3] K. Ošiņa, E. Rostoka, S. Isajevs, J. Sokolovska, T. Sjakste, and N. Sjakste, "Effects of an antimutagenic 1,4-dihydropyridine av-153 on expression of nitric oxide synthases and DNA repair-related enzymes and genes in kidneys of rats with a streptozotocin model of diabetes mellitus," *Basic and Clinical Pharmacology and Toxicology*, vol. 119, no. 5, pp. 458–463, 2016.

[4] M. G. Dekamin, S. Ilkhanizadeh, Z. Latifidoost, H. Daemi, Z. Karimi, and M. Barikani, "Alginic acid: a highly efficient

renewable and heterogeneous biopolymeric catalyst for one-pot synthesis of the Hantzsch 1,4-dihydropyridines," *RSC Advances*, vol. 4, no. 100, pp. 56658–56664, 2014.

[5] S. Zhaleh, N. Hazeri, M. R. Faghihi, and M. T. Maghsoodlou, "Chitosan: a sustainable, reusable and biodegradable organocatalyst for green synthesis of 1,4-dihydropyridine derivatives under solvent-free condition," *Research on Chemical Intermediates*, vol. 42, no. 12, pp. 8069–8081, 2016.

[6] M. Nasr-Esfahani, S. J. Hoseini, M. Montazerozohori, R. Mehrabi, and H. Nasrabadi, "Magnetic Fe_3O_4 nanoparticles: efficient and recoverable nanocatalyst for the synthesis of polyhydroquinolines and Hantzsch 1,4-dihydropyridines under solvent-free conditions," *Journal of Molecular Catalysis A: Chemical*, vol. 382, pp. 99–105, 2014.

[7] F. Saikh, R. De, and S. Ghosh, "Oxidative aromatization of Hantzsch 1,4-dihydropyridines by cupric bromide under mild heterogeneous condition," *Tetrahedron Letters*, vol. 55, no. 45, pp. 6171–6174, 2014.

[8] S. Paul, S. Sharma, M. Gupta, D. Choudhary, and R. Gupta, "Oxidative aromatization of Hantzsch 1,4-dihydropyridines by SiO2/P2O5-SeO2 under mild and heterogeneous conditions," *Bulletin of the Korean Chemical Society*, vol. 28, no. 2, pp. 336–338, 2007.

[9] M. M. Hashemi, Y. Ahmadibeni, and H. Ghafuri, "Aromatization of Hantzsch 1,4-dihydropyridines by hydrogen peroxide in the presence of cobalt(II) acetate," *Monatshefte fur Chemie*, vol. 134, no. 1, pp. 107–110, 2003.

[10] C.-B. Bai, N.-X. Wang, Y.-J. Wang, X.-W. Lan, Y. Xing, and J.-L. Wen, "A new oxidation system for the oxidation of Hantzsch-1,4-dihydropyridines and polyhydroquinoline derivatives under mild conditions," *RSC Advances*, vol. 5, no. 122, pp. 100531–100534, 2015.

[11] I. C. Cotterill, A. Y. Usyatinsky, J. M. Arnold et al., "Microwave assisted combinatorial chemistry. Synthesis of substituted pyridines," *Tetrahedron Letters*, vol. 39, no. 10, pp. 1117–1120, 1998.

[12] M. Nasr-Esfahani, B. Karami, and M. Behzadi, "A simple, efficient, one-pot three-component domino synthesis of Hantzsch pyridines under solvent-free condition," *Journal of Heterocyclic Chemistry*, vol. 46, no. 5, pp. 931–935, 2009.

[13] P. Guillermo, G. Olivia, F. Karina, H. Ofelia, and A. Cecilio, "A modification to the hantzsch method to obtain pyridines in a one pot reaction: use of a bentonitic clay in a dry medium," *Heterocyclic Communications*, vol. 2, no. 4, pp. 359-360, 1996.

[14] V. Jeena, S. Sithebe, and R. S. Robinson, "Copper-catalyzed synthesis of valuable heterocyclic compounds using a tandem oxidation process approach," *Synthetic Communications*, vol. 45, no. 12, pp. 1484–1491, 2015.

[15] J.-F. Soulé, H. Miyamura, and S. Kobayashi, "Direct amidation from alcohols and amines through a tandem oxidation process catalyzed by heterogeneous-polymer-incarcerated gold nanoparticles under aerobic conditions," *Chemistry - An Asian Journal*, vol. 8, no. 11, pp. 2614–2626, 2013.

[16] Z.-G. Wang, X.-H. Cao, Y. Yang, and M. Lu, "Green and efficient methods for one-pot aerobic oxidative synthesis of benzimidazoles from alcohols with TEMPO-PEG_{4000}-NHC-Cu(II) complex in water," *Synthetic Communications*, vol. 45, no. 12, pp. 1476–1483, 2015.

[17] Garima, V. P. Srivastava, and L. D. S. Yadav, "Biginelli reaction starting directly from alcohols," *Tetrahedron Letters*, vol. 51, no. 49, pp. 6436–6438, 2010.

[18] A. C. Cole, J. L. Jensen, I. Ntai et al., "Novel brønsted acidic ionic liquids and their use as dual solvent-catalysts," *Journal of the American Chemical Society*, vol. 124, no. 21, pp. 5962-5963, 2002.

[19] P. Sarma, S. Saikia, and R. Borah, "Studies on -SO_3H functionalized Brønsted acidic imidazolium ionic liquids (ILs) for one-pot, two-step synthesis of 2-styrylquinolines," *Synthetic Communications*, vol. 46, no. 14, pp. 1187–1196, 2016.

[20] X. B. Liu, M. Lu, T. T. Lu, and G. L. Gu, "Functionalized ionic liquid promoted aza-michael addition of aromatic amines," *Journal of the Chinese Chemical Society*, vol. 57, no. 6, pp. 1221–1226, 2010.

[21] Y.-L. Hu, X.-B. Liu, and D. Fang, "Efficient and convenient oxidation of sulfides to sulfones using H_2O_2 catalyzed by V_2O_5 in ionic liquid $[C_{12}min][HSO_4]$," *Catalysis Science & Technology*, vol. 4, no. 1, pp. 38–42, 2014.

[22] F. Dong, F. Zhenghao, and L. Zuliang, "Functionalized ionic liquid as the recyclable catalyst for Mannich-type reaction in aqueous media," *Catalysis Communications*, vol. 10, no. 8, pp. 1267–1270, 2009.

[23] R. S. Varma and D. Kumar, "Manganese triacetate mediated oxidation of Hantzsch 1,4- dihydropyridines to pyridines," *Tetrahedron Letters*, vol. 40, no. 1, pp. 21–24, 1999.

[24] M. Nasr-Esfahani, M. Moghadam, and G. Valipour, "Rapid and efficient aromatization of Hantzsch 1,4-dihydropyridines with potassium peroxomonosulfate catalyzed by manganese(III) Schiff base complexes," *Journal of the Iranian Chemical Society*, vol. 5, no. 2, pp. 244–251, 2008.

13

Study of the Deformation/Interaction Model: How Interactions Increase the Reaction Barrier

Zhiling Liang,[1] Houhe Liu,[1] Nianjun Su,[2] Dandan Song,[1] Yun Zhang,[1] Hong Huang,[1] Jianqi Zheng,[3] Cheng Zhong,[4] and Guodong Ye[1]

[1]*Guangdong Provincial Key Laboratory of Molecular Target & Clinical Pharmacology, School of Pharmaceutical Sciences, Guangzhou Medical University and The Fifth Affiliated Hospital of Guangzhou Medical University, Guangzhou 510700, China*
[2]*Department of Reproductive Health and Infertility, Guangdong Women and Children Hospital, Guangzhou 510010, China*
[3]*The First Clinical School of Guangzhou Medical University, Guangzhou 510275, China*
[4]*College of Chemistry and Molecular Sciences, Wuhan University, Wuhan 430072, China*

Correspondence should be addressed to Guodong Ye; gzhygd@gzhmu.edu.cn

Academic Editor: Arturo Espinosa Ferao

The interactions (including weak interactions) between dienophiles and dienes play an important role in the Diels-Alder reaction. To elucidate the influence of these interactions on the reactivity, a popular DFT functional and a variational DFT functional corrected with dispersion terms are used to investigate different substituent groups incorporated on the dienophiles and dienes. The bond order is used to track the trajectory of the cycloaddition reaction. The deformation/interaction model is used to obtain the interaction energy from the reactant complex to the inflection point until reaching the saddle point. The interaction energy initially increases with a decrease in the interatomic distance, reaching a maximum value, but then decreases when the dienophiles and dienes come closer. Reduced density gradient and chemical energy component analysis are used to analyse the interaction. Traditional transition state theory and variational transition state theory are used to obtain the reaction rates. The influence of tunneling on the reaction rate is also discussed.

1. Introduction

Interactions play an important role in reactions. Notably, the long-distance weak interaction force, which is known as the weak interaction, has been examined extensively in the study of multicomponent systems, like dimmers, clusters, and so on [1, 2]. However, many chemical reactions involve the breakage and formation of chemical bonds as well as three-dimensional conformational arrangement. Traditional examination methods are unable to cope with such increasing complication. The main reason for the complexity is that two molecules usually react from the reactant complex (RC) to the saddle point (SP) until the product complex (PC) is formed. For example, conformational distortion includes steric strain, torsion, and repulsions in dynamic systems, and the weak interaction also includes hydrogen bonding, van der Waals forces, salt bonding, and hydrophobic interaction forces. This requires expensive treatments (like topological analysis, basis set superposition error correction, etc. [3–8]) to track the trajectory of the reaction and reveal the relationship between the interaction energy (E_i) and the conformational change during the breakage and formation of chemical bonds. Luckily, the deformation/interaction model [9] seems to avoid this complication, compared with the above method, without losing accuracy in the description of E_i.

We can describe the change in chemical bonding during formation and dissociation using the Mayer bond order [10]. We can use the reduced density gradient (RDG) function isosurface to analyse the weak interaction region and determine the source of interaction [11, 12]. RDG is an extension of the atoms in molecules (AIM) theory, which gives clearer results. The relationship between energies and intermolecular forces was also elucidated by the energy parameters, according

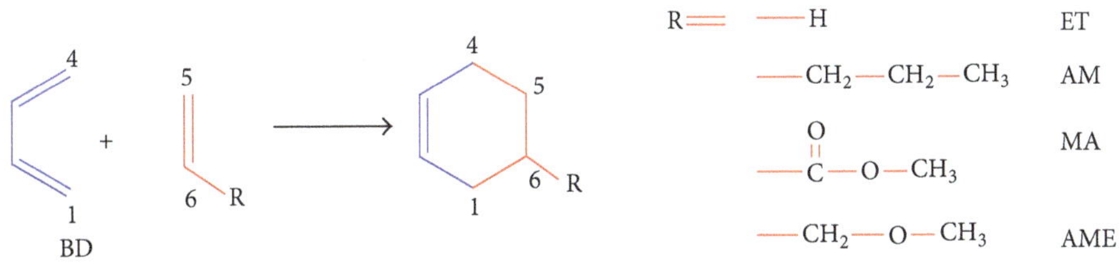

FIGURE 1: Schemes of the four Diels-Alder reactions.

to chemical energy component analysis (CECA) [13], and other kinetic parameters. To better match our experimental results, we used the hybrid DFT functional B3LYP and the variational DFT functional ωB97X-D to obtain the results. The latter contains an empirical dispersion correction to the ωB97X functional, as this provides long-range van der Waals interactions without additional computational cost. It also emerges that optimization of the ωB97 functional with empirical dispersion corrections leads to essentially zero dispersion correction.

The Diels-Alder reaction (DA) has a well-known secondary orbital interaction between the diene and the dienophile, which can influence the reaction [14]. The addition of dienophiles to dienes can proceed via a concerted mechanism, meaning that the two new CC bonds form "in concert" rather than sequentially. This popular reaction provides a good and standard example for us to examine the abovementioned interaction. We hope to find good correlation between the activation barrier and activation strain for many systems based on this exploration.

2. Calculation Methods

The four reactions with ethylene (ET), amylene (AM), methacrylate (MA), and allyl methyl ether (AME) as dienophiles with butadiene (BD) as the diene were studied. The structures are shown in Figure 1. The reactants, transition state (TS) and products were optimized by the Gaussian 09 suite of programs [15] using the B3LYP and ωB97X-D functionals and the 6-311++G (d,p) basis set. Graphical presentation of the weak interaction was achieved by Multiwfn [16], and E_i was decomposed into a variety of energy matrices by APEX4. Reaction rates and tunneling factors were calculated by KiSThelP [17].

3. Results and Discussion

3.1. Bond Order Analysis. Along the internal reaction coordinate (IRC), we simulated the change in bond order (n_{Mayer}) from the RC to the PC. Mayer defined n_{Mayer} based on Mulliken population analysis and linear combination of atomic orbitals [18], combining both to derive the formula for n_{Mayer} as

$$n_{Mayer} = \sum_{\mu \in A} \sum_{\nu \in B} (PS)_{\mu\nu} (PS)_{\nu\mu}, \quad (1)$$

where the off-diagonal matrix elements of PS are known as the Mulliken overlap populations.

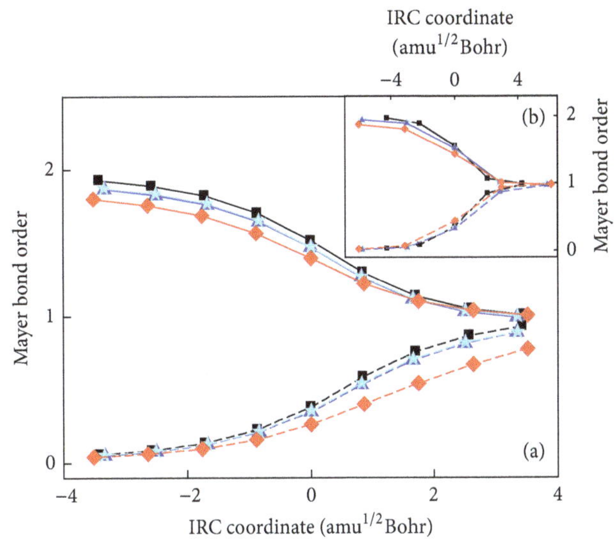

FIGURE 2: Analysis of the bond order. Bond order curves of BD + ET (black square), BD + AM (cyan star), BD + MA (red diamond), and BD + AME (blue triangle) from the RC to the PC. C_1-C_6 (- - -), C_5=C_6 (—). (a) B3LYP and (b) ωB97X-D.

Figure 2(a) shows that these trends are consistent with the empirical view. At the beginning, there is no bond formed between C_1 and C_6; the value of n_{Mayer} is close to 0. When the reaction proceeds to the PC, the value of n_{Mayer} between C_1 and C_6 increases to approximately 1, whereas when C_5=C_6 is changed to C_5-C_6, the corresponding value is reduced from 2 to 1. At the TS, the n_{Mayer} values of C_5-C_6 and C_1-C_6 are approximately 0.4 and 1.4, respectively (not 1.0). The changes in the n_{Mayer} value of C_5=C_6 in the four reactions are relatively similar, and the changes in the bond orders between C_1 and C_6 are different. This is due to their different structures; that is to say, the AM and AME are affected by the electron-donating group, whereas the MA is affected by the electron-withdrawing group. Meanwhile, the trend in Figure 2(b) is relatively similar.

3.2. Energy Analysis. We used the deformation/interaction model to analyse dynamic changes in the four DA reactions. The formula for the activation energy is

$$E_a = E_i + E_d, \quad (2)$$

where E_a is the activation energy, E_d is the energy required to distort the olefin reactants into their TS geometries, and E_i

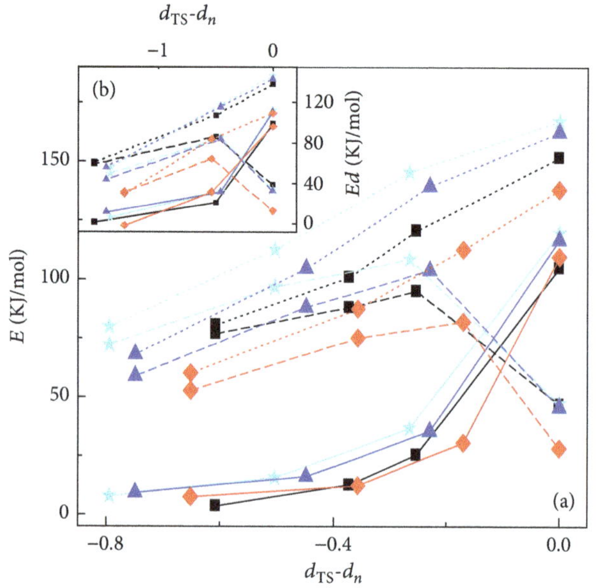

FIGURE 3: Energy analysis by the deformation/interaction model. Changes in the activation energy (...), deformation energy (—), and interaction energy of BD + ET (black square), BD + AM (cyan star), BD + MA (red diamond), and BD + AME (blue triangle) were analysed by the deformation/interaction model and projected onto the distance between C_1-C_6. (a) B3LYP and (b) ωB97X-D.

arises from a combination of closed-shell repulsion, charge transfer involving occupied and vacant orbital interactions, electrostatic interactions, and polarization effects [19].

As shown in Figures 3(a) and 3(b), although the energies obtained by ωB97X-D are smaller than those obtained by B3LYP, the two functionals showed similar trends in the calculated energy, and both E_a and E_d increased gradually, while E_i increased initially, reaching a maximum value, but then decreased at a smaller distance. There is an inflection point (IP) in the curve of E_i. Therefore, the process can be divided into two stages: the first stage is from RC to IP, and E_i plays an important role in the activation barrier at this stage, while the second stage is from IP to SP, and E_d exceeds the other chemical forces and governs the mechanism of the reaction. During the course of the reaction, E_a is much higher than E_d, therefore resulting in a positive E_i. This is different from S_N2 reactions, in which E_i is negative. It has been proven that the activation barrier decreases relative to E_d in all cases [20]. In particular, the values of E_a, E_i, E_d, and n_{Mayer} of BD + AM and BD + AME are similar and higher than those of BD + ET. This result confirms that the dienophile containing an electron-donating group is not conducive to the DA reaction, and while the E_a, E_i, and n_{Mayer} values of BD + MA are the smallest, they seem to be affected by the presence of an electron-withdrawing group, which is relatively favourable for the reaction. The symmetry or asymmetry of the structure often affects the E_d value of atoms, and the E_d values of BD in the four reactions are not the same. We found that the E_d value of diene is always greater than that of the dienophiles.

3.3. Analysis of the Weak Interaction. We used a new method to visualize the weak interaction, which depends on the RGD via the arithmetic expression as shown below:

$$\mathrm{RDG} = \frac{I_0}{2\sqrt[3]{3\pi^2}} * \frac{|\nabla\rho(r)|}{\sqrt[3]{\rho(r)^4}}. \quad (3)$$

This part only focuses on $|\nabla\rho(r)|/\sqrt[3]{\rho(r)^4}$, where ∇ is the gradient operator and $|\nabla\rho(r)|$ is the modulus of the electron density gradient. By combining the RDG and $\rho(r)$, we can determine which regions of the molecule are clearly associated with weak interactions. The RDG was used to analyse the interactions, which can be divided into medium- and long-range interactions based on the distance or into hydrogen bonding and electrostatic and van der Waals interactions according to the type of bond. A visualization of the weak interaction is shown in Figure 4, using BD + ET as an example. The points on the right side within the abscissa represent the nuclear area. The peak that extends to the upper left corner represents the molecular edge area. When it is farther away from the molecule, $|\nabla\rho(r)|$ and $\rho(r)$ are smaller, while $\rho(r)$ decreases faster, and the RDG value is stronger. The vertical bar, called a "spike" [11], on the leftmost side close to 0 indicates that there is a weak interaction (van der Waals interaction) between C_1-C_6 and C_4-C_5 in BD + ET. As is well known, the RDG value is closer to 0, the $\rho(r)$ value is smaller, and the weak interaction is stronger.

We also investigated the RDG values of different reactions from the RC to the IP until reaching the SP. The result is shown in Figure 5. When $\rho(r)$ is plotted along the X-axis, it gradually increases during the reaction. When the RDG values are plotted along the Y-axis, we found that the RDG values of C_1-C_6 and C_4-C_5 in BD + ET are similar. This is because the structure of BD + ET is symmetric; specifically, the RDG value begins to grow slightly in the first stage and decreases to almost the initial value in the second stage, as shown in Figure 5(a). However, the RDG value of C_1-C_6 in BD + AM and BD + AME decreases in the first stage and increases back to the original value in the second stage. At the same time, the RDG value of C_4-C_5 in BD + AM increases rapidly, while that of BD + AME increases slowly as the two atoms come closer. It should be noted that the two C_6 atoms here are substituted by electron-donating groups. For BD + MA, a continuous decrease in the RDG value between C_4-C_5 was observed, while the change in the RDG between C_1-C_6 is the same as that of BD + ET. It seems that the RDG value of C_1-C_6 changes irregularly, while that of C_4-C_5 changes regularly. It is noteworthy that C_5 is far away from the substituent based on our molecular design. We believed that the RDG value of C_5, as a distal end, is more susceptible to the substituent than that of C_6, as a near end. The electron-withdrawing group monotonically decreases the RDG value of C_4-C_5, while the electron-donating group produces the opposite effect.

In the calculation using the ωB97X-D functional, the RGD value of BD + ET increases in the first stage and decreases in the second stage, which is similar to the trend

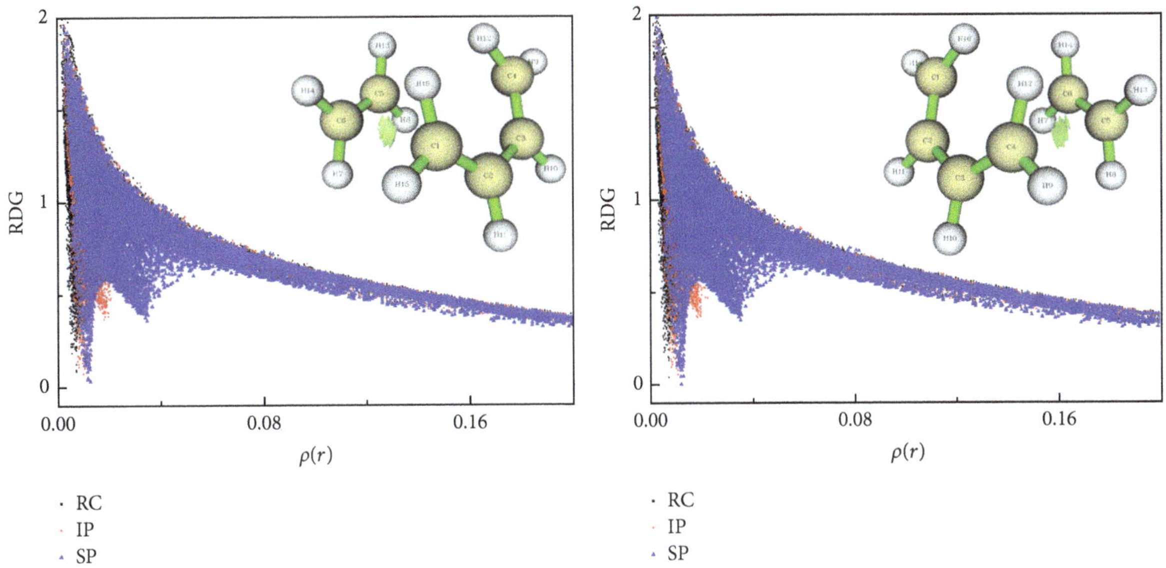

(a) (b)

FIGURE 4: The RDG region of BD + ET. RC (black square) to IP (red circle) to SP (blue triangle). C_1-C_6 (a) and C_4-C_5 (b).

obtained with the B3LYP functional. The RGD value of C_1-C_6 increases in the first stage and decreases to approximately the original value, while the RGD value of C_4-C_5 increases monotonously during the two stages. Compared with the results calculated by B3LYP, the calculation using the ωB97X-D functional accounts for the dispersion force to make the results more accurate. These results also prove that the RDG value of C_6, as a near end, is more susceptible to the substituent than that of C_5, as a distal end. It is noteworthy that the RGD values of C_4-C_5 in BD + AM and BD + AME are nearly the same. We found that the RGD value of C_4-C_5 in BD + MA also increases. This is in contrast to the results found using B3LYP and is different from the results of BD + AM and BD + AME. Its amplitude is obviously smaller than that of BD + AM and BD + AME. This validated our hypothesis that the influence of the electron-donating group on increasing the RDG value of C_4-C_5 is greater than that of the electron-withdrawing group. In brief, if the dienophile is substituted by an electron-withdrawing group, the spike value of a distal end determined by ωB97X-D increases slowly, while value determined by B3LYP shows a reverse trend from the RC to the IP until reaching the SP. This is in contrast to the spike value when the dienophile is substituted by an electron-donating group, which increases quickly as the reaction progresses.

3.4. Analysis of the Chemical Energy Component. To explain the above unusual RDG of the BD + MA system, we analysed the chemical energy component. Mayer proposed an energy decomposition method called CECA, in which the sum of the one- and two-center energy components should reproduce the total molecular HF energy. The interaction terms are combined and interpreted as electrostatic, exchange, overlapping, and kinetic energy contribution according to CECA [13]. Note that the two-centered energy here is not equivalent to E_i because it contains other energies (e.g., covalent components). However, we believe it can also elucidate the nature of the molecular interactions.

We compare the results of the two functional calculations in Figure 6. In Figures 6(a1), 6(b1), 6(c1), and 6(d1), the trends of the energies as the reaction proceeded are almost similar, except for the electrostatic contribution. The trends of C_1-C_6 are similar and negative, and the trends of C_4-C_5 are similar and positive in the four systems. However, we still found that BD + MA showed the largest difference (Figure 6(a1)). The electron-withdrawing/donating group influences the inductive effect. Thus, the electrostatic contribution was proven to be related to the n_{Mayer} and RDG. The contribution from the exchange and overlap terms continued to decline. We found that the exchange and overlap contributed a large portion of the two-center energy and had a dominating effect. In Figures 6(a2), 6(b2), 6(c2), and 6(d2), through calculation of the dispersion correction, we found that the trends of BD + MA were not consistent in Figures 6(b2), 6(c2), and 6(d2), but this did not affect the system as a whole.

3.5. Reaction Rate Results. In this part, the calculation results of the reaction rate constants (k) based on transition state theory (TST) and variational transition state theory (νTST) are discussed [21]. The temperature-dependent reaction rates, $k^{\text{TST}}(T)$, for bimolecular reactions were calculated by applying TST:

$$k^{\text{TST}}(T) = \lambda \frac{k_b T}{h} \cdot \frac{Q^{\neq}}{Q_A Q_B} \cdot \exp\left(-\frac{E}{RT}\right), \quad (4)$$

where $Q^{\neq}$, Q_A, and Q_B represent the total partition function of the TS and the reactants A and B, k_b represents Boltzmann's constant, R represents the gas constant, T represents the

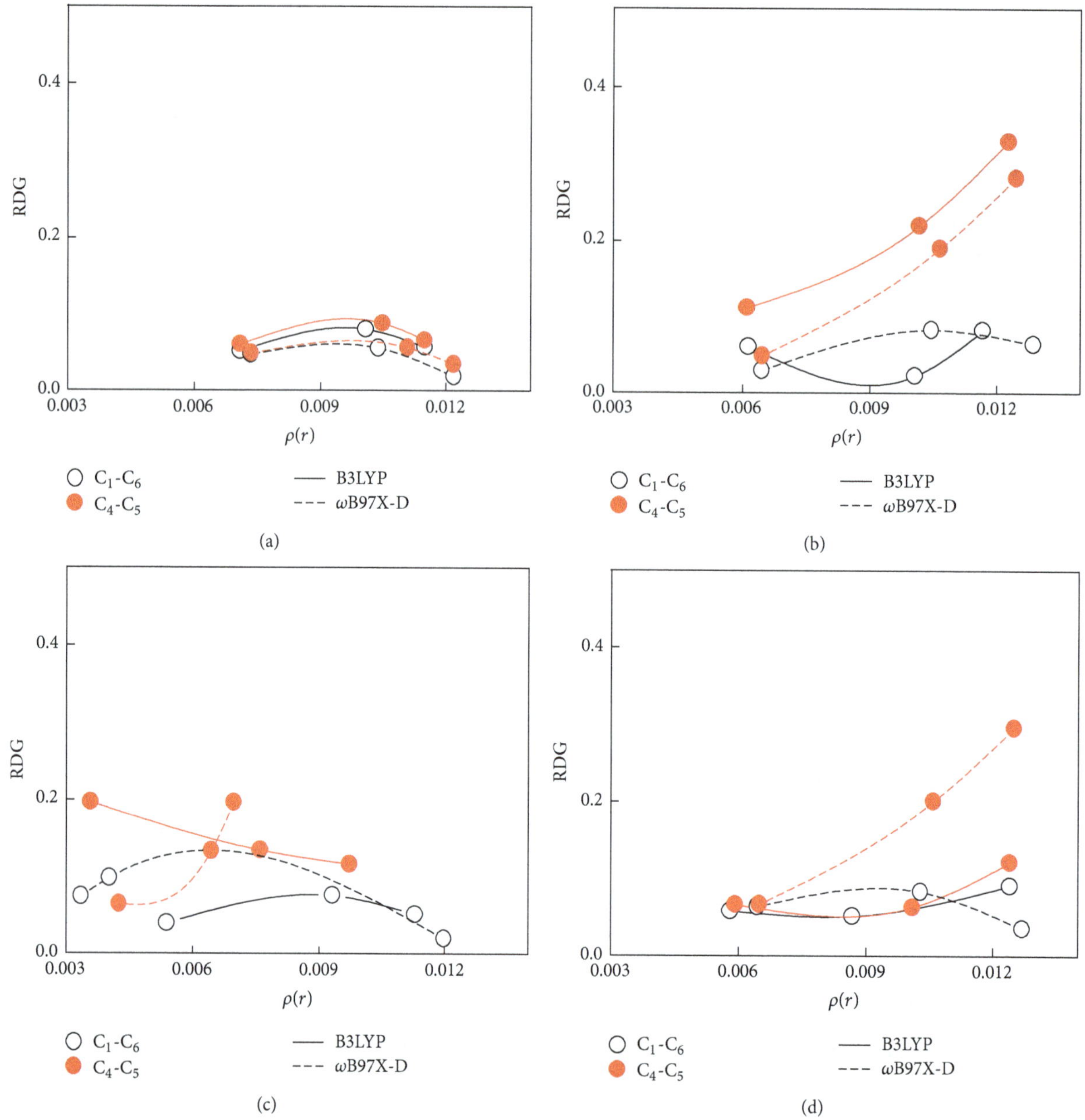

FIGURE 5: The change in the spike value among the IRC from the RC to the IP until reaching the SP for the four reactions. (a) BD + ET, (b) BD + AM, (c) BD + MA, (d) BD + AME, C_1-C_6 (white circle), C_4-C_5 (red circle). B3LYP (—), ωB97X-D (- - -).

temperature, h represents Planck's constant, and E is the activation energy. The constant expression of $k^{\nu\mathrm{TST}}(T)$ is

$$k^{\nu\mathrm{TST}}(T) = \frac{\sigma}{h\beta} \cdot \exp\left[-\beta \cdot \Delta G\left(T, s_*^{\nu\mathrm{TST}}\right)\right], \quad (5)$$

where σ represents the symmetry number and $\Delta G(T, s_*^{\nu\mathrm{TST}})$ coincides with the maximum free energy, $\beta = k_b/T$.

Since the tunneling factor $\kappa(T)$ is capable of affecting k, it is necessary to calculate it when k is considered in a precise way. In quantum chemistry, there is still a certain probability to achieve the reaction even when the energy is lower than the barrier. This may be due to penetration of a part of the particle through the middle of the barrier (also known as barrier penetration), which is represented by a correction term in the perturbative expansion of the parabolic barrier. $\kappa(T)$ can be obtained from the Wigner formula, which is shown below:

$$\kappa(T) = 1 + \frac{1}{24}\left(\frac{h\omega^{\neq}}{k_b T}\right)^2, \quad (6)$$

where $\omega^{\neq}$ is the imaginary frequency of the TS. Two virtual frequencies were found here because k_{TST} and $k_{\nu\mathrm{TST}}$ were calculated using different TSs.

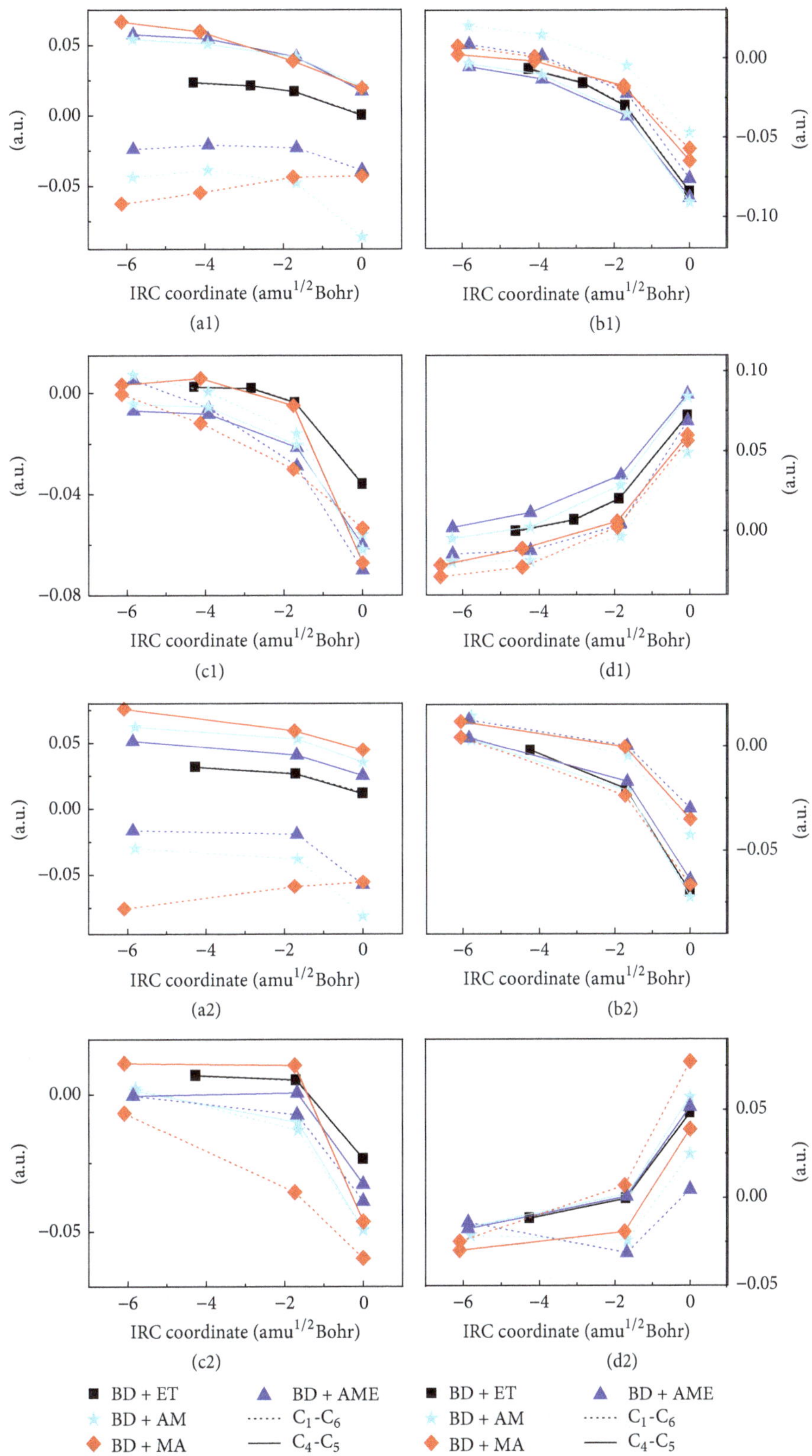

FIGURE 6: Analysis of the electrostatic (a), exchange (b), overlap (c), and kinetic energy (d). BD + ET (black square), BD + AM (cyan star), BD + MA (red diamond), BD + AME (blue triangle). C_1-C_6 (…), C_4-C_5 (—). (1) B3LYP; (2) ωB97X-D.

TABLE 1: The rate constants and tunneling factors.

Reactant	Functional	k_{TST} (cm^3/molec/s)	E_a kJ/mol	k_{vTST} (cm^3/molec/s)	$\omega^{\neq}$	$\kappa(T)$
BD + ET	B3LYP	5.52×10^{-34}	151.66	5.40×10^{-34}	−544.36	1.29
	ωB97X-D	4.46×10^{-32}	138.01	6.94×10^{-32}	−613.41	1.37
BD + AM	B3LYP	4.37×10^{-35}	167.16	4.69×10^{-35}	−540.58	1.28
	ωB97X-D	3.60×10^{-33}	146.69	4.23×10^{-33}	−612.82	1.36
BD + MA	B3LYP	2.70×10^{-31}	137.97	2.51×10^{-31}	−475.59	1.22
	ωB97X-D	2.62×10^{-27}	109.43	1.28×10^{-26}	−543.45	1.29
BD + AME	B3LYP	1.76×10^{-35}	162.37	1.44×10^{-35}	−534.06	1.28
	ωB97X-D	1.81×10^{-32}	143.03	2.07×10^{-32}	−594.03	1.34

Table 1 shows that the k_{TST} values were similar in magnitude to the k_{vTST} values. This means that the free and electron potential energy surfaces were nearly the same. However, the k obtained by ωB97X-D was greater than that obtained by B3LYP, which matches the results in Figure 3. The k_{TST} and k_{vTST} values of BD + MA were found to be the highest among the four systems. This result agreed with the result that BD + MA had the smallest E_a. However, the k_{TST} value is always of the same order as the k_{vTST} value but lower in the same reaction when determined by ωB97X-D, while the k_{TST} value is not always lower than the k_{vTST} value when determined by B3LYP. It seems that the results obtained by ωB97X-D become more regular through the use of the DFT-D dispersion correction. The $\kappa(T)$ of all reactions were found to range from 1.2 to 1.4. This increased k to almost the same level, which implies that $\kappa(T)$ does not significantly affect k. The reaction rate of cyclopentadiene with methyl vinyl ketone was 1.46×10^{-23} $cm^3 \cdot (molec^{-1} \cdot s^{-1})$ [22]. Experiments gave an Arrhenius activation energy for BD + ET of 115.5 kJ/mol at temperatures between 760 K and 921 K [23]. The E_a obtained by ωB97X-D was close to the experimental result.

4. Conclusions

The incorporation of an electron-withdrawing group affects the change in n_{Mayer}. E_i is the dominating energy when the cycloaddition reaction proceeds from the RC to the IP. The RDG analysis found that a dienophile substituted by an electron-withdrawing group decreased the spike value of the distal end. CECA also revealed that exchange and overlap terms provided the largest contributions to the two-center energy. Finally, k was related to E_a, but $\kappa(T)$ increased in equal proportion to k.

Conflicts of Interest

The authors declare that they have no conflicts of interest.

Authors' Contributions

Zhiling Liang, Houhe Liu, and Nianjun Su contributed to this work equally and should be regarded as co-first authors.

Acknowledgments

The authors' works are supported by the National Natural Science Foundation of China (Grant no. 21274032) and the Natural Science Foundation of Guangdong Province (Grant no. 2014A030313500). It was also funded by the project of the high level university of Guangdong province. The corresponding author, Guodong Ye, thanks the above funding sources.

References

[1] E. F. Kiefer and W. Gericke, "Nuclear magnetic resonance investigation of secondary valence forces. II. Intramolecular mercury-halogen coordination in 3-halopropylmercury compounds," *Journal of the American Chemical Society*, vol. 90, no. 19, pp. 5131–5136, 1968.

[2] X. An, H. Zhuo, W. Tian, Q. Li, W. Li, and J. Cheng, "Novel non-covalent interactions involved with the Al13M cluster (M = Li, Na, K, Cu, Ag, Au)," *Molecular Physics*, vol. 112, no. 22, pp. 2954–2962, 2014.

[3] R. Chaudret, B. De Courcy, J. Contreras-García et al., "Unraveling non-covalent interactions within flexible biomolecules: From electron density topology to gas phase spectroscopy," *Physical Chemistry Chemical Physics*, vol. 16, no. 21, pp. 9876–9891, 2014.

[4] D. J. R. Duarte, G. L. Sosa, and N. M. Peruchena, "Nature of halogen bonding. A study based on the topological analysis of the Laplacian of the electron charge density and an energy decomposition analysis," *Journal of Molecular Modeling*, vol. 19, no. 5, pp. 2035–2041, 2013.

[5] D. Mani and E. Arunan, "The X-C···Y (X = O/F, y = O/S/F/Cl/Br/N/P) 'carbon bond' and hydrophobic interactions," *Physical Chemistry Chemical Physics*, vol. 15, no. 34, pp. 14377–14383, 2013.

[6] D. Hugas, L. Guillaumes, M. Duran, and S. Simon, "Delocalization indices for non-covalent interaction: Hydrogen and DiHydrogen bond," *Computational and Theoretical Chemistry*, vol. 998, pp. 113–119, 2012.

[7] F. Fuster and S. J. Grabowski, "Intramolecular hydrogen bonds: The QTAIM and ELF characteristics," *The Journal of Physical Chemistry A*, vol. 115, no. 35, pp. 10078–10086, 2011.

[7] F. Fuster and S. J. Grabowski, "Intramolecular hydrogen bonds: The QTAIM and ELF characteristics," *The Journal of Physical Chemistry A*, vol. 115, no. 35, pp. 10078–10086, 2011.

[8] S. Grimme, C. Mück-Lichtenfeld, and J. Antony, "Noncovalent interactions between graphene sheets and in multishell (hyper)fullerenes," *The Journal of Physical Chemistry C*, vol. 111, no. 30, pp. 11199–11207, 2007.

[9] S. Osuna and K. N. Houk, "Cycloaddition reactions of butadiene and 1,3-dipoles to curved arenes, fullerenes, and nanotubes: Theoretical evaluation of the role of distortion energies on activation barriers," *Chemistry - A European Journal*, vol. 15, no. 47, pp. 13219–13231, 2009.

[10] A. J. Bridgeman, G. Cavigliasso, L. R. Ireland, and J. Rothery, "The Mayer bond order as a tool in inorganic chemistry," *Journal of the Chemical Society, Dalton Transactions*, no. 14, pp. 2095–2108, 2001.

[11] E. R. Johnson, S. Keinan, P. Mori-Sánchez, J. Contreras-García, A. J. Cohen, and W. Yang, "Revealing noncovalent interactions," *Journal of the American Chemical Society*, vol. 132, no. 18, pp. 6498–6506, 2010.

[12] N. Gillet, R. Chaudret, J. Contreras-García, W. Yang, B. Silvi, and J.-P. Piquemal, "Coupling quantum interpretative techniques: Another look at chemical mechanisms in organic reactions," *Journal of Chemical Theory and Computation*, vol. 8, no. 11, pp. 3993–3997, 2012.

[13] A. Hamza and I. Mayer, "Physical analysis of the diatomic "chemical" energy components," *Theoretical Chemistry Accounts*, vol. 109, no. 2, pp. 91–98, 2003.

[14] E. J. Corey and T. Loh, "First application of attractive intramolecular interactions to the design of chiral catalysts for highly enantioselective Diels-Alder reactions," *Journal of the American Chemical Society*, vol. 32, no. 23, pp. 6289–6292, 1991.

[15] M. J. Frisch, G. W. Trucks, and H. B. Schlegel, 2010, Gaussian 09. Gaussian Inc., Wallingford.

[16] T. Lu and F. Chen, "Multiwfn: a multifunctional wavefunction analyzer," *Journal of Computational Chemistry*, vol. 33, no. 5, pp. 580–592, 2012.

[17] S. Canneaux, F. Bohr, and E. Henon, "KiSThelP: A program to predict thermodynamic properties and rate constants from quantum chemistry results," *Journal of Computational Chemistry*, vol. 35, no. 1, pp. 82–93, 2014.

[18] I. Mayer, "On bond orders and valences in the Ab initio quantum chemical theory," *International Journal of Quantum Chemistry*, vol. 29, no. 1, pp. 73–84, 1986.

[19] C. A. Sader and K. N. Houk, "Distortion/interaction analysis of the reactivities and selectivities of halo- and methoxy-substituted carbenes with alkenes," *Arkivoc Archive for Organic Chemistry*, vol. 77, no. 77, pp. 843–850, 2014.

[20] D. H. Ess and K. N. Houk, "Theory of 1,3-dipolar cycloadditions: Distortion/interaction and frontier molecular orbital models," *Journal of the American Chemical Society*, vol. 130, no. 31, pp. 10187–10198, 2008.

[21] A. Fernandez-Ramos, B. A. Ellingson, B. C. Garrett, and D. G. Truhlar, "Variational transition state theory with multidimensional tunneling," *Reviews in Computational Chemistry*, vol. 23, pp. 125–232, 2007.

[22] L. L. Thomas, J. Tirado-Rives, and W. L. Jorgensen, "Quantum mechanical/molecular mechanical modeling finds diels alder reactions are accelerated less on the surface of water than in water," *Journal of the American Chemical Society*, vol. 132, no. 9, pp. 3097–3104, 2010.

[23] J. Sauer and R. Sustmann, "Mechanistic Aspects of Diels-Alder Reactions: A Critical Survey," *Angewandte Chemie International Edition*, vol. 19, no. 10, pp. 779–807, 1980.

14

MCM-41-Accelerated PWA Catalysis of Friedel-Crafts Reaction of Indoles and Isatins

Liuzhuang Xing,[1] YongHai Hui,[1] Jinghui Yang,[1] Xuejian Xing,[1] Yadong Hou,[1] Yang Wu,[1] Kui Fan,[1] and Wei Wang[2]

[1]*Key Laboratory of Oil & Gas Fine Chemicals, Ministry of Education & Xinjiang Uyghur Autonomous Region, College of Chemistry and Chemical Engineering, Xinjiang University, Urumqi 830046, China*
[2]*Department of Chemistry and Centre for Pharmacy, University of Bergen, 5020 Bergen, Norway*

Correspondence should be addressed to YongHai Hui; hyhai97@126.com and Wei Wang; wei.wang@uib.no

Academic Editor: Pasquale Longo

Ordered mesoporous siliceous material has been identified as one of the key elements of the catalysis concept. Here we report an efficient Friedel-Crafts reaction of indoles with isatins catalyzed by PWA/MCM-41, which got the di(indolyl)indolin-2-ones derivatives with high yield. Moreover, the catalysts were characterized by XRD and SEM/EDS, the EDS spectrum indicated that the catalyst used in this reaction also contains tungsten, and the proposed mechanism for the synthesis of 3,3-di(indolyl)indolin-2-ones was also discussed. Finally, the catalyst can be reused repeatedly for several times without obvious loss of activity.

1. Introduction

Friedel-Crafts reaction is a fundamental transformation for C-C bonds formation in organic synthesis [1]. Among the Friedel-Crafts reaction substrate of aromatic compounds, indoles and indole derivatives have received more and more attention [2] because of their privileged structure and quite pervasiveness in nature [3] as well as their excellent biological activities [4]. Isatin is one of the most important derivatives of indole with significant bioactive scaffold, containing double adjacent carbonyls with an indole unit and a γ-lactam moiety, which endows isatin with an amount of interesting synthetic applications [5].

The carbonyl group in 3-position of isatins readily makes isatins undergo electrophilic condensation reactions with indoles to produce 3,3-di(indolyl)indolin-2-ones, which also show excellent bioactivities [6–8]. Some literature sources revealed that the synthesis of 3,3-diaryloxindole skeleton can be achieved by coupling of indoles and isatins with acid [9, 10], I_2 [11], metal salts [12], ionic liquids [13], CAN [14], β-CD [15], Amberlyst-15 [16], and nanomaterials [17]. Despite these important contributions, developing and searching for a highly effective catalytic system is still challenging.

In the last few years, ordered mesoporous siliceous material supporting catalysts have been paid much attention because of their high surface areas, large pores, and ordered arrangement of structures [18]. Phosphotungstic acid (PWA) supported on heterogeneous materials has been widely used for Friedel-Crafts reaction [19, 20]. As a continuation of our interest in catalyst of supported mesoporous material in the organic synthesis [21, 22], herein, we reported isatins as the electrophiles to react with indoles using PWA/MCM-41 as catalyst at room temperature for the synthesis of 3,3-di(indolyl)indolin-2-ones.

2. Experimental

The purity of compounds was checked by thin layer chromatography (TLC) using ethyl acetate/petroleum ether (v/v) as eluent; IR spectrum was recorded on a Bruker Equinox 55 FT-IR spectrophotometer as KBr discs. The ^{1}H NMR and ^{13}C NMR spectra were recorded on an INOVA (400 MHz) FT-NMR spectrometer in $CDCl_3$ using TMS as an internal standard. The prepared catalysts were characterized by XRD, SEM, and EDS.

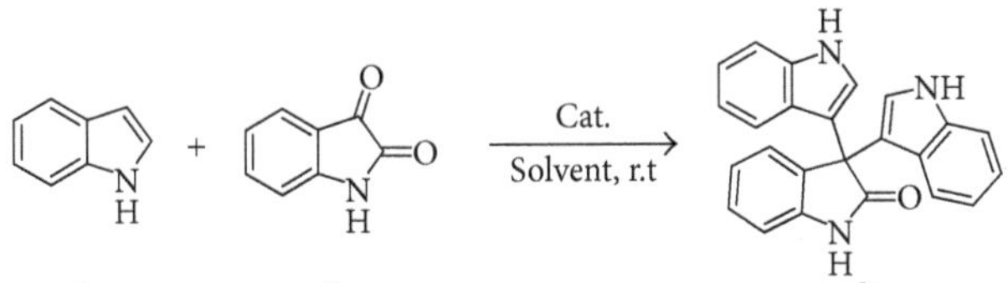

SCHEME 1: The Friedel-Crafts reaction of indole with isatin.

Indoles (0.20 mmol), isatins (0.10 mmol), and 60 wt% PWA/MCM-41 catalyst (0.0050 g) were stirred in THF (0.3 mL) at room temperature for 2.5 h and monitored by TLC. The product were purified by column chromatography on silica gel using petroleum ether/ethyl acetate (20 : 1, 10 : 1, 5 : 1, 2 : 1, 1 : 1) as the eluent. All compounds (**3a**–**3o**) were characterized by ^{1}H and ^{13}C NMR (400 MHz) spectral analysis. The catalyst was separated from the reaction mixture by centrifugalization, dried at 473 K for 4 h after washing with ethanol, then continued to be reused for the next reaction.

3. Results and Discussion

Initially, as a continuation of our work on the applications of mesoporous siliceous material catalysts for organic reaction, to optimize the reaction condition of the Friedel-Crafts reaction of indoles with isatins, we utilized isatin (0.10 mmol) and indole (0.20 mmol) as model substrates (Scheme 1), and the results are shown in Table 1. MCM-41 or PMoA as catalysts led to lower yields than PWA as catalyst (Table 1, entries (1)–(3)). To our delight, the yield was significantly improved with PWA/MCM-41 catalyzing the model reaction and the 60 wt% PWA/MCM-41 being the optimized catalyst that contrasted with (40–80) wt% and obtained up to 86% yield (Table 1, entries (4)–(8)). With the ratio of PWA/MCM-41 exceeding 60 wt%, the yields of the 3,3-di(indolyl)indolin-2-one decreased, because the MCM-41 pores might be blocked by the excessive amounts of PWA. Comparative experiments indicated that the pores indeed facilitated the Friedel-Crafts reaction [21].

The optimal catalyst was established, and further screening of the reaction conditions was focused on solvents, catalyst quantity, solvent quantity, and reaction time. The results revealed that the use of THF as a reaction medium was superior to other solvents (Table 1, entry (6) versus entries (9)–(15)). And with the amount of catalyst reduced from 0.0050 g to 0.0030 g, the yield of the reaction was significantly decreased (Table 1, entries (16) and (17)). However, the amount of solvent obviously influenced the formation of the 3,3-di(indolyl)indolin-2-one and the best outcome was obtained with 0.3 mL THF at 0.1 mmol scale of isatin (Table 1, entry (18)). To our delight, the reaction time was slightly decreased to 2.5 h and could also successfully deliver the desired product in excellent yield up to 99% (Table 1, entry (19)). Therefore, (0.005 g) 60 wt% PWA/MCM-41 catalyzed (0.10 mmol) isatins with (0.20 mmol) indoles with (0.3 mL) THF at room temperature for 2.5 h, which proved to be a robust method to build up the 3,3-di(indolyl)indolin-2-ones scaffold.

TABLE 1: Optimization of the reaction conditions[a].

Entry	Catalyst	Solvent	Time (h)	Yield (%)[b]
(1)	MCM-41	THF	3	31
(2)	PMoA	THF	3	39
(3)	PWA	THF	3	46
(4)	40 wt% PWA/MCM-41	THF	3	53
(5)	50 wt% PWA/MCM-41	THF	3	60
(6)	60 wt% PWA/MCM-41	THF	3	86
(7)	70 wt% PWA/MCM-41	THF	3	70
(8)	80 wt% PWA/MCM-41	THF	3	68
(9)	60 wt% PWA/MCM-41	EtOH	3	73
(10)	60 wt% PWA/MCM-41	Et_2O	3	54
(11)	60 wt% PWA/MCM-41	AcOEt	3	35
(12)	60 wt% PWA/MCM-41	acetone	3	47
(13)	60 wt% PWA/MCM-41	toluene	3	47
(14)	60 wt% PWA/MCM-41	CH_3CN	3	34
(15)	60 wt% PWA/MCM-41	CH_2Cl_2	3	24
(16)[c]	60 wt% PWA/MCM-41	THF	3	83
(17)[d]	60 wt% PWA/MCM-41	THF	3	43
(18)[e]	60 wt% PWA/MCM-41	THF	3	99
(19)[e]	60 wt% PWA/MCM-41	THF	2.5	99

[a]Reaction conditions (unless noted otherwise): 0.20 mmol of indole, 0.10 mmol of isatin, 0.0050 g of catalyst in 0.2 mL of solvent at room temperature for 3 h. [b]Isolated yield. [c]Use 0.0040 g of 60 wt% PWA/MCM-14 catalyst. [d]Use 0.0030 g of 60 wt% PWA/MCM-14 catalyst. [e]0.3 mL THF.

General procedure for synthesis of catalysts was exhibited in supporting information (see SI 2.1 S1). The catalysts were characterized using SEM equipped with EDS and XRD. The EDS spectrum and elemental contents of some zones are shown in Figure 1. It can be seen that the optimal catalyst (B zone in Figure 1(b)) in this system contains rich tungsten (X-ray energy 1.774 keV) compared with the pure MCM-41 [23] (A zone in Figure 1(a)). Meanwhile, the EDS spectrum indicated that the catalyst used in this reaction also contains tungsten and the catalyst could be used repeatedly.

The bands in the X-ray diffraction patterns (Figure 2) of pure MCM-41 (a), as well as the catalyst of PWA supported on MCM-41 (b) and the used catalyst PWA/MCM-41 (c) in the range of 1°–8° (Figure 2(a)), suggest that the hexagonal pore arrangement was kept [24]. The catalyst of 60 wt% PWA/MCM-41 exhibited clear diffractions at 10.2°, 20.7°, 22.5°, 29.6°, and 34.9° compared with pure MCM-41 that was associated with PWA (Figure 2(b)) [25], suggesting that PWA was supported on MCM-41 and the catalyst of PWA remained after catalyzing this reaction.

With the optimized parameters established, the substrate scope was investigated (Scheme 2) and the results were listed in Table 2. When the phenyl ring of indoles or isatins interacts with methyl substituent, the desired products **3b**–**3d** and **3k** are obtained in a moderate to good yield (Table 2, entries (2)–(4), (11)). However, product **3l**, the yield of 5-methyl-3,3-di(indolyl) oxindole, was excellent (Table 2, entry

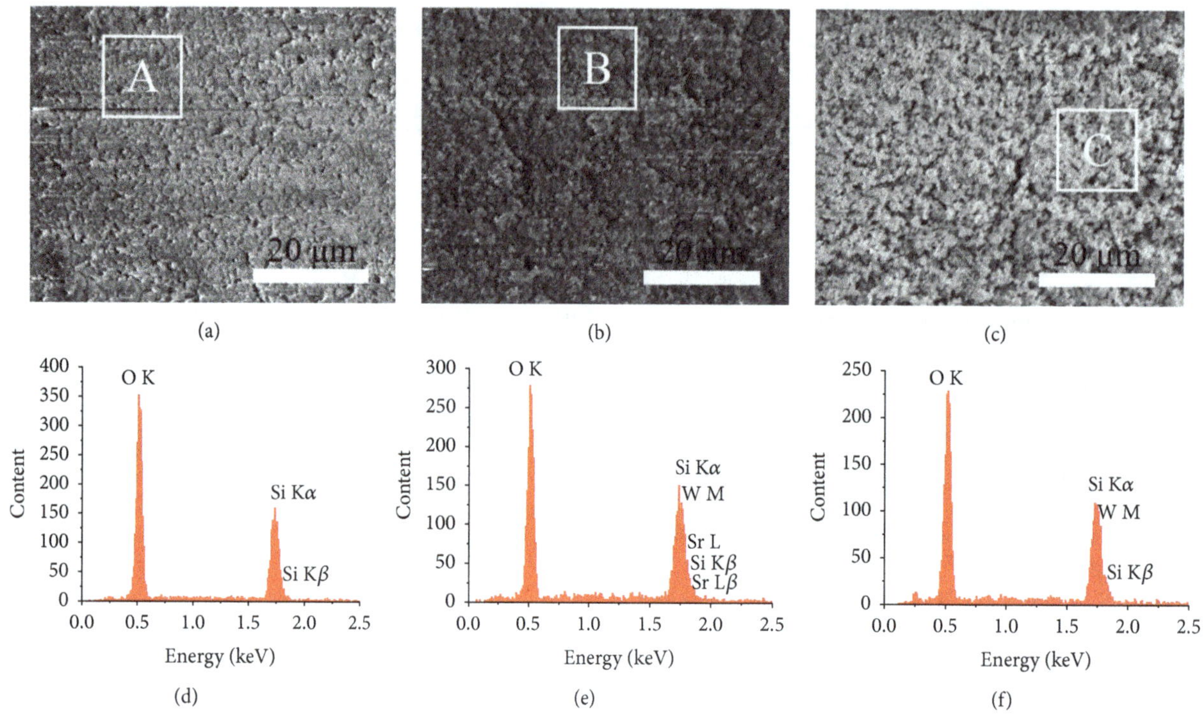

FIGURE 1: The SEM of pure MCM-41 (a), fresh catalyst 60 wt% PWA/MCM-41 (b), used catalyst 60 wt% PWA/MCM-41 (c), and EDS spectrum and elemental contents of the region A in (d), region B in (e), and region C in (f).

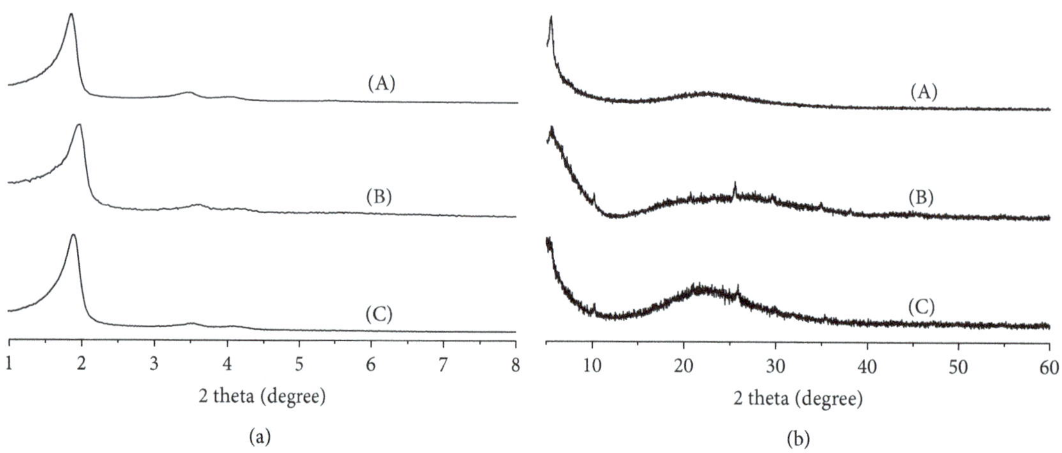

FIGURE 2: The small-angle range XRD (a) and wide-angle range XRD (b), pure MCM-41 (A), fresh catalyst 60 wt% PWA/MCM-41 (B), and used catalyst 60 wt% PWA/MCM-41 (C).

(12)). Wherever -F and -Cl groups are located in indoles or isatins of aromatic rings, they all could be tolerated well and gave the desired product derivatives **3e–3h**, **3m**, and **3o** in excellent yields (Table 2, entries (5)–(8), (13), and (15)). However, the benzene ring of indoles or isatins bearing sterically demanding groups, such as -Br and -NO_2, could only give corresponding products with lower yields, respectively (Table 2, entries (9), (10), and (14)). All compounds (**3a–3o**) were characterized by ^1H and ^{13}C NMR (400 MHz) spectral analysis (for all physical data of compounds, see SI 3 S2–S11).

On the basis of previous studies [9] and above results, a mechanism for this reaction was proposed. The first step is the activation of the 3-position carbonyl group of isatin (**2**) by 60 wt% PWA/MCM-41, which facilitates nucleophilic addition of indole (**1**) to form corresponding 3-hydroxy-3-indolyl oxindole (**5**). Then, hydrogen of indolyloxindole transferred to hydroxyl group and gave the intermediate (**6**), followed by its dehydration to (**7**). Finally, another indole reacted with (**7**) to provide the desired product 3,3-di(indolyl)indolin-2-one (**3**) and regenerate the catalyst for the next catalytic cycle (Scheme 3).

The study of reusability of catalytic is an important aspect of green chemistry. To investigate it, we used the above model reaction and optimized condition. After completion

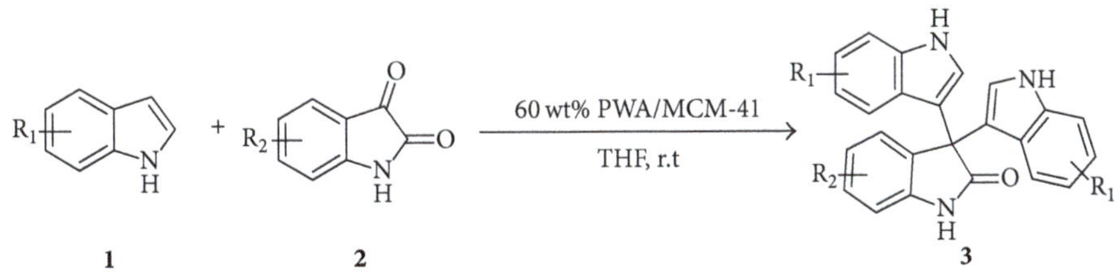

SCHEME 2: The Friedel-Crafts reaction of indoles with isatins.

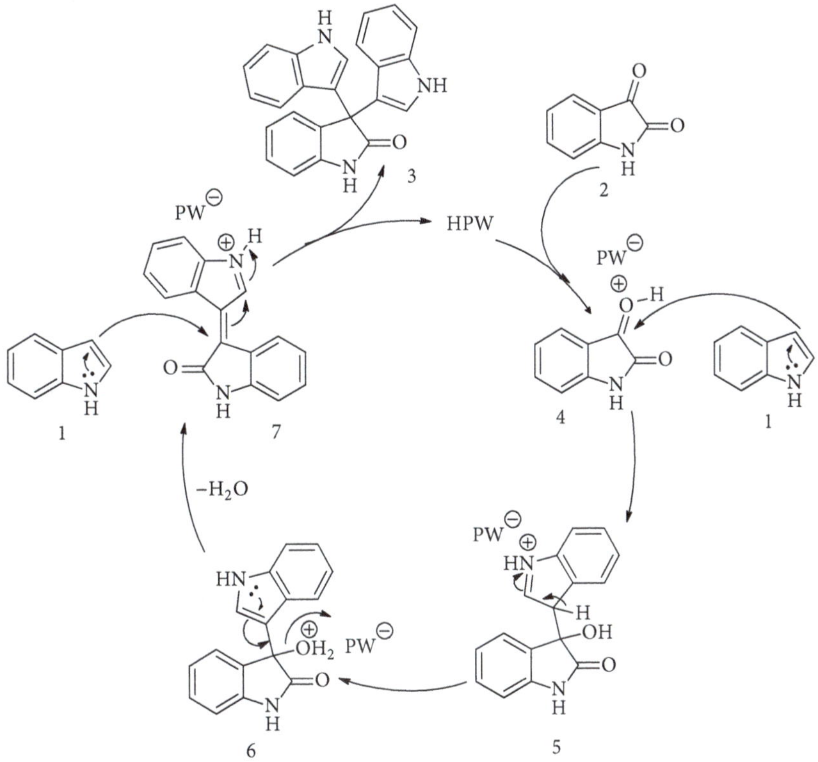

SCHEME 3: Proposed catalytic mechanism.

of the reaction, the catalyst was separated from the reaction mixture by centrifugalization, then washed with ethanol, and dried at 473 K for 4 h. The recovered catalyst was used for five consecutive reactions with no significant loss of activity (Figure 3) (for general cyclic procedure, see SI 2.3 S2).

4. Conclusion

In summary, we developed an efficient heterogeneous PWA/MCM-41 catalyzed Friedel-Crafts reaction of isatins with indoles, delivering a series of 3,3-di(indolyl)indolin-2-ones. The catalytic reaction could be carried out under mild conditions with broad substrate scope, good efficiency, and well functional group compatibility. Furthermore, the catalyst keeps the basic hexagonal pore arrangement and the components of PWA. The recovered catalyst was reused for five consecutive reactions with no significant loss of activity. Further investigation focusing on the scope of the currently synthetic applications is being conducted in our group.

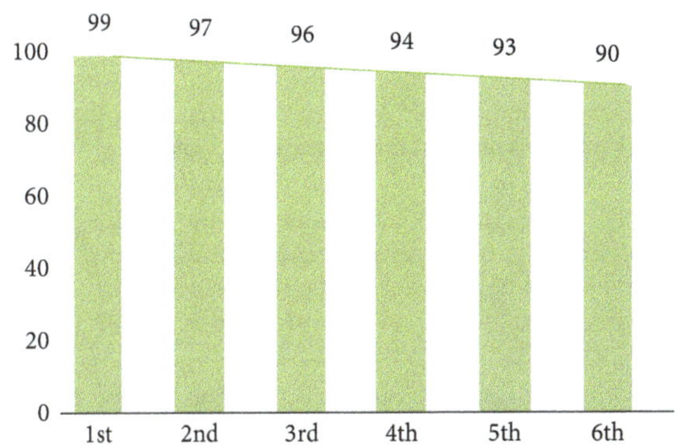

FIGURE 3: Reusability of the catalyst 60 wt% PWA/MCM-41.

TABLE 2: Substrate scope for the cyclocondensation recation[a].

Entry	R^1	R^2	Products	Yield (%)[b]	Yield (%)[c]	Ref.
(1)	H	H	**3a**	99	89	[10]
(2)	2-CH_3	H	**3b**	70	91	[9]
(3)	5-CH_3	H	**3c**	55	85	[14]
(4)	6-CH_3	H	**3d**	80	86	[14]
(5)	5-F	H	**3e**	96	87	[10]
(6)	6-F	H	**3f**	98	-	-
(7)	5-Cl	H	**3g**	98	89	[26]
(8)	6-Cl	H	**3h**	98	91	[26]
(9)	5-NO_2	H	**3i**	72	84	[14]
(10)	6-NO_2	H	**3j**	64	-	-
(11)	H	1-CH_3	**3k**	76	91	[10]
(12)	H	5-CH_3	**3l**	98	90	[9]
(13)	H	5-Cl	**3m**	99	86	[10]
(14)	H	4-Br	**3n**	60	90	[14]
(15)	H	7-F	**3o**	99	-	-

[a]Reaction conditions (unless noted otherwise): indoles (**1**, 0.20 mmol), isatins (**2**, 0.10 mmol), and 0.0050 g 60 wt% PWA/MCM-41 as catalyst in 0.3 mL THF at room temperature for 2.5 h. [b]Isolated yield. [c]Yield of reference.

Conflicts of Interest

All authors declare no conflicts of interest.

Acknowledgments

The authors acknowledge the National Natural Science Foundation of China (nos. 21362036 and 21162026) and the Thousand Talents Plan for financial support, Xinjiang University Analytical & Testing Center for instrumental analyses, and Adamas-beta Chemical Co. for all chemical reagents.

References

[1] G. Sartori and R. Maggi, "Use of solid catalysts in Friedel-Crafts acylation reactions," *Chemical Reviews*, vol. 111, no. 20, pp. 181–214, 2011.

[2] X. Liang, S. Z. Jiang, K. Wei, and Y. R. Yang, "Enantioselective total synthesis of (-)-Alstoscholarisine A," *Journal of the American Chemical Society*, vol. 138, no. 8, pp. 2560–2562, 2016.

[3] G. R. Pettit, J. D. Searcy, R. Tan et al., "Antineoplastic Agents. 585. Isolation of Bridelia ferruginea Anticancer Podophyllotoxins and Synthesis of 4-Aza-podophyllotoxin Structural Modifications," *Journal of Natural Products*, vol. 79, no. 3, pp. 507–518, 2016.

[4] H. Gurer-Orhan, C. Karaaslan, S. Ozcan et al., "Novel indole-based melatonin analogues: evaluation of antioxidant activity and protective effect against amyloid β-induced damage," *Bioorganic & Medicinal Chemistry*, vol. 24, no. 8, pp. 1658–1664, 2016.

[5] M. S. Shmidt, I. A. Perillo, A. Camelli, M. A. Fernández, and M. M. Blanco, "Polyfunctional 4-quinolinones. Synthesis of 2-substituted 3-hydroxy-4-oxo-1,4-dihydroquinolines," *Tetrahedron Letters*, vol. 57, no. 9, pp. 1022–1026, 2016.

[6] D. Sriram, P. Yogeeswari, N. S. Myneedu, and V. Saraswat, "Abacavir prodrugs: microwave-assisted synthesis and their evaluation of anti-HIV activities," *Bioorganic and Medicinal Chemistry Letters*, vol. 16, no. 8, pp. 2127–2129, 2006.

[7] T. Jiang, K. L. Kuhen, K. Wolff et al., "Design, synthesis and biological evaluations of novel oxindoles as HIV-1 non-nucleoside reverse transcriptase inhibitors. Part I," *Bioorganic & Medicinal Chemistry Letters*, vol. 16, no. 8, pp. 2105–2108, 2006.

[8] A. Kamal, Y. V. V. Srikanth, M. N. A. Khan, T. B. Shaik, and M. Ashraf, "Synthesis of 3,3-diindolyl oxyindoles efficiently catalysed by $FeCl_3$ and their in vitro evaluation for anticancer activity," *Bioorganic & Medicinal Chemistry Letters*, vol. 20, no. 17, pp. 5229–5231, 2010.

[9] J. Azizian, A. A. Mohammadi, N. Karimi, M. R. Mohammadizadeh, and A. R. Karimi, "Silica sulfuric acid a novel and heterogeneous catalyst for the synthesis of some new oxindole derivatives," *Catalysis Communications*, vol. 7, no. 10, pp. 752–755, 2006.

[10] G. Brahmachari and B. Banerjee, "Facile and one-pot access of 3,3-bis(indol-3-yl)indolin-2-ones and 2,2-bis(indol-3-yl)acenaphthylen-1(2H)-one derivatives via an eco- friendly pseudo-multicomponent reaction at room temperature using sulfamic acid as an organo-catalyst," *ACS Sustainable Chemistry & Engineering*, vol. 2, no. 12, pp. 2802–2812, 2014.

[11] P. Paira, A. Hazra, S. Kumar et al., "Efficient synthesis of 3,3-diheteroaromatic oxindole analogues and their in vitro evaluation for spermicidal potential," *Bioorganic & Medicinal Chemistry Letters*, vol. 19, no. 16, pp. 4786–4789, 2009.

[12] C. Praveen, A. Ayyanar, and P. T. Perumal, "Gold(III) chloride catalyzed regioselective synthesis of pyrano[3,4-b]indol-1(9H)-ones and evaluation of anticancer potential towards human cervix adenocarcinoma," *Bioorganic & Medicinal Chemistry Letters*, vol. 21, no. 14, pp. 4170–4173, 2011.

[13] K. Rad-Moghadam, M. Sharifi-Kiasaraie, and H. Taheri-Amlashi, "Synthesis of symmetrical and unsymmetrical 3,3-di(indolyl)indolin-2-ones under controlled catalysis of ionic liquids," *Tetrahedron*, vol. 66, no. 13, pp. 2316–2321, 2010.

[14] S. Y. Wang and S. J. Ji, "Facile synthesis of 3,3-di(heteroaryl)indolin-2-one derivatives catalyzed by ceric ammonium nitrate (CAN) under ultrasound irradiation," *Tetrahedron*, vol. 62, no. 7, pp. 1527–1535, 2006.

[15] Y. A. Tayade, D. R. Patil, Y. B. Wagh, A. D. Jangle, and D. S. Dalal, "An efficient synthesis of 3-indolyl-3-hydroxy oxindoles and 3,3-di(indolyl)indolin-2-ones catalyzed by sulfonated β-CD as a supramolecular catalyst in water," *Tetrahedron Letters*, vol. 56, no. 5, pp. 666–673, 2015.

[16] Y. Sarrafi, K. Alimohammadi, M. Sadatshahabi, and N. Norozipoor, "An improved catalytic method for the synthesis of 3,3-di(indolyl)oxindoles using Amberlyst 15 as a heterogeneous and reusable catalyst in water," *Monatshefte für Chemie*, vol. 143, no. 11, pp. 1519–1522, 2012.

[17] S. Z. Sayyed-Alangi and Z. Hossaini, "ZnO nanorods as an efficient catalyst for the green synthesis of indole derivatives using isatoic anhydride," *Chemistry of Heterocyclic Compounds*, vol. 51, no. 6, pp. 541–544, 2015.

[18] K. Sugino, N. Oya, N. Yoshie, and M. Ogura, "A simple modification creates a great difference: new solid-base catalyst using methylated N-substituted SBA-15," *Journal of the American Chemical Society*, vol. 133, no. 50, pp. 20030–20032, 2011.

[19] Suryani, C. M. Chang, Y. L. Liu, and Y. M. Lee, "Polybenzimidazole membranes modified with polyelectrolyte-functionalized multiwalled carbon nanotubes for proton exchange membrane fuel cells," *Journal of Materials Chemistry*, vol. 21, no. 20, pp. 7480–7486, 2011.

[20] X. Li, G. Qin, Y. Wang, and W. Wei, "Keggin-type phosphotungstic acid supported on mesoporous SiO_2–Al_2O_3 aerogel like beads and their application in the isopropylation of naphthalene," *Journal of Sol-Gel Science and Technology*, vol. 72, no. 2, pp. 405–414, 2014.

[21] K. Fan, Y. H. Hui, X. H. Hu, W. Shi, H. X. Pang, and Z. F. Xie, "PMoA/MCM-41 catalyzed aza-Michael reaction: special effects of mesoporous nanoreactor on chemical equilibrium and reaction rate through surface energy transformation," *New Journal of Chemistry*, vol. 39, no. 8, pp. 5916–5919, 2015.

[22] H. X. Pang, Y. H. Hui, K. Fan et al., "A catalysis study of mesoporous MCM-41 supported Schiff base and $CuSO_4·5H_2O$ in a highly regioselective synthesis of 4-thiazolidinone derivatives from cyclocondensation of mercaptoacetic acid," *Chinese Chemical Letters*, vol. 27, no. 3, pp. 335–339, 2016.

[23] D. E. Newbury and N. W. M. Ritchie, "Performing elemental microanalysis with high accuracy and high precision by scanning electron microscopy/silicon drift detector energy-dispersive X-ray spectrometry (SEM/SDD-EDS)," *Journal of Materials Science*, vol. 50, no. 2, pp. 493–518, 2014.

[24] C. T. Kresge, M. E. Leonowicz, W. J. Roth, J. C. Vartuli, and J. S. Beck, "Ordered mesoporous molecular sieves synthesized by a liquid-crystal template mechanism," *Nature*, vol. 359, no. 6397, pp. 710–712, 1992.

[25] S. H. Chai, H. P. Wang, Y. Liang, and B. Q. Xu, "Sustainable production of acrolein: gas-phase dehydration of glycerol over 12-tungstophosphoric acid supported on ZrO2 and SiO2," *Green Chemistry*, vol. 10, no. 10, pp. 1087–1093, 2008.

[26] S. J. Gharpure and S. K. Nanda, "Stereoselective synthesis of thiazino[4,3-: A] indoles using the thia-Pictet-Spengler reaction of indoles bearing N -tethered thiols and vinylogous thiocarbonates," *Organic & Biomolecular Chemistry*, vol. 14, no. 24, pp. 5586–5590, 2016.

15

Microwave-Assisted Synthesis of Tetrahydrocarbazole Linked 1,2-Diazoles in Aqueous-Brønsted Acid Catalytic System

Zubi Sadiq,[1] Erum Akbar Hussain,[1] Narjis Naz,[1] Ambreen Ghani,[2] Kausar Yasmeen,[1] and Fizzah Iqbal[1]

[1]*Department of Chemistry, Lahore College for Women University, Lahore 54000, Pakistan*
[2]*Department of Chemistry, University of Education, D.G. Khan Campus, Dera Ghazi Khan, Pakistan*

Correspondence should be addressed to Erum Akbar Hussain; erum.hussain@gmail.com

Academic Editor: Mohamed Afzal Pasha

A new series of tetrahydrocarbazole coupled 1,2-diazoles **4/5/6(a–o)** in moderate to excellent yield was synthesized successfully via multicomponent reaction approach by adopting Michael type-addition of hydrazines on in situ generated α,β-unsaturated ketones via aldol type strategy under synthetic microwave radiations and conventional heating. Structural confirmations of all the prepared compounds were achieved by spectroscopic techniques. The best results were achieved when equal amounts of water-acetic acid were used in microwave conditions in 4 minutes.

1. Introduction

Diazole ring is prominent moiety in wide range of bioactive compounds, agrochemicals, dyestuffs, photographic, and heat resistant polymers and attains distinct interest of researchers [1, 2]. Sufficient literature is available on its electronic applications as they are Intramolecular Charge Transfer (ICT) compounds and demonstrate fluorescent brightening characteristics. They possess distinct whole transport tendency in solar cells and electroluminescent devices [3]. Due to its synthetic versatility, it is one of the most important and stimulating units of high profiled molecules particularly with medicinal impact [4–7]. Moreover, 1,2-diazole having carbazole hybrids exhibit prominently enhanced material properties [8]. Tetrahydrocarbazole belongs to condensed heterocyclic system. Increasing demand of green protocols in organic synthesis provides microwave-assisted approach to environmentally undesirable procedures. Microwave mode is believed as a better lead in synthetic transformation with central usefulness such as accelerated rate of reaction, improved yield, clean setup, ecofriendly, economical, and/or high atom economy hence termed as "e-chemistry" [9]. Mostly one-pot multicomponent reactions (MCR) heteroannulation is successfully achieved on exposure to microwaves in one-pot assembly of neat reactants [10, 11].

Water as reaction solvent has become the prime choice due to ecofriendly nature [12]. This green solvent in combination with Brønsted acid like acetic acid is developed as catalytic system to afford desired 1,2-diazole entities in a benign way. This biodegradable catalytic system thus furnished mild conditions and sustainable reaction medium. In view of ongoing investigations on green protocols for preparing tetrahydrocarbazoles in our lab [13–15], we have modified conventional approach of condensation followed by Michael type-addition to obtain tetrahydrocarbazole linked 1,2-diazoles under synthetic microwave. Reaction time, yield, and purity of compounds have been compared in both processes individually. This effort is a contribution to designing innovative molecules with diverse properties in a clean and speedy way.

2. Materials and Methods

2.1. Materials. Melting points were recorded by open capillary method and are uncorrected. U-2800 Hitachi, UV-visible spectrophotometer was used for finding λ_{max}. KBr discs were used to record FTIR spectra on Midac Corporation FTIR spectrophotometer and values were reported in cm^{-1}. ^1H NMR spectra were measured on Bruker AXS 400 MHz spectrometer using TMS as an internal standard and DMSO-d_6

as solvent (chemical shift in δ, ppm). Mass spectra were recorded by GCMS-QP2010S Shimadzu Scientific Instruments, Inc. CHEM Discover 908010 focused microwave synthesis system was used to carry out all mentioned organic transformations. Crude products were identified on TLC (silica gel coated Al sheets, Merck 60 F_{254}, 0.2 mm thick) using UV light at 254/365 nm (CAMAG scientific Inc).

2.2. Synthesis

2.2.1. Tetrahydrocarbazole Linked 1,2-Diazoles Synthesis on Conventional Heating. To the solution of equimolar amount of substrate **(1a/1b/1c)** in dry methanol was added sodium methoxide (20 mol%) followed by addition of respective aldehydes **(2a–e)** (Table 1). After 30-minute reflux, hydrazine **(3a/3b/3c)** was added drop wise and reaction continued for further hours. When it got completed, monitored through TLC, reaction mixture was poured on crushed ice to get crude product precipitates. It was recrystallized in ethanol, chromatograph over silica (dichloromethane-methanol; 9.5 : 0.5), and yield was calculated.

2.2.2. Microwave Synthesis of Tetrahydrocarbazole Linked 1,2-Diazoles. 1.0 equiv of substrate **(1a/1b/1c)**, 1.0 equiv of aromatic aldehyde **(2a–e),** and 1.2 equiv of respective hydrazines **(3a/3b/3c)** were added in water-acetic acid (1:1) solvent in 10 mL sealed tube and irradiated at 300 W for appropriate time (Table 2). Standard microwave protocols were followed throughout reaction. After completion of the reaction, it was concentrated in vacuum; crude product was subjected to recrystallization with methanol to get pure product.

*2,3,4,9-Tetrahydro-1-(4,5-dihydro-5-phenyl-1H-pyrazol-3-yl)-1H-carbazole **(4a)**.* FTIR (KBr) cm^{-1}: 1578 (C=C), 1643 (C=N), 1043 (C-N), 2984 (C-H), 3064 (N-H); ^{1}H NMR (d_6-DMSO) δ: 10.33 (1H, s, H-9), 7.85–7.52 (5H, m, Ph), 7.48–7.32 (4H, m, Ph), 6.21 (1H, s, N-H pyra), 4.81 (1H, m, H_c), 3.68 (1H, dd, J = 18, 10 Hz, H_a), 3.43 (1H, dd, J = 18, 10 Hz, H_b), 1.85–1.76 (7H m, Cyclo). MS (EI): m/z = 315.174 $[M]^+$.

*2-(4,5-Dihydro-3-(6,7,8,9-tetrahydro-5H-carbazol-8-yl)-1H-pyrazol-5-yl)phenol **(4b)**.* FTIR (KBr) cm^{-1}: 1575 (C=C), 1642 (C=N), 1044 (C-N), 3061 (N-H), 1435 (O-H), 2982 (C-H); ^{1}H NMR (d_6-DMSO) δ: 10.35 (1H, s, H-9), 7.84–7.53 (4H, m, Ph), 7.45–7.34 (4H, m, Ph), 6.25 (1H, s, N-H pyra), 4.84 (1H, m, H_c), 3.63 (1H, dd, J = 18, 10 Hz, H_a), 3.45 (1H, dd, J = 18, 10 Hz, H_b), 1.82–1.74 (7H m, Cyclo). MS (EI): m/z = 331.168 $[M]^+$.

*2,3,4,9-Tetrahydro-1-(4,5-dihydro-5-(4-methoxyphenyl)-1H-pyrazol-3-yl)-1H-carbazole **(4c)**.* FTIR (KBr) cm^{-1}: 1577 (C=C), 1640 (C=N), 1047 (C-N), 3059 (N-H), 2987 (C-H), 1100 (C-OCH_3). ^{1}H NMR (d_6-DMSO) δ: 10.38 (1H, s, H-9), 7.86–7.44 (4H, m, Ph), 7.52–7.41 (4H, m, Ph), 6.26 (1H, s, N-H pyra), 4.82 (1H, m, H_c), 3.64 (1H, dd, J = 18, 10 Hz, H_a), 3.46 (1H, dd, J = 18, 10 Hz, H_b), 3.14 (3H, s, OCH_3), 1.80–1.72 (7H m, Cyclo). MS (EI): m/z = 345.184 $[M]^+$.

*1-(5-(4-Chlorophenyl)-4,5-dihydro-1H-pyrazol-3-yl)-2,3,4,9-tetrahydro-1H-carbazole **(4d)**.* FTIR (KBr) cm^{-1}: 1575 (C=C), 1638 (C=N), 1044 (C-N), 3064 (N-H), 2990 (C-H), 735 (C-Cl). ^{1}H NMR (d_6-DMSO) δ: 10.35 (1H, s, H-9), 7.84–7.53 (4H, m, Ph), 7.45–7.34 (4H, m, Ph), 6.25 (1H, s, N-H pyra), 4.85 (1H, m, H_c), 3.63 (1H, dd, J = 18, 10 Hz, H_a), 3.45 (1H, dd, J = 18, 10 Hz, H_b), 1.82–1.74 (7H m, Cyclo). MS (EI): m/z = 349.135 $[M]^+$.

*1-(5-(Furan-2-yl)-4,5-dihydro-1H-pyrazol-3-yl)-2,3,4,9-tetrahydro-1H-carbazole **(4e)**.* FTIR (KBr) cm^{-1}: 1572 (C=C), 1648 (C=N), 1045 (C-N), 3067 (N-H), 2994 (C-H), 1160 (C-O); ^{1}H NMR (d_6-DMSO) δ: 10.34 (1H, s, H-9), 7.88–7.73 (3H, m, Ph), 7.47–7.38 (4H, m, Ph), 6.27 (1H, s, N-H pyra), 4.83 (1H, m, H_c), 3.65 (1H, dd, J = 18, 10 Hz, H_a), 3.48 (1H, dd, J = 18, 10 Hz, H_b), 1.86–1.78 (7H m, Cyclo). MS (EI): m/z = 305.153 $[M]^+$.

*2,3,4,9-Tetrahydro-1-(4,5-dihydro-1-(2,4-dinitrophenyl)-5-phenyl-1H-pyrazol-3-yl)-1H-carbazole **(4f)**.* FTIR (KBr) cm^{-1}: 1577 (C=C), 1640 (C=N), 1047 (C-N), 2987 (C-H), 1450 (C-NO_2). ^{1}H NMR (d_6-DMSO) δ: 10.35 (1H, s, H-9), 8.30 (1H, dd, J = 8.8, 2.3 Hz, H-5DNPh), 7.84–7.53 (5H, m, Ph), 7.69 (1H, d, J = 8.8 Hz, H-6 DNPh), 7.61 (1H, d, J = 2.3 Hz, H-3DNPh), 7.45–7.34 (4H, m, Ph), 4.81 (1H, m, H_c), 3.63 (1H, dd, J = 18, 10 Hz, H_a), 3.45 (1H, dd, J = 18, 10 Hz, H_b), 1.82–1.74 (7H m, Cyclo). MS (EI): m/z = 481.175 $[M]^+$.

*2-(4,5-Dihydro-3-(6,7,8,9-tetrahydro-5H-carbazol-8-yl)-1-(2,4-dinitrophenyl)-1H-pyrazol-5-yl)phenol **(4g)**.* FTIR (KBr) cm^{-1}: 1574 (C=C), 1638 (C=N), 1049 (C-N), 2987 (C-H), 3300 (C-OH), 1398 (C-NO_2). ^{1}H NMR (d_6-DMSO) δ: 10.32 (1H, s, H-9), 8.27 (1H, dd, J = 8.8, 2.3 Hz, H-5DNPh), 7.80–7.48 (5H, m, Ph), 7.67 (1H, d, J = 8.8 Hz, H-6DNPh), 7.58 (1H, d, J = 8.8 Hz, H-3DNPh), 7.41–7.30 (4H, m, Ph), 4.84 (1H, m, H_c), 3.54 (1H, dd, J = 18, 10 Hz, H_a), 3.41 (1H, dd, J = 18, 10 Hz, H_b), 1.79–1.38 (7H m, Cyclo). MS (EI): m/z = 349.135 $[M]^+$.

*2,3,4,9-Tetrahydro-1-(4,5-dihydro-5-(4-methoxyphenyl)-1-(2,4-dinitrophenyl)-1H-pyrazol-3-yl)-1H-carbazole **(4h)**.* FTIR (KBr) cm^{-1}: 1578 (C=C), 1640 (C=N), 1049 (C-N), 2989 (C-H), 1545 (C-NO_2), 1101 (C-OCH_3). ^{1}H NMR (d_6-DMSO) δ: 10.34 (1H, s, H-9), 8.27 (1H, dd, J = 8.8, 2.3 Hz, H-5 DNPh), 7.79–7.46 (5H, m, Ph), 7.61 (1H, d, J = 8.8 Hz, H-6 DNPh), 7.53 (1H, d, J = 2.3 Hz, H-3 DNPh), 7.39–7.27 (4H, m, THCz), 4.82 (1H, m, H_c), 3.56 (1H, dd, J = 18, 10 Hz, H_a), 3.38 (1H, dd, J = 18, 10 Hz, H_b), 3.15 (3H, s, OCH_3), 1.77–1.67 (7H m, Cyclo). MS (EI): m/z 511.186 $[M]^+$.

*1-(5-(4-Chlorophenyl)-4,5-dihydro-1-(2,4-dinitrophenyl)-1H-pyrazol-3-yl)-2,3,4,9-tetrahydro-1H-carbazole **(4i)**.* FTIR (KBr) cm^{-1}: 1577 (C=C), 1648 (C=N), 1048 (C-N), 2990 (C-H), 1547 (C-NO_2), 1101 (C-Cl). ^{1}H NMR (d_6-DMSO) δ: 10.31 (1H, s, H-9), 8.26 (1H, dd, J = 8.8, 2.3 Hz, H-5DNPh), 7.74–7.58 (5H, m, Ph), 7.47 (1H, d, J = 8.8 Hz, H-6 DNPh), 7.56 (1H, d, J = 2.3 Hz, H-3 DNPh), 7.41–7.30 (4H, m, THCz), 4.82 (1H, m, H_c), 3.58 (1H, dd, J = 18, 10 Hz, H_a), 3.41 (1H, dd, J = 18, 10 Hz, H_b), 1.81–1.72 (7H m, Cyclo). MS (EI): m/z = 515.136 $[M]^+$.

*1-(5-(Furan-2-yl)-4,5-dihydro-1-(2,4-dinitrophenyl)-1H-pyrazol-3-yl)-2,3,4,9-tetrahydro-1H-carbazole **(4j)**.* FTIR (KBr) cm^{-1}: 1579 (C=C), 1640 (C=N), 1053 (C-N), 2997 (C-H), 1550

TABLE 1: An overview of reactants and diazoles **4/5/6(a–o)**.

Sr #		Aldehyde	Hydrazine unit	Product code
(1)		C_6H_5 **(2a)**	$H_2N{\cdot}NH_2{\cdot}H_2O$ **(3a)**	**4a**
(2)		*o*-OH-C_6H_4 **(2b)**	$H_2N{\cdot}NH_2{\cdot}H_2O$ **(3a)**	**4b**
(3)		*p*-OMe-C_6H_4 **(2c)**	$H_2N{\cdot}NH_2{\cdot}H_2O$ **(3a)**	**4c**
(4)		*p*-Cl-C_6H_4 **(2d)**	$H_2N{\cdot}NH_2{\cdot}H_2O$ **(3a)**	**4d**
(5)		C_4H_3O **(2e)**	$H_2N{\cdot}NH_2{\cdot}H_2O$ **(3a)**	**4e**
(6)	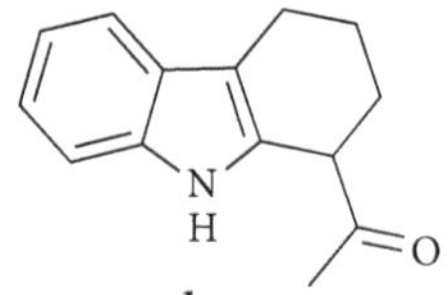	C_6H_5 **(2a)**	2,4-NO_2-C_6H_3 **(3b)**	**4f**
(7)		*o*-OH-C_6H_4 **(2b)**	2,4-NO_2-C_6H_3 **(3b)**	**4g**
(8)		*p*-OMe-C_6H_4 **(2c)**	2,4-NO_2-C_6H_3 **(3b)**	**4h**
(9)		*p*-Cl-C_6H_4 **(2d)**	2,4-NO_2-C_6H_3 **(3b)**	**4i**
(10)	**1a**	C_4H_3O **(2e)**	2,4-NO_2-C_6H_3 **(3b)**	**4j**
(11)		C_6H_5 **(2a)**	$CS(NH_2)NH{\cdot}NH_2$ **(3c)**	**4k**
(12)		*o*-OH-C_6H_4 **(2b)**	$CS(NH_2)NH{\cdot}NH_2$ **(3c)**	**4l**
(13)		*p*-OMe-C_6H_4 **(2c)**	$CS(NH_2)NH{\cdot}NH_2$ **(3c)**	**4m**
(14)		*p*-Cl-C_6H_4 **(2d)**	$CS(NH_2)NH{\cdot}NH_2$ **(3c)**	**4n**
(15)		C_4H_3O **(2e)**	$CS(NH_2)NH{\cdot}NH_2$ **(3c)**	**4o**
(16)		C_6H_5 **(2a)**	$H_2N{\cdot}NH_2{\cdot}H_2O$ **(3a)**	**5a**
(17)		*o*-OH-C_6H_4 **(2b)**	$H_2N{\cdot}NH_2{\cdot}H_2O$ **(3a)**	**5b**
(18)		*p*-OMe-C_6H_4 **(2c)**	$H_2N{\cdot}NH_2{\cdot}H_2O$ **(3a)**	**5c**
(19)		*p*-Cl-C_6H_4 **(2d)**	$H_2N{\cdot}NH_2{\cdot}H_2O$ **(3a)**	**5d**
(20)		C_4H_3O **(2e)**	$H_2N{\cdot}NH_2{\cdot}H_2O$ **(3a)**	**5e**
(21)	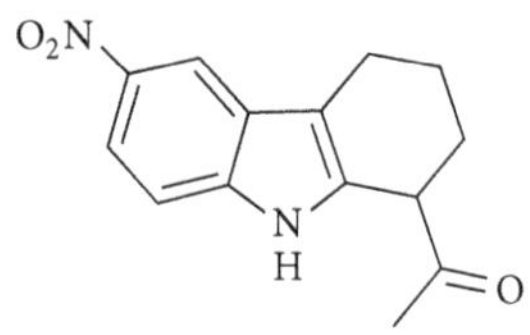	C_6H_5 **(2a)**	2,4-NO_2-C_6H_3 **(3b)**	**5f**
(22)		*o*-OH-C_6H_4 **(2b)**	2,4-NO_2-C_6H_3 **(3b)**	**5g**
(23)		*p*-OMe-C_6H_4 **(2c)**	2,4-NO_2-C_6H_3 **(3b)**	**5h**
(24)		4-Cl-C_6H_4 **(2d)**	2,4-NO_2-C_6H_3 **(3b)**	**5i**
(25)	**1b**	C_4H_3O **(2e)**	2,4-NO_2-C_6H_3 **(3b)**	**5j**
(26)		C_6H_5 **(2a)**	$CS(NH_2)NH{\cdot}NH_2$ **(3c)**	**5k**
(27)		*o*-OH-C_6H_4 **(2b)**	$CS(NH_2)NH{\cdot}NH_2$ **(3c)**	**5l**
(28)		*p*-OMe-C_6H_4 **(2c)**	$CS(NH_2)NH{\cdot}NH_2$ **(3c)**	**5m**
(29)		*p*-Cl-C_6H_4 **(2d)**	$CS(NH_2)NH{\cdot}NH_2$ **(3c)**	**5n**
(30)		C_4H_3O **(2e)**	$CS(NH_2)NH{\cdot}NH_2$ **(3c)**	**5o**
(31)		C_6H_5 **(2a)**	$H_2N{\cdot}NH_2{\cdot}H_2O$ **(3a)**	**6a**
(32)		*o*-OH-C_6H_4 **(2b)**	$H_2N{\cdot}NH_2{\cdot}H_2O$ **(3a)**	**6b**
(33)		*p*-OMe-C_6H_4 **(2c)**	$H_2N{\cdot}NH_2{\cdot}H_2O$ **(3a)**	**6c**
(34)		*p*-Cl-C_6H_4 **(2d)**	$H_2N{\cdot}NH_2{\cdot}H_2O$ **(3a)**	**6d**
(35)		C_4H_3O **(2e)**	$H_2N{\cdot}NH_2{\cdot}H_2O$ **(3a)**	**6e**
(36)	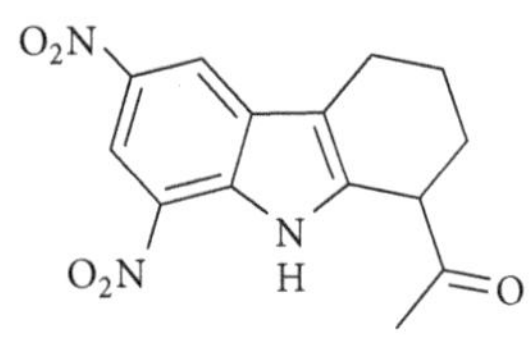	C_6H_5 **(2a)**	2,4-NO_2-C_6H_3 **(3b)**	**6f**
(37)		*o*-OH-C_6H_4 **(2b)**	2,4-NO_2-C_6H_3 **(3b)**	**6g**
(38)		*p*-OMe-C_6H_4 **(2c)**	2,4-NO_2-C_6H_3 **(3b)**	**6h**
(39)		*p*-Cl-C_6H_4 **(2d)**	2,4-NO_2-C_6H_3 **(3b)**	**6i**
(40)	**1c**	C_4H_3O **(2e)**	2,4-NO_2-C_6H_3 **(3b)**	**6j**
(41)		C_6H_5 **(2a)**	$CS(NH_2)NH{\cdot}NH_2$ **(3c)**	**6k**
(42)		*o*-OH-C_6H_4 **(2b)**	$CS(NH_2)NH{\cdot}NH_2$ **(3c)**	**6l**
(43)		*p*-OMe-C_6H_4 **(2c)**	$CS(NH_2)NH{\cdot}NH_2$ **(3c)**	**6m**
(44)		*p*-Cl-C_6H_4 **(2d)**	$CS(NH_2)NH{\cdot}NH_2$ **(3c)**	**6n**
(45)		C_4H_3O **(2e)**	$CS(NH_2)NH{\cdot}NH_2$ **(3c)**	**6o**

($C\text{-}NO_2$), 1159 (C-O-C). ^1H NMR (d_6-DMSO) δ: 10.35 (1H, s, H-9), 8.30 (1H, dd, J = 8.8, 2.3 Hz, H-5 DNPh), 7.79–7.48 (3H, m, Ph), 7.64 (1H, d, J = 8.8 Hz, H-6 DNPh), 7.55 (1H, d, J = 2.3 Hz, H-3 DNPh), 7.41–7.29 (4H, m, THCz), 4.84 (1H, m, H_c), 3.58 (1H, dd, J = 18, 10 Hz, H_a), 3.40 (1H, dd, J = 18, 10 Hz, H_b), 1.78–1.67 (7H m, Cyclo). MS (EI): m/z = 471.154 $[M]^+$.

*4,5-Dihydro-3-(6,7,8,9-tetrahydro-5H-carbazol-8-yl)-5-phenylpyrazole-1-carbothioamide **(4k)***. FTIR (KBr) cm^{-1}: 1578

TABLE 2: Summary of physical data of synthetic diazoles **4/5/6(a–o)**.

Entry	Product	Mol. formula	Color/state	Conventional		Microwave	
				Time (hr)	Yield (%)	Time (min)	Yield (%)
(1)	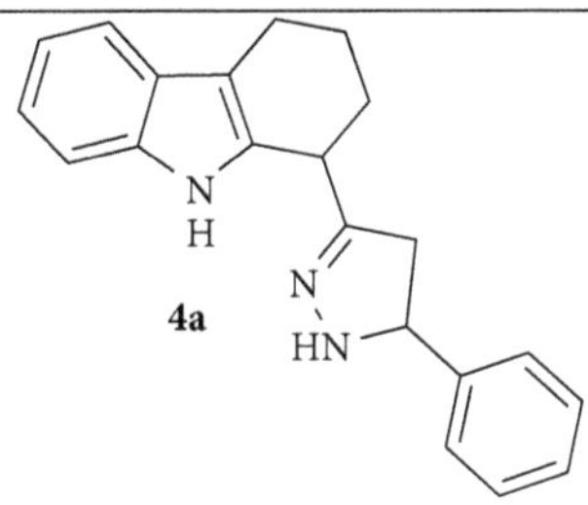 **4a**	$C_{21}H_{21}N_3$	Yellowish green solid	5	68	4	68
(2)	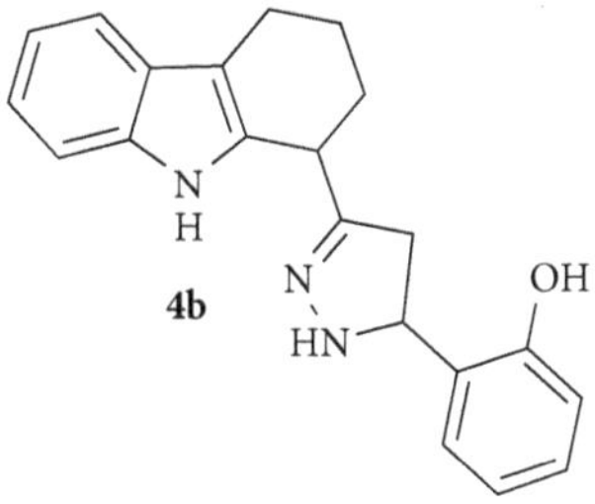 **4b**	$C_{21}H_{21}N_3O$	Yellowish green solid	6	72	5	72
(3)	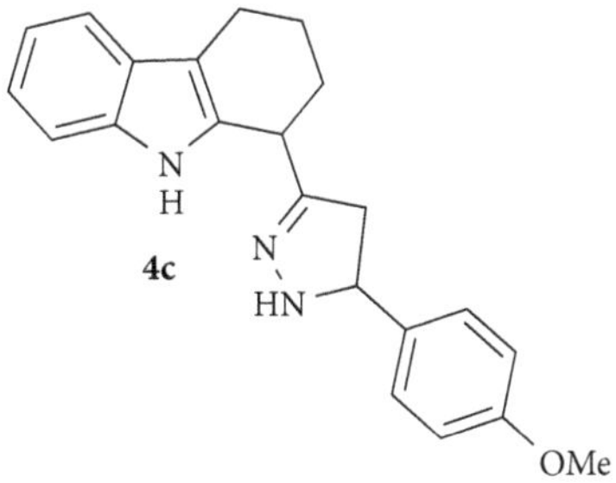 **4c**	$C_{27}H_{23}N_3O$	Yellowish green solid	5	82	5	83
(4)	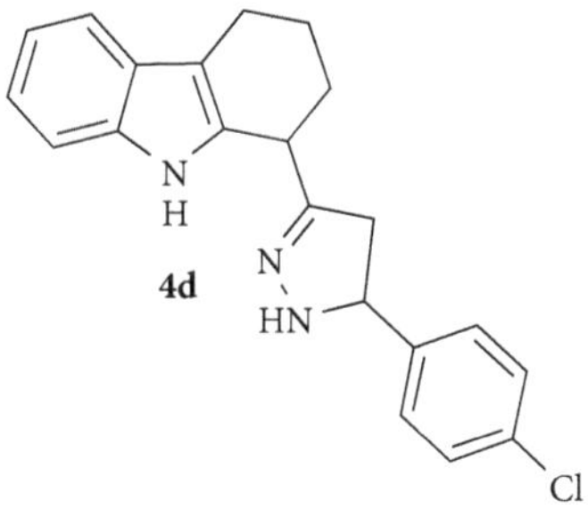 **4d**	$C_{21}H_{20}ClN_3$	Yellowish green solid	6	79	6	79
(5)	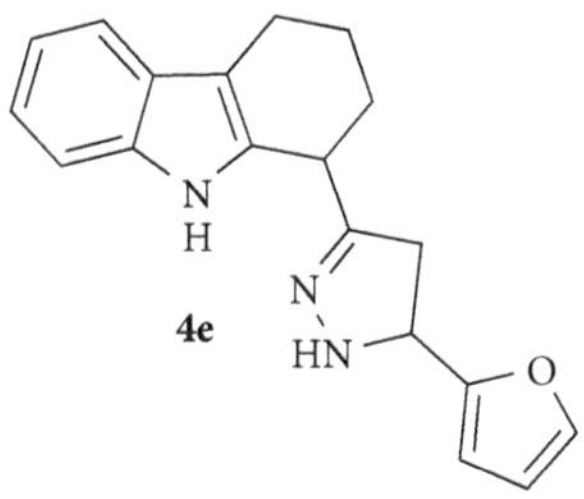 **4e**	$C_{19}H_{19}N_3O$	Yellowish green solid	4.5	80	6	80
(6)	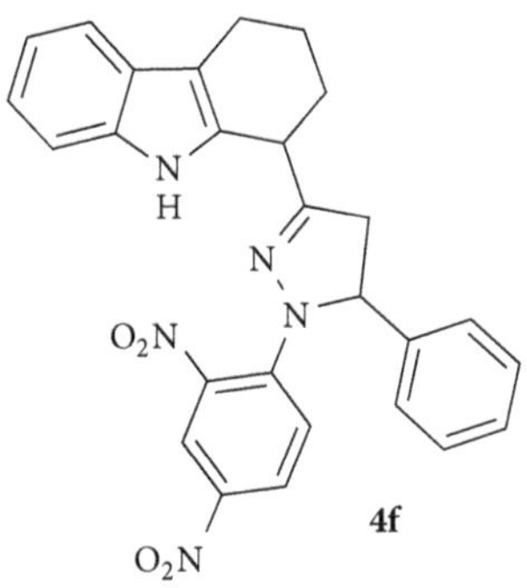 **4f**	$C_{27}H_{23}N_5O_4$	Yellowish green solid	4.5	71	7	71

Table 2: Continued.

Entry	Product	Mol. formula	Color/state	Conventional		Microwave	
				Time (hr)	Yield (%)	Time (min)	Yield (%)
(7)	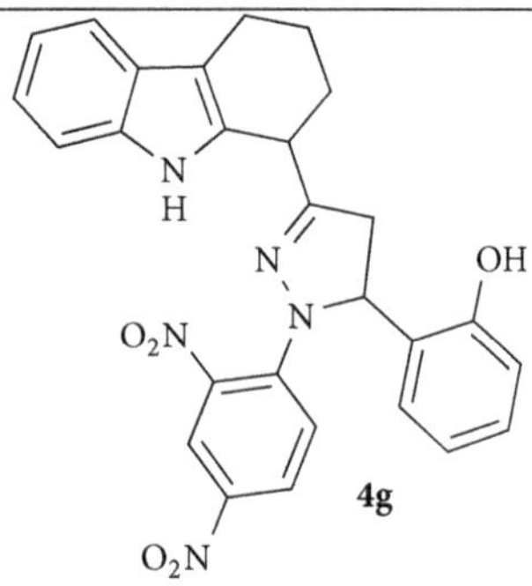 4g	$C_{27}H_{23}N_5O_5$	Yellowish green solid	5.5	84	5	82
(8)	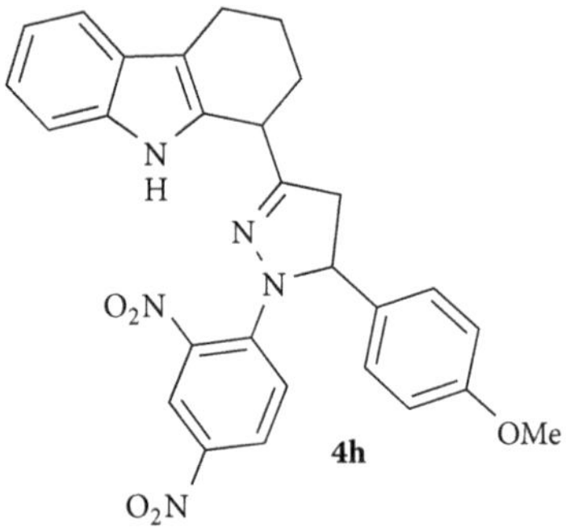 4h	$C_{28}H_{25}N_5O_5$	Yellowish green solid	4	83	4	83
(9)	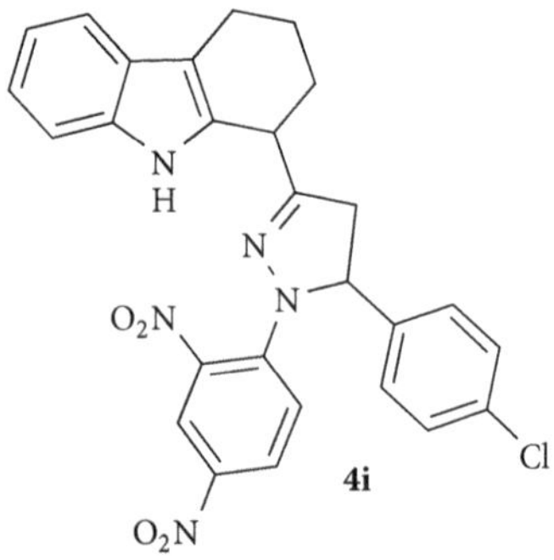 4i	$C_{27}H_{22}ClN_5O_4$	Yellowish green solid	5	77	6	77
(10)	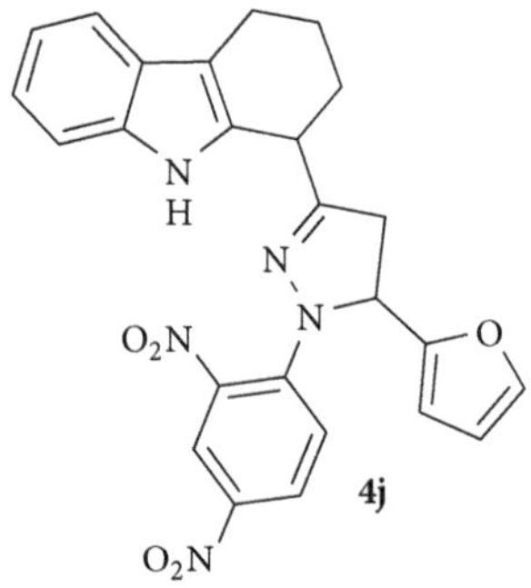 4j	$C_{25}H_{21}N_5O_5$	Yellowish green solid		84	6	81
(11)	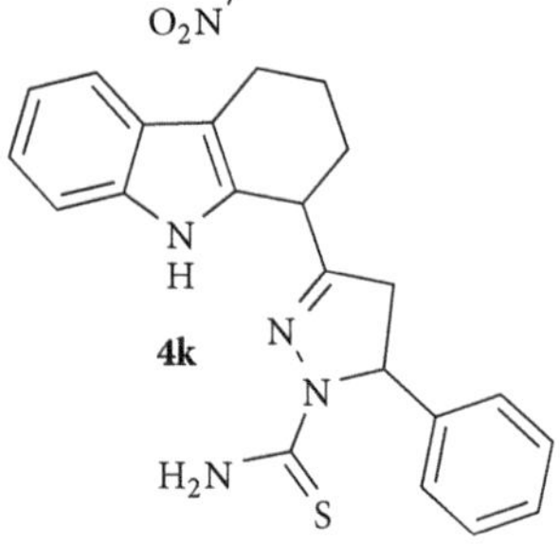 4k	$C_{27}H_{22}N_4S$	Yellowish green solid		75	6	75

Table 2: Continued.

Entry	Product	Mol. formula	Color/state	Conventional		Microwave	
				Time (hr)	Yield (%)	Time (min)	Yield (%)
(12)	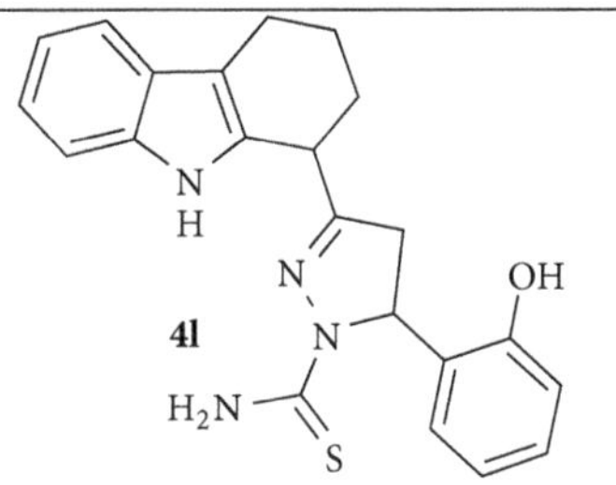 4l	$C_{27}H_{22}N_4OS$	Yellowish green solid	5.5	73	6	73
(13)	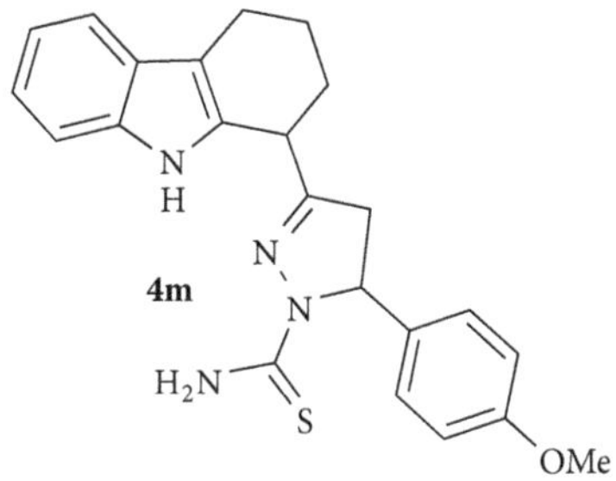 4m	$C_{23}H_{24}N_4OS$	Yellowish green solid	6.0	85	5	85
(14)	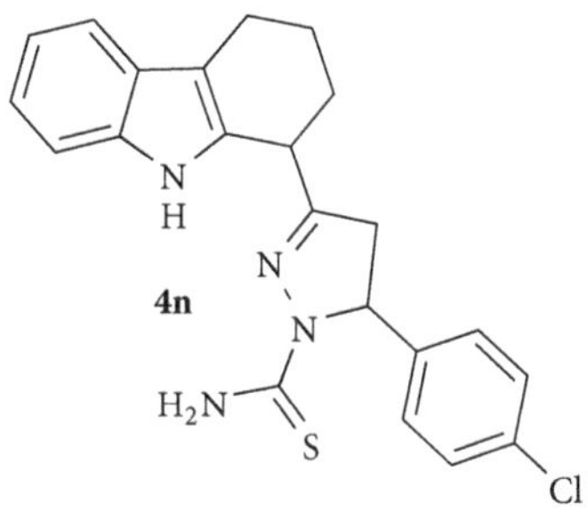 4n	$C_{27}H_{21}ClN_4S$	Yellowish green solid	6.5	87	6	87
(15)	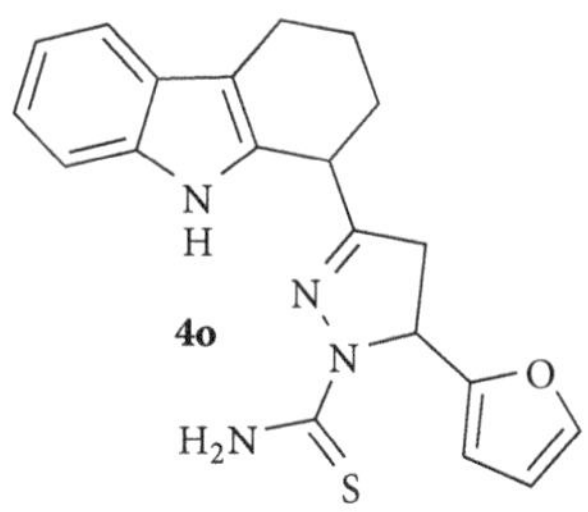 4o	$C_{20}H_{20}N_4OS$	Yellowish green solid	4.5	73	5	79
(16)	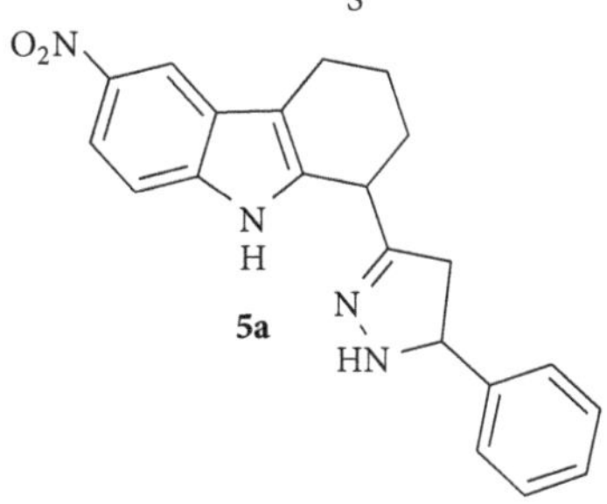 5a	$C_{21}H_{20}N_4O_2$	Reddish orange solid	5	76	6	76
(17)	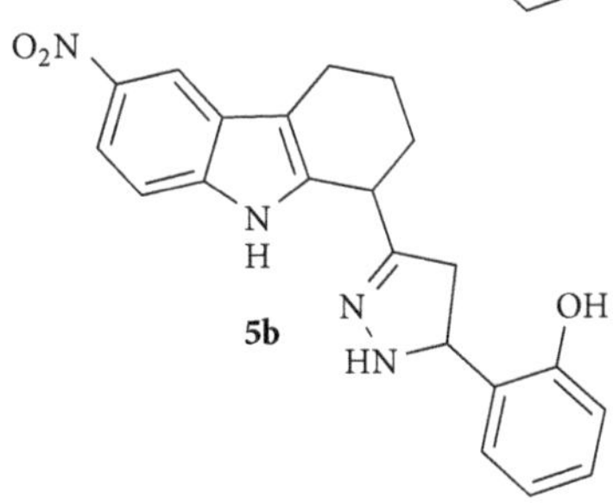 5b	$C_{21}H_{20}N_4O_3$	Reddish orange solid	5.5	75	6	75

Table 2: Continued.

Entry	Product	Mol. formula	Color/state	Conventional		Microwave	
				Time (hr)	Yield (%)	Time (min)	Yield (%)
(18)	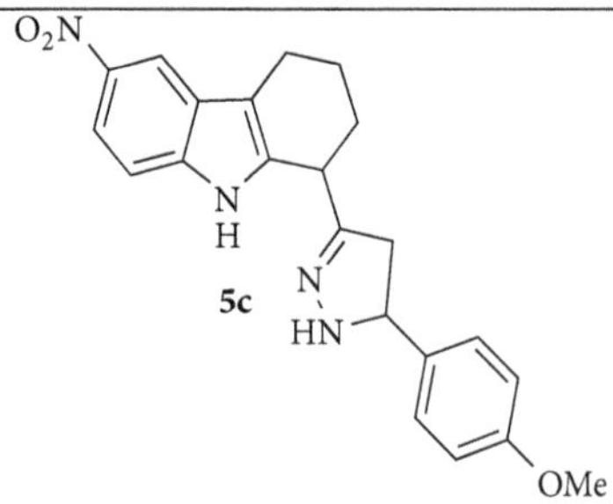 5c	$C_{27}H_{22}N_4O_3$	Reddish orange solid	4.5	78	5	78
(19)	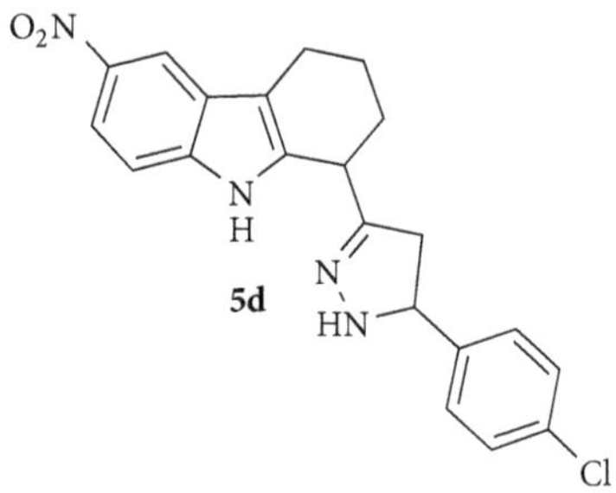 5d	$C_{21}H_{19}ClN_4O_2$	Reddish orange solid	5	84	5	84
(20)	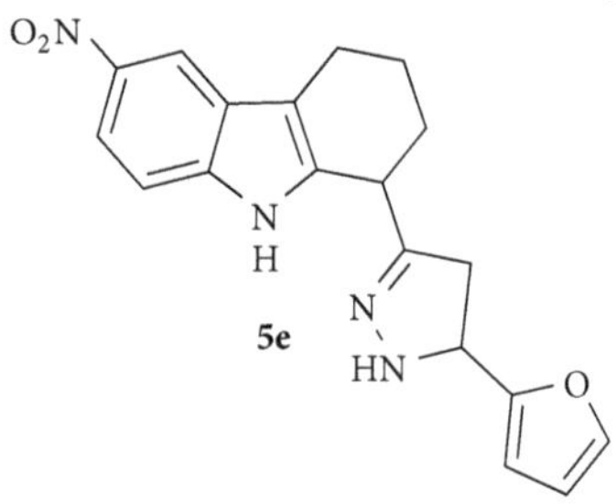 5e	$C_{19}H_{18}N_4O_3$	Reddish orange solid	5	84	6	84
(21)	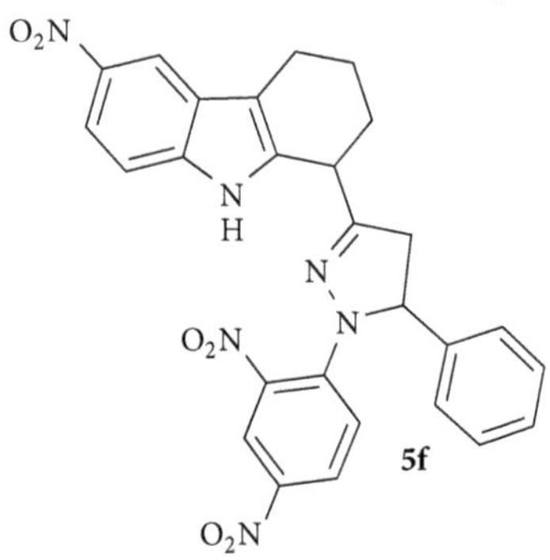 5f	$C_{27}H_{22}N_6O_6$	Reddish orange solid	4.75	74	5	74
(22)	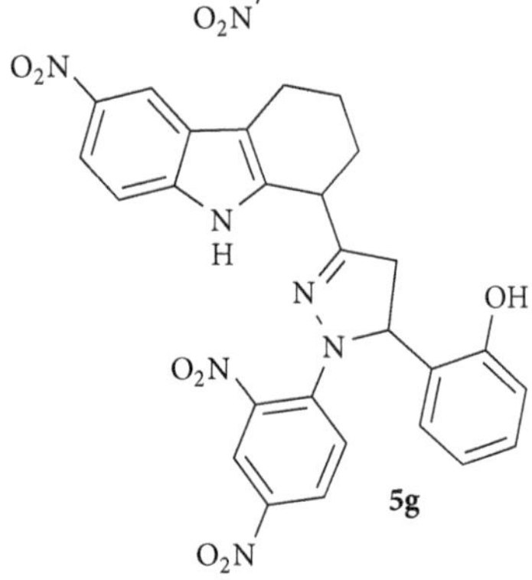 5g	$C_{27}H_{22}N_6O_7$	Reddish orange solid	6.5	75	6	75

TABLE 2: Continued.

Entry	Product	Mol. formula	Color/state	Conventional		Microwave	
				Time (hr)	Yield (%)	Time (min)	Yield (%)
(23)	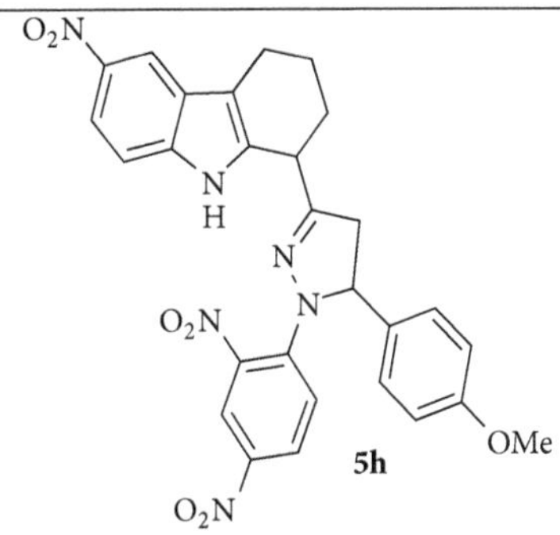 5h	$C_{28}H_{24}N_6O_7$	Reddish orange solid	5.75	84	6	84
(24)	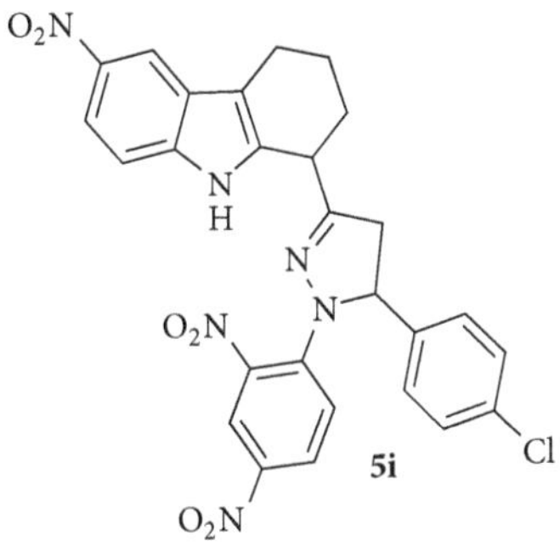 5i	$C_{27}H_{21}ClN_6O_6$	Reddish orange solid	5	78	5	78
(25)	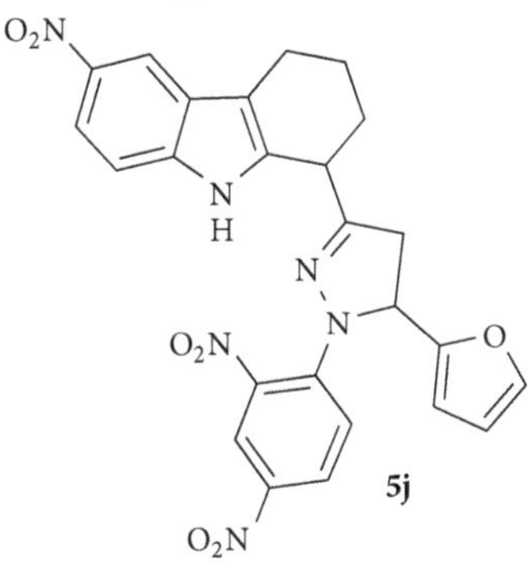 5j	$C_{25}H_{20}N_6O_7$	Reddish orange solid	4.5	80	5	80
(26)	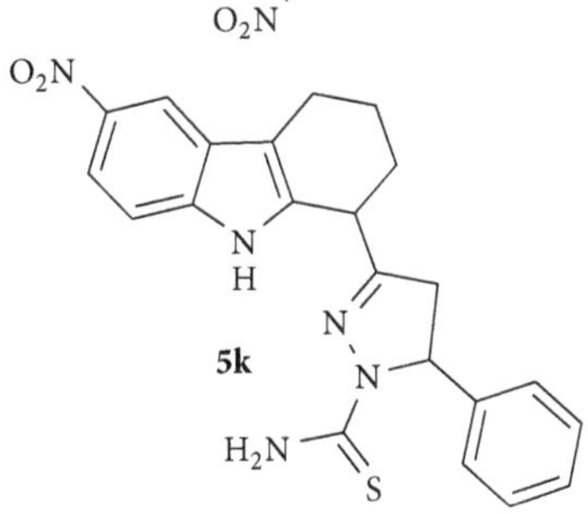 5k	$C_{27}H_{21}N_5O_2S$	Reddish orange solid	6.5	75	6	75
(27)	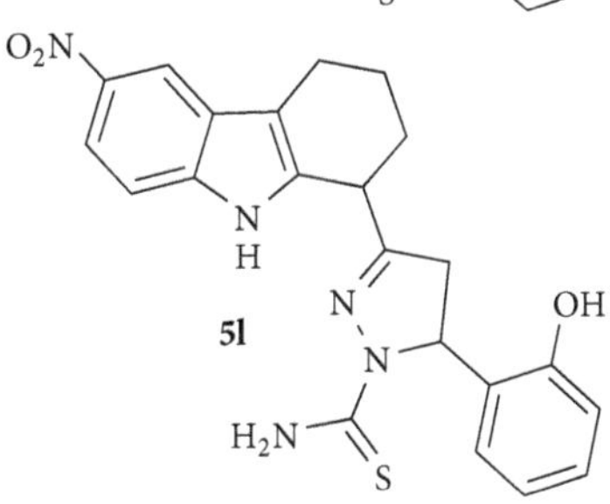 5l	$C_{27}H_{21}N_5O_3S$	Reddish orange solid	6	78	5	78

Table 2: Continued.

Entry	Product	Mol. formula	Color/state	Conventional		Microwave	
				Time (hr)	Yield (%)	Time (min)	Yield (%)
(28)	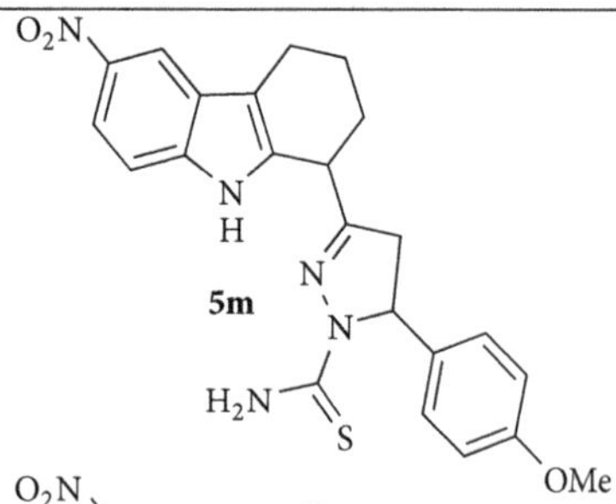 **5m**	$C_{23}H_{23}N_5O_3S$	Reddish orange solid	5.5	79	7	79
(29)	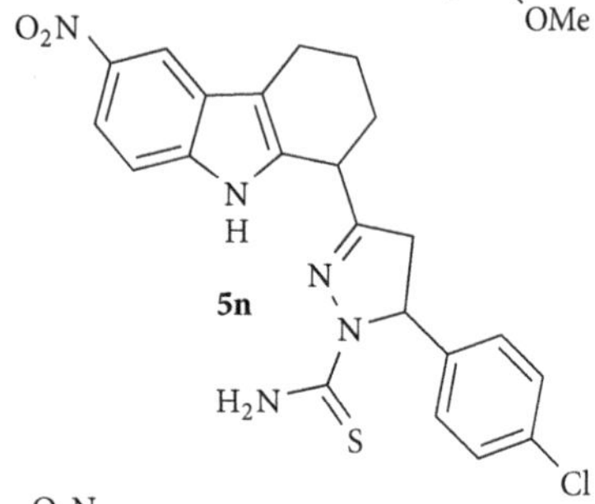 **5n**	$C_{27}H_{20}ClN_5O_2S$	Reddish orange solid	6	84	6	84
(30)	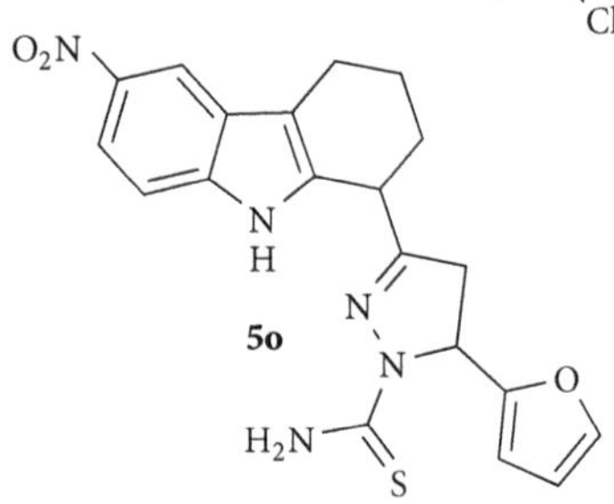 **5o**	$C_{20}H_{19}N_5O_3S$	Reddish orange solid	5.5	82	7	82
(31)	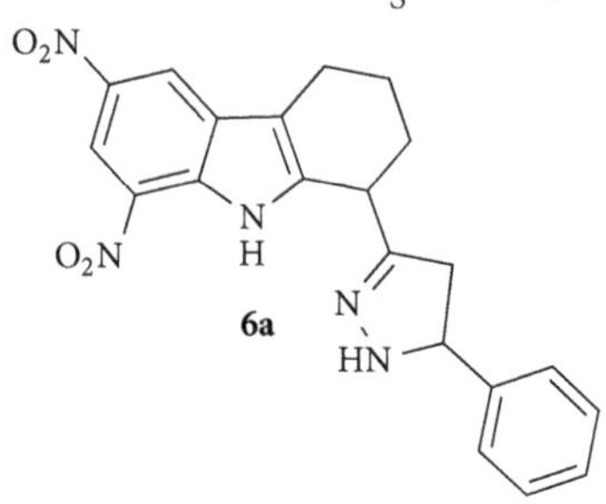 **6a**	$C_{21}H_{19}N_5O_4$	Yellow solid	2.5	75	5	88
(32)	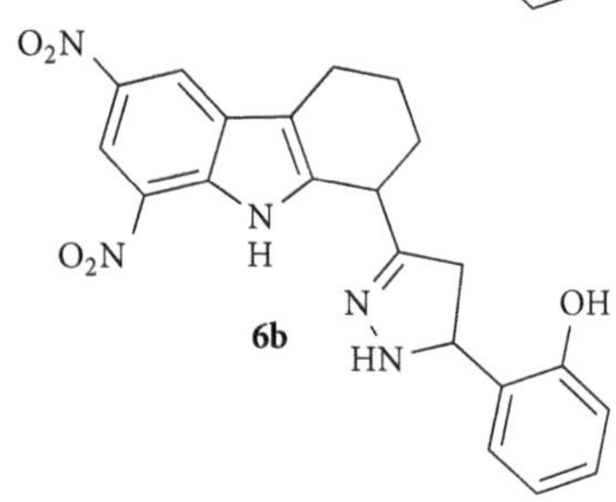 **6b**	$C_{21}H_{19}N_5O_5$	Yellow solid	3.5	62	7	85
(33)	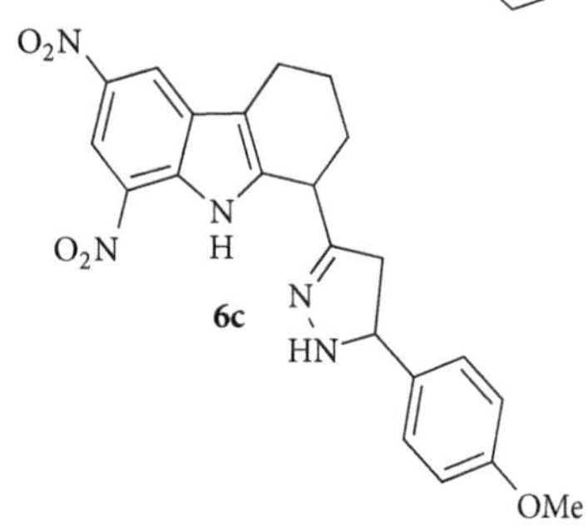 **6c**	$C_{27}H_{21}N_5O_5$	Yellow solid	3.75	65	8	79

Table 2: Continued.

Entry	Product	Mol. formula	Color/state	Conventional		Microwave	
				Time (hr)	Yield (%)	Time (min)	Yield (%)
(34)	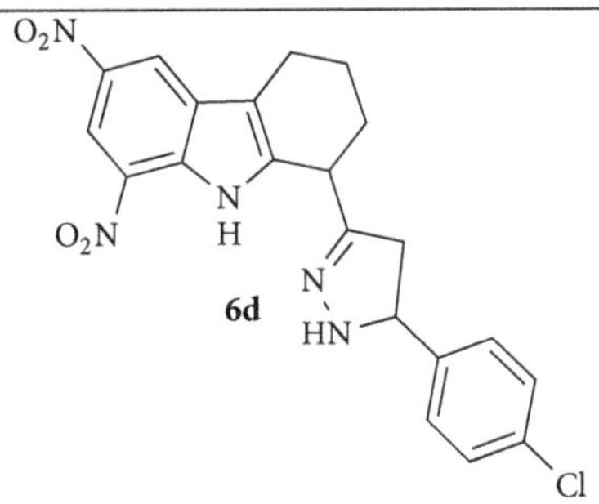 6d	$C_{21}H_{18}ClN_5O_4$	Yellow solid	3.5	60	6	86
(35)	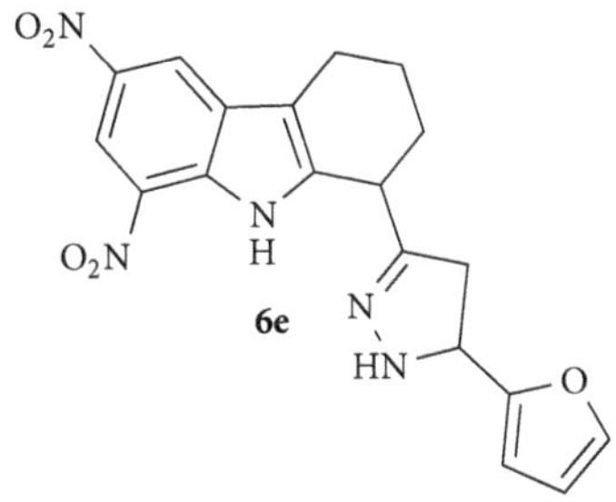 6e	$C_{19}H_{17}N_5O_5$	Yellow solid	4	65	8	78
(36)	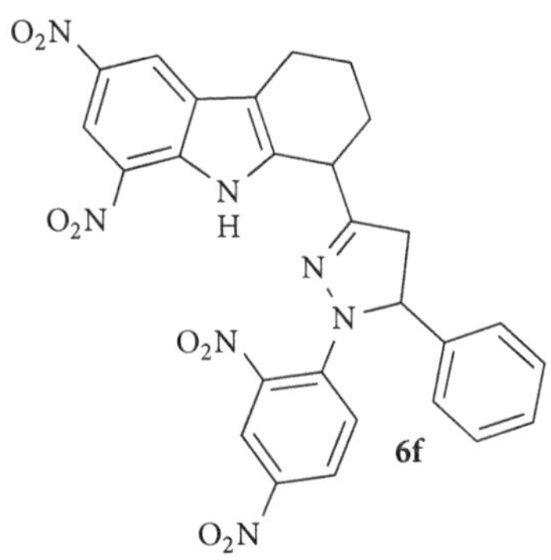 6f	$C_{27}H_{21}N_7O_8$	Yellowish brown gummy	3.75	24	6	76
(37)	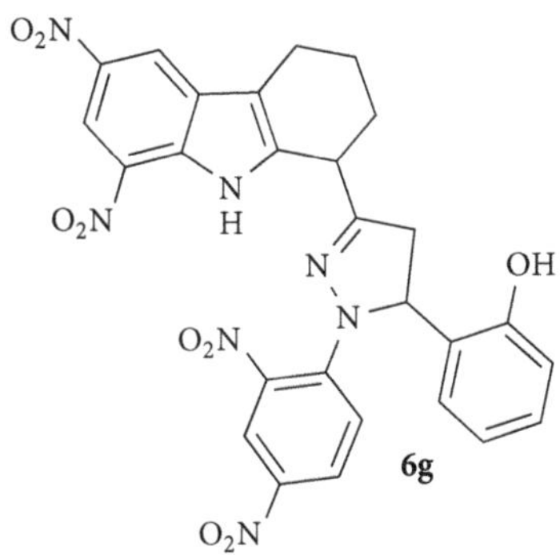 6g	$C_{27}H_{21}N_7O_9$	Reddish brown gummy	5.5	31	6	75
(38)	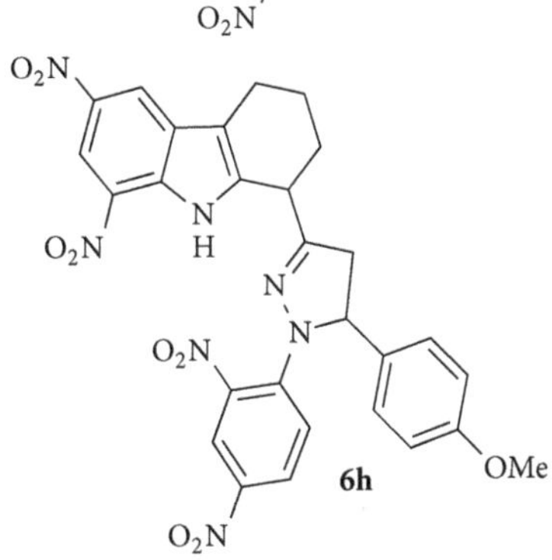 6h	$C_{28}H_{23}N_7O_9$	Reddish brown solid	6	39	5	78

Table 2: Continued.

Entry	Product	Mol. formula	Color/state	Conventional		Microwave	
				Time (hr)	Yield (%)	Time (min)	Yield (%)
(39)	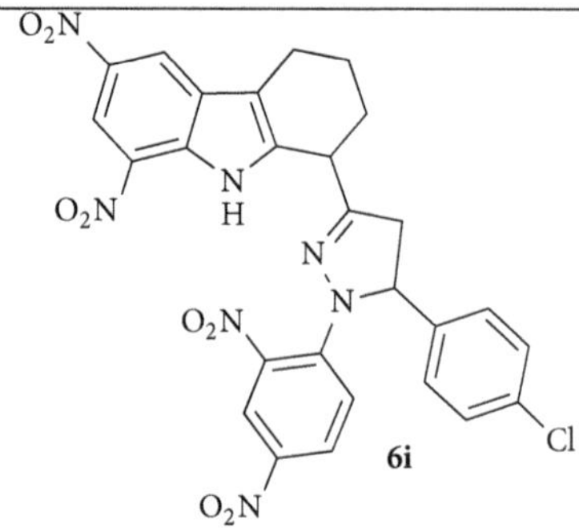 **6i**	$C_{27}H_{20}ClN_7O_8$	Dark brown solid	6.25	33	5	84
(40)	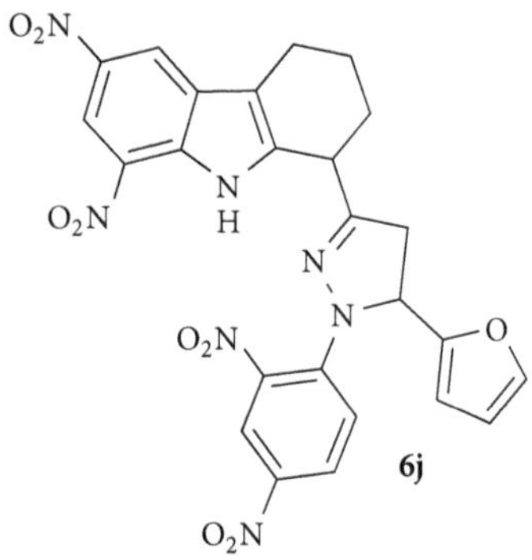 **6j**	$C_{29}H_{19}N_7O_9$	Reddish brown solid	5	36	6	80
(41)	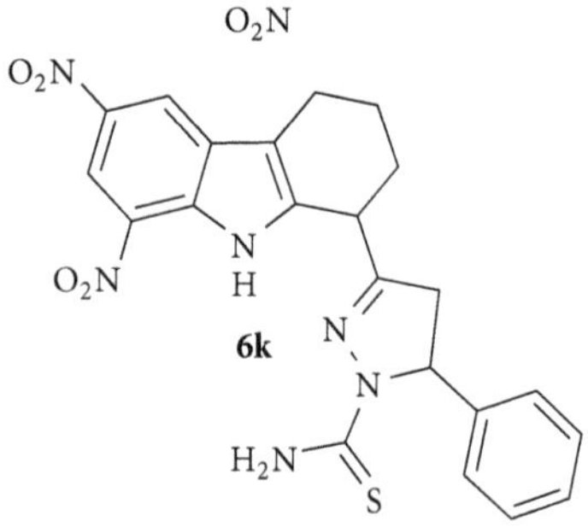 **6k**	$C_{27}H_{20}N_6O_4S$	Brownish Yellow solid	5.5	66	8	81
(42)	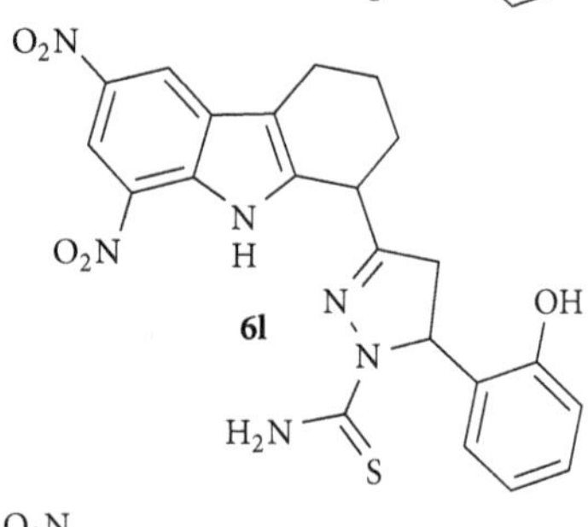 **6l**	$C_{22}H_{20}N_6O_5S$	Reddish yellow solid	4.75	63	5	82
(43)	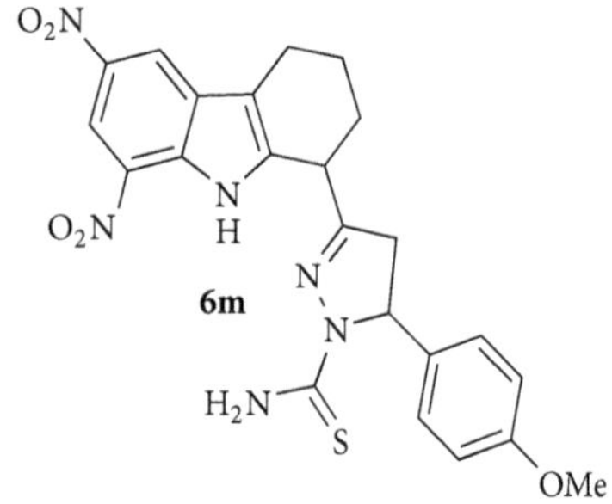 **6m**	$C_{23}H_{22}N_6O_5S$	Yellow solid	5	64	7	84

TABLE 2: Continued.

Entry	Product	Mol. formula	Color/state	Conventional Time (hr)	Conventional Yield (%)	Microwave Time (min)	Microwave Yield (%)
(44)	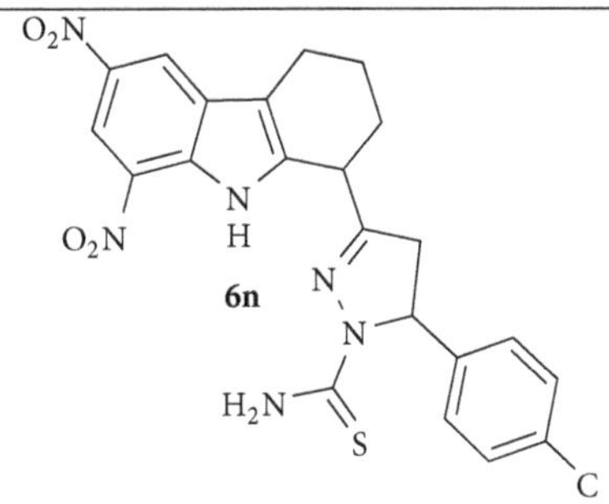 6n	$C_{22}H_{19}ClN_6O_4S$	Yellow solid	5.5	59	7	75
(45)	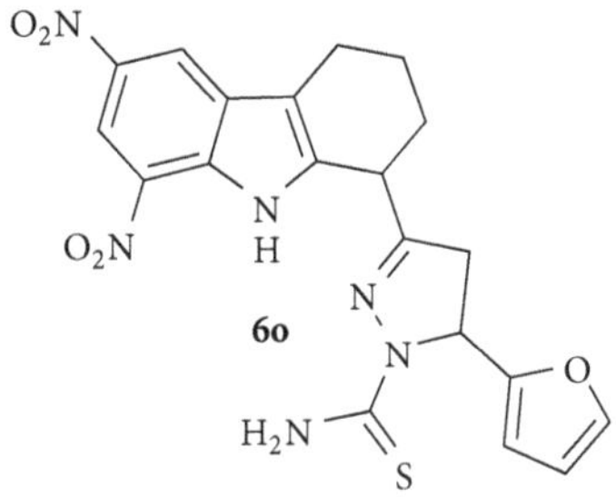 6o	$C_{20}H_{18}N_6O_5S$	Yellow solid	5.25	62	9	86

(C=C), 1640 (C=N), 1049 (C-N), 2989 (C-H), 3605 (C-NH_2), 1208 (C=S). ^{1}H NMR (d_6-DMSO) δ: 10.29 (1H, s, H-9), 7.77–7.48 (5H, m, Ph), 7.38–7.27 (4H, m, THCz), 4.85 (1H, m, H_c), 3.57 (1H, d, J = 18, 10 Hz, H_a), 3.42 (1H, d, J = 18, 10 Hz, H_b), 1.80–1.72 (7H m, Cyclo). MS (EI): m/z = 497.170 $[M]^+$.

*4,5-Dihydro-3-(6,7,8,9-tetrahydro-5H-carbazol-8-yl)-5-(2-hydroxyphenyl)pyrazole-1-carbothioamide **(4l)***. FTIR (KBr) cm^{-1}: 1579 (C=C), 1635 (C=N), 1054 (C-N), 2979 (C-H), 3300 (C-OH), 3598 (C-NH_2), 1200 (C=S). ^{1}H NMR (d_6-DMSO) δ: 10.33 (1H, s, H-9), 7.81–7.48 (4H, m, Ph), 7.41–7.29 (4H, m, THCz), 4.81 (1H, m, H_c), 3.59 (1H, dd, J = 18, 10 Hz, H_a), 3.41 (1H, dd, J = 18, 10 Hz, H_b), 1.78–1.70 (7H m, Cyclo). MS (EI): m/z = 390.151 $[M]^+$.

*4,5-Dihydro-3-(6,7,8,9-tetrahydro-5H-carbazol-8-yl)-5-(4-methoxyphenyl)pyrazole-1-carbothioamide **(4m)***. FTIR (KBr) cm^{-1}: 1579 (C=C), 1640 (C=N), 1053 (C-N), 2997 (C-H), 1101 (C-OCH_3), 3599 (C-NH_2), 1201 (C=S). ^{1}H NMR (d_6-DMSO) δ: 10.29 (1H, s, H-9), 7.78–7.49 (4H, m, Ph), 7.38–7.27 (4H, m, THCz), 4.84 (1H, m, H_c), 3.57 (1H, dd, J = 18, 10 Hz, H_a), 3.39 (1H, dd, J = 18, 10 Hz, H_b), 3.11 (3H, s, OCH_3), 1.78–1.71 (7H m, Cyclo). MS (EI): m/z = 404.167 $[M]^+$.

*5-(4-Chlorophenyl)-4,5-dihydro-3-(6,7,8,9-tetrahydro-5H-carbazol-8-yl)pyrazole-1-carbothioamide **(4n)***. FTIR (KBr) cm^{-1}: 1582 (C=C), 1635 (C=N), 1051 (C-N), 2987 (C-H), 740 (C-Cl), 3595 (C-NH_2), 1200 (C=S). ^{1}H NMR (d_6-DMSO) δ: 10.31 (1H, s, H-9), 7.82–7.50 (4H, m, Ph), 7.42–7.29 (4H, m, THCz), 4.80 (1H, m, H_c), 3.60 (1H, dd, J = 18, 10 Hz, H_a), 3.42 (1H, dd, J = 18, 10 Hz, H_b), 1.80–1.68 (7H m, Cyclo). MS (EI): m/z = 408.118 $[M]^+$.

*5-(Furan-2-yl)-4,5-dihydro-3-(6,7,8,9-tetrahydro-5H-carbazol-8-yl)pyrazole-1-carbothioamide **(4o)***. FTIR (KBr) cm^{-1}: 1579 (C=C), 1640 (C=N), 1053 (C-N), 2997 (C-H), 1159 (C-O-C), 3598 (C-NH_2), 1200 (C=S). ^{1}H NMR (d_6-DMSO) δ: 10.34 (1H, s, H-9), 7.85–7.55 (3H, m, Ph), 7.48–7.37 (4H, m, THCz), 4.85 (1H, m, H_c), 3.66 (1H, dd, J = 18, 10 Hz, H_a), 3.49 (1H, dd, J = 18, 10 Hz, H_b), 1.85–1.79 (7H m, Cyclo). MS (EI): m/z = 364.136 $[M]^+$.

*2,3,4,9-Tetrahydro-1-(4,5-dihydro-5-phenyl-1H-pyrazol-3-yl)-6-nitro-1H-carbazole **(5a)***. FTIR (KBr) cm^{-1}: 1574 (C=C), 1638 (C=N), 1049 (C-N), 3056 (N-H), 2987 (C-H), 1540 (C-NO_2). ^{1}H NMR (d_6-DMSO) δ: 10.86 (1H, s, H-9), 8.20 (1H, dd, J = 8.8, 2.3 Hz, H-7NO_2THCz), 7.68 (1H, d, J = 8.8 Hz, H-8 NO_2THCz), 7.47–7.38 (5H, m, Ph), 7.23 (1H, d, J = 2.3 Hz, H-5 NO_2-THCz), 6.34 (s, 1H, N-H pyra), 5.33 (1H, m, H_c), 3.55 (1H, dd, J = 18, 12 Hz, H_a), 3.21 (1H, dd, J = 18, 12 Hz, H_b), 1.75–1.67 (m, 7H, Cyclo). MS (EI): m/z = 360.159 $[M]^+$.

*2-(4,5-Dihydro-3-(6,7,8,9-tetrahydro-3-nitro-5H-carbazol-8-yl)-1H-pyrazol-5-yl)phenol **(5b)***. FTIR (KBr) cm^{-1}: 1572 (C=C), 1630 (C=N), 1043 (C-N), 3054 (N-H), 298 (C-H), 3300 (C-OH), 1551 (C-NO_2). ^{1}H NMR (d_6-DMSO) δ: 10.63 (1H, s, H-9), 8.26 (1H, dd, J = 8.8, 2.3 Hz, H-7NO_2THCz), 7.68 (1H, d, J = 8.8 Hz, H-8 NO_2THCz), 7.52–7.41 (4H, m, Ph), 7.25 (1H, d, J = 2.3 Hz, H-5 NO_2-THCz), 6.37 (s, 1H, N-Hpyra), 5.38 (1H, m, H_c), 3.58 (1H, dd, J = 18, 12 Hz, H_a), 3.25 (1H, dd, J = 18, 12 Hz, H_b), 1.78–1.69 (m, 7H, Cyclo). MS (EI): m/z = 376.154 $[M]^+$.

*2,3,4,9-Tetrahydro-1-(4,5-dihydro-5-(4-methoxyphenyl)-1H-pyrazol-3-yl)-6-nitro-1H-carbazole **(5c)***. FTIR (KBr) cm^{-1}:

1574 (C=C), 1638 (C=N), 1049 (C-N), 3056 (N-H), 2987 (C-H), 1100 (C-OCH_3), 1549 (C-NO_2). ^1H NMR (d_6-DMSO) δ: 10.62 (1H, s, H-9), 8.26 (1H, dd, J = 8.8, 2.3 Hz, H-7NO_2THCz), 7.68 (1H, d, J = 8.8 Hz, H-8 NO_2THCz), 7.47–7.34 (4H, m, Ph), 7.22 (1H, d, J = 2.1 Hz, H-5 NO_2THCz), 6.36 (s, 1H, N-Hpyra), 5.38 (1H, m, H_c), 3.52 (1H, dd, J = 18, 12 Hz, H_a), 3.29 (1H, dd, J = 18, 12 Hz, H_b), 3.12 (3H, s, OCH_3), 1.76–1.64 (m, 7H, Cyclo). MS (EI): m/z = 390.169 $[M]^+$.

*1-(5-(4-Chlorophenyl)-4,5-dihydro-1H-pyrazol-3-yl)-2,3,4,9-tetrahydro-6-nitro-1H-carbazole **(5d)**.* FTIR (KBr) cm^{-1}: 1574 (C=C), 1638 (C=N), 1049 (C-N), 3056 (N-H), 2987 (C-H), 730 (C-Cl), 1550 (C-NO_2). ^1H NMR (d_6-DMSO) δ: 10.69 (1H, s, H-9), 8.26 (1H, dd, J = 8.4, 2.1 Hz, H-7NO_2THCz), 7.68 (1H, d, J = 8.4 Hz, H-8 NO_2THCz), 7.47–7.35 (4H, m, Ph), 7.22 (1H, d, J = 2.1 Hz, H-5 NO_2THCz), 6.32 (s, 1H, N-Hpyra), 5.26 (m, 1H, H_c), 3.54 (1H, dd, J = 18, 12 Hz, H_a), 3.27 (1H, dd, J = 18, 12 Hz, H_b), 1.76–1.64 (m, 7H, Cyclo). MS (EI): m/z = 390.120 $[M]^+$.

*1-(5-(Furan-2-yl)-4,5-dihydro-1H-pyrazol-3-yl)-2,3,4,9-tetrahydro-6-nitro-1H-carbazole **(5e)**.* FTIR (KBr) cm^{-1}: 1578 (C=C), 1645 (C=N), 1054 (C-N), 3045 (N-H), 2985 (C-H), 1164 (C-O-C), 1339 (C-NO_2). ^1H NMR (d_6-DMSO) δ: 10.69 (1H, s, H-9), 8.26 (1H, dd, J = 8.8, 2.3 Hz, H-7NO_2THCz), 7.68 (1H, d, J = 8.8 Hz, H-8 NO_2THCz), 7.44–7.31 (3H, m, Ph), 7.22 (1H, d, J = 2.3 Hz, H-5 NO_2THCz), 6.34 (s, 1H, N-Hpyra), 5.27 (1H, m, H_c), 3.52 (1H, dd, J = 18, 12 Hz, H_a), 3.25 (1H, dd, J = 18, 12 Hz, H_b), 1.76–1.64 (m, 7H, Cyclo). MS (EI): m/z = 350.138 $[M]^+$.

*2,3,4,9-Tetrahydro-1-(4,5-dihydro-1-(2,4-dinitrophenyl)-5-phenyl-1H-pyrazol-3-yl)-6-nitro-1H-carbazole **(5f)**.* FTIR (KBr) cm^{-1}: 1574 (C=C), 1639 (C=N), 1056 (C-N), 3046 (N-H), 2987 (C-H), 1550 (C-NO_2). ^1H NMR (d_6-DMSO) δ: 10.68 (1H, s, H-9), 8.34 (1H, dd, J = 8.8, 2.3 Hz, H-7 NO_2THCz), 8.28 (1H, d, J = 2.3 Hz, H-5 NO_2THCz), 7.67 (1H, d, J = 8.8 Hz, H-5 DNPh), 7.58 (1H, d, J = 8.8 Hz, H-5DNTHCz), 7.49–7.37 (5H, m, Ph), 7.18 (1H, d, J = 8.8 Hz, H-8 NO_2THCz), 7.12 (1H, d, J = 2.3 Hz, H-3DNPh), 5.24 (1H, m, H_c), 3.53 (1H, dd, J = 18, 12 Hz, H_a), 3.29 (1H, dd, J = 18, 12 Hz, H_b), 1.76–1.64 (m, 7H, Cyclo). MS (EI): m/z = 526.160 $[M]^+$.

*2-(4,5-Dihydro-3-(6,7,8,9-tetrahydro-3-nitro-5H-carbazol-8-yl)-1-(2,4-dinitrophenyl)-1H-pyrazol-5-yl)phenol **(5g)**.* FTIR (KBr) cm^{-1}: 1569 (C=C), 1640 (C=N), 1049 (C-N), 2987 (C-H), 3400 (C-OH), 1445 (C-NO_2). ^1H NMR (d_6-DMSO) δ: 10.76 (1H, s, H-9), 8.33 (1H, dd, J = 8.8, 2.1 Hz, H-5 DNPh), 8.28 (1H, dd, J = 8.8, 2.3 Hz, H-7 NO_2THCz), 7.62 (1H, d, J = 8.8 Hz, H-6 DNPh), 7.55 (1H, dd, J = 8.8 Hz, H-3 DNPh), 7.46–7.34 (4H, m, Ph), 7.24 (1H, d, J = 2.3 Hz, H-5, NO_2THCz), 7.15 (1H, d, J = 8.8 Hz, H-8 NO_2THCz), 5.18 (1H, m, H_c), 3.45 (1H, dd, J = 18, 12 Hz, H_a), 3.21 (1H, dd, J = 18, 12 Hz, H_b), 1.76–1.64 (m, 7H, Cyclo). MS (EI): m/z = 542.155 $[M]^+$.

*2,3,4,9-Tetrahydro-1-(4,5-dihydro-5-(4-methoxyphenyl)-1-(2,4-dinitrophenyl)-1H-pyrazol-3-yl)-6-nitro-1H-carbazole **(5h)**.* FTIR (KBr) cm^{-1}: 1578 (C=C), 1644 (C=N), 1044 (C-N), 2980 (C-H), 1058 (C-OCH_3), 1354 (C-NO_2). ^1H NMR (d_6-DMSO) δ: 10.44 (1H, s, H-9), 8.35 (1H, dd, J = 8.8, 2.3 Hz, H-7 NO_2THCz), 8.27 (1H, d, J = 2.3 Hz, H-5 NO_2THCz), 7.57 (1H, d, J = 8.8 Hz, H-8 NO_2THCz), 7.49 (1H, dd, J = 8.8, 2.1 Hz, H-5 DNPh), 7.41–7.32 (4H, m, Ph), 7.11 (1H, d, J = 8.8, 2.1 Hz, H-6 DNPh), 6.92 (1H, d, J = 2.3 Hz, H-3 DNPh), 5.25 (1H, m, H_c), 3.47 (1H, dd, J = 18, 12 Hz, H_a), 3.20 (1H, dd, J = 18, 12 Hz, H_b), 3.08 (3H, s, OCH_3), 1.72–1.58 (m, 7H, Cyclo). MS (EI): m/z = 556.171 $[M]^+$.

*4,5-Dihydro-3-(6,7,8,9-tetrahydro-3-nitro-5H-carbazol-8-yl)-5-(2-hydroxyphenyl)pyrazole-1-carbothioamide **(5i)**.* FTIR (KBr) cm^{-1}: 1574 (C=C), 1646 (C=N), 1055 (C-N), 2987 (C-H), 745 (C-Cl), 1450 (C-NO_2). ^1H NMR (d_6-DMSO) δ: 10.54 (1H, s, H-9), 8.32 (1H, dd, J = 8.9, 2.1 Hz, H-7 NO_2THCz), 8.26 (1H, d, J = 2.1 Hz, H-5NO_2THCz), 7.66 (1H, d, J = 8.9, 2.4 Hz, H-5 DNPh), 7.56 (1H, d, J = 8.9 Hz, H-8 NO_2THCz), 7.44–7.32 (4H, m, Ph), 7.28 (1H, d, J = 2.3 Hz, H-3, DNPh), 7.18 (1H, d, J = 8.9 Hz, H-6 DNPh), 5.34 (1H, m, H_c), 3.52 (1H, dd, J = 18, 12 Hz, H_a), 3.29 (1H, dd, J = 18, 12 Hz, H_b), 3.12 (3H, s, OCH_3), 1.76–1.64 (m, 7H, Cyclo). MS (EI): m/z = 560.121 $[M]^+$.

*1-(5-(Furan-2-yl)-4,5-dihydro-1-(2,4-dinitrophenyl)-1H-pyrazol-3-yl)-2,3,4,9-tetrahydro-6-nitro-1H-carbazole **(5j)**.* FTIR (KBr) cm^{-1}: 1569 (C=C), 1637 (C=N), 1049 (C-N), 2980 (C-H), 1164 (C-O-C), 1490 (C-NO_2). ^1H NMR (d_6-DMSO) δ: 10.42 (1H, s, H-9), 8.30 (1H, dd, J = 8.9, 2.1 Hz, H-7 NO_2THCz), 8.28 (1H, d, J = 2.1 Hz, H-5 NO_2THCz), 7.69 (1H, dd, J = 8.9, 2.1 Hz, H-5 DNPh), 7.61 (1H, d, J = 8.8 Hz, H-8 NO_2THCz), 7.23 (1H, d, J = 2.3 Hz, H-3, DNPh), 7.16 (1H, d, J = 2.3 Hz, H-6 DNPh), 5.28 (1H, m, H_c), 3.54 (1H, dd, J = 18, 12 Hz, H_a), 3.31 (1H, dd, J = 18, 12 Hz, H_b), 1.78–1.69 (m, 7H, Cyclo). MS (EI): m/z = 516.139 $[M]^+$.

*4,5-Dihydro-3-(6,7,8,9-tetrahydro-3-nitro-5H-carbazol-8-yl)-5-phenylpyrazole-1-carbothioamide **(5k)**.* FTIR (KBr) cm^{-1}: 1572 (C=C), 1630 (C=N), 1043 (C-N), 2980 (C-H), 3587 (C-NH_2), 1200 (C=S), 1490 (C-NO_2). ^1H NMR (d_6-DMSO) δ: 10.49 (1H, s, H-9), 8.11 (1H, dd, J = 8.8, 2.3 Hz, H-7NO_2THCz), 7.62 (1H, d, J = 8.8 Hz, H-8 NO_2THCz), 7.41–7.34 (5H, m, Ph), 7.24 (1H, d, J = 2.3 Hz, H-5 NO_2THCz), 5.22 (1H, m, H_c), 3.37 (1H, dd, J = 18, 12 Hz, H_a), 3.18 (1H, dd, J = 18, 12 Hz, H_b), 1.73–1.67 (m, 7H, Cyclo). MS (EI): m/z = 419.142 $[M]^+$.

*4,5-Dihydro-3-(6,7,8,9-tetrahydro-3-nitro-5H-carbazol-8-yl)-5-(2-hydroxyphenyl)pyrazole-1-carbothioamide **(5l)**.* FTIR (KBr) cm^{-1}: 1568 (C=C), 1625 (C=N), 1039 (C-N), 2979 (C-H), 3300 (C-OH), 3604 (C-NH_2), 1500 (C-NO_2), 1202 (C=S). ^1H NMR (d_6-DMSO) δ: 10.55 (1H, s, H-9), 8.15 (1H, dd, J = 8.8, 2.3 Hz, H-7 NO_2THCz), 7.58 (1H, d, J = 8.8 Hz, H-8 NO_2THCz), 7.37–7.28 (5H, m, Ph), 7.19 (1H, d, J = 2.3 Hz, H-5 NO_2THCz), 5.21 (1H, m, H_c), 3.32 (1H, dd, J = 18, 12 Hz, H_a), 3.14 (1H, dd, J = 18, 12 Hz, H_b), 1.69–1.43 (m, 7H, Cyclo). MS (EI): m/z = 435.137 $[M]^+$.

*4,5-Dihydro-3-(6,7,8,9-tetrahydro-3-nitro-5H-carbazol-8-yl)-5-(4-methoxyphenyl)pyrazole-1-carbothioamide **(5m)**.* FTIR (KBr) cm^{-1}: 1578 (C=C), 1640 (C=N), 1049 (C-N), 2989

(C-H), 3592 (C-NH_2), 1535 (C-NO_2), 1204 (C=S), 1058 (C-OCH_3). ^{1}H NMR (d_6-DMSO) δ: 10.57 (1H, s, H-9), 8.13 (1H, dd, *J* = 8.8, 2.3 Hz, H-7 NO_2THCz), 7.49 (1H, d, *J* = 8.8 Hz, H-8 NO_2THCz), 7.31–7.17 (5H, m, Ph), 7.08 (1H, d, *J* = 2.3 Hz, H-5 NO_2THCz), 5.24 (1H, m, H_c), 3.30 (1H, d, *J* = 18, 12 Hz, H_a), 3.21 (1H, d, *J* = 18, 12 Hz, H_b), 3.08 (3H, s, OCH_3), 1.65–1.46 (m, 7H, Cyclo). MS (EI): *m/z* = 449.152 $[M]^+$.

5-(4-Chlorophenyl)-4,5-dihydro-3-(6,7,8,9-tetrahydro-3-nitro-5H-carbazol-8-yl)pyrazole-1-carbothioamide ***(5n)***. FTIR (KBr) cm^{-1}: 1578 (C=C), 1640 (C=N), 1049 (C-N), 2987 (C-H), 3480 (C-NH_2), 1490 (C-NO_2), 1200 (C=S), 725 (C-Cl). ^{1}H NMR (d_6-DMSO) δ: 10.42 (1H, s, H-9), 8.13 (1H, dd, *J* = 8.8, 2.3 Hz, H-7 NO_2THCz), 7.64 (1H, d, *J* = 8.8 Hz, H-8 NO_2-THCz), 7.47–7.29 (5H, m, Ph), 7.35 (1H, d, *J* = 2.3 Hz, H-5 NO_2THCz), 5.19 (1H, m, H_c), 3.38 (1H, dd, *J* = 18, 12 Hz, H_a), 3.19 (1H, dd, *J* = 18, 12 Hz, H_b), 1.75–1.58 (m, 7H, Cyclo). MS (EI): *m/z* = 453.103 $[M]^+$.

5-(Furan-2-yl)-4,5-dihydro-3-(6,7,8,9-tetrahydro-3-nitro-5H-carbazol-8-yl)pyrazole-1-carbothioamide ***(5o)***. FTIR (KBr) cm^{-1}: 1578 (C=C), 1640 (C=N), 1049 (C-N), 2987 (C-H), 3595 (C-NH_2), 1148 (C-O-C), 1537 (C-NO_2), 1200 (C=S). ^{1}H NMR (d_6-DMSO) δ: 10.38 (1H, s, H-9), 7.98 (1H, dd, *J* = 8.8, 2.3 Hz, H-7 NO_2THCz), 7.52 (1H, d, *J* = 8.8 Hz, H-8 NO_2THCz), 7.35–7.21 (5H, m, Ph), 7.17 (1H, d, *J* = 2.3 Hz, H-5 NO_2THCz), 5.25 (1H, m, H_c), 3.58 (1H, dd, *J* = 18, 12 Hz, H_a), 3.29 (1H, dd, *J* = 18, 12 Hz, H_b), 1.55–1.38 (m, 7H, Cyclo). MS (EI): *m/z* = 409.121 $[M]^+$.

6,7,8,9-Tetrahydro-8-(4,5-dihydro-5-phenyl-1H-pyrazol-3-yl)-1,3-dinitro-5H-carbazole ***(6a)***. FTIR (KBr) cm^{-1}: 1576 (C=C), 1639 (C=N), 1045 (C-N), 3069 (N-H), 2998 (C-H), 1550 (C-NO_2). ^{1}H NMR (d_6-DMSO) δ: 10.52 (1H, s, H-9), 8.31 (1H, d, *J* = 2.3 Hz, H-7 DNTHCz), 7.52 (1H, d, *J* = 2.3 Hz, H-5 DNTHCz), 7.38–7.31 (5H, m, Ph), 6.20 (1H, s, N-H Pyra), 5.15 (1H, m, H_c), 3.53 (1H, dd, *J* = 18, 9.6 Hz, H_a), 3.26 (1H, dd, *J* = 18, 9.6 Hz, H_b), 1.72–1.61 (m, 7H, Cyclo). MS (EI): *m/z* = 405.144 $[M]^+$.

2-(4,5-Dihydro-3-(6,7,8,9-tetrahydro-1,3-dinitro-5H-carbazol-8-yl)-1H-pyrazol-5-yl)phenol ***(6b)***. FTIR (KBr) cm^{-1}: 1576 (C=C), 1637 (C=N), 1044 (C-N), 3068 (N-H), 2996 (C-H), 3329 (C-OH), 1495 (C-NO_2). ^{1}H NMR (d_6-DMSO) δ: 10.53 (1H, s, H-9), 8.28 (1H, d, *J* = 2.3 Hz, H-7 DNTHCz), 7.53 (1H, d, *J* = 2.3 Hz, H-5 DNTHCz), 7.38–7.29 (4H, m, Ph), 6.22 (1H, s, N-H Pyra), 5.18 (1H, m, H_c), 3.48 (1H, dd, *J* = 18, 9.6 Hz, H_a), 3.25 (1H, dd, *J* = 18, 9.6 Hz, H_b), 1.71–1.63 (m, 7H, Cyclo). MS (EI): *m/z* = 421.139 $[M]^+$.

6,7,8,9-Tetrahydro-8-(4,5-dihydro-5-(4-methoxyphenyl)-1H-pyrazol-3-yl)-1,3-dinitro-5H-carbazole ***(6c)***. FTIR (KBr) cm^{-1}: 1578 (C=C), 1640 (C=N), 1043 (C-N), 3069 (N-H), 2998 (C-H), 1100 (C-OCH_3), 1500 (C-NO_2). ^{1}H NMR (d_6-DMSO) δ: 10.54 (1H, s, H-9), 8.32 (1H, d, *J* = 2.3 Hz, H-7 DNTHCz), 7.56 (1H, d, *J* = 2.3 Hz, H-5 DNTHCz), 7.41–7.34 (5H, m, Ph), 6.21 (1H, s, N-H Pyra), 5.21 (1H, m, H_c), 3.52 (1H, dd, *J* = 18, 9.6 Hz, H_a), 3.29 (1H, dd, *J* = 18, 9.7 Hz, H_b), 3.14 (3H, s, OCH_3), 1.76–1.64 (m, 7H, Cyclo). MS (EI): *m/z* = 435.154 $[M]^+$.

8-(5-(4-Chlorophenyl)-4,5-dihydro-1H-pyrazol-3-yl)-6,7,8,9-tetrahydro-1,3-dinitro-5H-carbazole ***(6d)***. FTIR (KBr) cm^{-1}: 1574 (C=C), 1639 (C=N), 1042 (C-N), 3070 (N-H), 2996 (C-H), 740 (C-Cl) 1498 (C-NO_2). ^{1}H NMR (d_6-DMSO) δ: 10.52 (1H, s, H-9), 8.31 (1H, d, *J* = 2.4 Hz, H-7 DNTHCz), 7.67 (1H, d, *J* = 2.4 Hz, H-5 DNTHCz), 7.44–7.38 (4H, m, Ph), 6.24 (1H, s, N-H Pyra), 4.95 (1H, m, H_c), 3.56 (1H, dd, *J* = 18, 9.8 Hz, H_a), 3.32 (1H, dd, *J* = 18, 9.8 Hz, H_b), 1.77–1.68 (m, 7H, Cyclo). MS (EI): *m/z* = 439.105 $[M]^+$.

8-(5-(Furan-2-yl)-4,5-dihydro-1H-pyrazol-3-yl)-6,7,8,9-tetrahydro-1,3-dinitro-5H-carbazole ***(6e)***. FTIR (KBr) cm^{-1}: 1567 (C=C), 1646 (C=N), 1048 (C-N), 3078 (N-H), 2988 (C-H), 1145 (C-O-C), 1437 (C-NO_2). ^{1}H NMR (d_6-DMSO) δ: 10.54 (1H, s, H-9), 8.32 (1H, d, *J* = 2.3 Hz, H-7 DNTHCz), 7.56 (1H, d, *J* = 2.3 Hz, H-5 DNTHCz), 7.41–7.34 (3H, m, Ph), 6.21 (1H, s, N-H Pyra), 4.81 (1H, m, H_c), 3.52 (1H, dd, *J* = 18, 9.6 Hz, H_a), 3.29 (1H, dd, *J* = 18, 9.6 Hz, H_b), 1.76–1.64 (m, 7H, Cyclo). MS (EI): *m/z* = 395.123 $[M]^+$.

6,7,8,9-Tetrahydro-8-(4,5-dihydro-1-(2,4-dinitrophenyl)-5-phenyl-1H-pyrazol-3-yl)-1,3-dinitro-5H-carbazole ***(6f)***. FTIR (KBr) cm^{-1}: 1570 (C=C), 1645 (C=N), 1046 (C-N), 2990 (C-H), 1550 (C-NO_2). ^{1}H NMR (d_6-DMSO) δ: 10.51 (1H, s, H-9), 8.26 (1H, d, *J* = 2.1 Hz, H-7 DNTHCz), 7.62 (1H, d, *J* = 8.9, 2.6 Hz, H-5 DNPh), 7.48 (1H, d, *J* = 2.1 Hz, H-5 DNTHCz), 7.41 (1H, d, *J* = 8.9 Hz, H-6 DNPh), 7.34–7.21 (5H, m, Ph), 7.45 (1H, d, *J* = 2.6 Hz, H-3 DNPh), 4.89 (1H, m, H_c), 3.47 (1H, dd, *J* = 18, 9.6 Hz, H_a), 3.23 (1H, dd, *J* = 18, 9.8 Hz, H_b), 1.72–1.58 (m, 7H, Cyclo). MS (EI): *m/z* = 571.145 $[M]^+$.

2-(4,5-Dihydro-3-(6,7,8,9-tetrahydro-1,3-dinitro-5H-carbazol-8-yl)-1-(2,4-dinitrophenyl)-1H-pyrazol-5-yl)phenol ***(6g)***. FTIR (KBr) cm^{-1}: 1576 (C=C), 1648 (C=N), 1045 (C-N), 2998 (C-H), 3329 (C-OH), 1447 (C-NO_2). ^{1}H NMR (d_6-DMSO) δ: 10.48 (1H, s, H-9), 8.27 (1H, d, *J* = 2.1 Hz, H-7 DNTHCz), 7.57 (1H, dd, *J* = 8.1, 2.6 Hz, H-5 DNPh), 7.46 (1H, d, *J* = 2.1 Hz, H-5 DNTHCz), 7.35 (1H, d, *J* = 8.1 Hz, H-6 DNPh), 7.27–7.18 (4H, m, Ph), 7.11 (1H, d, *J* = 2.6 Hz, H-3 DNPh), 4.87 (1H, m, H_c), 3.48 (1H, dd, *J* = 18, 9.6 Hz, H_a), 3.26 (1H, dd, *J* = 18, 9.6 Hz, H_b), 1.75–1.61 (m, 7H, Cyclo). MS (EI): *m/z* = 587.140 $[M]^+$.

6,7,8,9-Tetrahydro-8-(4,5-dihydro-5-(4-methoxyphenyl)-1-(2,4-dinitrophenyl)-1H-pyrazol-3-yl)-1,3-dinitro-5H-carbazole ***(6h)***. FTIR (KBr) cm^{-1}: 1568 (C=C), 1645 (C=N), 1048 (C-N), 2995 (C-H), 1170 (C-OCH_3), 1449 (C-NO_2). ^{1}H NMR (d_6-DMSO) δ: 10.47 (1H, s, H-9), 8.27 (1H, d, *J* = 2.1 Hz, H-7 DNTHCz), 7.59 (1H, dd, *J* = 8.1, 2.6 Hz, H-5 DNPh), 7.47 (1H, d, *J* = 2.1 Hz, H-5 DNTHCz), 7.38 (1H, d, *J* = 8.1 Hz, H-6 DNPh), 7.36–7.24 (4H, m, Ph), 7.19 (1H, d, *J* = 2.6 Hz, H-3 DNPh), 4.81 (1H, m, H_c), 3.49 (1H, dd, *J* = 18, 9.6 Hz, H_a), 3.25 (1H, dd, *J* = 18, 9.8 Hz, H_b), 3.08 (3H, s, OCH_3), 1.75–1.64 (m, 7H, Cyclo). MS (EI): *m/z* = 601.156 $[M]^+$.

8-(5-(4-Chlorophenyl)-4,5-dihydro-1-(2,4-dinitrophenyl)-1H-pyrazol-3-yl)-6,7,8,9-tetrahydro-1,3-dinitro-5H-carbazole ***(6i)***. FTIR (KBr) cm^{-1}: 1569 (C=C), 1639 (C=N), 1044 (C-N), 2989 (C-H), 735 (C-Cl), 1450 (C-NO_2). ^1H NMR (d_6-DMSO) δ: 10.54 (1H, s, H-9), 8.32 (1H, d, J = 2.1 Hz, H-7 DNTHCz), 7.68 (1H, dd, J = 8.4, 2.6 Hz, H-5 DNPh), 7.57 (1H, d, J = 2.1 Hz, H-5 DNTHCz), 7.49 (1H, d, J = 8.4 Hz, H-6 DNPh), 7.37–7.26 (4H, m, Ph), 7.21 (1H, d, J = 2.6 Hz, H-3 DNPh), 4.82 (1H, m, H_c), 3.52 (1H, dd, J = 18, 9.6 Hz, H_a), 3.29 (1H, dd, J = 18, 9.6 Hz, H_b), 1.76–1.65 (m, 7H, Cyclo). MS (EI): m/z = 605.106 $[M]^+$.

8-(5-(Furan-2-yl)-4,5-dihydro-1-(2,4-dinitrophenyl)-1H-pyrazol-3-yl)-6,7,8,9-tetrahydro-1,3-dinitro-5H-carbazole ***(6j)***. FTIR (KBr) cm^{-1}: 1578 (C=C), 1647 (C=N), 1045 (C-N), 2996 (C-H), 1159 (C-O-C), 1438 (C-NO_2). ^1H NMR (d_6-DMSO) δ: 10.54 (1H, s, H-9), 8.32 (1H, d, J = 2.1 Hz, H-7 DNTHCz), 7.67 (1H, d, J = 8.1 Hz, H-5 DNPh), 7.58 (1H, d, J = 2.1 Hz, H-5 DNTHCz), 7.49 (1H, dd, J = 8.1, 2.6 Hz, H-6 DNPh), 7.38–7.31 (3H, m, Ph), 7.23 (1H, d, J = 2.6 Hz, H-3 DNPh), 4.87 (1H, m, H_c), 3.54 (1H, dd, J = 18, 9.7 Hz, H_a), 3.27 (1H, dd, J = 18, 9.7 Hz, H_b), 1.78–1.66 (m, 7H, Cyclo). MS (EI): m/z = 561.124 $[M]^+$.

4,5-Dihydro-3-(6,7,8,9-tetrahydro-1,3-dinitro-5H-carbazol-8-yl)-5-phenylpyrazole-1-carbothioamide ***(6k)***. FTIR (KBr) cm^{-1}: 1576 (C=C), 1639 (C=N), 1045 (C-N), 2988 (C-H), 3610 (C-NH_2), 1450 (C-NO_2) 1202 (C=S). ^1H NMR (d_6-DMSO) δ: 10.52 (1H, s, H-9), 8.26 (1H, d, J = 2.7 Hz, H-7 DNTHCz), 7.57–7.48 (5H, m, Ph), 7.41 (1H, d, J = 2.7 Hz, H-5 DNTHCz), 4.86 (1H, m, H_c), 3.49 (1H, dd, J = 18, 9.8 Hz, H_a), 3.31 (1H, dd, J = 18, 9.8 Hz, H_b), 1.78–1.67 (m, 7H, Cyclo). MS (EI): m/z = 464.127 $[M]^+$.

4,5-Dihydro-3-(6,7,8,9-tetrahydro-1,3-dinitro-5H-carbazol-8-yl)-5-(2-hydroxyphenyl)pyrazole-1-carbothioamide ***(6l)***. FTIR (KBr) cm^{-1}: 1569 (C=C), 1639 (C=N), 1044 (C-N), 2994 (C-H), 3300 (C-OH), 3590 (C-NH_2), 1445 (C-NO_2). ^1H NMR (d_6-DMSO) δ: 10.48 (1H, s, H-9), 8.28 (1H, d, J = 2.7 Hz, H-7 DNTHCz), 7.65–7.51 (4H, m, Ph), 7.39 (1H, d, J = 2.7 Hz, H-5 DNTHCz), 4.92 (1H, m, H_c), 3.52 (1H, dd, J = 18, 9.6 Hz, H_a), 3.27 (1H, dd, J = 18, 9.6 Hz, H_b), 1.76–1.64 (m, 7H, Cyclo). MS (EI): m/z = 480.122 $[M]^+$.

4,5-Dihydro-3-(6,7,8,9-tetrahydro-1,3-dinitro-5H-carbazol-8-yl)-5-(4-methoxyphenyl)pyrazole-1-carbothioamide ***(6m)***. FTIR (KBr) cm^{-1}: 1578 (C=C), 1637 (C=N), 1043 (C-N), 2998 (C-H), 1058 (C-OCH_3), 3600 (C-NH_2), 1436 (C-NO_2), 1204 (C=S). ^1H NMR (d_6-DMSO) δ: 10.51 (1H, s, H-9), 8.32 (1H, d, J = 2.4 Hz, H-7 DNTHCz), 7.69–7.53 (5H, m, Ph), 7.39 (1H, d, J = 2.4 Hz, H-5 DNTHCz), 4.95 (1H, m, H_c), 3.54 (1H, dd, J = 18, 9.8 Hz, H_a), 3.29 (1H, dd, J = 18, 9.8 Hz, H_b), 3.11 (3H, s, OCH_3), 1.76–1.64 (m, 7H, Cyclo). MS (EI): m/z = 491.137 $[M]^+$.

5-(4-Chlorophenyl)-4,5-dihydro-3-(6,7,8,9-tetrahydro-1,3-dinitro-5H-carbazol-8-yl)pyrazole-1-carbothioamide ***(6n)***. FTIR (KBr) cm^{-1}: 1576 (C=C), 1648 (C=N), 1045 (C-N), 2992 (C-H), 760 (C-Cl), 3550 (C-NH_2), 1447 (C-NO_2), 1206 (C=S). ^1H NMR (d_6-DMSO): δ: 10.54 (1H, s, H-9), 8.37 (1H, d, J = 2.6 Hz, H-7 DNTHCz), 7.62–7.54 (4H, m, Ph), 7.43 (1H, d, J = 2.6 Hz, H-5 DNTHCz), 4.93 (1H, m, H_c), 3.55 (1H, dd, J = 18, 9.6 Hz, H_a), 3.31 (1H, dd, J = 18, 9.6 Hz, H_b), 1.78–1.66 (m, 7H, Cyclo). MS (EI): m/z = 498.088 $[M]^+$.

5-(Furan-2-yl)-4,5-dihydro-3-(6,7,8,9-tetrahydro-1,3-dinitro-5H-carbazol-8-yl)pyrazole-1-carbothioamide ***(6o)***. FTIR (KBr) cm^{-1}: 1576 (C=C), 1639 (C=N), 1042 (C-N), 2994 (C-H), 1160 (C-O-C), 3580 (C-NH_2), 1449 (C-NO_2), 1202 (C=S). ^1H NMR (d_6-DMSO) δ: 10.57 (1H, s, H-9), 8.36 (1H, d, J = 2.4 Hz, H-7 DNTHCz), 7.74–7.57 (3H, m, Ph), 7.41 (1H, d, J = 2.4 Hz, H-5 DNTHCz), 4.97 (1H, m, H_c), 3.57 (1H, dd, J = 18, 9.6 Hz, H_b), 3.32 (1H, dd, J = 18, 9.6 Hz, H_b), 1.80–1.67 (m, 7H, Cyclo). MS (EI): m/z = 454.106 $[M]^+$.

3. Results and Discussion

Tetrahydrocarbazole linked 1,2-diazole have become increasingly useful in many fields; prior to the synthesis of title compounds, ketone **1(a–c)** was prepared by reported method [16]. Initially, this reaction was carried out in two steps, that is, formation of α,β-unsaturated ketone by the action of ketone **1(a–c)** and aromatic aldehydes **2(a–e)** in the presence of catalyst. In the later step, addition of hydrazine resulted in the diazole formation and the yield was considerably decreased. In search of better results in terms of yield, reaction time, and easy workup, MCR approach (Scheme 1) had been employed for the synthesis of tetrahydrocarbazole based 1,2-diazoles **4–6(a–o)** where **1(a–c)**, **2(a–e),** and **3(a–c)** in alkaline medium using 20 mol% sodium methoxide in methanol were stirred together (Table 1).

The color of reaction mixture gradually amplified upon addition of base which indicated the formation of enolate ion. Slow addition of aldehyde with continuous stirring led in situ to α,β-unsaturated ketone generation indicated by TLC and FTIR which was consumed by hydrazine to furnish titled diazoles in excellent yield. The product was achieved by pouring reaction mixture on crushed ice.

Resultant precipitates were filtered, vacuum dried, and subjected to column chromatography using dichloromethane-methanol as solvent system. Mechanistically, attack of hydrazine took place via Michael type-addition on α,β-unsaturated ketone followed by removal of water which furnishes pyrazoline moiety. Initially, the reaction was carried out in acetic acid : water system but it remained incomplete even after longer time of more than 24 hours. This is why basic conditions were opted for conventional method. Reaction conditions were optimized for conventional heating method using model reaction. **1c**, **2a**, and **3a** were reacted in the presence of different amounts of sodium methoxide as catalyst. Various concentrations were explored where 20 mol% was found to be ideal. Sodium methoxide was most effective in abstracting the proton from sterically hindered ketone (**1c**) relative to typical sodium hydroxide or potassium hydroxide catalysts. Its small amount was sufficient to provide product in 2.5 hours in terms of yield; increase in catalyst amount decreased the product yield (Table 3).

TABLE 3: Optimization of reaction conditions.

Entry	Conventional heating			Microwave irradiation		
	Catalyst* (mol%)	Time (hr)	Yield (%)	Catalyst** ratio	Time (min)	Yield (%)
(1)	10	5	48	2 : 1	15	32
(2)	20	2.5	75	1 : 1	4	88
(3)	30	4	43	1 : 1.5	6	46
(4)	40	3.5	30	1 : 2	12	38

*Sodium methoxide; **acetic-acid-water.

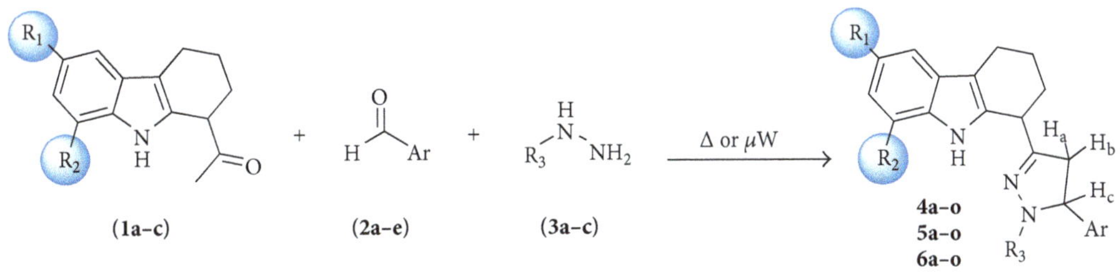

SCHEME 1: Preparation of tetrahydrocarbazoles linked 1,2-diazoles.

To improve and make this approach more benign, clean, and high-yielding, we have attempted to prepare diazoles under microwave radiations. Therefore, to attain benchmark reaction conditions in microwave, ketone **(1c)**, benzaldehyde **(2a)**, and hydrazine hydrate **(3a)** in 1 : 1 : 1.2 equivalents were exposed to microwave radiations at 300 W. It has been observed that microwave-assisted Aza-Michael reaction proceeds more speedily in acetic acid-water catalytic system than in water alone. This led to exploring various combinations of catalytic system [17].

The aqueous-Brønsted acid system (1 : 1) was used as green solvent and catalyst that demonstrated excellent results while reaction progress was monitored by TLC (Table 3). We were contented on the observation that the reaction proceeded smoothly and almost all the reactants were consumed within 4 minutes to afford the desired product **(6a)** in satisfactory yield (Table 3, Entry 2). Reaction was also performed on less power levels (200–250 W) but all results led to longer time with incomplete reactants consumption. To inspect the scope and efficacy of the current procedure, reactions between different substrates **(1a–c)**, various aromatic aldehydes with electron donating/withdrawing groups **(2a–e)**, and different hydrazines **(3a–c)** were carried out and results are summarized in Table 3. From these results, it has been observed that all the reactions proceeded smoothly in very short reaction time to afford the corresponding tetrahydrocarbazole linked 1,2-diazoles in good to excellent (Table 2). Diazoles bearing electron donating groups (-OH, $-OCH_3$, -furfural) on aromatic ring of carbinol carbon demonstrated remarkable yield. It might be helpful to envision that these groups forcing the electrons towards electrophilic aldehyde group make it tend to be less contented and more reactive. Reaction time reduced significantly from hours to few minutes by applying microwave radiations.

The structure confirmation was achieved with the help of FTIR, ^{1}H NMR, and mass spectroscopic analysis of the compounds. The protons associated with tetrahydrocarbazole (THCz), 6-nitro tetrahydrocarbazole (NO_2 THCz) and 6,8-dinitrotetrahydrocarbazole (DNTHCz), and aromatic moieties were matched with that of reported in literature [13, 15, 16]. In ^{1}H NMR, a multiplet at δ 4.85–4.80, 5.38–5.18, and 5.19–4.91 was assigned to enantiotopic proton H_c in **4a–o, 5a–o,** and **6a–o**, respectively. A doublet resonated at δ 3.68–3.56, 3.58–3.34, and 3.57–3.47 attributed to diastereotopic proton H_a while another doublet at δ 3.38–3.48, 3.31–3.14, 3.32–3.23 corresponds to H_b, in **4a–o, 5a–o,** and **6a–o**, subsequently confirming 1,2-diazole ring. The singlet responsible for N-H pyrazoline ring was observed at δ 6.27–6.21, 6.37–6.32, and 6.24–6.20 in **4a–e, 5a–e,** and **6a–e** [18]. Further confirmation of heterocyclic ring formation was endorsed by absorption bands in ranges from 3078 to 3045 cm^{-1} in FTIR spectrum.

4. Conclusion

In the present study, we have successfully established a rapid, efficient, one-pot, and highly selective synthesis of a new series of tetrahydrocarbazole linked 1,2-diazoles by employing Fischer indole, aldol condensation, and Michael type cycloaddition reactions. Both alkaline and acidic conditions were found suitable to afford desired molecules. Key advantages of described procedure are operational simplicity, shorter reaction time, convenient workup, and high yields. We are optimistic to construct novel structural leads in the domains of heterocyclic chemistry. This new synthetic series is remarkable as both key units THCz and 1,2-diazole has renowned biological profile. We believe that our procedure is user-friendly and economically cheap due to inexpensive catalyst and simple synthetic protocols.

Conflicts of Interest

The authors declare no conflicts of interest regarding this work.

Acknowledgments

The authors are grateful to Higher Education Commission, Pakistan, for providing the generous grant to carry out this project and analysis.

References

[1] N. Harikrishna, A. M. Isloor, K. Ananda, A. Obaid, and H.-K. Fun, "Synthesis, and antitubercular and antimicrobial activity of 1′-(4-chlorophenyl)pyrazole containing 3,5-disubstituted pyrazoline derivatives," *New Journal of Chemistry*, vol. 40, no. 1, pp. 73–76, 2016.

[2] M. E. Rahman, H. M. Faidallah, A. K. Kuerban, and A. Naveed, "Synthesis of pyrazoles and bipyrazoles- a review," *American Journal of PharmTech Research*, vol. 5, no. 1, pp. 89–112, 2015.

[3] T. Sano, Y. Nishio, Y. Hamada, H. Takahashi, T. Usuki, and K. Shibata, "Design of conjugated molecular materials for optoelectronics," *Journal of Materials Chemistry*, vol. 10, no. 1, pp. 157–161, 2000.

[4] R. N. Doughty and H. D. White, "Carvedilol: Use in chronic heart failure," *Expert Review of Cardiovascular Therapy*, vol. 5, no. 1, pp. 21–31, 2007.

[5] R. P. Schleimer and E. Benjamini, "Effects of prostaglandin synthesis inhibition on the immune response," *Immunopharmacology*, vol. 3, no. 3, pp. 205–219, 1981.

[6] S. Elz and W. L. Heil, "Synthesis, biological in vitro evaluation and stereoselectivity of ondansetron analogues: novel 5-HT2A receptor antagonists," *Bioorganic and Medicinal Chemistry Letters*, vol. 5, no. 7, pp. 667–672, 1995.

[7] T. Taj, R. R. Kamble, T. M. Gireesh, R. K. Hunnur, and S. B. Margankop, "One-pot synthesis of pyrazoline derivatised carbazoles as antitubercular, anticancer agents, their DNA cleavage and antioxidant activities," *European Journal of Medicinal Chemistry*, vol. 46, no. 9, pp. 4366–4373, 2011.

[8] C.-J. Lu, H. Li, Q.-F. Xu, Q.-H. Xu, and J.-M. Lu, "Synthesis and Morphology of Two Carbazole-Pyrazoline-Containing Polymer Systems and Their Electrical Memory Performance," *ChemPlusChem*, vol. 80, no. 8, pp. 1354–1362, 2015.

[9] A. A. El-Kateb, N. M. Abd El-Rahman, T. S. Saleh, M. H. Ali, A. S. Elhaddad, and A. Y. El-Dosoky, "Microwave mediated facile synthesis of some novel pyrazole, pyrimidine, pyrazolo[1,5-a]pyrimidine, triazolo[1,5-a]pyrimidine and pyrimido[1,2-a] benzimidazole derivatives under solventless condition," *Nature and Science*, vol. 10, no. 11, pp. 77–86, 2012.

[10] P. Ali, N. Pathan, and T. Ben Hadda, "Benign Methodology and Efficient Catalysis for the One-Pot Multicomponent Synthesis of Dihydropyrimidinones and Thiones: A New Key for Old Lock," *Journal of Applied Chemistry*, vol. 2014, pp. 1–6, 2014.

[11] K. Manna and Y. K. Agrawal, "Microwave assisted synthesis of new indophenazine 1,3,5-trisubstruted pyrazoline derivatives of benzofuran and their antimicrobial activity," *Bioorganic and Medicinal Chemistry Letters*, vol. 19, no. 10, pp. 2688–2692, 2009.

[12] D. Bandyopadhyay, S. Mukherjee, L. C. Turrubiartes, and B. K. Banik, "Ultrasound-assisted aza-Michael reaction in water: A green procedure," *Ultrasonics Sonochemistry*, vol. 19, no. 4, pp. 969–973, 2012.

[13] E. A. Hussain, A. Yasmeen, Z. Sadiq, N. Naz, and H. I. Shaheen, "Thermal/microwave/DES mediated synthesis of iminotetrahydrocarbazoles and their biological evaluation," *International Journal of Biology, Pharmacy and Allied Sciences*, vol. 5, no. 12, pp. 3331–3344, 2016.

[14] G. Yaqub, A. Hannan, E. Akbar et al., "Synthesis, antibacterial, and antifungal activities of novel pyridazino carbazoles," *Journal of Chemistry*, vol. 2013, Article ID 818739, 7 pages, 2013.

[15] Z. Sadiq, E. Akbar, and N. Naz, "Synthesis and antimicrobial assessment of efficacious carbazole derivatives," *International Journal of Pharma and Bio Sciences*, vol. 5, no. 1, pp. P455–P464, 2014.

[16] N. A. M. M. Shmeiss, M. M. F. Ismail, A. M. Soliman, and H. I. El-Diwani, "Synthesis of novel 1-substituted and 1,9-disubstituted-1,2,3,4-tetrahydro- 9H-carbazole derivatives as potential anticancer agents," *Molecules*, vol. 5, no. 10, pp. 1101–1112, 2000.

[17] B. Sharifzadeh, N. O. Mahmoodi, M. Mamaghani, K. Tabatabaeian, A. S. Chirani, and I. Nikokar, "Facile regioselective synthesis of novel bioactive thiazolyl-pyrazoline derivatives via a three-component reaction and their antimicrobial activity," *Bioorganic and Medicinal Chemistry Letters*, vol. 23, no. 2, pp. 548–551, 2013.

[18] Z. N. Siddiqui and T. Khan, "An efficient synthesis of novel bis-chalcones and bis-pyrazolines in the presence of cellulose sulfuric acid as biodegradable catalyst under solvent-free conditions," *Journal of the Brazilian Chemical Society*, vol. 25, no. 6, pp. 1002–1011, 2014.

16

Synthesis, Characterization, and Relative Study on the Catalytic Activity of Zinc Oxide Nanoparticles Doped $MnCO_3$, $-MnO_2$, and $-Mn_2O_3$ Nanocomposites for Aerial Oxidation of Alcohols

Mohamed E. Assal, Mufsir Kuniyil, Mohammed Rafi Shaik, Mujeeb Khan, Abdulrahman Al-Warthan, Mohammed Rafiq H. Siddiqui, and Syed Farooq Adil

Department of Chemistry, College of Science, King Saud University, P.O. Box 2455, Riyadh 11451, Saudi Arabia

Correspondence should be addressed to Syed Farooq Adil; sfadil@ksu.edu.sa

Academic Editor: Mohamed Afzal Pasha

Zinc oxide nanoparticles doped manganese carbonate catalysts [X% ZnO_x–$MnCO_3$] (where $X = 0$–7) were prepared via a facile and straightforward coprecipitation procedure, which upon different calcination treatments yields different manganese oxides, that is, [X% ZnO_x–MnO_2] and [X% ZnO_x–Mn_2O_3]. A comparative catalytic study was conducted to evaluate the catalytic efficiency between carbonates and oxides for the selective oxidation of secondary alcohols to corresponding ketones using molecular oxygen as a green oxidizing agent without using any additives or bases. The prepared catalysts were characterized by different techniques such as SEM, EDX, XRD, TEM, TGA, BET, and FTIR spectroscopy. The 1% ZnO_x–$MnCO_3$ calcined at 300°C exhibited the best catalytic performance and possessed highest surface area, suggesting that the calcination temperature and surface area play a significant role in the alcohol oxidation. The 1% ZnO_x–$MnCO_3$ catalyst exhibited superior catalytic performance and selectivity in the aerial oxidation of 1-phenylethanol, where 100% alcohol conversion and more than 99% product selectivity were obtained in only 5 min with superior specific activity (48 $mmol \cdot g^{-1} \cdot h^{-1}$) and 390.6 turnover frequency (TOF). The specific activity obtained is the highest so far (to the best of our knowledge) compared to the catalysts already reported in the literatures used for the oxidation of 1-phenylethanol. It was found that ZnO_x nanoparticles play an essential role in enhancing the catalytic efficiency for the selective oxidation of alcohols. The scope of the oxidation process is extended to different types of alcohols. A variety of primary, benzylic, aliphatic, allylic, and heteroaromatic alcohols were selectively oxidized into their corresponding carbonyls with 100% convertibility without overoxidation to the carboxylic acids under base-free conditions.

1. Introduction

Selective catalytic oxidation of alcohols to corresponding carbonyl compounds is one of the most pivotal functional group transformations in organic chemistry [1–6]. The oxidation products such as aldehyde and ketone derivatives are important precursors and intermediates in the cosmetics, perfumery, flame-retardants, insecticides, confectionary, biofuels, and pharmaceutical industries [7–11]. Usually, conventional oxidation of alcohols was conducted using toxic and expensive stoichiometric oxidizing agents such as CrO_3 $KMnO_4$, NaClO, $Na_2Cr_2O_7$, SeO_2, and Br_2 and leads to producing harmful by-products [12, 13]. In contrast, using clean and low cost oxidizing agents, such as aqueous hydrogen peroxide and particularly molecular oxygen to produce water, is the only side product and has gained growing interest from the sustainable and green chemistry point of views [14, 15]. In this context, noble metals such as gold [16–20], palladium [21], platinum [22–24], ruthenium [25, 26], and rhodium [27, 28] have been used as a heterogeneous catalyst for the alcohol oxidation with high catalytic activities and selectivities. Beside their high costs, these precious metals also have serious toxicity issues and difficulty in preparation and rarity of these noble metals makes these catalysts impractical for industrial applications [29, 30]. Therefore, huge efforts have been made in order to replace these expensive noble metal catalysts with cheaper and plentiful nonnoble metals, for example, copper [31, 32], cobalt [33, 34], nickel [35–37], iron [38, 39], vanadium [40], silver [41, 42], chromium [43], molybdenum [44, 45], rhenium [46], and zirconium [47], for selective oxidation of alcohols. The metal, metal oxide, and mixed metal oxide nanoparticles catalysts were also found to be highly effective for the catalytic oxidation of alcohols. In addition, it has been widely reported that the catalytic activity of mixed metal oxide nanoparticles catalysts was remarkably enhanced upon doping with other metals, maybe because the metal nanoparticles possess huge surface area [48, 49].

Manganese carbonate ($MnCO_3$) has being considered one of the inexpensive and most stable metal oxides with high catalytic performance. Because $MnCO_3$ has various leads such as being most active, being stable, having low cost, and being environmental friendly catalyst, $MnCO_3$ and manganese oxides have been widely castoff as a catalyst or catalyst support for the catalytic oxidation of alcohols into carbonyls [50–53]. Moreover, it has several features, such as being highly active, stable, inexpensive, and ecofriendly catalyst [51–56]. However, different types of manganese oxide, mixed manganese oxide, and noble metal doped/supported Mn oxides were widely employed for the catalytic oxidation of numerous organic substrates, for instance, oxidation of naphthalene to carbon dioxide [57], oxidation of carbon monoxide to CO_2 [58, 59], oxidation of toluene to CO_2 [60], oxidation of ethylene and propylene to CO_2 [61], oxidation of cyclohexane to cyclohexanol and cyclohexanone [62], oxidation of alkyl aromatics to ketones [63], oxidation of 4-tert-butyltoluene to 4-tert-butylbenzaldehyde [64], and oxidation of formaldehyde to CO_2 [65].

In continuation of our efforts on the use of different mixed metal oxide nanoparticles as an efficient catalyst for the selective oxidation of alcohols in presence of molecular O_2 [18, 32, 41, 48, 49], we demonstrate herein a facile and straightforward procedure for the preparation of ZnO_x NPs doped manganese carbonates or oxides employed for oxidation of secondary alcohols with environmentally friendly oxidizing agent such molecular O_2 which produces water as the only by-product under base-free conditions. The reaction circumstances have been optimized with different weight percentages of ZnO_x, reaction times, calcination temperatures, reaction temperatures, and catalyst amounts using oxidation of 1-phenylethanol into acetophenone as a reaction model. The as-prepared catalyst was employed for oxidation of different types of alcohols. It was found that most of alcohols that were used in this study were completely converted to their corresponding aldehydes or ketones in extremely short reaction time without overoxidation to the acids. Moreover, the prepared catalysts have been characterized by several types of techniques, like scanning electron microscopy (SEM), energy dispersive X-ray spectroscopy (EDX), X-ray diffraction (XRD), transmission electron microscopy (TEM), Fourier transform infrared spectroscopy (FTIR), thermogravimetric analysis (TGA), and Brunauer-Emmett-Teller (BET) surface area measurement spectroscopies.

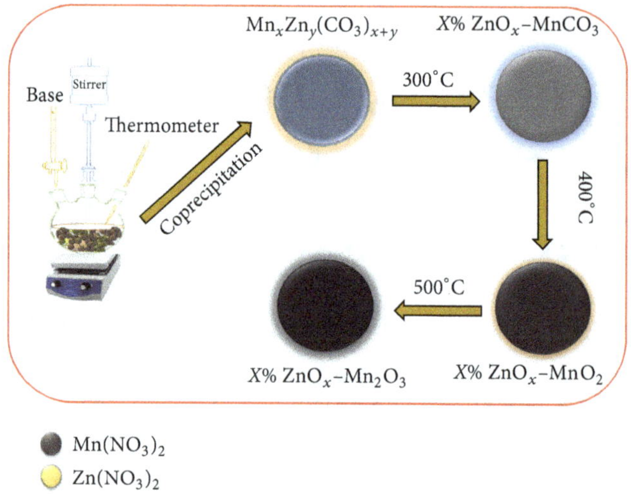

SCHEME 1: Graphical illustration of the synthesis of the X% ZnO_x–$MnCO_3$ oxide and other products formed.

2. Experimental Section

2.1. Preparation of the Catalysts. ZnO_x NPs doped $MnCO_3$ catalysts of the type X% ZnO_x–$MnCO_3$ (where $X = 0$, 1, 3, 5, and 7) were prepared by coprecipitation method where % X denotes w/w%. Stoichiometric amount of manganese (II) nitrate tetrahydrate ($Mn(NO_3)_2{\cdot}4H_2O$) and zinc nitrate hexahydrate ($Zn(NO_3)_2{\cdot}6H_2O$) was dissolved in distilled water. About 100 mL of the mixture of solutions was taken in a glass three-necked round-bottom flask. The resulting mixture was heated to 80°C, while stirring using a mechanical stirrer and 0.5 M solution of sodium hydrogen carbonate ($NaHCO_3$) was added dropwise until the solution attained a pH 9. The solution was continued to stir at the same temperature for about 3 hours and then left on stirring over night at room temperature. The solution was filtered by centrifugation and the solid product obtained was dried at 70°C overnight; this upon calcination at different temperatures yielded different catalysts, which were used for comparative study of catalytic properties. A pictorial illustration is given in Scheme 1.

2.2. Characterization Techniques. Scanning electron microscopy (SEM) and elemental analysis (energy dispersive X-ray analysis: EDX) were carried out using Jeol SEM model JSM 6360A (Japan). This was used to determine the morphology of nanoparticles and its elemental composition. Fourier transform infrared (FTIR) spectra were registered on a Perkin-Elmer 1000 FTIR instrument (Waltham, MA, USA); the samples were prepared in KBr pellets. Transmission electron microscopy (TEM) was carried out using Jeol TEM model JEM-1101 (Japan), which was used to determine the shape and size of nanoparticles. Powder X-ray diffraction studies were carried out using Altima IV [Make: Regaku] X-ray diffractometer. BET surface area was measured on a NOVA 4200e surface area and pore size analyzer. Thermogravimetric analysis was carried out using Perkin-Elmer Thermogravimetric Analyzer 7.

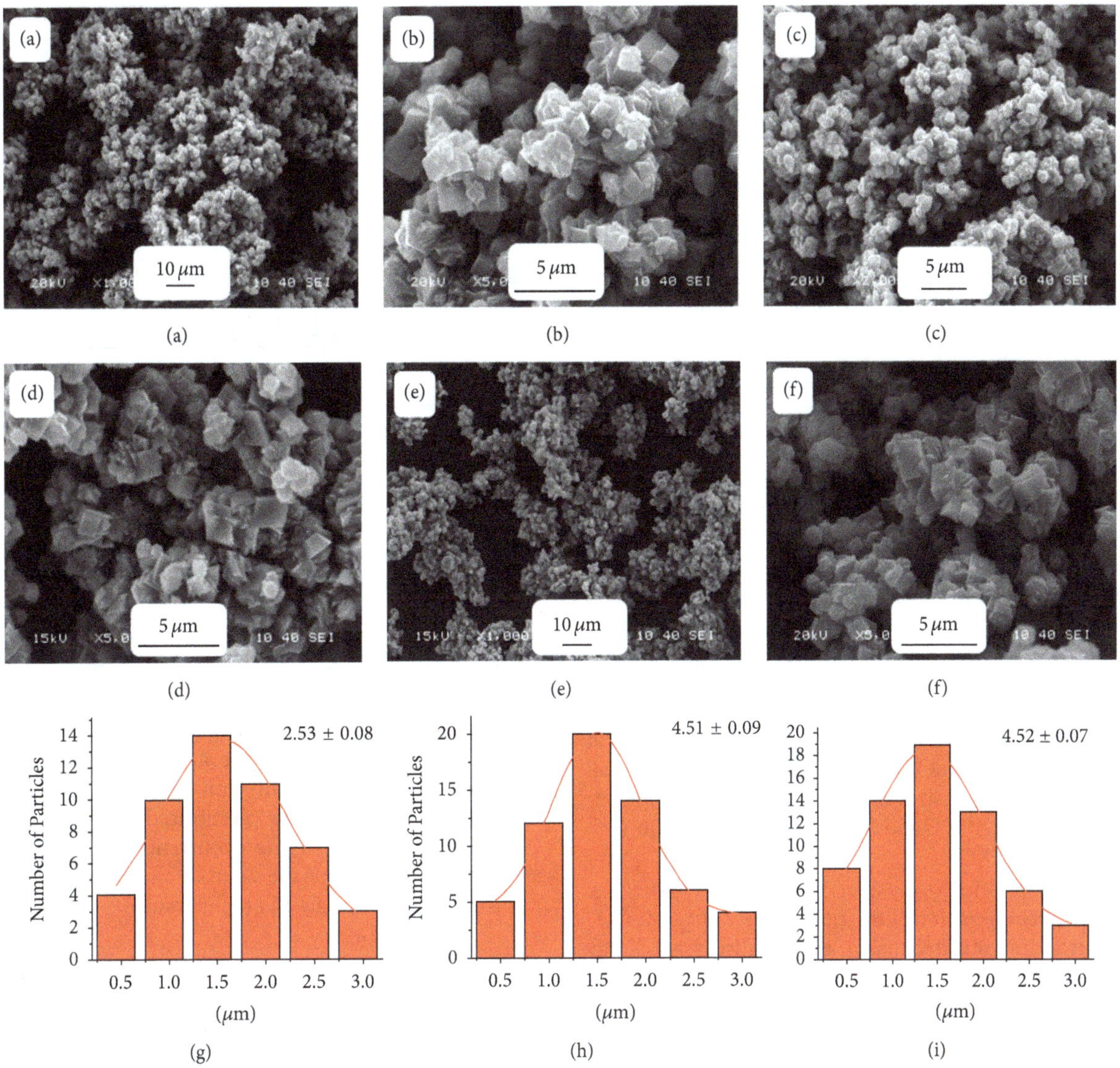

FIGURE 1: SEM analysis of the catalysts calcined at (a)-(b) 300°C; (c)-(d) 400°C; (e)-(f) 500°C; (a)-(b) overview image for as-prepared 1% ZnO_x–$MnCO_3$; (c)-(d) overview image of 1% ZnO_x–MnO_2; (e)-(f) overview image of 1% ZnO_x–Mn_2O_3; (g) particle size distribution of 1% ZnO_x–$MnCO_3$; (h) particle size distribution of 1% ZnO_x–MnO_2; (i) particle size distribution of 1% ZnO_x–Mn_2O_3.

2.3. Typical Procedure for Alcohol Oxidation. Liquid-phase oxidation of 1-phenylethanol was performed in glass flask equipped with a magnetic stirrer, reflux condenser, and thermometer. In a typical experiment, a mixture of the 1-phenylethanol (2 mmol), toluene (10 mL), and the catalyst (0.3 g) was transferred in a glass three-necked round-bottomed flask (100 mL); the resulting mixture was then heated to desired temperature with vigorous stirring. The oxidation experiment was started by bubbling oxygen gas at a flow rate of 20 mL/min into the reaction mixture. After the reaction, the solid catalyst was filtered off by centrifugation and the liquid products were analyzed by gas chromatography to determine the conversion of the alcohol and product selectivity by GC, 7890A, Agilent Technologies Inc., equipped with a flame ionization detector (FID) and a 19019S-001 HP-PONA column.

3. Results and Discussion

3.1. Characterization of the Catalysts. The as-synthesized catalysts surface morphology was analyzed by using scanning electron microscopy (SEM). The SEM images of the as-synthesized catalysts calcined at different temperatures, 1% ZnO_x–$MnCO_3$ at 300°C, 1% ZnO_x–MnO_2 at 400°C, and 1% ZnO_x–Mn_2O_3 at 500°C, exhibited morphology as shown in Figures 1(a)–1(e). The SEM image of the catalysts discloses well defined and spherical shaped morphology. Furthermore, it is observed that the morphology of the catalysts did not vary with the different calcination temperatures but it varies with size of the particles; as the calcination temperature increases ZnO_x NPs size increases and also the spherical shaped morphology of all the catalysts clearly exhibited in Figures 1(a)–1(e). The particle size distribution of SEM micrographs

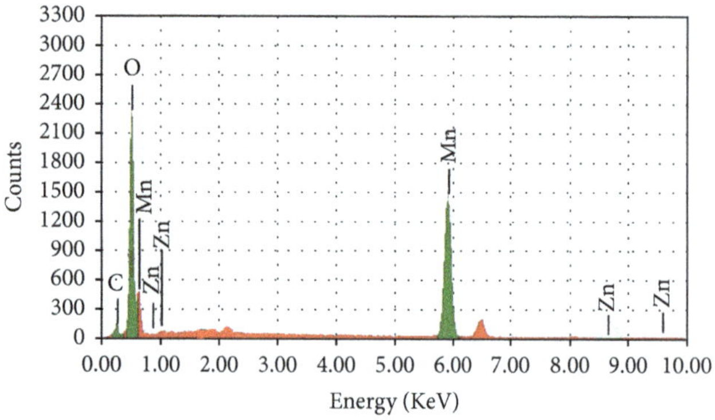

FIGURE 2: EDX analysis of the as-synthesized catalysts calcined 1% ZnO_x–$MnCO_3$.

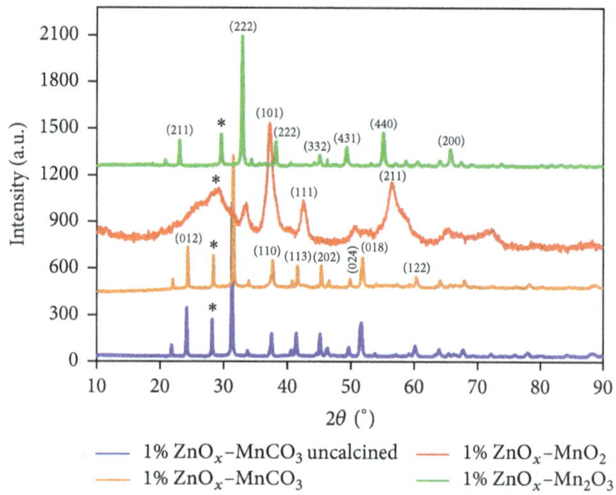

FIGURE 3: XRD pattern of the catalyst 1% ZnO_x–$MnCO_3$ uncalcined and catalyst calcined at different temperatures 1% ZnO_x–$MnCO_3$; 1% ZnO_x–MnO_2; and 1% ZnO_x–Mn_2O_3.

of the as-synthesized catalysts, 1% ZnO_x–$MnCO_3$ calcined at different temperatures, 1% ZnO_x–$MnCO_3$ at 300°C, 1% ZnO_x–MnO_2 at 400°C, and 1% ZnO_x–Mn_2O_3 at 500°C, was assessed by using the program software J as shown in Figures 1(g)–1(i). The particle size was found to be approximately equal to 2.53, 4.51, and 4.79 μm in the catalyst precalcined at 300, 400, and 500°C correspondingly. From the attained results, it is noticed that while there is a variation in particle sizes among the precalcined catalysts 1% ZnO_x–$MnCO_3$ at 300°C and 1% ZnO_x–MnO_2 at 400°C, there is a merely small difference in the particle sizes among the precalcined catalysts 1% ZnO_x–MnO_2 at 400°C and 1% ZnO_x–Mn_2O_3 at 500°C.

3.2. Energy Dispersive X-Ray Spectroscopy (EDX) Analysis. The elemental composition of the catalyst 1% ZnO_x–$MnCO_3$ at 300°C was investigated using energy dispersive X-ray spectroscopy (EDX). The intense signal at 5.5–6 keV strongly recommends that "Mn" was the major element, which has an optical absorption in this range due to the surface plasmon resonance (SPR). The signals at 0.8, 1.1, 8.6, and 9.6 strongly endorse that "Zn" was the element. It was also prominent that the other signals were also attained in the range 0.0–0.5 keV which denotes the typical absorption of carbon and oxygen. The found values of composition percentage of the Zn, Mn, and O_2 present in the catalyst were nearly identical to the values of the composition castoff in preparing the catalyst shown in Figure 2.

3.3. X-Ray Diffraction (XRD) Analysis. The as-prepared catalysts crystal structure was analyzed by using X-ray diffraction (XRD). The XRD images of the catalysts 1% ZnO_x–$MnCO_3$ uncalcined and calcined at different temperatures, 1% ZnO_x–$MnCO_3$ at 300°C, 1% ZnO_x–MnO_2 at 400°C, and 1% ZnO_x–Mn_2O_3 at 500°C are shown in Figure 3. The found diffraction signals confirmed that as-synthesized catalysts are in crystalline nature. The calcined catalysts conceded various phases of manganese oxide alongside with the traces of ZnO_x. The uncalcined catalyst shows patterns which were obtained to be in agreement with the standard XRD data of rhodochrosite syn-$MnCO_3$ (JCPDS number 44-1472). With the calcination at 300°C the XRD pattern exhibits the existence of rhodochrosite D-$MnCO_3$ (JCPDS number 1-0981). Subsequently when the calcination temperature was further increased to 400°C the XRD spectra attained were in agreement with pyrolusite MnO_2 (JCPDS number 24-0735). When the catalyst was calcined at 500°C the XRD pattern obtained was in agreement with bixbyite-Mn_2O_3 (JCPDS number 2-0909). The reflections noticed with symbol asterisk (∗) might be owing to the occurrence of ZnO_x. The as-synthesized nanocomposite catalysts have been matched with known compounds stated in the literature.

3.4. Fourier Transform Infrared Spectroscopy (FTIR) Analysis. The FTIR spectrum was used to classify the functional groups on the surface of the as-synthesized catalyst. The FTIR spectra of the catalysts at different calcination temperatures (300, 400, and 500°C), that is, 1% ZnO_x–$MnCO_3$, 1% ZnO_x–MnO_2, and 1% ZnO_x–Mn_2O_3, correspondingly are displayed in Figure 4. High wavenumber zone displays stretching vibrations of the surface hydroxyl group and physisorbed water. The characteristic bands of νOH located around 3475 cm^{-1} were found to be in the 300, 400, and 500°C calcined catalyst. The intensity of these bands decreases as the calcination temperature increase demonstrates the decrease of the existence of OH group and H_2O on the catalyst surface. The absorption band at around 1635 cm^{-1} resembles fingerprint bending vibration modes of hydroxyl groups [83]. Therefore, it can be said that OH group in the catalyst surface acts a significant part in the 1-phenylethanol oxidation, in which the 1% ZnO_x–$MnCO_3$ catalyst displays the highest catalytic performance associated with the other calcination temperatures [84]. In case of 300°C calcination temperature, the presence of a sharp peak at nearly 1356 cm^{-1} is a fingerprint for surface carbonate (CO_3^{-2}) group which confirmed the occurrence of carbonate group in the surface of the catalyst and these results are in good agreement with EDX and XRD results [85, 86]. In case of 400 and 500°C, that is, 1% ZnO_x–MnO_2 and 1% ZnO_x–Mn_2O_3 correspondingly, the absorption peak

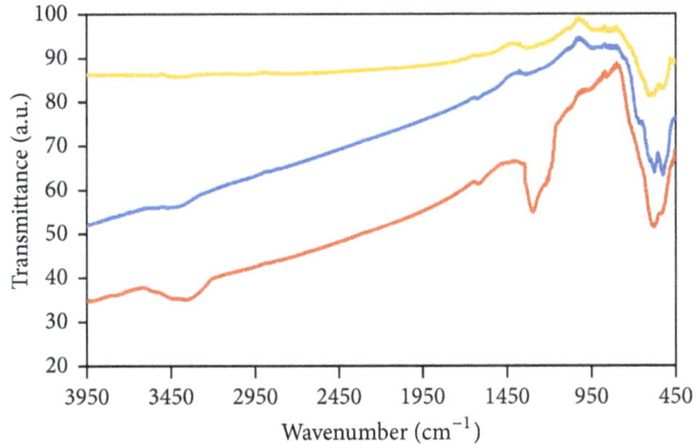

FIGURE 4: FTIR spectra of the catalyst calcined at different temperatures 1% ZnO_x–$MnCO_3$; 1% ZnO_x–MnO_2; and 1% ZnO_x–Mn_2O_3.

at around 1356 cm^{-1} representing carbonate group almost disappears, maybe owing to manganese carbonate converted to manganese oxides at elevated temperature. Sharp peaks are attained in the range of 520–610 cm^{-1} typical for different oxides of manganese [87, 88].

3.5. High-Resolution Transmission Electron Microscopy (HRTEM) Analysis. The HRTEM images of catalysts calcined at different temperatures, 1% ZnO_x–$MnCO_3$ at 300°C, 1% ZnO_x–MnO_2 at 400°C, and 1% ZnO_x–Mn_2O_3 at 500°C, are displayed in Figures 5(a)–5(c). Moreover, the HRTEM image reveals that the calcination temperature plays significant role in particle size. The particles size and agglomeration of the as-synthesized catalyst increase with increase of calcination temperature. It might also be owing to the decomposition of $MnCO_3$ to manganese oxide. The interplanar distance estimated from the HRTEM image of the catalysts calcined at 300°C (Figure 5(a)) revealed d-spacing of 0.29 nm and 0.23 nm resembles the (104) and the (110) planes of rhombohedral $MnCO_3$. The sample calcined at 400°C revealed d-spacing of 0.25 nm resembled the (201) planes of MnO_2 (Figure 5(b)). For the sample calcined at 500°C (Figure 5(c)), the most noticeable lattice spacing resembles the (200) plane of α-Mn_2O_3.

3.6. Thermal Gravimetric Analysis. For the as-synthesized catalyst calcined at different temperatures, 1% ZnO_x–$MnCO_3$ at 300°C, 1% ZnO_x–MnO_2 at 400°C, and 1% ZnO_x–Mn_2O_3 at 500°C, thermal stability was investigated by thermogravimetric analysis (TGA) as displayed in Figure 6. The thermogram of the as-synthesized catalyst calcined at 300°C demonstrates that the catalyst is steady up to 415°C with very insignificant percentage of weight loss that is less than 3%, due to the exclusion of volatile surface-absorbed water and volatile impurities. The highest loss of weight percentage was noticed by nearly 10%–12% in the temperature range of about 420–600°C which appears to be due to the loss of CO_2 of $MnCO_3$ to form MnO_2 and further oxidation of

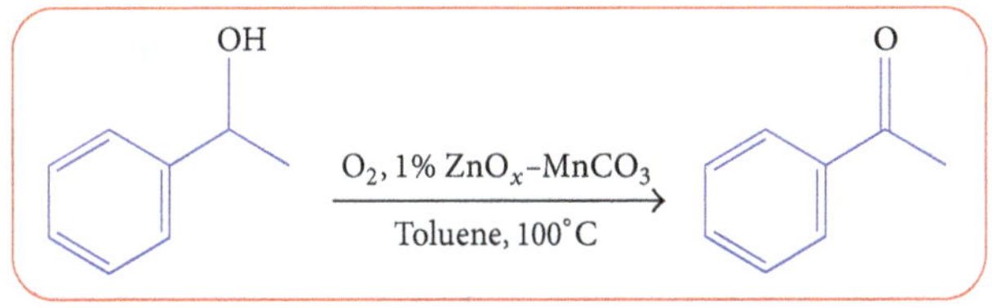

SCHEME 2: Aerial oxidation of 1-phenylethanol into acetophenone.

MnO_2 to Mn_2O_3 as indicated from the literature by Zhu et al. [89]. When the temperature is increased up to 800°C, the highest weight loss is noticed around 14%. Furthermore, the as-synthesized catalyst calcined at 400°C exhibited the maximum total weight loss percentage around 9% when the catalyst was heated up to 800°C, whereas in the case of the catalyst calcined at 500°C it displayed weight loss percentage of around 4%. Consequently, the catalysts displayed a thermal stability up to around 415°C and a marginally thermal stability decrease attained with raise in temperature.

3.7. Surface Area Measurements. The surface area of the as-synthesized catalysts was studied using BET surface area analysis, in order to determine the surface area and to understand the relationship between surface area and catalytic activity of the as-prepared catalyst for the oxidation of 1-phenylethanol. Table 1 displayed that the specific surface area of the prepared catalyst calcined at different calcination temperature such as 300°C, 400°C, and 500°C, that is, 1% ZnO_x–$MnCO_3$, 1% ZnO_x–MnO_2, and 1% ZnO_x–Mn_2O_3, respectively, was about 120.26, 69.83, and 21.79 m^2·g^{-1}, respectively. It can be observed that the catalyst with composition 1% ZnO_x–$MnCO_3$ calcined at 300°C has higher specific surface area than the catalyst obtained by calcining the prepared material at higher temperatures, that is, 400°C and 500°C. The results obtained suggest that at higher calcination temperatures there is a decrease in the surface area due to the sintering. Thus, this may partially be responsible for high catalytic performance in case of the catalyst calcined at 300°C, while in case of 400 and 500°C calcination treatment, there is a considerable decrease in the specific surface area, maybe due to agglomeration of ZnO_x NPs, which has negative effect on the catalyst functioning, which leads to poor alcohol conversion. From the above findings, it can be said that both calcination treatment and the surface area of the synthesized catalyst play a crucial role in the surface area of the synthesized catalyst, which in turn affects the working of the catalyst.

3.8. Catalytic Evaluation. In order to get effective results, the various parameters such as various percentages weight of ZnO_x, reaction time, calcination temperature, catalyst amount, and reaction temperature were optimized as shown in Tables 1–4. For this purpose, the oxidation of 1-phenylethanol was chosen as a substrate model (Scheme 2).

3.8.1. Influence of Calcination Temperature. The effect of calcination temperature on the oxidation of 1-phenylethanol

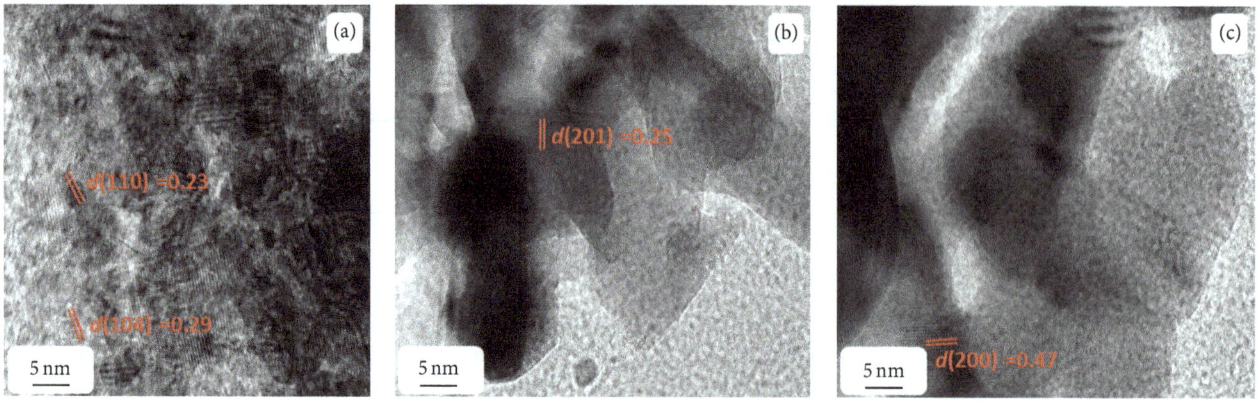

(a) (b) (c)

FIGURE 5: The HRTEM images of catalyst calcined at different temperatures: (a) 1% ZnO_x–$MnCO_3$; (b) 1% ZnO_x–MnO_2; and (c) 1% ZnO_x–Mn_2O_3.

TABLE 1: Catalytic oxidation of 1-phenylethanol in different calcination temperatures of the prepared catalysts.[a]

Entry	Catalyst	T (°C)	SA ($m^2\ g^{-1}$)	Conv. (%)	Sp. activity ($mmol{\cdot}g^{-1}{\cdot}h^{-1}$)	TON	TOF (h^{-1})	Sel. (%)
1	1% ZnO_x–$MnCO_3$	300	120.26	100.00	22.22	54.25	180.83	>99
2	1% ZnO_x–MnO_2	400	69.83	81.49	18.11	44.21	147.37	>99
3	1% ZnO_x–Mn_2O_3	500	21.79	55.94	12.43	30.35	101.16	>99

[a] Reaction conditions: 2 mmol of 1-phenylethanol, 300 mg of catalyst, oxygen with rate 20 mL/min, reaction temperature at 100°C, 10 mL of toluene, and 18 min of reaction time.

TABLE 2: Effect of different weight % of ZnO_x on the 1-phenylethanol oxidation.[a]

Entry	Catalyst	Conv. (%)	Sp. activity ($mmol{\cdot}g^{-1}{\cdot}h^{-1}$)	TON	TOF (h^{-1})	Sel. (%)
1	0% ZnO_x–$MnCO_3$	78.31	17.40	/	/	>99
2	1% ZnO_x–$MnCO_3$	100.00	22.22	54.25	180.83	>99
3	3% ZnO_x–$MnCO_3$	92.43	20.54	18.08	60.27	>99
4	5% ZnO_x–$MnCO_3$	72.29	16.06	10.85	36.17	>99
5	7% ZnO_x–$MnCO_3$	60.14	13.36	7.70	25.67	>99

[a] Reaction conditions: 2 mmol of 1-phenylethanol, 300 mg of catalyst, calcination temperature at 300°C, oxygen with rate 20 mL/min, reaction temperature at 100°C, 10 mL of toluene, and 18 min of reaction time.

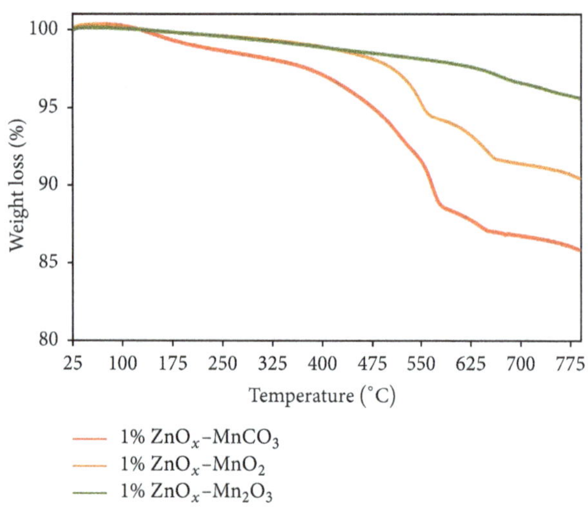

FIGURE 6: TGA of as-synthesized catalyst calcined at different calcination temperatures 300°C, 400°C, and 500°C; 1% ZnO_x–$MnCO_3$; 1% ZnO_x–MnO_2; and 1% ZnO_x–Mn_2O_3.

has been examined. The synthesized catalyst calcined at different temperatures such as 300°C, 400°C, and 500°C. The catalysts showed high product selectivity during all oxidation experiments (above 99%). Notwithstanding the oxidation of 1-phenylethanol is strongly affected by calcination temperature. The obtained results concluded that the applied calcination treatment has a marked inhibiting effect on the oxidation process by decreasing the alcohol conversion [90, 91]. For example, the catalyst calcined at 300°C, that is, 1% ZnO_x–$MnCO_3$, exhibits the highest alcohol conversion (100%) along with 22.22 $mmol{\cdot}g^{-1}{\cdot}h^{-1}$ specific activity and 180.83 h^{-1} TOF, within 18 min of the reaction (Table 1, entry 1), while catalysts calcined at 400 and 500°C, that is, 1% ZnO_x–MnO_2 and 1% ZnO_x–Mn_2O_3, respectively, gave a lower conversion of 81.49% and 55.94%, respectively, under similar circumstances (Table 1, entries 2, 3). It is worth mentioning that these observations correlate with the BET surface area results of the prepared catalyst. 1% ZnO_x–$MnCO_3$, 1% ZnO_x–MnO_2, and 1% ZnO_x–Mn_2O_3 were 120.26, 69.83, and

TABLE 3: Effect of reaction temperature on the oxidation of 1-phenylethanol using 1% ZnO_x–$MnCO_3$ as the catalyst.[a]

Entry	T (°C)	Conv. (%)	Sp. activity ($mmol·g^{-1}·h^{-1}$)	TON	TOF (h^{-1})	Sel. (%)
1	20	36.07	8.01	19.57	65.23	>99
2	40	52.84	11.74	28.67	95.57	>99
3	60	69.12	15.36	37.50	125.0	>99
4	80	84.69	18.82	45.94	153.13	>99
5	100	100.0	22.22	54.25	180.83	>99

[a]Reaction conditions: 2 mmol of 1-phenylethanol, 300 mg of catalyst, calcination temperature at 300°C, oxygen with rate 20 mL/min, 10 mL of toluene, and 18 min of reaction time.

TABLE 4: The effect of the catalyst 1% ZnO_x–$MnCO_3$ amount on the catalytic efficiency.[a]

Entry	Catalyst amount (mg)	Conv. (%)	Sp. activity ($mmol·g^{-1}·h^{-1}$)	TON	TOF (h^{-1})	Sel. (%)
1	100	40.58	97.39	66.05	792.60	>99
2	200	56.13	67.36	45.67	548.04	>99
3	300	70.98	56.78	38.51	462.12	>99
4	400	86.01	51.61	34.96	419.52	>99
5	500	100	48.0	32.55	390.6	>99

[a]Reaction conditions: 2 mmol of 1-phenylethanol, calcination temperature at 300°C, oxygen with rate 20 mL/min, reaction temperature at 100°C, 10 mL of toluene, and 5 min of reaction time.

$21.79\ m^2·g^{-1}$, respectively. As we mentioned, the prepared catalyst 1% ZnO_x–$MnCO_3$ exhibited the highest alcohol conversion and possessed the highest surface area among all other calcination temperatures. In contrast, the catalyst calcined at 400, that is, 1% ZnO_x–MnO_2, and 500°C, that is, 1% ZnO_x–Mn_2O_3, provided lesser activity and lower surface area. Therefore, it can be deduced that the catalytic activity is strongly influenced by calcination temperature of the catalyst. So we chose to use the catalyst calcined at 300°C, that is, 1% ZnO_x–$MnCO_3$, as the best calcination temperature to optimize other parameters. The results including 1-phenylethanol conversion, surface area, specific activity, turnover number (TON), turnover frequency (TOF), and acetophenone selectivity over the prepared catalyst were summarized in Table 1 and plotted in Figure 7.

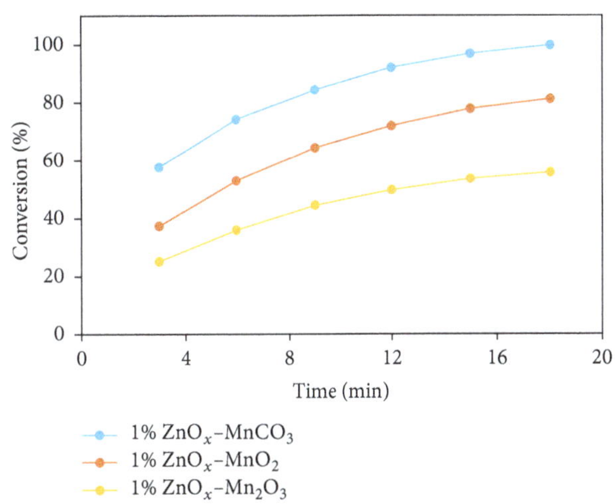

FIGURE 7: Graphical illustration of 1-phenylethanol oxidation using catalyst calcined at different calcination temperature.

3.8.2. Influence of the Amount of ZnO_x Promotor. In order to get the best catalytic performance, the influence of % ZnO_x on the reaction yield was studied by varying load of ZnO_x on $MnCO_3$ support ranging from 0 to 7% and the graphical representation of the results has been shown in Figure 8 and listed in Table 2. The results showed that undoped $MnCO_3$ catalyst, that is, 0% ZnO_x–$MnCO_3$, gives about 78.31% conversion of 1-phenylethanol along with specific activity of $17.40\ mmol·g^{-1}·h^{-1}$ within 18 min (Table 2, entry 1). However, after ZnO_x NPs doping on $MnCO_3$, the catalytic efficiency has been remarkably enhanced, where the 1% ZnO_x–$MnCO_3$ catalyst exhibits a full alcohol conversion under identical reaction conditions and the calculated specific activity was found to be $22.22\ mmol·g^{-1}·h^{-1}$ and $180.83\ h^{-1}$ TOF (Table 2, entry 2), whereas as the weight percentage of ZnO_x is further increased, the catalysts 3% ZnO_x–$MnCO_3$, 5% ZnO_x–$MnCO_3$, and 7% ZnO_x–$MnCO_3$ provided a lower alcohol conversion of 92.43, 72.29, and 60.14%, respectively. Additionally, the specific activity, TON, and TOF were found to decrease with increased ZnO_x content in the catalyst (Table 2, entries 3–5), although the selectivity towards acetophenone was more than 99% for all reactions. As illustrated in Figure 8, the ZnO_x NPs play a fundamental role in improving the catalytic performance for the selective oxidation of 1-phenylethanol. Moreover, as % ZnO_x increases the conversion of 1-phenylethanol decreases, probably due to the agglomeration of ZnO_x NPs in the catalyst surface or to a blocking of active sites as indicated by the results of the BET analysis. The results indicate that 1% ZnO_x–$MnCO_3$ catalyst was the best among all catalysts synthesized. Hence, we choose to use 1% ZnO_x–$MnCO_3$ in the further studies.

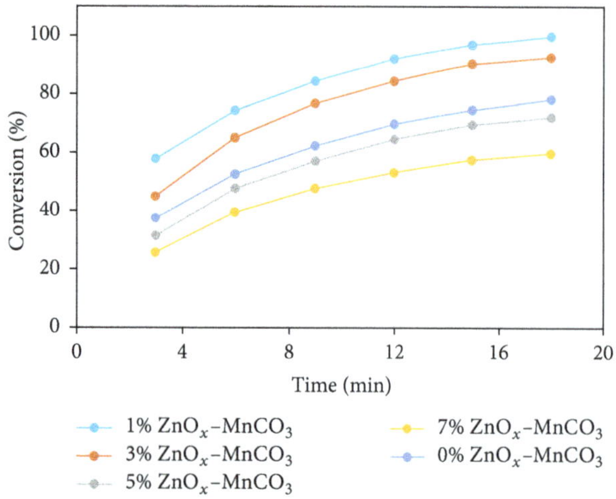

FIGURE 8: The effect of different weight % ZnO_x on the 1-phenylethanol oxidation.

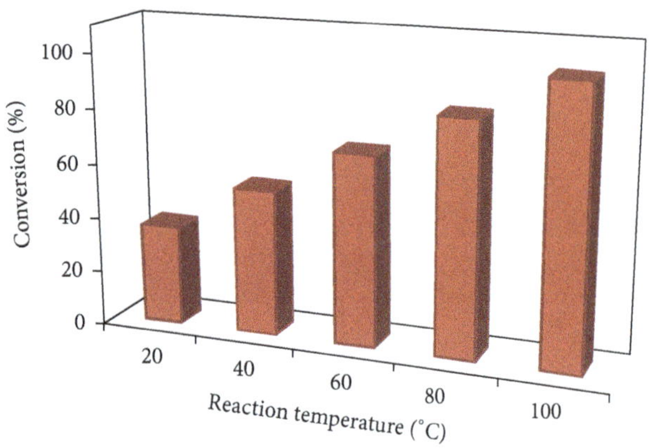

FIGURE 9: Catalytic performance of 1% ZnO_x–$MnCO_3$ catalyst as a function of reaction temperature.

3.8.3. Influence of Temperature. The effect of reaction temperature on the conversion of 1-phenylethanol and the acetophenone selectivity was investigated by considering five different temperatures in the range of 20–100°C. The results including alcohol conversion, specific activity, TON, TOF, and product selectivity have been summarized in Table 3 and plotted in Figure 9. The acetophenone selectivity was almost unchanged (>99%), whereas the conversion of 1-phenylethanol strongly depends on the reaction temperature. For instance, at low reaction temperature of 20°C, a relatively low alcohol conversion of 36.07% has been obtained (Table 3, entry 1). As expected, the high temperature contributed to a higher oxidation rate and led to the remarkable improvement of catalytic activity of the prepared catalyst. A full conversion of 1-phenylethanol was achieved at 100°C within relatively short reaction time (18 min) under identical reaction conditions (Table 3, entry 5). In addition, when the reaction temperature increased, the specific activity, TON, and TOF were also increased as shown in Table 3. Therefore, 100°C was chosen to be an optimum reaction temperature in this study to achieve highest catalytic performance.

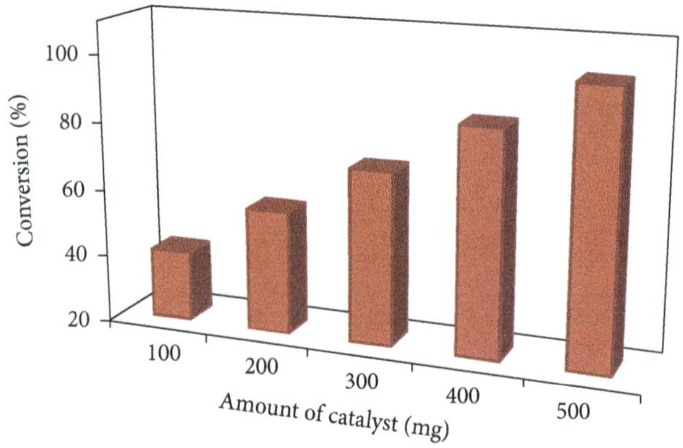

FIGURE 10: Catalytic activity of 1% ZnO_x–$MnCO_3$ catalyst as a function of catalyst amount.

3.8.4. Influence of Catalyst Amount. The 1-phenylethanol conversion and the selectivity of acetophenone using different catalyst amounts were also evaluated, and the results are illustrated in Table 4 and Figure 10. The oxidation reaction was carried out in presence of 100, 200, 300, 400, and 500 mg of 1% ZnO_x–$MnCO_3$ catalyst under the optimized reaction conditions. It is noteworthy that the catalytic activity of the as-synthesized catalyst is directly proportional to the catalyst amounts. According to Table 4, the alcohol conversion increased from 40.58% to 100% by increasing the catalyst amount from 100 to 500 mg within short reaction time (5 min), whereas the product selectivity was almost unchanged during all oxidation processes (above 99%). Moreover, a blank reaction has also been examined in the absence of the catalyst. It was found that no formation of acetophenone was obtained, indicating that the catalyst is necessary for the oxidation of 1-phenylethanol.

Furthermore, in order to show the superior catalytic activity of the present catalytic system, the effectiveness of our catalytic system has been compared with the ones already reported in the scientific literature in the oxidation of 1-phenylethanol (Table 5). Obviously, the as-synthesized 1% ZnO_x–$MnCO_3$ is the most efficient catalyst among all the listed catalysts. Herein, the prepared catalyst has been used for 1-phenylethanol oxidation and exhibited a full conversion and selectivity within extremely short reaction time of 5 min at 100°C and highest specific activity (48 $mmol{\cdot}g^{-1}{\cdot}h^{-1}$) compared to all other listed catalysts. In contrast, all listed catalysts take a very long reaction time to completely oxidize 1-phenylethanol and require higher reaction temperature. For instance, Hosseini-Sarvari et al. [76] have prepared Ag nanoparticles doped ZnO support (Ag/ZnO) and used it as a catalyst for selective oxidation of alcohols to their corresponding carbonyls. The nano Ag/ZnO catalyst provided 90% conversion of 1-phenylethanol and more than 99% selectivity of acetophenone along with 30 $mmol{\cdot}g^{-1}{\cdot}h^{-1}$ specific activity after long reaction time of 6 h at 100°C. In another example, Nepak and Darbha [70] reported liquid-phase selective alcohol oxidation over Au–Pd NPs doped in sodium titanate nanotubes (NaTNTs) as a catalyst. The Au–Pd/NaTNT catalyst exhibited 1-phenylethanol conversion of 84% and product

TABLE 5: A comparison between performance of our catalytic system and earlier reported catalysts for 1-phenylethanol oxidation.

Catalyst	Conv. (%)	Sel. (%)	T (°C)	Time	Sp. activity ($mmol{\cdot}g^{-1}{\cdot}h^{-1}$)	Ref.
1% ZnO_x–$MnCO_3$	100	>99	100	5 min	48.0	This work
Cu/AlO(OH)	99	97	27	5 h	8.08	[66]
$VOSO_4$/TEMPO	99	96	80	2 h	37.78	[67]
SSDT	90	>99	reflux	1 h	1.8	[68]
Au/MgO	97	>99	130	2 h	19.4	[69]
Au–Pd/NaTNT	84	86	120	10 h	42.0	[70]
CdS-MTA	40	>99	5	1.5 h	0.60	[71]
Au–Pd/LDH	>99	>99	80	1.5 h	6.67	[72]
Au/TiO_2	100	>99	110	24 h	0.83	[73]
$Fe(NO_3)_3{\cdot}9H_2O$/ABNO	87	>99	RT	4 h	5.44	[74]
$FeCl_3$-imine@SiO_2	90	>99	80	6 h	3.26	[75]
Ag/ZnO	90	>99	100	6 h	30.0	[76]
CoTE4PyP-MT	92	89	70	2 h	30.67	[77]
Mn(III)-salen	46.9	>99	20	40 min	5.77	[78]
5 wt% Ir/TiO_2	97	95	80	48 h	0.40	[79]
Ru/CaO–ZrO_2	>99	>98	40	6 h	1.67	[80]
$CeCrO_3$	100	100	90	6 h	15.15	[62]
Ru/Mg–LaO	96	>99	80	4 h	2.4	[81]
$CoAl_2O_4$	63.45	83.25	80	8 h	0.79	[34]
Fe_3O_4	76	>99	80	18 h	1.4	[82]

selectivity of about 86%, and the specific activity of this conversion was about 42 $mmol{\cdot}g^{-1}{\cdot}h^{-1}$ within long reaction time of 10 h at higher reaction temperature of 120°C. As a result, the prepared 1% ZnO_x–$MnCO_3$ catalyst has been found to be the best choice for the alcohol oxidation.

3.9. Recycling Studies. The reusability of the catalyst has considerable importance from both commercial and academic point of view. In order to evaluate the recyclability and the stability, the oxidation of 1-phenylethanol was performed 5 times under optimum reaction conditions with the recycling of 1% ZnO_x–$MnCO_3$ catalyst and results are shown in Figure 11. After the first use of the catalyst in the oxidation of 1-phenylethanol to give acetophenone, the toluene was evaporated and fresh toluene was added and the mixture was filtered by centrifugation to recover the catalyst. The filtered catalyst was washed sequentially with toluene. The results revealed that the 1% ZnO_x–$MnCO_3$ catalyst was reused for at least 5 cycles with no appreciable decrease in its activity being observed and selectivity remained unchanged throughout all runs. During the five recycling reactions, the conversion of 1-phenylethanol decreased gradually from 100% to 92.37%, possibly due to the mass loss during the filtration method [92, 93]. Therefore, results indicate an excellent recyclability and stability of the as-prepared catalyst.

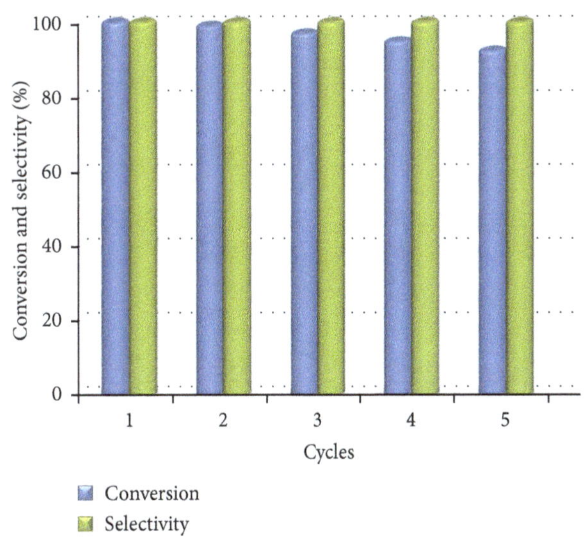

FIGURE 11: Recyclability of 1% ZnO_x–$MnCO_3$ for the aerial oxidation of 1-phenylethanol (reaction conditions: 2 mmol of 1-phenylethanol, calcination temperature at 300°C, oxygen with rate 20 mL/min, 0.5 g of catalyst, 10 mL of toluene, reaction temperature at 100°C, and 5 min of reaction time).

3.10. Oxidation of Different Types of Alcohols over 1% ZnO_x–$MnCO_3$ Catalyst. For evaluation of the catalyst capabilities, a variety of alcohols including secondary, benzylic aliphatic, primary allylic, and heteroatomic alcohols were subjected to oxidation with molecular O_2 under the optimized circumstances (Table 6, entries 1–26). The optimal reaction conditions include 2 mmol of alcohol in 10 mL toluene with 20 mL/min oxygen flow rate and 100°C reaction temperature in presence of 1% ZnO_x–$MnCO_3$ catalyst (0.5 g) calcined at 300°C. The results are summarized in Table 6. The results showed that all secondary aromatic alcohols were selectively oxidized into their corresponding ketones with 100% convertibility under optimized circumstances (Table 6,

Table 6: Oxidation of different types of alcohols over 1% ZnO_x–$MnCO_3$ catalyst.[a]

R. number	Reactants	Products	Time (min)	Conv. (%)	Sel. (%)
1	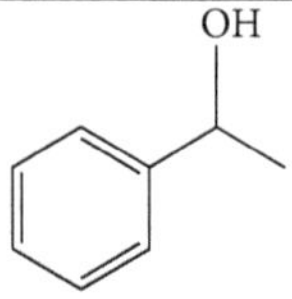	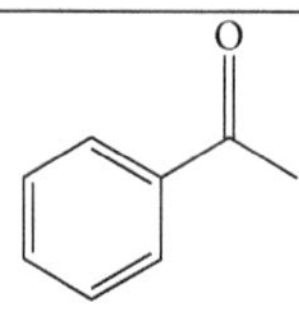	5	100	>99
2	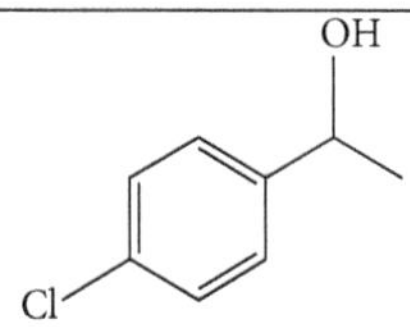	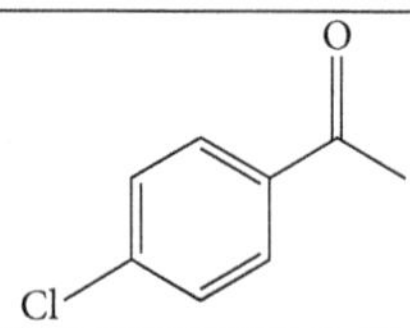	8	100	>99
3	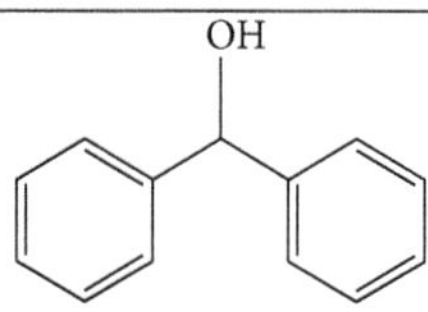	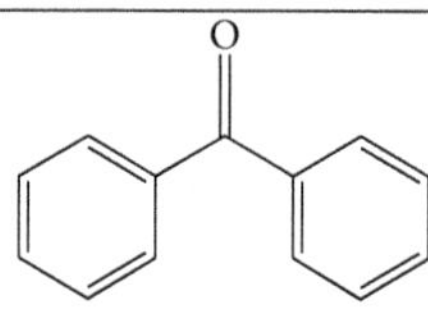	4	100	>99
4	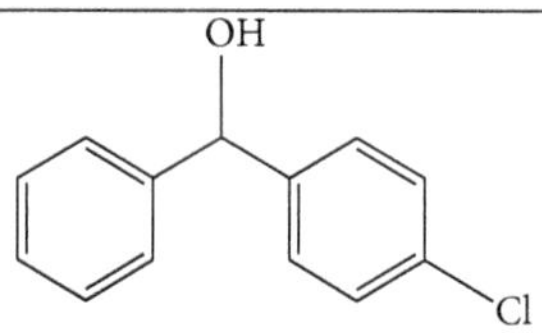	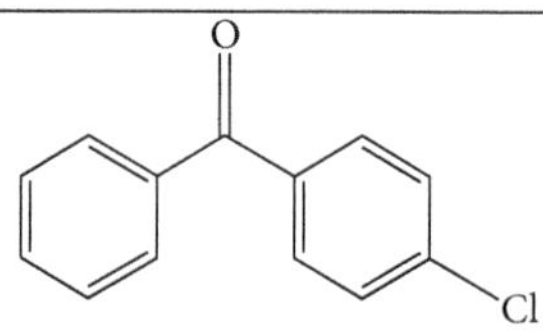	6	100	>99
5	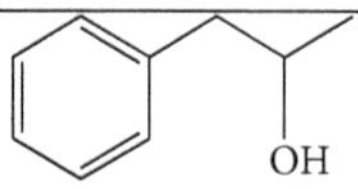	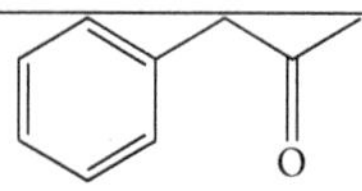	10	100	>99
6	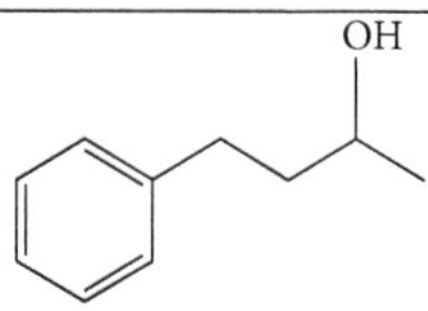	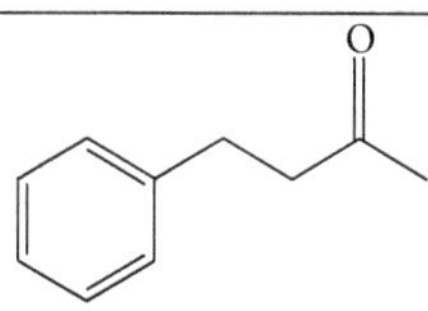	13	100	>99
7	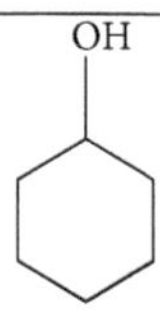	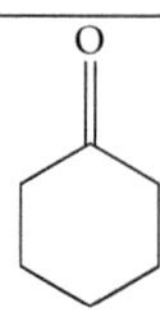	30	100	>99
8	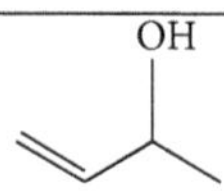	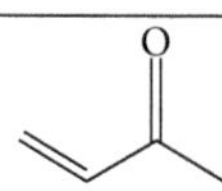	80	100	>99
9	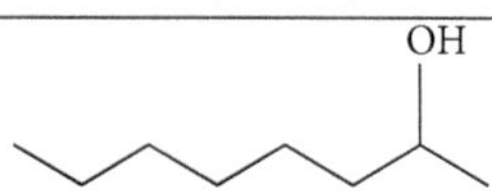	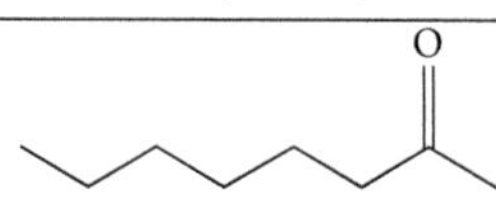	85	100	>99
10	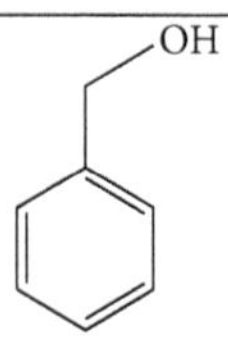	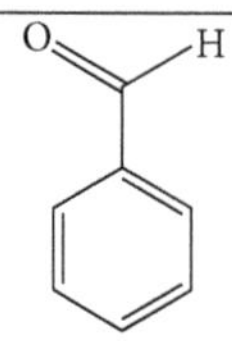	4	100	>99
11	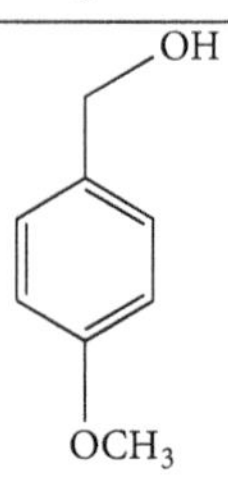	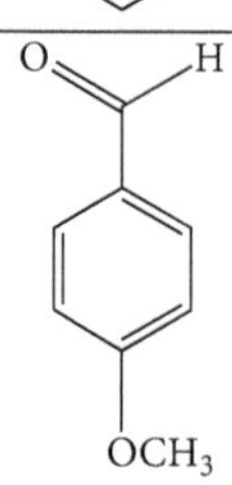	4	100	>99

TABLE 6: Continued.

R. number	Reactants	Products	Time (min)	Conv. (%)	Sel. (%)
12	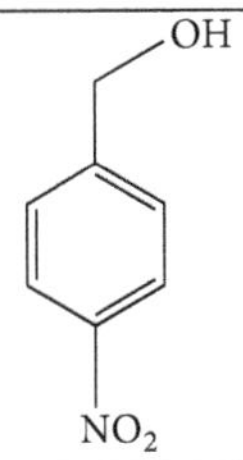	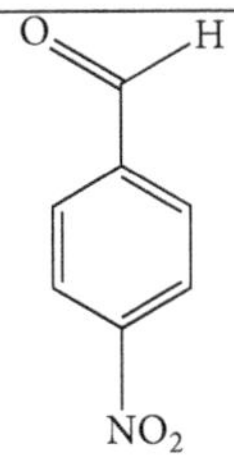	7	100	>99
13	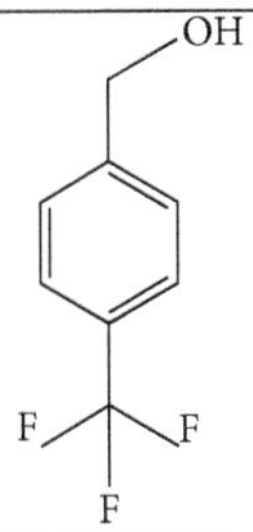	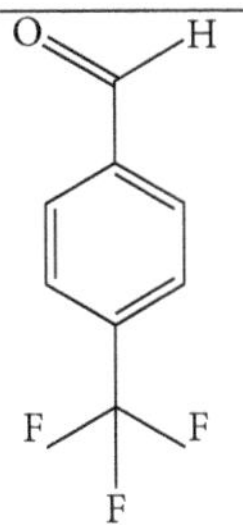	8	100	>99
14	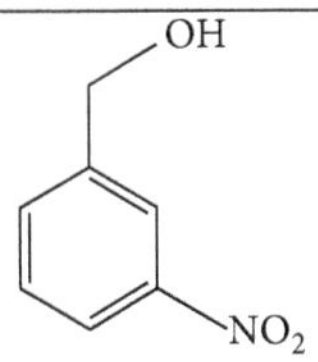	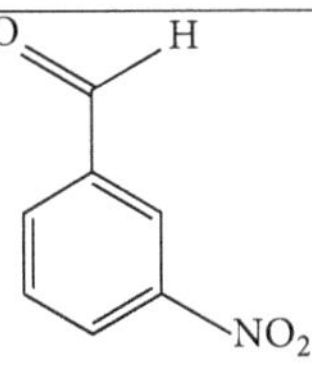	8	100	>99
15	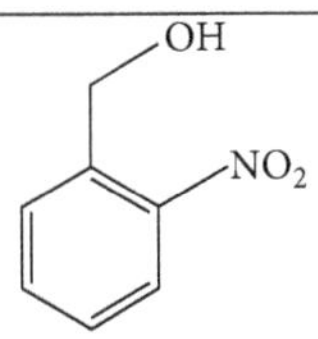	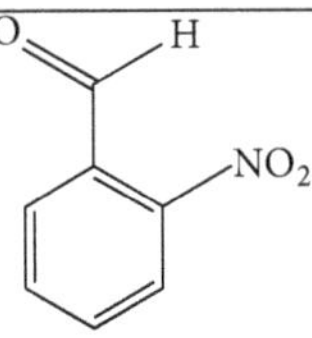	11	100	>99
16	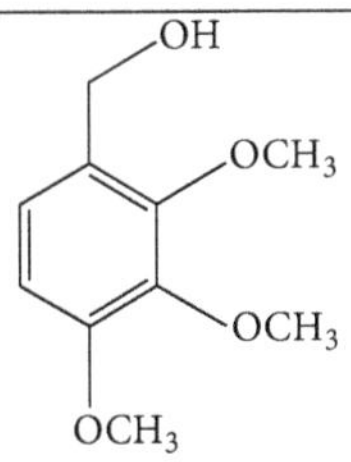	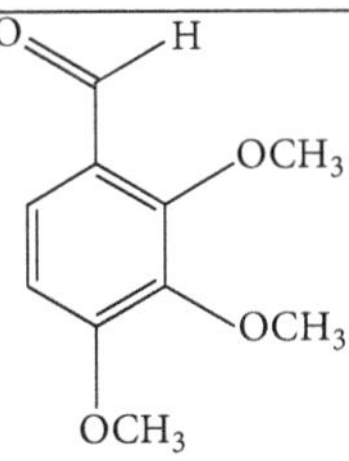	9	100	>99
17	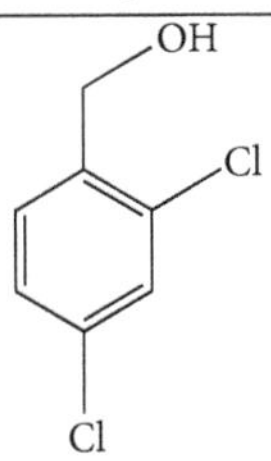	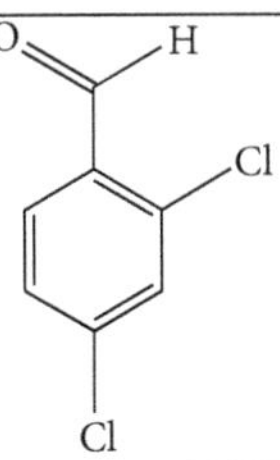	12	100	>99
18	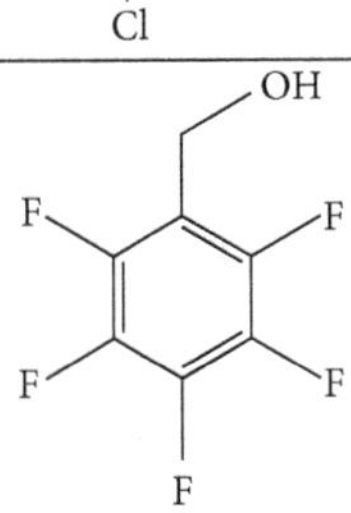	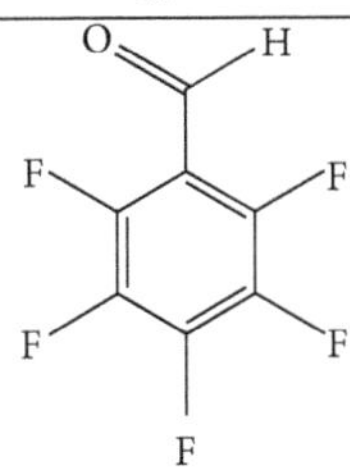	15	100	>99
19	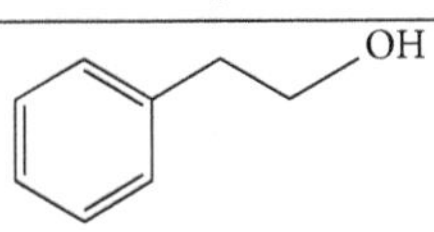	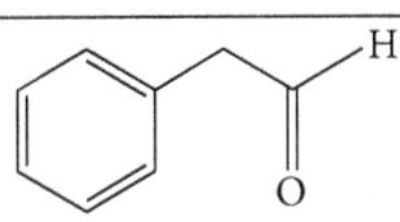	10	100	>99

TABLE 6: Continued.

R. number	Reactants	Products	Time (min)	Conv. (%)	Sel. (%)
20	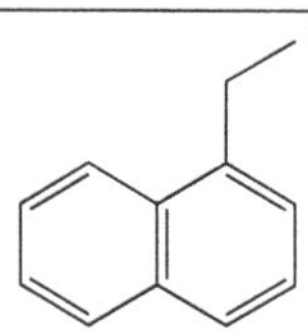	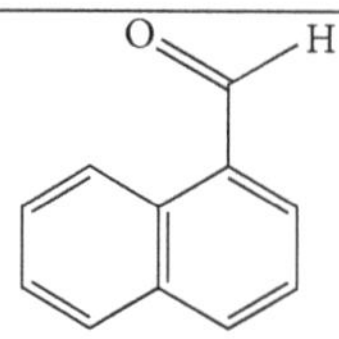	20	100	>99
21	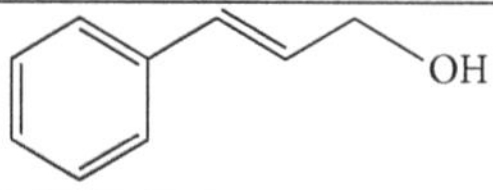	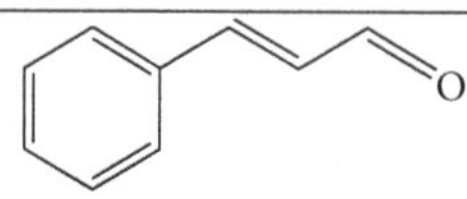	7	100	>99
22	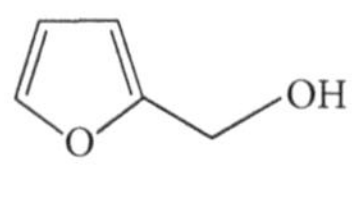	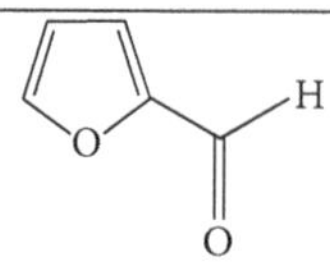	20	100	>99
23	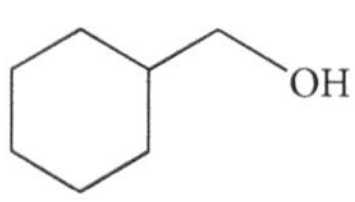	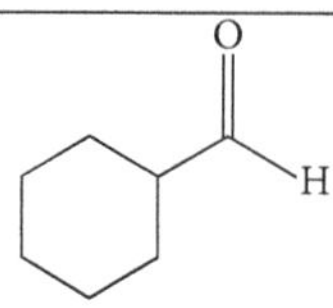	35	100	>99
24	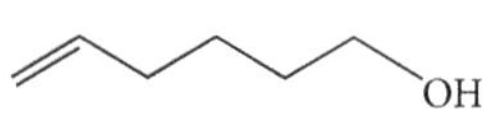	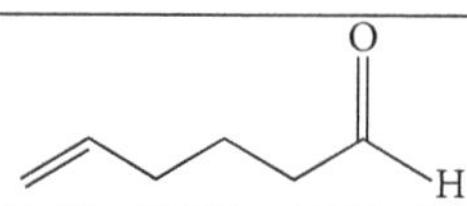	70	100	>99
25	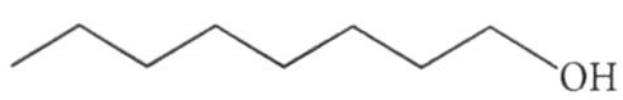	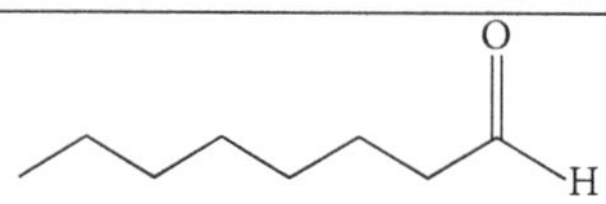	75	100	>99
26	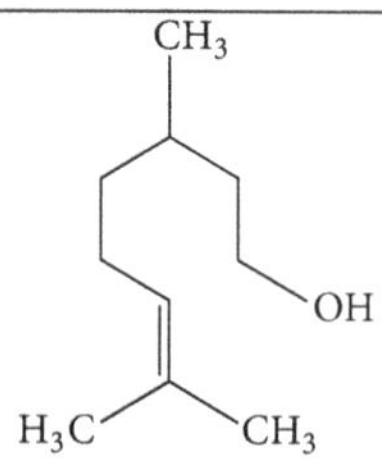	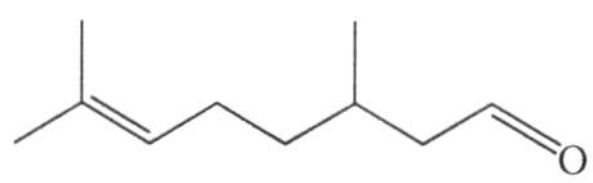	90	100	>99

[a]Reaction conditions: 2 mmol of alcohol, 0.5 g of catalyst, calcination temperature at 300°C, oxygen with rate 20 mL/min, reaction temperature at 100°C, and 10 mL of toluene.

entries 1–6). Additionally, an excellent selectivity (above 99%) towards ketones has been obtained throughout all oxidation processes and no by-products were found in the reaction mixture. It is noteworthy that the benzhydrol was demonstrated to be the most reactive among all secondary aromatic alcohols and gave 100% conversion after 4 min, while the 4-chlorobenzhydrol gave 100% conversion after longer reaction time, maybe because 4-chlorobenzhydrol contains electron-withdrawing group that deactivate the phenyl ring by decreasing the electron density (Table 6, entries 3 and 4). Furthermore, 1-phenylethanol and its derivatives also afforded complete conversion and more than 99% selectivity in relatively short reaction times (Table 6, entries 1 and 2). Commonly, the oxidation of aliphatic alcohols is much more difficult than aromatic ones [94–96]. Indeed, our catalytic protocol was also found to be effective for oxidation of secondary aliphatic alcohols to their corresponding ketones under the similar reaction conditions (Table 6, entries 7–9). As expected, it was necessary to increase reaction time, because the oxidation of secondary aliphatic alcohols is more difficult than secondary aromatic alcohols; for example, 2-octanol required relatively longer reaction time (85 min) for complete oxidation to 2-octanone (Table 6, entry 9).

In order to exploit the applicability of the present catalytic system and to generalize the scope of the oxidation process, the oxidation reaction has extended to various types of alcohol such as primary aromatic, allylic, heteroaromatic, and aliphatic alcohols. Generally, all primary aromatic alcohols could be oxidized rapidly to the corresponding aldehydes with complete conversions within very short reaction times (Table 6, entries 10–22). Moreover, more than 99% selectivity to corresponding aldehydes has been achieved in most of oxidation reactions and no other products were detected in the reaction mixture because no overoxidation to carboxylic acids occurred. The oxidation of the primary benzylic alcohols was strongly influenced by the electronic properties of the substituents on the benzyl group. In general, the oxidation rate of aromatic alcohols with electron-releasing groups was relatively higher than that of alcohols with electron-withdrawing group [13, 97, 98]. For instance, oxidation of an aromatic alcohol containing electron-donating group such 4-methoxylbenzyl alcohol gave 100% conversion only in 4 min (Table 6, entry 11), while oxidation of alcohol bearing electron-withdrawing group like 4-(trifluoromethyl)benzyl alcohol afforded 100% conversion within longer reaction time (8 min) (Table 6, entry 13). On the other hand, *para*-substituted alcohols have higher activities relative to *ortho*- and *meta*-substituent, probably because the *para*-position has minimum steric hindrance compared to other positions [99, 100]. For an example, *para*-nitrobenzyl alcohol was fully oxidized into its corresponding aldehyde within 7 min (Table 6, entry 12), while *meta*- and *ortho*-nitrobenzyl alcohol were completely oxidized after longer times, 8 and 11 min, respectively (Table 6, entries 14 and 15). Steric hindrance is another essential factor that affects the rate of the oxidation processes; the bulky groups such as 4-CF_3, 2,4-DiCl, 2,3,4-TriOMe, and 2,3,4,5,6-pentaflouro attached to benzyl alcohol decrease the performance of the oxidation reaction, maybe because of the steric resistance that impedes the oxidation of the alcohols bearing bulky substituents (Table 6, entries 13, 16–18). Cinnamyl alcohol, an example of allylic alcohols, has been completely oxidized to cinnamaldehyde with more than 99% cinnamaldehyde selectivity within 7 min (Table 6, entry 21). A heteroaromatic alcohol such as furfuryl alcohol has been converted to furfural within 20 min, with complete conversion and product selectivity (Table 6, entry 22). Furthermore, compared to primary aromatic alcohols, the oxidation of primary aliphatic alcohols exhibited relatively low reactivity towards oxidation reaction (Table 6, entries 23–26). In this regard, the oxidation of cyclohexanemethanol, 5-Hexen-1-ol, octan-1-ol, and citronellol into their corresponding aldehydes occurs in slightly longer reaction times (Table 6, entries 23–26). As a result, it can be said that the aerobic oxidation of alcohols catalyzed by 1% ZnO_x–$MnCO_3$ has been affected by two factors, electronic and steric effects.

4. Conclusions

In conclusion, ZnO_x NPs doped $MnCO_3$ was found to be an efficient catalyst towards the catalytic oxidation of alcohols to carbonyls with molecular O_2 as a clean oxidant without using any additive or base. Interestingly, the ZnO_x NPs play a fundamental role in the catalytic performance because the oxidation rate has remarkably improved after doping ZnO_x NPs on $MnCO_3$. A complete 1-phenylethanol conversion can be accomplished within extremely short reaction time of only 5 min along with 48 $mmol·g^{-1}·h^{-1}$ specific activity and 390.6 TOF. The obtained specific activity is much higher than that reported in earlier literatures. Additionally, wide range of benzylic, aliphatic, allylic, primary, and secondary alcohols has been studied for aerobic oxidation into their corresponding aldehydes and ketones with excellent alcohols conversion within relatively short reaction times under mild reaction conditions. Furthermore, the very high product selectivity (>99%) has been achieved for most of alcohols used in this work. The as-prepared catalyst can be recycled and reused for five times without any obvious loss of catalytic activity and the selectivity remained almost unchanged. This catalytic system offers several advantages including complete alcohol conversions and extremely high specific activities and selectivity towards products in extremely short reaction times by using ecofriendly, low cost, selective, facile synthesis, and reusable heterogeneous catalyst will make this system useful and applicable for the selective oxidation of alcohols.

Conflicts of Interest

The authors declare that they have no conflicts of interest.

Acknowledgments

The authors extend their appreciation to the Deanship of Scientific Research at King Saud University for funding this work through the Research Group Project no. RG–1436–032.

References

[1] G.-J. ten Brink, I. W. C. E. Arends, and R. A. Sheldon, "Green, catalytic oxidation of alcohols in water," *Science*, vol. 287, no. 5458, pp. 1636–1639, 2000.

[2] M. Pagliaro, S. Campestrini, and R. Ciriminna, "Ru-based oxidation catalysis," *Chemical Society Reviews*, vol. 34, no. 10, pp. 837–845, 2005.

[3] K. Mori, T. Hara, T. Mizugaki, K. Ebitani, and K. Kaneda, "Hydroxyapatite-supported palladium nanoclusters: a highly active heterogeneous catalyst for selective oxidation of alcohols by use of molecular oxygen," *Journal of the American Chemical Society*, vol. 126, no. 34, pp. 10657–10666, 2004.

[4] I. E. Markó, P. R. Giles, M. Tsukazaki, S. M. Brown, and C. J. Urch, "Copper-catalyzed oxidation of alcohols to aldehydes and ketones: an efficient, aerobic alternative," *Science*, vol. 274, no. 5295, pp. 2044–2046, 1996.

[5] T. Mallat and A. Baiker, "Oxidation of alcohols with molecular oxygen on solid catalysts," *Chemical Reviews*, vol. 104, no. 6, pp. 3037–3058, 2004.

[6] J. Ahmad, P. Figiel, M. Räisänen, M. Leskelä, and T. Repo, "Aerobic oxidation of benzylic alcohols with bis(3,5-di-terr-butylsalicylaldimine)copper(II) complexes," *Applied Catalysis A: General*, vol. 371, no. 1-2, pp. 17–21, 2009.

[7] C. Parmeggiani and F. Cardona, "Transition metal based catalysts in the aerobic oxidation of alcohols," *Green Chemistry*, vol. 14, no. 3, pp. 547–564, 2012.

[8] Z. Guo, B. Liu, Q. Zhang, W. Deng, Y. Wang, and Y. Yang, "Recent advances in heterogeneous selective oxidation catalysis for sustainable chemistry," *Chemical Society Reviews*, vol. 43, no. 10, pp. 3480–3524, 2014.

[9] G. Palmisano, E. García-López, G. Marcì et al., "Advances in selective conversions by heterogeneous photocatalysis," *Chemical Communications*, vol. 46, no. 38, pp. 7074–7089, 2010.

[10] S. Higashimoto, N. Suetsugu, M. Azuma, H. Ohue, and Y. Sakata, "Efficient and selective oxidation of benzylic alcohol by O_2 into corresponding aldehydes on a TiO_2 photocatalyst under visible light irradiation: effect of phenyl-ring substitution on the photocatalytic activity," *Journal of Catalysis*, vol. 274, no. 1, pp. 76–83, 2010.

[11] M. Zhang, Q. Wang, C. Chen, L. Zang, W. Ma, and J. Zhao, "Oxygen atom transfer in the photocatalytic oxidation of alcohols by TiO_2: oxygen isotope studies," *Angewandte Chemie*, vol. 121, no. 33, pp. 6197–6200, 2009.

[12] S. Yurdakal, G. Palmisano, V. Loddo, V. Augugliaro, and L. Palmisano, "Nanostructured rutile TiO_2 for selective photocatalytic oxidation of aromatic alcohols to aldehydes in water," *Journal of the American Chemical Society*, vol. 130, no. 5, pp. 1568-1569, 2008.

[13] D. I. Enache, J. K. Edwards, P. Landon et al., "Solvent-free oxidation of primary alcohols to aldehydes using Au-Pd/TiO_2 catalyst," *Science*, vol. 311, no. 5759, pp. 362–365, 2006.

[14] M. M. Dell'Anna, M. Mali, P. Mastrorilli, P. Cotugno, and A. Monopoli, "Oxidation of benzyl alcohols to aldehydes and ketones under air in water using a polymer supported palladium catalyst," *Journal of Molecular Catalysis A: Chemical*, vol. 386, pp. 114–119, 2014.

[15] B.-Z. Zhan and A. Thompson, "Recent developments in the aerobic oxidation of alcohols," *Tetrahedron*, vol. 60, no. 13, pp. 2917–2935, 2004.

[16] J. Yu, J. Li, H. Wei, J. Zheng, H. Su, and X. Wang, "Hydrotalcite-supported gold catalysts for a selective aerobic oxidation of benzyl alcohol driven by visible light," *Journal of Molecular Catalysis A: Chemical*, vol. 395, pp. 128–136, 2014.

[17] Y. Li, Y. Gao, and C. Yang, "A facile and efficient synthesis of polystyrene/gold-platinum composite particles and their application for aerobic oxidation of alcohols in water," *Chemical Communications*, vol. 51, no. 36, pp. 7721–7724, 2015.

[18] S. Alabbad, S. F. Adil, M. E. Assal, M. Khan, A. Alwarthan, and M. R. H. Siddiqui, "Gold & silver nanoparticles supported on manganese oxide: synthesis, characterization and catalytic studies for selective oxidation of benzyl alcohol," *Arabian Journal of Chemistry*, vol. 7, no. 6, pp. 1192–1198, 2014.

[19] R. H. Adnan, G. G. Andersson, M. I. J. Polson, G. F. Metha, and V. B. Golovko, "Factors influencing the catalytic oxidation of benzyl alcohol using supported phosphine-capped gold nanoparticles," *Catalysis Science and Technology*, vol. 5, no. 2, pp. 1323–1333, 2015.

[20] T. Wang, X. Yuan, S. Li, L. Zeng, and J. Gong, "CeO_2-modified Au@ SBA-15 nanocatalysts for liquid-phase selective oxidation of benzyl alcohol," *Nanoscale*, vol. 7, no. 17, pp. 7593–7602, 2015.

[21] Y. Yan, X. Jia, and Y. Yang, "Palladium nanoparticles supported on CNT functionalized by rare-earth oxides for solvent-free aerobic oxidation of benzyl alcohol," *Catalysis Today*, vol. 259, pp. 292–302, 2014.

[22] Y. Hong, X. Yan, X. Liao et al., "Platinum nanoparticles supported on Ca(Mg)-zeolites for efficient room-temperature alcohol oxidation under aqueous conditions," *Chemical Communications*, vol. 50, no. 68, pp. 9679–9682, 2014.

[23] T. Wang, H. Shou, Y. Kou, and H. Liu, "Base-free aqueous-phase oxidation of non-activated alcohols with molecular oxygen on soluble Pt nanoparticles," *Green Chemistry*, vol. 11, no. 4, pp. 562–568, 2009.

[24] K. Kon, S. M. A. Hakim Siddiki, and K.-I. Shimizu, "Size- and support-dependent Pt nanocluster catalysis for oxidant-free dehydrogenation of alcohols," *Journal of Catalysis*, vol. 304, pp. 63–71, 2013.

[25] Z.-W. Yang, X. Zhao, T.-J. Li et al., "Catalytic properties of palygorskite supported Ru and Pd for efficient oxidation of alcohols," *Catalysis Communications*, vol. 65, pp. 34–40, 2015.

[26] D. Canseco-Gonzalez and M. Albrecht, "Wingtip substituents tailor the catalytic activity of ruthenium triazolylidene complexes in base-free alcohol oxidation," *Dalton Transactions*, vol. 42, no. 20, pp. 7424–7432, 2013.

[27] A. Wusiman and C.-D. Lu, "Selective oxidation of benzylic, allylic and propargylic alcohols using dirhodium(II) tetraamidinate as catalyst and aqueous tert-butyl hydroperoxide as oxidant," *Applied Organometallic Chemistry*, vol. 29, no. 4, pp. 254–258, 2015.

[28] J. Olguín, H. Müller-Bunz, and M. Albrecht, "Springloaded porphyrin NHC hybrid rhodium(III) complexes: carbene dissociation and oxidation catalysis," *Chemical Communications*, vol. 50, no. 26, pp. 3488–3490, 2014.

[29] M. Xie, X. Dai, S. Meng, X. Fu, and S. Chen, "Selective oxidation of aromatic alcohols to corresponding aromatic aldehydes using In_2S_3 microsphere catalyst under visible light irradiation," *Chemical Engineering Journal*, vol. 245, pp. 107–116, 2014.

[30] Q. Cao, L. M. Dornan, L. Rogan, N. L. Hughes, and M. J. Muldoon, "Aerobic oxidation catalysis with stable radicals," *Chemical Communications*, vol. 50, no. 35, pp. 4524–4543, 2014.

[31] P. Cruz, Y. Pérez, I. Del Hierro, and M. Fajardo, "Copper, copper oxide nanoparticles and copper complexes supported on mesoporous SBA-15 as catalysts in the selective oxidation of benzyl alcohol in aqueous phase," *Microporous and Mesoporous Materials*, vol. 220, pp. 136–147, 2016.

[32] R. Ali, M. E. Assal, and A. Al-Warthan, "Selective oxidation of benzylic alcohols with molecular oxygen catalyzed by copper-manganese oxide nanoparticles," *Asian Journal of Chemistry*, vol. 25, no. 9, article 4815, 2013.

[33] S. M. Seyedi, R. Sandaroos, and G. H. Zohuri, "Novel cobalt(II) complexes of amino acids-Schiff bases catalyzed aerobic oxidation of various alcohols to ketones and aldehyde," *Chinese Chemical Letters*, vol. 21, no. 11, pp. 1303–1306, 2010.

[34] C. Ragupathi, J. Judith Vijaya, S. Narayanan, S. K. Jesudoss, and L. John Kennedy, "Highly selective oxidation of benzyl alcohol to benzaldehyde with hydrogen peroxide by cobalt aluminate catalysis: a comparison of conventional and microwave methods," *Ceramics International*, vol. 41, no. 2, pp. 2069–2080, 2015.

[35] S. R. Ali, P. Chandra, M. Latwal, S. K. Jain, V. K. Bansal, and S. P. Singh, "Synthesis of nickel hexacyanoferrate nanoparticles and their potential as heterogeneous catalysts for the solvent-free oxidation of benzyl alcohol," *Chinese Journal of Catalysis*, vol. 32, no. 11-12, pp. 1844–1849, 2011.

[36] A. R. Hajipour, H. Karimi, and A. Koohi, "Selective oxidation of alcohols over nickel zirconium phosphate," *Chinese Journal of Catalysis*, vol. 36, no. 7, pp. 1109–1116, 2015.

[37] H.-B. Ji, T.-T. Wang, M.-Y. Zhang, Q.-L. Chen, and X.-N. Gao, "Green oxidation of alcohols by a reusable nickel catalyst in the presence of molecular oxygen," *Reaction Kinetics and Catalysis Letters*, vol. 90, no. 2, pp. 251–257, 2007.

[38] R. Naik, A. Nizam, A. Siddekha, and M. A. Pasha, "An efficient sonochemical oxidation of benzyl alcohols into benzaldehydes by $FeCl_3/HNO_3$ in acetone," *Ultrasonics Sonochemistry*, vol. 18, no. 5, pp. 1124–1127, 2011.

[39] R. Cang, B. Lu, X. Li, R. Niu, J. Zhao, and Q. Cai, "Iron-chloride ionic liquid immobilized on SBA-15 for solvent-free oxidation of benzyl alcohol to benzaldehyde with H_2O_2," *Chemical Engineering Science*, vol. 137, pp. 268–275, 2015.

[40] G. C. Behera and K. M. Parida, "Liquid phase catalytic oxidation of benzyl alcohol to benzaldehyde over vanadium phosphate catalyst," *Applied Catalysis A: General*, vol. 413-414, pp. 245–253, 2012.

[41] S. F. Adil, M. E. Assal, M. Khan, A. Al-Warthan, and M. H. Rafiq Siddiqui, "Nano silver-doped manganese oxide as catalyst for oxidation of benzyl alcohol and its derivatives: synthesis, characterisation, thermal study and evaluation of catalytic properties," *Oxidation Communications*, vol. 36, no. 3, pp. 778–791, 2013.

[42] M. Deng, G. Zhao, Q. Xue, L. Chen, and Y. Lu, "Microfibrous-structured silver catalyst for low-temperature gas-phase selective oxidation of benzyl alcohol," *Applied Catalysis B: Environmental*, vol. 99, no. 1-2, pp. 222–228, 2010.

[43] N. Noshiranzadeh, R. Bikas, K. Ślepokura, M. Mayeli, and T. Lis, "Synthesis, characterization and catalytic activity of new Cr(III) complex in oxidation of primary alcohols to aldehydes," *Inorganica Chimica Acta*, vol. 421, pp. 176–182, 2014.

[44] P. S. N. Rao, K. T. V. Rao, P. S. Sai Prasad, and N. Lingaiah, "The role of vanadia for the selective oxidation of benzyl alcohol over heteropolymolybdate supported on alumina," *Chinese Journal of Catalysis*, vol. 32, no. 11, pp. 1719–1726, 2011.

[45] A. V. Biradar, M. K. Dongare, and S. B. Umbarkar, "Selective oxidation of aromatic primary alcohols to aldehydes using molybdenum acetylide oxo-peroxo complex as catalyst," *Tetrahedron Letters*, vol. 50, no. 24, pp. 2885–2888, 2009.

[46] S. C. A. Sousa, J. R. Bernardo, P. R. Florindo, and A. C. Fernandes, "Efficient and selective oxidation of alcohols catalyzed by oxo-rhenium complexes," *Catalysis Communications*, vol. 40, pp. 134–138, 2013.

[47] T. W. Goh, C. Xiao, R. V. Maligal-Ganesh, X. Li, and W. Huang, "Utilizing mixed-linker zirconium based metal-organic frameworks to enhance the visible light photocatalytic oxidation of alcohol," *Chemical Engineering Science*, vol. 124, pp. 45–51, 2015.

[48] M. R. H. Siddiqui, I. Warad, S. F. Adil, R. M. Mahfouz, and A. Al-Arifi, "Nano-gold supported nickel manganese oxide: synthesis, characterisation and evaluation as oxidation catalyst," *Oxidation Communications*, vol. 35, no. 2, article 476, 2012.

[49] S. F. Adil, S. Alabbad, M. Kuniyil et al., "Vanadia supported on nickel manganese oxide nanocatalysts for the catalytic oxidation of aromatic alcohols," *Nanoscale Research Letters*, vol. 10, no. 1, article 52, 2015.

[50] Q. Tang, X. Gong, P. Zhao, Y. Chen, and Y. Yang, "Copper-manganese oxide catalysts supported on alumina: physicochemical features and catalytic performances in the aerobic oxidation of benzyl alcohol," *Applied Catalysis A: General*, vol. 389, no. 1-2, pp. 101–107, 2010.

[51] F. Arena, B. Gumina, A. F. Lombardo et al., "Nanostructured MnO_x catalysts in the liquid phase selective oxidation of benzyl alcohol with oxygen: part I. Effects of Ce and Fe addition on structure and reactivity," *Applied Catalysis B: Environmental*, vol. 162, pp. 260–267, 2015.

[52] M. Yang, Q. Ling, R. Rao et al., "Mn_3O_4-NiO-Ni/CNTs catalysts prepared by spontaneous redox at high temperature and their superior catalytic performance in selective oxidation of benzyl alcohol," *Journal of Molecular Catalysis A: Chemical*, vol. 380, pp. 61–69, 2013.

[53] Y. Chen, H. Zheng, Z. Guo et al., "Pd catalysts supported on $MnCeO_x$ mixed oxides and their catalytic application in solvent-free aerobic oxidation of benzyl alcohol: support composition and structure sensitivity," *Journal of Catalysis*, vol. 283, no. 1, pp. 34–44, 2011.

[54] S. Mandal, C. Santra, K. K. Bando et al., "Aerobic oxidation of benzyl alcohol over mesoporous Mn-doped ceria supported Au nanoparticle catalyst," *Journal of Molecular Catalysis A: Chemical*, vol. 378, pp. 47–56, 2013.

[55] M. Yang, Q. Ling, H. Yang, C. Li, and A. Zhang, "Enhanced catalytic activity of K-birnessite MnO_2 confined in carbon nanotubes for selective oxidation of benzyl alcohol," *Catalysis Communications*, vol. 46, pp. 238–241, 2014.

[56] V. Mahdavi and M. Mardani, "Preparation of manganese oxide immobilized on SBA-15 by atomic layer deposition as an efficient and reusable catalyst for selective oxidation of benzyl alcohol in the liquid phase," *Materials Chemistry and Physics*, vol. 155, pp. 136–146, 2015.

[57] T. J. Clarke, S. A. Kondrat, and S. H. Taylor, "Total oxidation of naphthalene using copper manganese oxide catalysts," *Catalysis Today*, vol. 258, pp. 610–615, 2015.

[58] X. Guo, J. Li, and R. Zhou, "Catalytic performance of manganese doped $CuO-CeO_2$ catalysts for selective oxidation of CO in hydrogen-rich gas," *Fuel*, vol. 163, pp. 56–64, 2016.

[59] M. Tepluchin, S. Kureti, M. Casapu, E. Ogel, S. Mangold, and J.-D. Grunwaldt, "Study on the hydrothermal and SO_2 stability of Al_2O_3-supported manganese and iron oxide catalysts for lean CO oxidation," *Catalysis Today*, vol. 258, pp. 498–506, 2015.

[60] S. C. Kim, Y.-K. Park, and J. W. Nah, "Property of a highly active bimetallic catalyst based on a supported manganese oxide for the complete oxidation of toluene," *Powder Technology*, vol. 266, pp. 292–298, 2014.

[61] M. Piumetti, D. Fino, and N. Russo, "Mesoporous manganese oxides prepared by solution combustion synthesis as catalysts for the total oxidation of VOCs," *Applied Catalysis B: Environmental*, vol. 163, pp. 277–287, 2015.

[62] Z. Feng, Y. Xie, F. Hao, P. Liu, and H. Luo, "Catalytic oxidation of cyclohexane to KA oil by zinc oxide supported manganese 5,10,15,20-tetrakis(4-nitrophenyl)porphyrin," *Journal of Molecular Catalysis A: Chemical*, vol. 410, pp. 221–225, 2015.

[63] A. S. Burange, S. R. Kale, and R. V. Jayaram, "Oxidation of alkyl aromatics to ketones by tert-butyl hydroperoxide on manganese dioxide catalyst," *Tetrahedron Letters*, vol. 53, no. 24, pp. 2989–2992, 2012.

[64] W. H. Yu, C. H. Zhou, D. S. Tong, and T. N. Xu, "Aerobic oxidation of 4-tert-butyltoluene over cobalt and manganese supported hexagonal mesoporous silicas as heterogeneous catalysts," *Journal of Molecular Catalysis A: Chemical*, vol. 365, pp. 194–202, 2012.

[65] J. Pei, X. Han, and Y. Lu, "Performance and kinetics of catalytic oxidation of formaldehyde over copper manganese oxide catalyst," *Building and Environment*, vol. 84, pp. 134–141, 2015.

[66] S. G. Babu, P. A. Priyadarsini, and R. Karvembu, "Copper on boehmite: a simple, selective, efficient and reusable heterogeneous catalyst for oxidation of alcohols with periodic acid in water at room temperature," *Applied Catalysis A: General*, vol. 392, no. 1-2, pp. 218–224, 2011.

[67] Z. Du, J. Ma, H. Ma, M. Wang, Y. Huang, and J. Xu, "Vanadyl sulfate: a simple catalyst for oxidation of alcohols with molecular oxygen in combination with 2,2,6,6-tetramethyl-piperidyl-1-oxyl," *Catalysis Communications*, vol. 11, no. 8, pp. 732–735, 2010.

[68] F. M. Moghaddam, N. Masoud, B. K. Foroushani, S. Saryazdi, N. Ghonouei, and E. Daemi, "Silica-supported DABCO-tribromide: a recoverable reagent for oxidation of alcohols to the corresponding carbonyl compounds," *Scientia Iranica*, vol. 20, no. 3, pp. 598–602, 2013.

[69] V. V. Costa, M. Estrada, Y. Demidova et al., "Gold nanoparticles supported on magnesium oxide as catalysts for the aerobic oxidation of alcohols under alkali-free conditions," *Journal of Catalysis*, vol. 292, pp. 148–156, 2012.

[70] D. Nepak and S. Darbha, "Selective aerobic oxidation of alcohols over Au–Pd/sodium titanate nanotubes," *Catalysis Communications*, vol. 58, pp. 149–153, 2015.

[71] I. Tamiolakis, I. N. Lykakis, and G. S. Armatas, "Mesoporous CdS-sensitized TiO_2 nanoparticle assemblies with enhanced photocatalytic properties: selective aerobic oxidation of benzyl alcohols," *Catalysis Today*, vol. 250, pp. 180–186, 2015.

[72] Y. Shi, H. Yang, X. Zhao et al., "Au–Pd nanoparticles on layered double hydroxide: highly active catalyst for aerobic oxidation of alcohols in aqueous phase," *Catalysis Communications*, vol. 18, pp. 142–146, 2012.

[73] F. Vindigni, S. Dughera, F. Armigliato, and A. Chiorino, "Aerobic oxidation of alcohols on Au/TiO_2 catalyst: new insights on the role of active sites in the oxidation of primary and secondary alcohols," *Monatshefte fur Chemie*, vol. 147, no. 2, pp. 391–403, 2016.

[74] L. Wang, S. Shang, G. Li, L. Ren, Y. Lv, and S. Gao, "Iron/ABNO-catalyzed aerobic oxidation of alcohols to aldehydes and ketones under ambient atmosphere," *Journal of Organic Chemistry*, vol. 81, no. 5, pp. 2189–2193, 2016.

[75] D. Sahu, A. R. Silva, and P. Das, "A novel iron(III)-based heterogeneous catalyst for aqueous oxidation of alcohols using molecular oxygen," *RSC Advances*, vol. 5, no. 96, pp. 78553–78560, 2015.

[76] M. Hosseini-Sarvari, T. Ataee-Kachouei, and F. Moeini, "A novel and active catalyst Ag/ZnO for oxidant-free dehydrogenation of alcohols," *Materials Research Bulletin*, vol. 72, pp. 98–105, 2015.

[77] X. Zhou and H. Ji, "Cobalt porphyrin immobilized on montmorillonite: a highly efficient and reusable catalyst for aerobic oxidation of alcohols to carbonyl compounds," *Chinese Journal of Catalysis*, vol. 33, no. 11-12, pp. 1906–1912, 2012.

[78] Y. Zhang, Q. Zhou, W. Ma, and J. Zhao, "Enantioselective oxidation of racemic secondary alcohols catalyzed by chiral Mn(III)-salen complex with sodium hypochlorite as oxidant," *Catalysis Communications*, vol. 45, pp. 114–117, 2014.

[79] A. Yoshida, Y. Takahashi, T. Ikeda, K. Azemoto, and S. Naito, "Catalytic oxidation of aromatic alcohols and alkylarenes with molecular oxygen over Ir/TiO_2," *Catalysis Today*, vol. 164, no. 1, pp. 332–335, 2011.

[80] T. Yasu-eda, S. Kitamura, N.-O. Ikenaga, T. Miyake, and T. Suzuki, "Selective oxidation of alcohols with molecular oxygen over Ru/CaO–ZrO_2 catalyst," *Journal of Molecular Catalysis A: Chemical*, vol. 323, no. 1-2, pp. 7–15, 2010.

[81] M. L. Kantam, R. S. Reddy, U. Pal et al., "Ruthenium/magnesium–lanthanum mixed oxide: an efficient reusable catalyst for oxidation of alcohols by using molecular oxygen," *Journal of Molecular Catalysis A: Chemical*, vol. 359, pp. 1–7, 2012.

[82] M. Shaikh, M. Satanami, and K. V. S. Ranganath, "Efficient aerobic oxidation of alcohols using magnetically recoverable catalysts," *Catalysis Communications*, vol. 54, pp. 91–93, 2014.

[83] W. Gac, "The influence of silver on the structural, redox and catalytic properties of the cryptomelane-type manganese oxides in the low-temperature CO oxidation reaction," *Applied Catalysis B: Environmental*, vol. 75, no. 1-2, pp. 107–117, 2007.

[84] R. K. Kunkalekar and A. V. Salker, "Low temperature carbon monoxide oxidation over nanosized silver doped manganese dioxide catalysts," *Catalysis Communications*, vol. 12, no. 3, pp. 193–196, 2010.

[85] L. Chen, B. Sun, X. Wang, F. Qiao, and S. Ai, "2D ultrathin nanosheets of Co–Al layered double hydroxides prepared in l-asparagine solution: enhanced peroxidase-like activity and colorimetric detection of glucose," *Journal of Materials Chemistry B*, vol. 1, no. 17, pp. 2268–2274, 2013.

[86] A. C. Tas, "Calcium metal to synthesize amorphous or cryptocrystalline calcium phosphates," *Materials Science and Engineering C*, vol. 32, no. 5, pp. 1097–1106, 2012.

[87] D. P. Dubal, D. S. Dhawale, R. R. Salunkhe, S. M. Pawar, and C. D. Lokhande, "A novel chemical synthesis and characterization of Mn_3O_4 thin films for supercapacitor application," *Applied Surface Science*, vol. 256, no. 14, pp. 4411–4416, 2010.

[88] E. N. Maslen, V. A. Streltsov, N. R. Streltsova, and N. Ishizawa, "Electron density and optical anisotropy in rhombohedral carbonates. III. Synchrotron X-ray studies of $CaCO_3$, $MgCO_3$ and $MnCO_3$," *Acta Crystallographica Section B*, vol. 51, no. 6, pp. 929–939, 1995.

[89] C. Zhu, G. Saito, and T. Akiyama, "A new $CaCO_3$-template method to synthesize nanoporous manganese oxide hollow structures and their transformation to high-performance $LiMn_2O_4$ cathodes for lithium-ion batteries," *Journal of Materials Chemistry A*, vol. 1, no. 24, pp. 7077–7082, 2013.

[90] G. Zhan, Y. Hong, V. T. Mbah et al., "Bimetallic Au–Pd/MgO as efficient catalysts for aerobic oxidation of benzyl alcohol: a green bio-reducing preparation method," *Applied Catalysis A: General*, vol. 439-440, pp. 179–186, 2012.

[91] G. Wu, Y. Gao, F. Ma et al., "Catalytic oxidation of benzyl alcohol over manganese oxide supported on MCM-41 zeolite," *Chemical Engineering Journal*, vol. 271, pp. 14–22, 2015.

[92] E. G. Rodrigues, J. J. Delgado, X. Chen, M. F. R. Pereira, and J. J. M. Órfão, "Selective oxidation of glycerol catalyzed by gold supported on multiwalled carbon nanotubes with different surface chemistries," *Industrial and Engineering Chemistry Research*, vol. 51, no. 49, pp. 15884–15894, 2012.

[93] X. Yu, Y. Huo, J. Yang, S. Chang, Y. Ma, and W. Huang, "Reduced graphene oxide supported Au nanoparticles as an efficient catalyst for aerobic oxidation of benzyl alcohol," *Applied Surface Science*, vol. 280, pp. 450–455, 2013.

[94] E. Assady, B. Yadollahi, M. Riahi Farsani, and M. Moghadam, "Zinc polyoxometalate on activated carbon: an efficient catalyst for selective oxidation of alcohols with hydrogen peroxide," *Applied Organometallic Chemistry*, vol. 29, no. 8, pp. 561–565, 2015.

[95] S. Hasannia and B. Yadollahi, "Zn–Al LDH nanostructures pillared by Fe substituted Keggin type polyoxometalate: synthesis, characterization and catalytic effect in green oxidation of alcohols," *Polyhedron*, vol. 99, pp. 260–265, 2015.

[96] J. Albadi, A. Alihoseinzadeh, and A. Razeghi, "Novel metal oxide nanocomposite of Au/CuO–ZnO for recyclable catalytic aerobic oxidation of alcohols in water," *Catalysis Communications*, vol. 49, pp. 1–5, 2014.

[97] Z. Nadealian, V. Mirkhani, B. Yadollahi, M. Moghadam, S. Tangestaninejad, and I. Mohammadpoor-Baltork, "Selective oxidation of alcohols to aldehydes using inorganic–organic hybrid catalyst based on zinc substituted polyoxometalate and ionic liquid," *Journal of Coordination Chemistry*, vol. 65, no. 6, pp. 1071–1081, 2012.

[98] K. P. Peterson and R. C. Larock, "Palladium-catalyzed oxidation of primary and secondary allylic and benzylic alcohols," *Journal of Organic Chemistry*, vol. 63, no. 10, pp. 3185–3189, 1998.

[99] M. M. Kadam, K. B. Dhopte, N. Jha, V. G. Gaikar, and P. R. Nemade, "Synthesis, characterization and application of γ-MnO_2/graphene oxide for the selective aerobic oxidation of benzyl alcohols to corresponding carbonyl compounds," *New Journal of Chemistry*, vol. 40, no. 2, pp. 1436–1442, 2016.

[100] M. Heidari-Golafzani, M. Rabbani, R. Rahimi, and A. Azad, "Catalytic oxidation of primary and secondary alcohols over a novel TCPP/Zn–Fe_2O_4@ZnO catalyst," *RSC Advances*, vol. 5, no. 121, pp. 99640–99645, 2015.

17

1-(2-Chlorobenzyloxy)-3-[1,2,3]triazol-1-yl-propan-2-ol Derivatives: Synthesis, Characterization, and DFT-Based Descriptors Analysis

Eloisa Román-Maldonado,[1] Horacio Reyes,[2] Miguel A. Sanchez-Carmona,[1] Nelly González-Rivas,[1] and Erick Cuevas-Yañez[1]

[1]*Centro Conjunto de Investigación en Química Sustentable UAEM-UNAM, Universidad Autónoma del Estado de México, Carretera Toluca-Atlacomulco Km 14.5, 50200 Toluca, MEX, Mexico*
[2]*División de Ingeniería Química, Instituto de Estudios Superiores de Jocotitlán, Carretera Toluca-Atlacomulco Km 44.8, Ejido de San Juan y San Agustin, Jocotitlán, MEX, Mexico*

Correspondence should be addressed to Erick Cuevas-Yañez; ecuevasy@uaemex.mx

Academic Editor: Anton Kokalj

A novel series of 1-(2-chlorobenzyloxy)-3-[1,2,3]triazol-1-yl-propan-2-ol derivatives was designed and synthesized using copper catalyzed alkyne-azide cycloaddition in the key step. Theoretical investigation of molecular and electronic properties by means of global and local reactivity indexes of the synthetized compounds was carried out, using DFT (Density Functional Theory) at PBEPBE/6-31++G** level.

1. Introduction

Depression and similar psychological disorders affect about 350 million people worldwide and represent one of the leading causes of disability and a major contributor to the burden of suicide and ischemic heart diseases [1, 2]. This illness is associated with a decreased production of serotonin in conjunction with a decreased sensitivity of postsynaptic receptors to available serotonin [3]. Therefore, one of the most extended current therapies for this kind of diseases involve the use of commercially available drugs which work as selective serotonin reuptake inhibitors (SSRI) such as fluoxetine among other analogous compounds [4, 5]. Although all these compounds are effective as antidepressants, some side effects are observed during their use. This reason has motivated the design and development of diverse molecules, which display good reuptake inhibiting activity [6–8].

A particular molecule that has attracted us is a derivative from chlorobenzyloxy-3,6-dihydropyridinyl propan-2-ol named DJLDU-3-79, **1** [9]. This compound was designed from a hybridization of two active reuptake inhibitory molecules in order to improve the above-mentioned properties. However, a synthetic challenge in the preparation of this molecule is the tetrahydropyridine moiety prepared from 4-piperidone analogs, which in turn are obtained through several reaction steps [10, 11]. In this regard, we propose a synthesis of a novel family of compounds **2** analogs to molecule **1** based on a substitution of tetrahydropyridine ring by 1,2,3-triazole which is easily synthesized by copper catalyzed alkyne-azide cycloaddition (CuAAC), the most powerful click reaction (Scheme 1).

On the other hand, DFT-based methods have already proven to be useful in determining molecular structure of organic compounds, as well as in elucidating their electronic structure and reactivity [12]. There are two kinds of DFT-based reactivity indexes: global and local quantities. The global parameters measure the electronic features of the molecules as a whole and the local quantity reflects the reactivity of particular regions or atoms. The properties calculated in this study have been successfully applied in to explain the reactivity of organic and inorganic molecules, proteins, solids, and others [13].

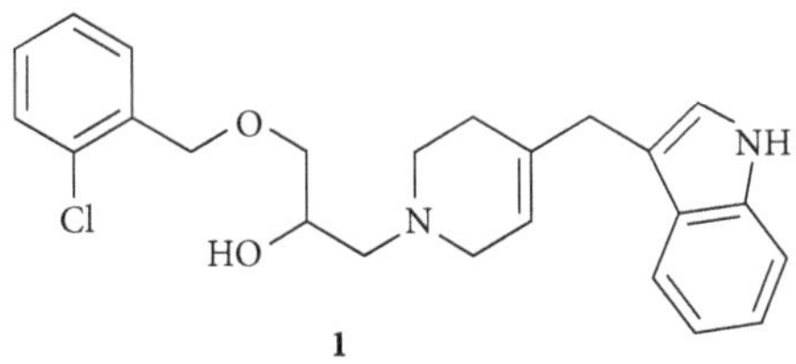

1

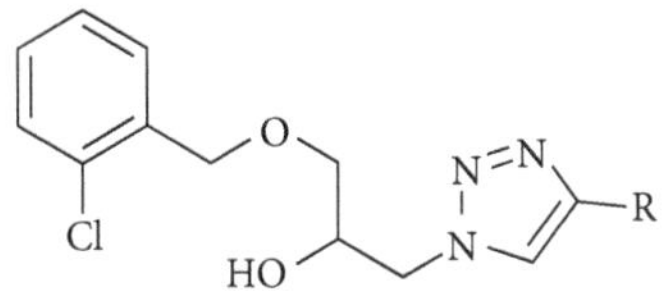

2

Scheme 1: Structure of DJLDU-3-79 (**1**) and general structure for molecules **2** proposed in this work.

Several studies have confirmed closed relationships between the biological performance of a family of compounds and their chemical reactivity. The pharmacokinetic and pharmacodynamic properties of SSRI molecules are strongly influenced by their interactions to the body proteins [9]. Therefore, the electron transfer properties in protein-ligand binding could be studied by using DFT reactivity indexes. To accomplish this goal, the use of global chemical reactivity parameters, chemical potential (I), electron affinity (A), chemical potential (μ), softness (S), and electrophilicity (ω), could aid the understanding of the interaction process between SSRI and biological molecules. Also, the local parameters could give some light about which molecular moieties are most involved in the binding structure

This report summarizes our most recent findings in the click chemistry area aiming to search and develop new molecules with high SSRI activity; also the theoretical characterization of the obtained molecules was carried out.

2. Experimental Section

2.1. Synthesis. The starting materials were purchased from Aldrich Chemical Co. and were used without further purification. Solvents were distilled before use. Silica plates of 0.20 mm thickness were used for thin layer chromatography. Melting points were determined with a Krüss Optronic melting point apparatus and they are uncorrected. ^{1}H and ^{13}C NMR spectra were recorded using a Bruker AVANCE 300; the chemical shifts (δ) are given in ppm relative to TMS as internal standard (0.00). For analytical purposes the mass spectra were recorded on a Shimadzu GCMS-QP2010 Plus in the EI mode, 70 eV, 200°C via direct inlet probe. Only the molecular and parent ions (m/z) are reported. IR spectra were recorded on a Bruker TENSOR 27 FT instrument.

*2.1.1. 2-Chlorophenyl-methanol (**4**).* A solution of 2-chlorobenzaldehyde (1.125 mL, 1.40 g, 10.03 mmol) in ethanol (15 mL) was added to a suspension of sodium borohydride (0.189 g, 4.97 mmol) in absolute ethanol (6 mL) maintaining the temperature below 25°C. The resulting mixture was heated at 50°C during 1.5 h. The solvent was removed in vacuo, and water (20 mL) was added; the mixture was acidified with diluted HCl (10%) to pH = 5. The product was extracted with CH_2Cl_2 (3 × 15 mL), the organic phases were joined and dried over Na_2SO_4, and the solvent was removed in vacuo to yield a white solid (1.27 g, 90%), which was used without additional purification, m.p. 50°C; ^{1}H NMR (300 MHz, $CDCl_3$) δ 7.49 (dd, J = 7.3, 2.0 Hz, 1H), 7.37 (dd, J = 7.5, 1.8 Hz, 1H), 7.32–7.20 (m, 2H), 4.79 (d, J = 6.0 Hz, 1H), 1.69 (s, 1H); ^{13}C NMR (75 MHz, $CDCl_3$) δ 137.1, 131.6, 128.2, 127.7, 127.6, 125.9, 61.7; IR (ATR, cm^{-1}): 3302, 3216, 2912, 2859, 1439; MS [EI+] m/z (%): 142 $[M]^+$ (1), 107 $[C_7H_7O]^+$ (35), 77 $[C_6H_5]^+$ (100), 51 $[C_4H_3]^+$ (91).

*2.1.2. 2-(2-Chloro-benzyloxymethyl)-oxirane (**5**).* A mixture of 40% aqueous solution of NaOH (6.5 mL) (2-chlorophenyl)-methanol **4** (0.71 g, 5 mmol) and tetrabutylammonium iodide (0.09 g, 0.25 mmol) was vigorously stirred at room temperature for 24 h. The mixture was cooled to 0°C and epichlorohydrin (1.55 mL, 1.83 g, and 19.9 mmol) was added dropwise. After completion, the reaction mixture was stirred at 0°C for 30 minutes and at room temperature overnight. The mixture was poured into ice cold water (20 mL). The product was extracted with CH_2Cl_2 (3 × 15 mL), the organic phases were joined and dried over Na_2SO_4, and the solvent was removed in vacuo to yield a pale yellow oil (1.27 g, 90%), which was used without additional purification; ^{1}H NMR (300 MHz, $CDCl_3$) δ 7.50 (dd, J = 7.4, 2.1 Hz, 1H), 7.35 (dd, J = 7.4, 1.8 Hz, 1H), 7.31–7.19 (m, 2H), 4.68 (d, J = 3.9 Hz, 2H), 3.49–3.87 (dd, J = 11.4, 5.8 Hz, 2H), 3.27–3.18 (m, 1H), 2.65–2.82 (dd, J = 5.0, 2.7 Hz, 2H); ^{13}C NMR (75 MHz, $CDCl_3$) δ 135.4, 132.6, 129.0, 128.7, 128.5, 126.5, 71.1, 69.9, 50.5, 43.9; IR (ATR, cm^{-1}): 3061, 2998, 2923, 2864, 1250, 1096; MS [EI+] m/z (%): 198 $[M]^+$ (5), 56 $[C_3H_5O]^+$ (100). HRMS (EI): calcd. For $C_{10}H_{11}ClO_2$: 198.0448; found: 198.0442.

*2.1.3. 1-Azido-3-(2-chloro-benzyloxy)-propan-2-ol (**6**).* 2-(2-Chloro-benzyloxymethyl)-oxirane **5** (0.965 g, 4.86 mmol), NH_4Cl (0.529 g, 9.89 mmol), and NaN_3 (2.652 g, 40.8 mmol) were added successively to a mixture of MeOH (8 mL) and H_2O (1 mL) and the reaction mixture was stirred at room temperature for 24 h. The solvent was removed in vacuo and the product was extracted with AcOEt (3 × 15 mL); the organic phases were joined and dried over Na_2SO_4 and the solvent was removed in vacuo to yield a pale yellow oil (1.2981 g, 74.12%), which was used without additional purification; ^{1}H NMR (300 MHz, $CDCl_3$) δ 7.40 (dd, J = 7.2, 2.2 Hz, 1H), 7.35–7.30 (m, 1H), 7.29–7.16 (m, 2H), 4.60 (s, 2H), 3.94 (ddd, J = 10.8, 5.9, 4.9 Hz, 2H), 3.54 (s, broad, 1H), 3.52–3.34 (ddd, J = 5.4, 2.3, 1.3 Hz, 2H), 3.22 (s, 1H); ^{13}C NMR (75 MHz, $CDCl_3$) δ 135.3, 133.2, 129.4, 129.4, 129.1, 126.9, 71.85, 70.6, 69.6, 53.5; IR (ATR, cm^{-1}): 3426, 2097, 1092, 751; MS [EI+] m/z (%): 241 $[M]^+$ (5), 149 $[M\text{-}C_7H_6ClO]^+$ (100), 125 $[M\text{-}C_7H_6Cl]^+$ (90), 107 [M- $C_3H_6N_3O$]+ (90), 89 $[M\text{-}C_2H_4N_3O]^+$ (60), 43 $[M\text{-}N_3]^+$ (65). HRMS (EI): calcd. For $C_{10}H_{12}ClN_3O_2$: 241.0618; found: 241.0622.

2.2. General Procedure for the Synthesis of 1-(2-Chlorobenzyloxy)-3-[1,2,3]triazol-1-yl-propan-2-oles. The appropriate alkyne (1 mol) was added in one portion to a solution of 1-azido-3-(2-chloro-benzyloxy)-propan-2-ol **6** (0.241 g, 1 mmol), CuI (0.009 g, 0.05 mmol), and DIPEA (0.17 mL, 0.129 g, 1 mmol) in CH_2Cl_2 (15 mL). The resulting mixture was stirred at room temperature for 24 h. An aqueous solution of 2% EDTA (30 mL) was added and the stirring was continued for additional 24 h. The product was extracted with CH_2Cl_2 (3 × 15 mL), the organic phases were joined and dried over Na_2SO_4, the solvent was removed in vacuo, and the final product was purified by crystallization.

2.2.1. 1-(2-Chlorobenzyloxy)-3-(4-phenyl-[1,2,3]triazol-1-yl)-propan-2-ol (7). White solid; yield: 85%; mp 58°C; ^{1}H NMR (300 MHz, $CDCl_3$) δ 7.86 (s, 1H), 7.78–7.65 (m, 2H), 7.47–7.30 (m, 5H), 7.32 (s, 1H), 7.30–7.20 (m, 2H), 4.65 (s, 2H), 4.64–4.57 (m), 4.49 (dd, J = 8.7, 7.0 Hz, 1H), 4.48–4.36 (m, 1H), 3.59–3.52 (m, 1H), 3.44–3.37 (m, 1H), 2.62 (s, broad, 1H); ^{13}C NMR (75 MHz, $CDCl_3$) δ 135.3, 133.2, 129.4, 129.4, 129.1, 126.9, 71.8, 70.6, 69.6, 53.5; IR (ATR, cm^{-1}): 3227, 3061, 2866, 1440, 750; MS [EI+] *m/z* (%): 343 $[M]^+$ (5), 125 $[C_7H_6Cl]^+$ (100). HRMS (EI): calcd. For $C_{18}H_{18}ClN_3O_2$: 343.1088; found: 343.1091.

*2.2.2. 1-(2-Chlorobenzyloxy)-3-(4-p-tolyloxymethyl-[1,2,3]triazol-1-yl)-propan-2-ol (**8**).* White solid; yield: 99%; mp 25°C; ^{1}H NMR (300 MHz, $CDCl_3$) δ 7.70 (s, 1H), 7.46 (dd, J = 7.6, 1.7 Hz, 1H), 7.41 (dd, J = 3.8, 1.7 Hz, 1H), 7.27 (dd, J = 12.2, 1.8 Hz, 1H), 7.20–7.13 (m, 1H), 7.05 (d, J = 8.5 Hz, 2H), 6.87 (d, J = 2.3 Hz, 1H), 6.83 (d, J = 2.1 Hz, 1H), 5.08 (s, 2H), 4.57 (s, 2H), 4.52–4.36 (dddd, J = 13.9, 14.0, 3.8, 7.1 Hz, 2H), 4.20 (dd, J = 9.2, 5.2 Hz, 1H), 3.53–3.46 (ddd, J = 9.7, 5.4, 4.9 Hz, 2H), 2.49 (s, broad, 1H), 2.25 (s, 3H); ^{13}C NMR (75 MHz, $CDCl_3$) δ 155.1, 143.0, 134.7, 134.2, 132.3, 129.5, 128.5, 128.5, 127.9, 125.8, 123.4, 113.8, 70.6, 69.5, 68.1, 60.93, 52.1, 19.5; IR (ATR, cm^{-1}): 3312, 2925, 2860, 1269, 1120; MS [EI+] *m/z* (%): 387 $[M]^+$ (5), 125 $[C_7H_6Cl]^+$ (100). HRMS (EI): calcd. For $C_{20}H_{22}ClN_3O_3$: 387.1350; found: 387.1351.

*2.2.3. 1-(2-Chlorobenzyloxy)-3-[4-(4-methoxy-phenoxymethyl)-[1,2,3]triazol-1-yl]-propan-2-ol (**9**).* White solid; yield: 46%; mp 76°C; ^{1}H NMR (300 MHz, $CDCl_3$) δ 7.78 (s, 1H), 7.75–7.68 (m, 2H), 7.39 (dd, J = 10.7, 4.2 Hz, 2H), 7.31–7.20 (m, 2H), 6.95 (d, J = 8.4 Hz, 2H), 4.65 (s, 1H), 4.60–4.46 (dddd, J = 14.2, 3.6, 14.0, 6.8 Hz, 2H), 4.36–4.26 (m, 1H), 3.84 (s, 3H), 3.63–3.53 (ddd J = 9.8, 4.9, 9.7, 5.7 Hz, 2H), 2.04 (s, broad, 1H); ^{13}C NMR (75 MHz, $CDCl_3$) δ 159.6, 147.5, 135.0, 133.5 de C3, 129.6, 129.6, 129.3, 126.9, 126.9, 126.9, 120.4, 114.2, 71.3, 70.8, 69.3, 55.3, 52.9; IR (ATR, cm^{-1}): 3256, 3141, 2958, 2924, 2854, 1360, 1249; MS [EI+] *m/z* (%): 403 $[M]^+$ (28), 125 $[C_7H_6Cl]^+$ (100).

*2.2.4. 1-(2-Chlorobenzyloxy)-3-(4-p-tolyl-[1,2,3]triazol-1-yl)-propan-2-ol (**10**).* White solid; yield: 51%; mp 120°C; ^{1}H NMR (300 MHz, $CDCl_3$) δ 7.81 (s, 1H), 7.64–7.55 (m, 2H), 7.42 (dd, J = 6.4, 2.9 Hz, 1H), 7.29–7.14 (m, 5H), 4.64 (s, 2H), 4.58–4.45 (dddd, J = 14.0, 7.2, 3.8, 4.8 Hz, 2H), 4.44–4.38 (m, 1H), 3.61–3.57 (m, 2H), 2.36 (s, 3H), 1.26 (s, broad, 1H); ^{13}C NMR (75 MHz, $CDCl_3$) δ 155.1, 143.0, 134.7, 134.22, 132.3, 129.5, 128.5, 128.8, 127.9, 125.8, 123.4, 113.8, 70.6, 69.5, 68.1, 60.9, 52.1, 19.5; IR (ATR, cm^{-1}): 3392, 2918, 2865, 1442, 1107; MS [EI+] *m/z* (%): 357 $[M]^+$ (5), 125 $[C_7H_6Cl]^+$ (100). HRMS (EI): calcd. For $C_{19}H_{20}ClN_3O_2$: 357.1244; found: 357.1247.

*2.2.5. 2-{1-[3-(2-Chlorobenzyloxy)-2-hydroxy-propyl]-1H-[1,2,3]triazol-4-ylmethyl}-2-methyl-cyclopentane-1,3-dione (**11**).* Colorless oil; yield: 66%; ^{1}H NMR (300 MHz, $CDCl_3$) δ 7.44 (s, 1H), 7.48–7.41 (m, 1H), 7.36–7.31 (m, 1H), 7.26–7.19 (m, 2H), 4.60 (s, 2H), 4.46–4.29 (dddd, J = 14.0, 13.9, 7.3, 3.6 Hz, 2H), 4.14 (dq, J = 9.4, 5.3 Hz, 1H), 3.50 (d, J = 5.2 Hz, 2H), 3.05 (s, 2H), 2.80 (d, J = 3.2 Hz, 4H), 2.01 (s, broad, 1H), 1.17 (s, 3H); ^{13}C NMR (75 MHz, $CDCl_3$) δ 212.2, 202.7, 137.5, 130.5, 128.0, 124.5, 124.4, 124.1, 122.02, 118.20, 66.9, 65.6, 64.1, 50.3, 48.1, 30.1, 26.0, 25.9, 25.3; IR (ATR, cm^{-1}): 3304, 2925, 2869, 1442, 1361, 1219; MS [EI+] *m/z* (%): 391 $[M]^+$ (10), 125 $[C_7H_6Cl]^+$ (100). HRMS (EI): calcd. For $C_{19}H_{22}ClN_3O_4$: 391.1299; found: 391.1303.

*2.2.6. 3-{1-[3-(2-Chlorobenzyloxy)-2-hydroxy-propyl]-[1,2,3]triazol-4-yl}-propan-1-ol (**12**).* Colorless oil; yield: 84%; ^{1}H NMR (300 MHz, $CDCl_3$) δ 7.44 (s, 1H), 7.43–7.35 (m, 1H), 7.30–7.21 (m, 3H), 4.78 (s, broad, 1H), 4.64 (s, 2H), 4.53–4.38 (dddd, J = 14.0, 14.0, 7.1, 3.6 Hz), 4.28–4.20 (m, 1H), 3.66 (t, J = 6.4 Hz, 2H), 3.58–3.51 (dddd, J = 9.7, 9.7, 5.5, 5.0 Hz, 2H), 2.79 (t, J = 7.4 Hz, 2H), 2.17 (s, broad, 1H), 1.89 (p, J = 6.6 Hz, 2H); ^{13}C NMR (75 MHz, $CDCl_3$) δ 147.0, 134.8, 133.1, 129.3, 129.2, 126.6, 122.3, 71.2, 70.5, 68.8, 61.3, 52.6, 31.5, 21.7; IR (ATR, cm^{-1}): 3487, 2912, 2862, 1716, 1446, 1219; MS [EI+] *m/z* (%): 325 $[M]^+$ (10), 126 $[C_5H_8N_3O]^+$ (100). HRMS (EI): calcd. For $C_{15}H_{20}ClN_3O_3$: 325.1193; found: 325.1195.

*2.2.7. 1-(2-Chlorobenzyloxy)-3-[4-(4-nitro-phenoxymethyl)-[1,2,3]triazol-1-yl]-propan-2-ol (**13**).* White solid; yield: 55%; mp 85°C; ^{1}H NMR (300 MHz, $CDCl_3$) δ 8.20 (d, J = 8.2 Hz, 2H), 8.17 (d, J = 2.1 Hz, 2H), 7.81 (s, 1H), 7.43–7.23 (m, 4H), 7.08 (m, 2H), 4.63 (d, J = 1.1 Hz, 2H), 4.62 (dddd, J = 14.1, 14.0 7.1, 4.6 Hz, 2H), 4.30 (m, 1H), 3.64–3.49 (dddd, J = 9.7, 9.7, 5.0, 4.7 Hz, 2H), 2.16 (s, 1H); ^{13}C NMR (75 MHz, $CDCl_3$) δ 162.7, 142.1, 141.4, 134.6, 132.9, 129.2, 129.2, 128.9, 126.5, 125.5, 124.4, 114.5, 71.0, 70.4, 68.75, 52.7; IR (ATR, cm^{-1}): 3253, 2878, 2856, 1336; MS [EI+] *m/z* (%): 418 $[M]^+$ (5), 125 $[C_7H_6Cl]^+$ (100). HRMS (EI): calcd. For $C_{19}H_{19}ClN_4O_5$: 418.1044; found: 418.1048.

*2.2.8. 1-(2-Chlorobenzyloxy)-3-[4-(4-chlorophenoxymethyl)-[1,2,3]triazol-1-yl]-propan-2-ol (**14**).* White solid; yield: 89%; mp 67-68°C; ^{1}H NMR (300 MHz, $CDCl_3$) δ 7.73 (s, 1H), 7.41 (dd, J = 9.6, 4.8 Hz, 1H), 7.30–7.20 (m, 5H), 6.94–6.89 (m, 2H), 5.16 (s, 2H), 4.63 (s, 2H), 4.60–4.40 (m, 2H), 4.29 (dd, J = 4.5, 3.7 Hz, 1H), 3.62–3.46 (m, 2H), 2.95 (s, broad, 1H); ^{13}C NMR (75 MHz, $CDCl_3$) δ 156.1, 143.0, 134.3, 132.8, 128.9, 128.9, 128.8, 128.7, 128.7, 126.2, 123.6, 115.4, 70.5, 70.1, 68.5, 61.5, 52.2; IR (ATR, cm^{-1}): 3168, 2921, 2869, 1489, 1237; MS [EI+] *m/z* (%): 407 $[M]^+$ (2), 185 $[C_9H_{10}ClO]^+$ (80), 127 $[C_6H_4ClO]^+$ (100). HRMS (EI): calcd. For $C_{19}H_{19}Cl_2N_3O_3$: 407.0803; found: 407.0808.

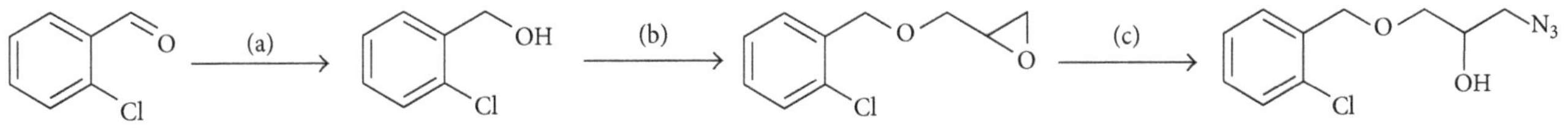

3 4 5 6

SCHEME 2: Reagents and conditions: (a) $NaBH_4$ and MeOH. (b) Epichlorohydrin, NaOH, and N-Bu_4NBr. (c) NaN_3, NH_4Cl, and MeOH-H_2O.

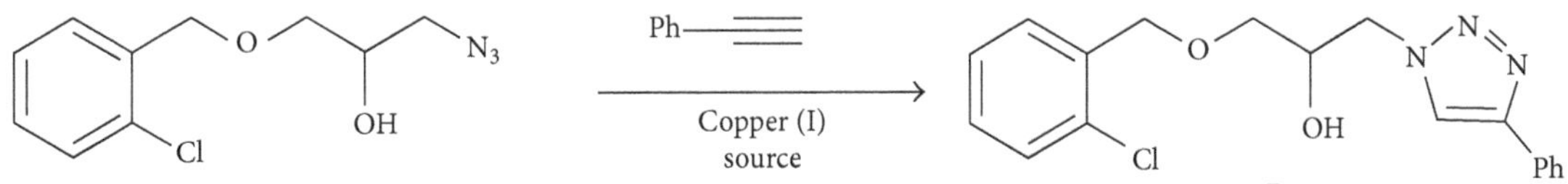

6 7

SCHEME 3: Reaction of azido alcohol **6** with phenylacetylene.

*2.2.9. 1-(2-Chlorobenzyloxy)-3-[4-(6-methoxynaphthalen-2-yl)-[1,2,3]triazol-1-yl]-propan-2-ol (**15**).* White solid; yield: 94%; mp 100°C; ^1H NMR (300 MHz, $CDCl_3$) δ 8.10 (s, 1H), 7.90 (s, 1H), 7.79 (dt, J = 8.5, 1.8 Hz, 1H), 7.72 (s, 1H), 7.68 (dd, J = 8.8, 1.8 Hz, 1H), 7.40 (dd, J = 17.1, 9.3 Hz), 7.29–7.20 (m, 3H), 7.16–7.09 (m, 2H), 4.66 (s, 2H), 4.60–4.46 (ddd, J = 3.5, 13.9, 6.7 Hz, 2H), 4.36 (s, 1H), 3.92 (s, 3H), 3.65–3.57 (dddd, J = 9.7, 4.9, 9.7, 5.5 Hz, 2H), 2.16 (s, broad, 1H); ^{13}C NMR (75 MHz, $CDCl_3$) δ 157.9, 147.6, 135.1, 134.3, 133.4, 129.7, 129.6, 129.5, 129.2, 128.8, 127.3, 126.9, 125.5, 124.9, 121.2, 119.2, 105.7, 71.5, 70.8, 69.2, 55.3, 53.3; IR (ATR, cm^{-1}): 3391, 3131, 2921, 2852, 1606, 1216; MS [EI+] m/z (%): 423 $[M]^+$ (10), 226 $[C_{13}H_{10}N_3O]^+$ (80), 185 $[C_9H_{10}ClO]^+$ (35), 125 $[C_7H_6Cl]^+$ (100). HRMS (EI): calcd. For $C_{23}H_{22}ClN_3O_3$: 423.1350; found: 423.1356.

*2.2.10. 1-(2-Chlorobenzyloxy)-3-[4-(4-fluorophenyl)-[1,2,3]triazol-1-yl]-propan-2-ol (**16**).* Colorless oil; yield: 66%; ^1H NMR (300 MHz, $CDCl_3$) δ 7.82 (s, 1H), 7.77–7.70 (m, 2H), 7.53 (dd, J = 5.7, 3.3 Hz, 1H), 7.38 (td, J = 7.4, 2.2 Hz, 1H), 7.26 (ddt, J = 7.6, 2.2, 1.1 Hz, 1H), 7.30–7.21 (m, 2H), 7.09 (t, J = 7.4 Hz, 1H), 4.65 (d, J = 0.8 Hz, 2H), 4.61–4.46 (dd, J = 14.0, 7.0 Hz, 2H), 4.31 (p, J = 5.2 Hz, 1H), 3.64–3.54 (dd, J = 9.7, 5.7 Hz, 2H), 2.17 (s, 1H); ^{13}C NMR (75 MHz, $CDCl_3$) δ 164.1, 160.8, 146.5, 134.8, 133.2, 129.5, 129.4, 129.1, 127.2, 127.1, 126.7, 126.4, 120.8, 115.7, 115.5, 71.2, 70.6, 69.0, 52.9; IR (ATR, cm^{-1}): 3365, 2922, 2864, 1497, 1224; MS [EI+] m/z (%): 361 $[M]^+$ (25), 164 $[C_8H_7FN_3]^+$ (100). HRMS (EI): calcd. For $C_{18}H_{17}ClFN_3O_2$: 361.0993; found: 361.0997.

*2.2.11. 1-(2-Chlorobenzyloxy)-3-[4-(2,4-difluorophenyl)-[1,2,3] triazol-1-yl]-propan-2-ol (**17**).* White solid; yield: 67%; mp 77°C; ^1H NMR (300 MHz, $CDCl_3$) δ 8.31–8.15 (m, 1H), 8.00 (s, 1H), 7.47–7.33 (m, 2H), 7.26 (dd, J = 5.9, 3.5 Hz, 2H), 6.99 (td, J = 8.4, 2.6 Hz, 1H), 6.88 (ddd, J = 11.1, 8.9, 2.6 Hz, 1H), 4.66 (s, 2H), 4.61–4.49 (ddd, J = 14.1, 6.9, 3.7 Hz, 2H), 4.37–4.27 (m, 1H), 3.65–3.55 (dddd, J = 9.7, 9.6, 5.6, 4.6 Hz, 2H), 1.65 (s, broad, 1H); ^{13}C NMR (75 MHz, $CDCl_3$) δ 163.4, 160.1, 139.6, 134.9, 134.3, 132.6, 128.8, 128.5, 126.1, 123.21, 114.2, 111.1, 103.6, 70.74, 70.1, 68.5, 52.4; IR (ATR, cm^{-1}): 3194, 3151, 2908, 1361; MS [EI+] m/z (%): 379 $[M]^+$ (10), 239 $[C_{10}H_{11}F_2N_3O]^+$ (100). HRMS (EI): calcd. For $C_{18}H_{16}ClF_2N_3O_2$: 379.0889; found: 379.0891.

*2.2.12. 1-(2-Chlorobenzyloxy)-3-[4-(2-trifluoromethylphenyl)-[1,2,3]triazol-1-yl]-propan-2-ol (**18**).* Colorless oil; yield: 68%; ^1H NMR (300 MHz, $CDCl_3$) δ 7.91 (s, 1H), 7.90 (d, J = 6.9 Hz, 1H), 7.73 (dd, J = 8.0, 1.4 Hz, 1H), 7.47 (dd, J = 7.9, 1.5 Hz, 1H), 7.42 (ddd, J = 6.9, 3.7, 1.6 Hz, 1H), 7.37–7.32 (m, 1H), 7.23 (td, J = 6.5, 5.7, 3.7 Hz, 2H), 4.63 (d, J = 1.2 Hz, 2H), 4.63–4.50 (dddd, J = 14.1, 14.0, 6.8, 3.9 Hz, 2H), 4.30 (dq, J = 9.8, 5.3 Hz, 1H), 3.60–3.51 (dddd, J = 9.8, 9.7, 5.6, 4.9 Hz, 2H), 1.26 (s, broad, 1H); ^{13}C NMR (75 MHz, $CDCl_3$) δ 163.4, 160.1, 139.6, 134.9, 134.3, 132.6, 128.8, 128.5, 126.1, 123.21, 114.2, 111.1, 103.6, 70.74, 70.1, 68.5, 52.4; IR (ATR, cm^{-1}): 3344, 1354, 1168.86, 1107; MS [EI+] m/z (%): 411 $[M]^+$ (5), 125 $[C_7H_6Cl]^+$ (100). HRMS (EI): calcd. For $C_{19}H_{17}ClF_3N_3O_2$: 411.0961; found: 411.0965.

2.3. Theoretical Calculations. All the electronic structure computations were carried out with NWChem package [14]. A frequency analysis was applied to the optimized geometries to verify the corresponding stability criterion. The PBEPBE [15] exchange and correlation functional was used with the 6-311++g** Pople's basis set. All the reactivity parameters were computed from a finite differences approach.

3. Results and Discussion

3.1. Chemistry. The first experiments were carried out with the objective to prepare 1-azido-3-(2-chloro-benzyl-oxy)-propan-2-ol, **6** (Scheme 2). Initially, 2-chlorophenyl methanol **4** prepared from reduction of 4-chlorobenzaldehyde **3** was reacted with epichlorohydrin to give the aryl epoxide **5** [9, 15] which was opened with sodium azide to afford azido alcohol **6** in a 62% global yield from 4-chlorobenzaldehyde **3**.

Azido alcohol **6** was the starting material for the following CuAAC experiments. In a model reaction, compound **6** was treated with phenylacetylene in presence of diverse Cu(I) catalytic sources according to that described by Rostovtsev et al. [16], Tornøe et al. [17] and our group [18, 19] (Scheme 3), obtaining compound **7** as only reaction product. From these experiments presented in Table 1, we selected CuI-DIPEA for CuAAC reaction because a best yield (85%) was reached using this catalytic system.

Taking these optimized conditions, a synthetic procedure was developed and a series of (2-chlorobenzyloxy)-1,2,3-triazolyl-propan-2-ol derivatives was prepared from azido

Table 1: Effect of catalyst, reducing agent, base, and solvent.

Entry	Copper salt	Reducing agent	Base	Solvent	% yield
7	$CuSO_4$	Glucose	NaOH	$MeOH-H_2O$	34
8	$CuSO_4$	Sodium ascorbate	none	$tBuOH-H_2O$	28
9	CuI	None	NaOH	$MeOH-H_2O$	40
10	CuI	None	DIPEA	MeOH	35
11	CuI	None	DIPEA	CH_2Cl_2	85

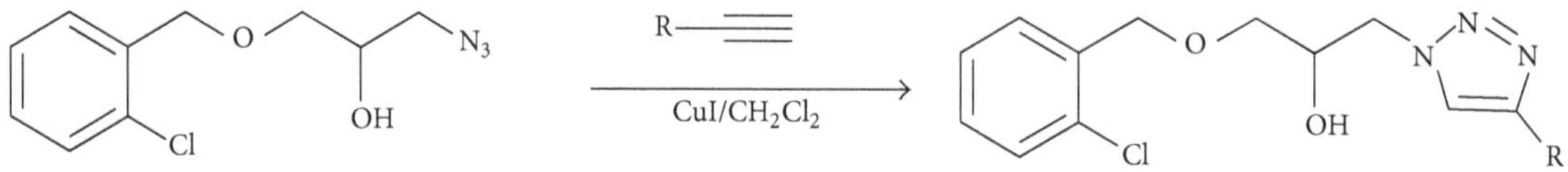

6

Scheme 4

Table 2: 1,2,3-Triazoles prepared via CuAAC.

Compound	R^1	% yield
7	Ph	85
8	$CH_2O(4\text{-}CH_3)C_6H_4$	99
9	$CH_2O(4\text{-}OCH_3)C_6H_4$	46
10	$4\text{-}CH_3C_6H_4$	51
11	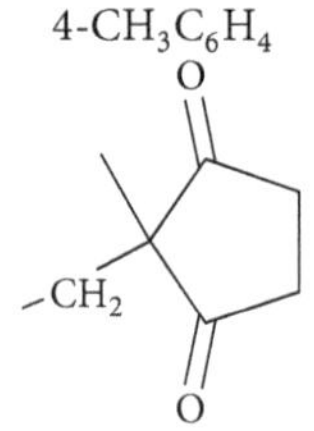	66
12	$(CH_2)_3OH$	84
13	$CH_2O(4\text{-}NO_2)C_6H_4$	55
14	$CH_2O(4\text{-}Cl)C_6H_4$	89
15	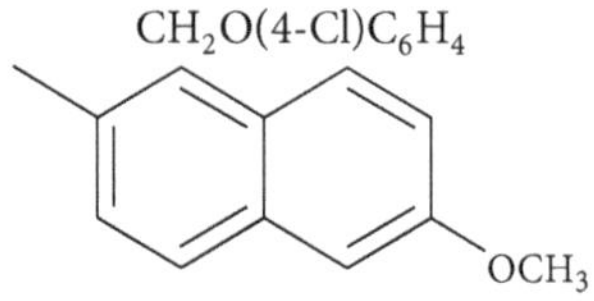	94
16	$4\text{-}FC_6H_4$	66
17	$2,4\text{-}FC_6H_3$	67
18	$2\text{-}CF_3C_6H_4$	68

alcohol **6** with different alkynes. The results in Table 2 show that this process occurs with high efficiency and functional group tolerance; see Scheme 4.

3.2. DFT Characterization. In order to compare triazoles 7–**18** with compound **1** and to establish possible SSRI candidates, a theoretical study was carried out. The global reactivity indexes of studied compounds are presented in Table 3, electronic affinity (A), ionization potential (I), chemical potential (μ), chemical softness (S), and electrophilicity (ω). The obtained results are gathered in Table 3.

Ionization potential (I) is defined as the energy needed to remove an electron from a molecule. According to obtained results, the molecule **12** showed the highest ionization potential value, and the molecule **15** has the closest value to **1**, showing that electron-rich aromatic rings are necessary to mimic the properties of molecule **1**. On the other hand, the presence of electron attracting groups in the aromatics increases the value of I; the aliphatic moieties have the same effect.

Electron affinity (EA) measures the ability of a molecule to accept electrons and form anions. There are some EA negative in the obtained values, because the molecules released energy in the process of electron acceptance. In a general view, only the half of the molecules has positive EA values like one. Nevertheless, the computed values are different compared with **1**, with the molecules **16** with a behavior closest to **1**.

Chemical potential (μ) measures the escaping tendency of an electronic cloud in the ground state system. The obtained results shown that molecule **1** has higher value (less negative) than synthesized triazolyl derivatives, making compound **1** a better electron donor than obtained compounds. Nevertheless, almost all the compounds have pretty close values to **1**, except for compound **7** and in minor degree **12**.

Chemical softness (S) is related to polarizability. The values S for compounds **14** and **18** are remarkably close to the value of compound **1**, suggesting a similar facility to charge transfer; other compounds with similar values are **15**–**17**, respectively. As matter of fact, the trifluorobenzyl group in compound **18** shows similar characteristics to indolyl moiety in compound **1**.

Electrophilicity (ω) gives an idea of the stabilization energy when the system acquires electrons from the environment up to saturation. Since the electrophilicity power is computed through ($\mu^2/2\eta$), the differences in electrophilicity power (ω) between the compounds are mainly due to differences in their chemical potential. In this regard, the results suggest that triazole moiety is a better nucleophile for compounds 7–**18** than tetrahydropyridine system in compound **1**. In a general way all the parameters obtained by DFT calculations show the necessity of having an electron-rich aromatic bond to mimic the indolyl moiety in compound **1**.

TABLE 3: Global reactivity indexes in eV.

	A	I	μ	S	ω
1	0.113	6.546	−3.216	0.155	0.804
7	−1.042	7.717	−8.879	0.056	2.219
8	−0.162	7.593	−3.878	0.128	0.969
9	−0.129	7.250	−3.689	0.135	0.922
10	−0.105	7.510	−3.808	0.131	0.952
12	−0.158	8.362	−4.260	0.117	1.065
13	0.911	8.253	−3.671	0.136	0.917
14	0.347	7.312	−3.482	0.143	0.870
15	0.526	6.619	−3.044	0.164	0.761
16	0.271	7.397	−3.562	0.140	0.890
17	0.371	7.537	−3.583	0.139	0.895
18	0.568	7.688	−3.559	0.140	0.889

$\mu = 1/2(\varepsilon_I + \varepsilon_A)$; $\eta = \varepsilon_A - \varepsilon_I$; $S = 1/$; $\omega = \mu^2/2\eta$.

TABLE 4: Local reactivity indexes for selected atoms.

Compound		f^+	f^-	s^+	s^-
	N1	0.01	0.04	0.01	0.06
7	N2	0.02	**0.08**	0.03	0.12
	N3	0.02	0.04	0.03	0.06
	N1	0.00	0.02	0.00	0.07
8	N2	0.01	**0.03**	0.03	0.10
	N3	0.02	0.01	0.07	0.03
	N1	0.00	0.04	0.00	0.14
9	N2	0.02	**0.08**	0.07	0.29
	N3	0.01	0.04	0.03	0.14
	N1	0.01	0.04	0.03	0.14
10	N2	0.02	**0.08**	0.07	0.28
	N3	0.02	0.03	0.07	0.10
	N1	0.00	0.03	0.00	0.11
12	N2	0.01	**0.10**	0.03	0.37
	N3	0.01	0.04	0.03	0.14
	N1	0.01	0.03	0.04	0.12
13	N2	0.02	**0.06**	0.08	0.23
	N3	0.00	0.03	0.00	0.12
	N1	0.02	0.01	0.08	0.13
14	N2	0.05	**0.03**	0.22	0.27
	N3	0.03	0.02	0.13	0.13
	N1	0.01	0.03	0.04	0.11
15	N2	0.03	**0.05**	0.11	0.19
	N3	0.02	0.03	0.07	0.11
	N1	0.02	0.02	0.07	0.07
16	N2	0.02	**0.03**	0.07	0.11
	N3	0.03	0.04	0.11	0.15
	N1	0.00	0.04	0.00	0.15
17	N2	0.02	**0.07**	0.07	0.27
	N3	0.02	0.04	0.07	0.17
	N1	0.02	0.02	0.08	0.08
18	N2	0.02	**0.04**	0.08	0.16
	N3	0.03	0.03	0.12	0.12

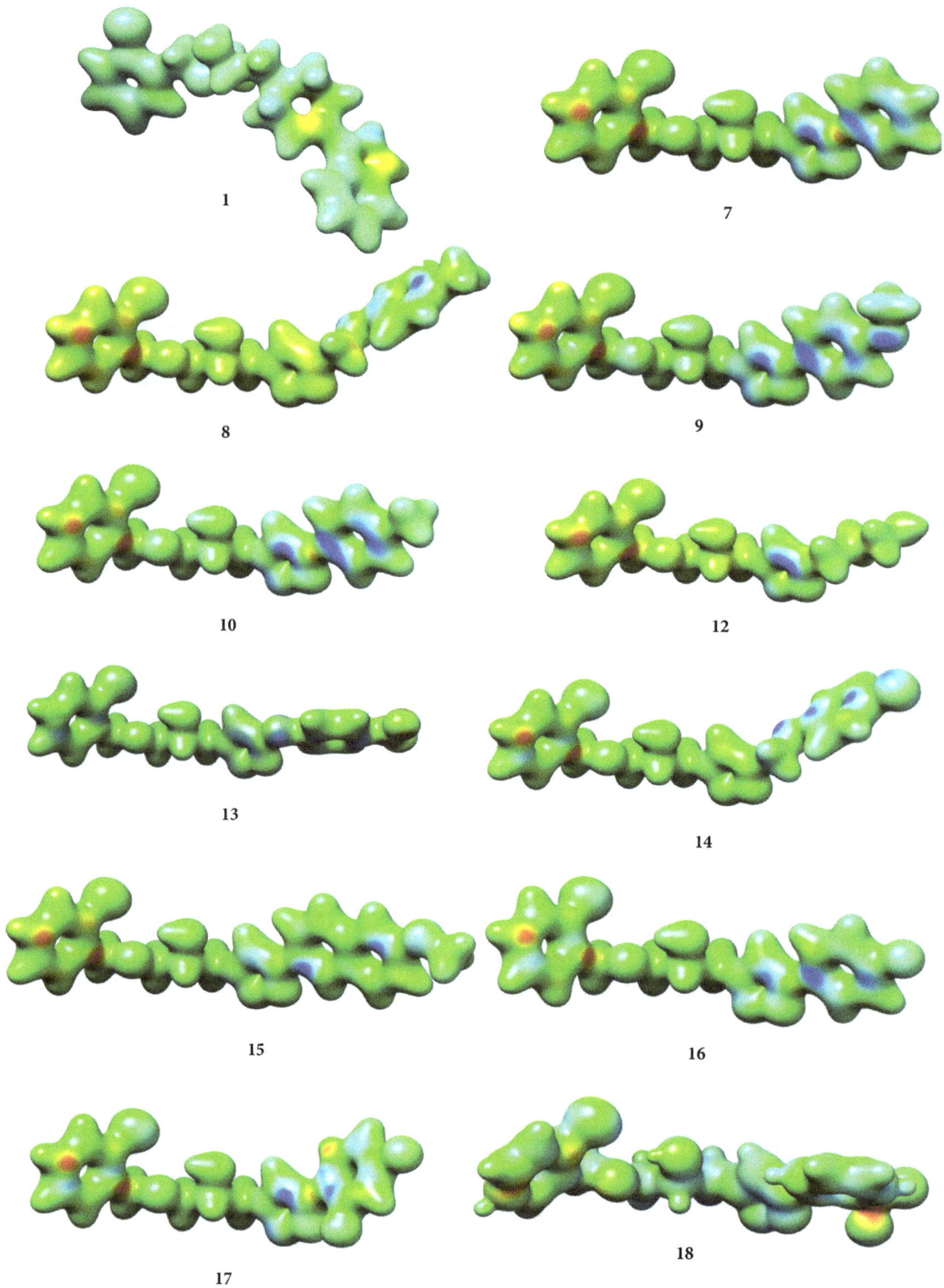

FIGURE 1: Dual descriptor $f^{\pm}$ for the molecules studied in this work.

The local properties of synthesized compounds are hard to analyze due to the diverse nature of substituent groups used in this study. For this reason, the three nitrogen atoms in triazole ring were used as a probe for the local properties in the prepared molecules. As shown in Table 4, the highest values of f^- can be observed for N2 atom in all molecules, indicating that this atom is the most susceptible to a nucleophilic attack. In addition, a visible tendency for the case of f^+ was not found. The local softness (s) followed the same tendencies as Fukui functions; nevertheless, this function could be comparable between molecules, being higher in compounds **14–18** on N2 atom.

Other important index is known as dual descriptor $f^{\pm}$ which contains information about the chemical philicity (electrophilicity/nucleophilicity). In this case, the graphic maps are generally sufficient for rationalizing the chemical reactivity. The maps of dual descriptor for the set of molecules studied in this work are represented in Figure 1, displaying the most probable sites for a possible nucleophilic-electrophilic attack. In a general way, the most electrophilic sites are found on the chlorobenzyl moiety, while the most nucleophilic sites are located on the opposite side of the molecule independent of the electronic nature of substituent group. These elements reveal some similarities between compound **1** and synthesized compounds with nearest values for compounds **14–18**. The calculated parameters also allow establishing interaction sites in these molecules.

4. Conclusions

A new class of derivatives of chlorobenzyloxy triazolyl-propan-2-ol was obtained from CuAAC reaction using commercially available raw materials through simple and efficient methods. Theoretical calculations indicate that a high similarity can be found between these products and the leading molecule **1**; consequently, a good SSRI ability would be expected for these compounds, suggesting more studies about this topic. Also, DFT descriptors show that the presence of an electron-rich aromatic ring is necessary to mimic the indolyl moiety present in compound **1**.

Conflicts of Interest

The authors declare that they have no conflicts of interest.

Acknowledgments

Financial support from CONACYT (Project no. 135053) is gratefully acknowledged. The authors would like to thank Signa S. A. de C. V. for some graciously donated solvents and reagents and to N. Zavala, A. Nuñez, and L. Triana for the technical support.

References

[1] M. Reddy, "Depression: the disorder and the burden," *Indian Journal of Psychological Medicine*, vol. 32, no. 1, pp. 1–2, 2010.

[2] M. Marcus, M. T. Yasamy, M. Ommeren, D. Chisholm, and S. Saxena, *Depression: A Global Public Health Concern*, World Health Organization, 2012, http://www.who.int/mental_health/management/depression/who_paper_depression_wfmh_2012.pdf.

[3] K. Sangkuhl, T. Klein, and R. Altman, "Selective Serotonin Reuptake Inhibitors (SSRI) pathway," *Pharmacogenetics Genomics*, vol. 19, no. 11, pp. 907–909, 2009.

[4] M. Vaswani, F. K. Linda, and S. Ramesh, "Role of selective serotonin reuptake inhibitors in psychiatric disorders: a comprehensive review," *Progress in Neuro-Psychopharmacology and Biological Psychiatry*, vol. 27, no. 1, pp. 85–102, 2003.

[5] D. J. Nutt, S. Forshall, C. Bell et al., "Mechanisms of action of selective serotonin reuptake inhibitors in the treatment of psychiatric disorders," *European Neuropsychopharmacology*, vol. 9, supplement 3, pp. S81–S86, 1999.

[6] O. M. Ghoneim, D. A. Ibrahim, I. M. El-Deeb, S. H. Lee, and R. G. Booth, "A novel potential therapeutic avenue for autism: design, synthesis and pharmacophore generation of SSRIs with dual action," *Bioorganic and Medicinal Chemistry Letters*, vol. 21, no. 22, pp. 6714–6723, 2011.

[7] J. I. Andrés, J. Alcázar, J. M. Alonso et al., "Synthesis of 7-amino-3a,4-dihydro-3H-[1]benzopyrano[4,3-c]isoxazole derivatives displaying combined α2-adrenoceptor antagonistic and 5-HT reuptake inhibiting activities," *Bioorganic and Medicinal Chemistry*, vol. 14, no. 13, pp. 4361–4372, 2006.

[8] J. M. Dorsey, M. G. Miranda, N. V. Cozzi, and K. G. Pinney, "Synthesis and biological evaluation of 2-(4-fluorophenoxy)-2-phenyl-ethyl piperazines as serotonin-selective reuptake inhibitors with a potentially improved adverse reaction profile," *Bioorganic and Medicinal Chemistry*, vol. 12, no. 6, pp. 1483–1491, 2004.

[9] T. L. Nolan, D. J. Lapinsky, J. N. Talbot et al., "Identification of a novel selective serotonin reuptake inhibitor by coupling monoamine transporter-based virtual screening and rational molecular hybridization," *ACS Chemical Neuroscience*, vol. 2, no. 9, pp. 544–552, 2011.

[10] J. Barluenga, F. Aznar, C. Valdés, and M.-P. Cabal, "Stereoselective synthesis of 4-piperidone and 4-aminotetrahydropyridine derivatives by the imino Diels-Alder reaction of 2-amino-1,3-butadienes," *Journal of Organic Chemistry*, vol. 58, no. 12, pp. 3391–3396, 1993.

[11] M. M. Faul, M. E. Kobierski, and M. E. Kopach, "Green chemistry approach to the synthesis of N-substituted piperidones," *Journal of Organic Chemistry*, vol. 68, no. 14, pp. 5739–5741, 2003.

[12] P. Geerlings, F. De Proft, and W. Langenaeker, "Conceptual density functional theory," *Chemical Reviews*, vol. 103, no. 5, pp. 1793–1873, 2003.

[13] F. Aparicio, N. González-Rivas, J. Ireta et al., "Soft-Soft interactions in the protein-protein recognition process: the K^+ channel-charybdotoxin case," *International Journal of Quantum Chemistry*, vol. 112, no. 22, pp. 3618–3623, 2012.

[14] E. Apra, T. L. Windus, T. P. Straatsma et al., *NWChem, A Computational Chemistry Package for Parallel Computers, Version 5.1*, Pacific Northwest National Laboratory, Richland, Wash, USA, 2005.

[15] C. Adamo and V. Barone, "Toward reliable density functional methods without adjustable parameters: the PBE0 model," *Journal of Chemical Physics*, vol. 110, no. 13, pp. 6158–6170, 1999.

[16] V. V. Rostovtsev, L. G. Green, V. V. Fokin, and K. B. Sharpless, "A stepwise huisgen cycloaddition process: copper(I)-catalyzed regioselective 'ligation' of azides and terminal alkynes," *Angewandte Chemie - International Edition*, vol. 41, no. 14, pp. 2596–2599, 2002.

[17] C. W. Tornøe, C. Christensen, and M. Meldal, "Peptidotriazoles on solid phase: [1,2,3]-Triazoles by regiospecific copper(I)-catalyzed 1,3-dipolar cycloadditions of terminal alkynes to azides," *Journal of Organic Chemistry*, vol. 67, no. 9, pp. 3057–3064, 2002.

[18] M. A. García, Z. G. Ríos, J. González et al., "The use of glucose as alternative reducing agent in copper-catalyzed alkyne-azide cycloaddition," *Letters in Organic Chemistry*, vol. 8, no. 10, pp. 701–706, 2011.

[19] J. González, V. M. Pérez, D. O. Jiménez, G. Lopez-Valdez, D. Corona, and E. Cuevas-Yañez, "Effect of temperature on triazole and bistriazole formation through copper-catalyzed alkyne-azide cycloaddition," *Tetrahedron Letters*, vol. 52, no. 27, pp. 3514–3517, 2011.

18

Synthesis, Molecular Docking Studies, and Antifungal Activity Evaluation of New Benzimidazole-Triazoles as Potential Lanosterol 14α-Demethylase Inhibitors

Nafiz Öncü Can,[1,2] Ulviye Acar Çevik,[2,3] Begüm Nurpelin Sağlık,[2,3] Serkan Levent,[2,3] Büşra Korkut,[4] Yusuf Özkay,[2,3] Zafer Asım Kaplancıklı,[2] and Ali Savaş Koparal[5]

[1]*Department of Analytical Chemistry, Faculty of Pharmacy, Anadolu University, 26470 Eskişehir, Turkey*
[2]*Department of Pharmaceutical Chemistry, Faculty of Pharmacy, Anadolu University, 26470 Eskişehir, Turkey*
[3]*Doping and Narcotic Compounds Analysis Laboratory, Faculty of Pharmacy, Anadolu University, 26470 Eskişehir, Turkey*
[4]*Department of Pharmaceutical Toxicology, Faculty of Pharmacy, Anadolu University, 26470 Eskişehir, Turkey*
[5]*Department of Environmental Engineering, Faculty of Engineering, Anadolu University, Eskişehir, Turkey*

Correspondence should be addressed to Yusuf Özkay; yozkay@anadolu.edu.tr

Academic Editor: Gabriel Navarrete-Vazquez

Due to anticandidal importance of azole compounds, a new series of benzimidazole-triazole derivatives (**5a–5s**) were designed and synthesized as ergosterol inhibitors. The chemical structures of the target compounds were characterized by spectroscopic methods. The final compounds were screened for antifungal activity against *Candida glabrata* (ATCC 90030), *Candida krusei* (ATCC 6258), *Candida parapsilosis* (ATCC 22019), and *Candida albicans* (ATCC 24433). Compounds **5i** and **5s** exhibited significant inhibitory activity against *Candida* strains with MIC_{50} values ranging from 0.78 to 1.56 μg/mL. Cytotoxicity results revealed that IC_{50} values of compounds **5i** and **5s** against NIH/3T3 are significantly higher than their MIC_{50} values. Effect of the compounds **5i** and **5s** against ergosterol biosynthesis was determined by LC-MS-MS analysis. Both compounds caused a significant decrease in the ergosterol level. The molecular docking studies were performed to investigate the interaction modes between the compounds and active site of lanosterol 14-α-demethylase (CYP51), which is as a target enzyme for anticandidal azoles. Theoretical ADME predictions were also calculated for final compounds.

1. Introduction

Recently, the incidence of systemic fungal infection has become an important complication and a major cause of morbidity and mortality in immune compromised individuals such as patients undergoing anticancer chemotherapy or organ transplants and patients with AIDS [1, 2]. According to the mechanism of action, there are six classes of antifungal agents: fungal ergosterol synthesis inhibitors (azoles: ketoconazole, fluconazole, and voriconazole), glucan synthesis inhibitors (echinocandins and caspofungin), ergosterol disruptors (polyenes antibiotics: amphotericin B), squalene epoxidase inhibitors (terbinafine and naftifine), chitin synthesis inhibitors (nikkomycin), and nucleic acid synthesis inhibitors (5-fluorocytosine) (Figure 1) [3]. Due to its high therapeutic index, azoles are first-line drugs for the treatment of invasive fungal infections [4, 5]. Most therapies, designed to treat fungal infections, target the ergosterol biosynthesis pathway or its end product [6].

In fungi, lanosterol 14-α-demethylase (CYP51) belongs to a superfamily of monooxygenases called cytochrome P450, which catalyzes the oxidative removal of the 14-methyl group (C-32) of lanosterol to give 14,15-desaturated intermediates in ergosterol biosynthesis. CYP51 is one of the key enzymes of sterol biosynthesis in different biological kingdoms and serves the metabolic function such as membrane permeability, membrane fluidity, enzyme activity, cell morphology, and cell cycle progression [7–9]. This enzyme is found in all

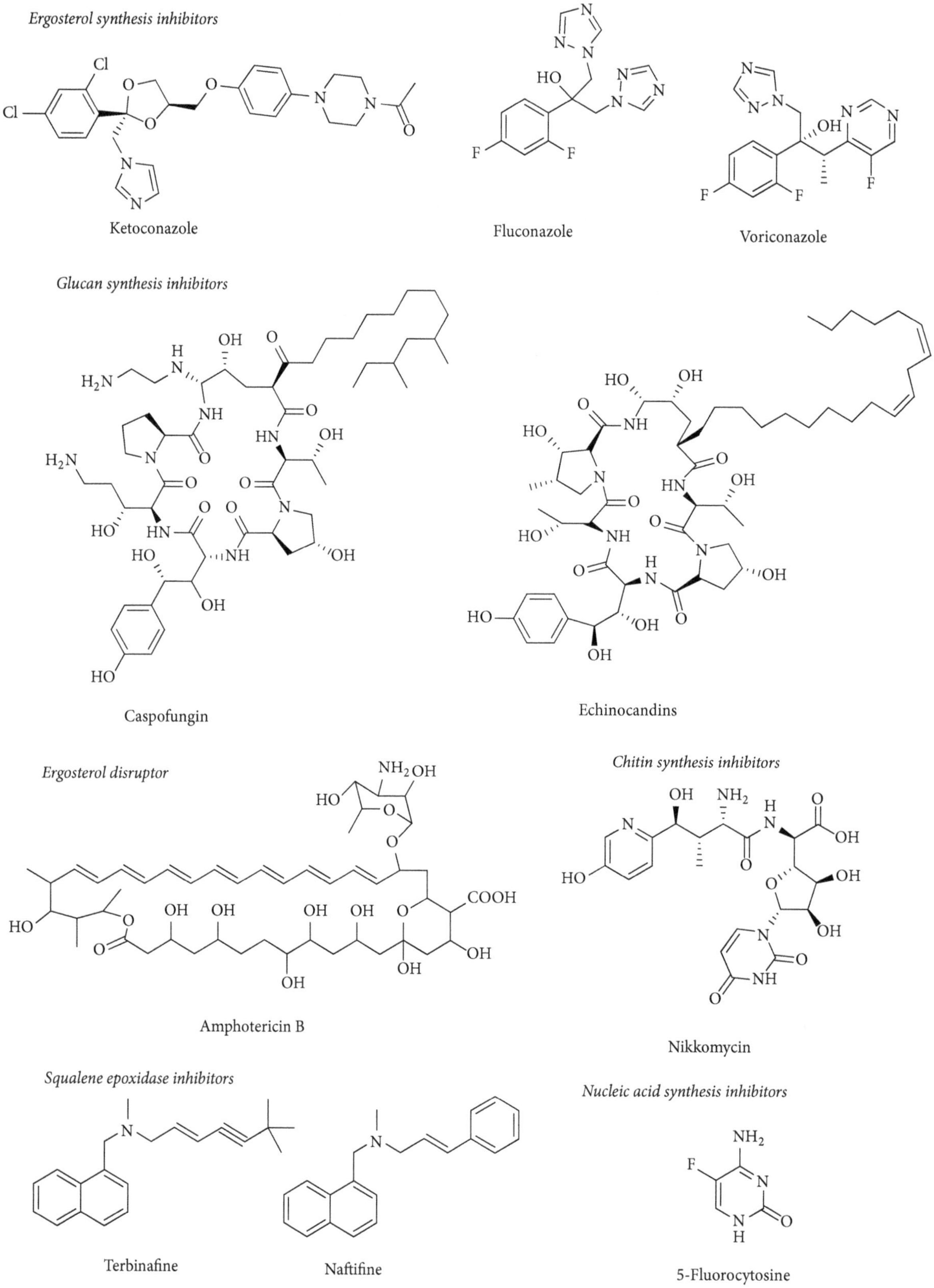

FIGURE 1: Chemical structures of some antifungal drugs which are in clinical use.

eukaryotes (including humans) and because the azoles interact also with other cytochrome P450-dependent enzymes (CYP3A4), a selective inhibition of the enzyme is essential for an increased therapeutic index [10–12].

In general, the active site of CYP51 is divided into four subsites: a coordination bond with iron of the heme group, the hydrophobic region, the hydrophilic H-bonding region, and the narrow hydrophobic cleft formed. The affinity of azole antifungals to the lanosterol 14a-demethylase is determined not only by the coordination binding of the nitrogen of azole ring to the heme iron in the active side (N-4 of triazole and N-3 of imidazole) but also by the affinity of N-1 substituent for the apoprotein part of the enzyme. The remaining part of the azole antifungal fits in the similar way like lanosterol in the hydrophobic groove of lanosterol 14-α-demethylase [13–17].

Based on these properties of the active site of CYP51, we are provided with an attempt to design new azole derivatives to find potent systemic antifungal agents that have a broad antifungal spectrum but with less potential to develop drug resistance. Several studies have been performed, and many of them have supported the finding that benzimidazole and triazole rings were significant structural moieties for antifungal activity [18–28]. In the present study, prompted from anticandidal potential of azole compounds, new benzimidazole-triazole hybrid compounds were synthesized and evaluated for antifungal activity.

2. Experimental

2.1. Chemistry. Entire chemicals used in the study were purchased from either Sigma-Aldrich (Sigma-Aldrich Corp., St. Louis, MO, USA) or Merck (Merck KGaA, Darmstadt, Germany) and used without further chemical or biological purification. Microwave syntheses were realized by using a Monowave 300 high-performance microwave reactor (Anton-Paar, Austria). Melting points of synthesized compounds were determined by using an automatic melting point determination system (MP90 series, Mettler-Toledo, OH, USA) and were presented as uncorrected. ^{1}H and ^{13}C NMR spectra were recorded in DMSO-d_6 by a Bruker digital FT-NMR spectrometer (Bruker Bioscience, MA, USA) at 300 MHz and 75 MHz, respectively. The IR spectra of the compounds were recorded using an IRAffinity-1S Fourier transform IR (FTIR) spectrometer (Shimadzu, Kyoto, Japan). High resolution mass spectrometric studies were performed using an LCMS-IT-TOF system (Shimadzu, Kyoto, Japan). Chemical purities of the compounds were checked by classical TLC applications performed on silica gel 60 F_{254} (Merck KGaA, Darmstadt, Germany); LCMS-IT-TOF chromatograms were also used for the same purpose.

*2.1.1. Methyl 4-(5(6)-Methyl-1H-benzimidazol-2-yl)benzoate **(1)**.* In a vial (30 mL) of microwave synthesis reactor (Anton-Paar, Monowave 300, Austria) equipped with a magnetic stirrer, a solution of methyl-4-formyl benzoate (2.4 g, 0.015 mol) in DMF (10 mL) and sodium disulfite (2.85 g, 0.015 mol) were heated under conditions of 240°C and 10 bar for 5 min. After this stage, 5-methyl-1,2-phenylenediamine (1.83 g, 0.015 mol) was added and then reaction mixture was kept under the same reaction conditions in microwave reactor. The mixture was cooled and poured into iced-water. The obtained solid was filtered, washed with water, dried, and recrystallized from ethanol to gain product with 86% yields.

*2.1.2. 4-(5(6)-Methyl-1H-benzimidazol-2-yl)benzohydrazide **(2)**.* Methyl 4-(5-methyl-*1H*-benzimidazol-2-yl)benzoate **(1)** (2.66 g, 0.01 mol) and hydrazine hydrate (5 mL) in ethanol (15 mL) were heated under the conditions of 150°C and 10 bar for 10 min in microwave synthesis reactor (Anton-Paar Monowave 300, Austria). The mixture was cooled and poured into iced-water. The obtained solid was filtered, washed with water, dried, and recrystallized from ethanol to gain product with 92% yields.

*2.1.3. N-Alkyl-2-[4-(5(6)-methyl-1H-benzimidazol-2-yl)benzoyl]hydrazine-1-carbothioamide **(3a, 3b)**.* 4-(5-Methyl-1*H*-benzimidazol-2-yl)benzohydrazide **(2)** (2.66 g, 0.01 mol) and methyl isothiocyanate or ethyl isothiocyanate (0.012 mol) in ethanol were refluxed for 2 h. The precipitated product was filtered, washed with ethanol, and dried. Compounds **3a** and **3b** were obtained with yields of 78% and 81%, respectively.

*2.1.4. 4-Alkyl-5-[4-(5(6)-methyl-1H-benzimidazol-2-yl)phenyl)]-4H-1,2,4-triazole-3-thiol **(4a-4b)**.* *N*-Alkyl-2-[4-(5(6)-methyl-*1H*-benzimidazol-2-yl)benzoyl]hydrazine-1-carbothioamide **(3a, 3b)** (0.001 mol) in ethanol was refluxed under stirring for 2 h in the presence of NaOH (0.012 mol). After completion of reaction, the solution was acidified with HCl 37%; the precipitate was filtered, washed with water, dried, and then recrystallized from ethanol. Compounds **4a** and **4b** were obtained with yields of 71% and 73%, respectively.

*2.1.5. 2-(4-(4-Alkyl-5-(2-(substitutedphenyl)-2-oxo-ethylthio)-4H-1,2,4-triazol-3yl)-phenyl)-5(6)-methyl-1H-benzimidazole **(5a–5s)**.* A solution of **4a** or **4b** (0.001 mol) in acetone (10 mL), an appropriate substituted 2-bromoacetophenone derivative (0.001 mol), and potassium carbonate (0.138 g, 0.001 mol) were refluxed at 40°C for 12 h. The solvent was evaporated; residue was washed with water, dried, and recrystallized from ethanol.

*2.1.6. 2-(4-(4-Methyl-5-(2-phenyl-2-oxo-ethylthio)-4H-1,2,4-triazol-3yl)-phenyl)-5(6)-methyl-1H-benzimidazole **(5a)**.* Yield: 81%. M.p. 247-248°C. FTIR (ATR, cm^{-1}): 3211 (N-H), 1676 (C=O), 848 (1,4-disubstituted benzene C-H). ^{1}H-NMR (300 MHz, DMSO-d_6): δ = 2.44 (3H, s, -CH_3), 3.72 (3H, s, -CH_3), 4.96 (2H, s, -CH_2-), 7.07 (1H, d, J = 8.19 Hz, Benzimidazole C-H), 7.42 (1H, s, Benzimidazole C-H), 7.51–7.60 (3H, m, Benzimidazole C-H, Monosubstituted Benzene C-H), 7.70 (1H, t, J = 7.88 Hz, Monosubstituted Benzene C-H), 7.89 (2H, d, J = 8.35 Hz, 1,4-disubstituted benzene C-H), 8.04 (2H, d, J = 7.88 Hz, Monosubstituted Benzene C-H), 8.32 (2H, d, J = 8.35 Hz, 1,4-disubstituted benzene C-H), 13.15 (1H, s, Benzimidazole -NH). ^{13}C-NMR (75 MHz, DMSO-d_6): δ (ppm): 21.81, 32.49, 41.27, 115.76, 124.54, 126.83, 127.18, 128.42, 128.93, 129.22, 129.31, 130.37, 131.62, 132.37, 134.26, 135.73, 150.37, 151.08, 155.30, 192.85. $[M+H]^+$ calcd for $C_{25}H_{21}N_5OS$: 440.1540; found: 440.1527.

2.1.7. 2-(4-(4-Methyl-5-(2-(4-cyanophenyl)-2-oxo-ethylthio)-4H-1,2,4-triazol-3yl)-phenyl)-5(6)-methyl-1H-benzoimidazole ***(5b)***. Yield: 83%. M.p. 239-240°C. FTIR (ATR, cm^{-1}): 3093 (N-H), 1680 (C=O), 849 (1,4-disubstituted benzene C-H). ^{1}H-NMR (300 MHz, DMSO-d_6): δ = 2.49 (3H, s, -CH_3), 3.74 (3H, s, -CH_3), 4.99 (2H, s, -CH_2-), 7.29 (1H, d, J = 8.28 Hz, Benzimidazole C-H), 7.57 (1H, s, Benzimidazole C-H), 7.67 (1H, d, J = 8.28 Hz, Benzimidazole C-H), 8.02 (2H, d, J = 8.45 Hz, 1,4-disubstituted benzene C-H), 8.06 (2H, d, J = 8.40, 4-cyanophenyl C-H), 8.19 (2H, d, J = 8.40, 4-cyanophenyl C-H), 8.36 (2H, d, J = 8.45 Hz, 1,4-disubstituted benzene C-H), 13.32 (1H, s, Benzimidazole -NH). ^{13}C-NMR (75 MHz, DMSO-d_6): δ (ppm): 21.73, 32.59, 41.13, 114.31, 114.84, 116.03, 118.56, 126.77, 127.59, 128.17, 129.47, 129.54, 130.16, 133.32, 135.06, 135.36, 139.04, 149.01, 151.26, 154.96, 193.43. $[M+H]^+$ calcd for $C_{26}H_{20}N_6OS$: 465.1492; found: 465.1485.

2.1.8. 2-(4-(4-Methyl-5-(2-(4-fluorophenyl)-2-oxo-ethylthio)-4H-1,2,4-triazol-3yl)-phenyl)-5(6)-methyl-1H-benzimidazole ***(5c)***. Yield: 77%. M.p. 255-256°C. FTIR (ATR, cm^{-1}): 2968 (N-H), 1681 (C=O), 843 (1,4-disubstituted benzene C-H). ^{1}H-NMR (300 MHz, DMSO-d_6): δ = 2.43 (3H, s, -CH_3), 3.71 (3H, s, -CH_3), 4.94 (2H, s, -CH_2-), 7.05 (1H, d, J = 8.13 Hz, Benzimidazole C-H), 7.37–7.43 (3H, m, 4-Fluorophenyl C-H, Benzimidazole C-H), 7.51 (1H, d, J = 8.13 Hz, Benzimidazole C-H), 7.88 (2H, d, J = 8.31 Hz, 1,4-disubstituted benzene C-H), 8.11–8.17 (2H, m, 4-Fluorophenyl C-H), 8.32 (2H, d, J = 8.31 Hz, 1,4-disubstituted benzene C-H), 12.11 (1H, s, Benzimidazole -NH). ^{13}C-NMR (75 MHz, DMSO-d_6): δ (ppm): 21.81, 32.48, 41.10, 116.36 (d, J = 21.8 Hz), 124.42, 127.14, 128.29, 129.19, 130.04, 131.38, 131.51, 131.85, 132.02 (d, J = 9.0 Hz), 132.22, 132.49, 132.51 (d, J = 3.0 Hz), 150.44, 150.99, 155.33, 165.79 (d, J = 251.3 Hz), 192.53. $[M+H]^+$ calcd for $C_{25}H_{20}FN_5OS$: 458.1445; found: 458.1443.

2.1.9. 2-(4-(4-Methyl-5-(2-(4-bromophenyl)-2-oxo-ethylthio)-4H-1,2,4-triazol-3yl)-phenyl)-5(6)-methyl-1H-benzimidazole ***(5d)***. Yield: 75%. M.p. 259-260°C. FTIR (ATR, cm^{-1}): 3201 (N-H), 1680 (C=O), 850 (1,4-disubstituted benzene C-H). ^{1}H-NMR (300 MHz, DMSO-d_6): δ = 2.43 (3H, s, -CH_3), 3.71 (3H, s, -CH_3), 4.92 (2H, s, -CH_2-), 7.05 (1H, s, Benzimidazole C-H), 7.34–7.56 (2H, m, Ar-H), 7.78–7.95 (6H, m, Ar-H), 8.31 (2H, s, Ar-H), 12.91 (1H, s, Benzimidazole -NH). ^{13}C-NMR (75 MHz, DMSO-d_6): δ (ppm): 21.81, 32.48, 41.10, 111.63, 119.13, 123.99, 124.85, 127.08, 128.39, 129.20, 130.92, 131.33, 131.96, 132.36, 132.78, 134.77, 142.46, 150.31, 150.90, 155.35, 193.23. $[M+H]^+$ calcd for $C_{25}H_{20}BrN_5OS$: 518.0645; found: 518.0635.

2.1.10. 2-(4-(4-Methyl-5-(2-(4-methylphenyl)-2-oxo-ethylthio)-4H-1,2,4-triazol-3yl)-phenyl)-5(6)-methyl-1H-benzimidazole ***(5e)***. Yield: 78%. M.p. 261-262°C. FTIR (ATR, cm^{-1}): 3197 (N-H), 1674 (C=O), 850 (1,4-disubstituted benzene C-H). ^{1}H-NMR (300 MHz, DMSO-d_6): δ = 2.39 (3H, s, -CH_3), 2.44 (3H, s, -CH_3), 3.70 (3H, s, -CH_3), 4.91 (2H, s, -CH_2-), 7.07 (1H, d, J = 8.13 Hz, Benzimidazole C-H), 7.37 (2H, d, J = 8.12, 4-methylphenyl C-H), 7.41 (1H, br.s, Benzimidazole C-H), 7.51 (1H, br.s, Benzimidazole C-H), 7.88 (2H, d, J = 8.37 Hz, 1,4-disubstituted benzene C-H), 7.93 (2H, d, J = 8.12, 4-methylphenyl C-H), 8.31 (2H, d, J = 8.37 Hz, 1,4-disubstituted benzene C-H), 12.92 (1H, s, Benzimidazole -NH). ^{13}C-NMR (75 MHz, DMSO-d_6): δ (ppm): 21.68, 21.82, 32.48, 41.25, 111.75, 119.21, 124.28, 127.09, 128.28, 129.05, 129.20, 129.84, 130.77, 131.94, 132.80, 133.22, 135.86, 144.79, 150.47, 151.06, 155.31, 193.39. $[M+H]^+$ calcd for $C_{26}H_{23}N_5OS$: 454.1696; found: 454.1683.

2.1.11. 2-(4-(4-Methyl-5-(2-(4-chlorophenyl)-2-oxo-ethylthio)-4H-1,2,4-triazol-3yl)-phenyl)-5(6)-methyl-1H-benzimidazole ***(5f)***. Yield: 76%. M.p. 226-227°C. FTIR (ATR, cm^{-1}): 3190 (N-H), 1680 (C=O), 850 (1,4-disubstituted benzene C-H). ^{1}H-NMR (300 MHz, DMSO-d_6): δ = 2.44 (3H, s, -CH_3), 3.71 (3H, s, -CH_3), 4.93 (2H, s, -CH_2-), 7.04 (1H, d, J = 8.55 Hz, Benzimidazole C-H), 7.45 (1H, br.s, Benzimidazole C-H), 7.57 (1H, br.s, Benzimidazole C-H), 7.64 (2H, d, J = 8.55, 4-chlorophenyl C-H), 7.88 (2H, d, J = 8.33 Hz, 1,4-disubstituted benzene C-H), 8.05 (2H, d, J = 8.55, 4-chlorophenyl C-H), 8.31 (2H, d, J = 8.33 Hz, 1,4-disubstituted benzene C-H), 12.93 (1H, s, Benzimidazole -NH). ^{13}C-NMR (75 MHz, DMSO-d_6): δ (ppm): 21.86, 32.49, 41.12, 111.62, 119.12, 123.99, 124.85, 127.07, 127.13, 129.20, 129.41, 130.86, 131.97, 132.78, 134.45, 139.15, 142.45, 144.67, 150.90, 155.35, 193.12. $[M+H]^+$ calcd for $C_{25}H_{20}ClN_5OS$: 474.1150; found: 474.1149.

2.1.12. 2-(4-(4-Methyl-5-(2-(2,4-dichlorophenyl)-2-oxo-ethylthio)-4H-1,2,4-triazol-3yl)-phenyl)-5(6)-methyl-1H-benzimidazole ***(5g)***. Yield: 74%. M.p. 159-160°C. FTIR (ATR, cm^{-1}): 3305 (N-H), 1685 (C=O), 854 (1,4-disubstituted benzene C-H). ^{1}H-NMR (300 MHz, DMSO-d_6): δ = 2.45 (3H, s, -CH_3), 3.68 (3H, s, -CH_3), 4.81 (2H, s, -CH_2-), 7.10 (1H, d, J = 8.22 Hz, Benzimidazole C-H), 7.43 (1H, s, Benzimidazole C-H), 7.54 (1H, d, J = 7.98 Hz, Benzimidazole C-H), 7.61 (1H, dd, J = 8.38 Hz–2.01 Hz Dichlorophenyl C-H), 7.78 (1H, d, J = 1.98 Hz, Dichloropheyl), 7.90 (3H, m, 1,4-disubstituted benzene C-H, Dichlorophenyl C-H), 8.33 (2H, d, J = 8.43 Hz, 1,4-disubstituted benzene C-H), 13.28 (1H, s, Benzimidazole -NH). ^{13}C-NMR (75 MHz, DMSO-d_6): δ (ppm): 21.81, 32.09, 43.05, 111.51, 119.19, 124.00, 124.88, 127.24, 128.05, 128.32, 129.25, 130.56, 131.98, 132.07, 132.11, 133.41, 134.14, 135.82, 137.27, 143.97, 150.22, 154.89, 194.94. $[M+H]^+$ calcd for $C_{25}H_{19}Cl_2N_5OS$: 508.0760; found: 508.0755.

2.1.13. 2-(4-(4-Methyl-5-(2-(2,4-difluorophenyl)-2-oxo-ethylthio)-4H-1,2,4-triazol-3yl)-phenyl)-5(6)-methyl-1H-benzimidazole ***(5h)***. Yield: 82%. M.p. 258-259°C. FTIR (ATR, cm^{-1}): 3196 (N-H), 1676 (C=O), 850 (1,4-disubstituted benzene C-H). ^{1}H-NMR (300 MHz, DMSO-d_6): δ = 2.44 (3H, s, -CH_3), 3.71 (3H, s, -CH_3), 4.83 (2H, s, -CH_2-), 7.05 (1H, d, J = 8.40 Hz, Benzimidazole C-H), 7.25–7.34 (2H, m, Ar-H), 7.43–7.58 (2H, m, Ar-H), 7.89 (2H, d, J = 8.33 Hz, 1,4-disubstituted benzene C-H), 7.98–8.06 (1H, m, Ar-H), 8.31 (2H, d, J = 8.33 Hz, 1,4-disubstituted benzene C-H), 12.92 (1H, s, Benzimidazole -NH). ^{13}C-NMR (75 MHz, DMSO-d_6): δ (ppm): 21.86, 32.45, 44.29, 105.79 (t, J = 26.81 Hz), 111.62, 113.03 (dd, J = 24.98 Hz–3.31 Hz), 119.12, 123.99, 124.85, 127.07, 128.22, 129.21, 131.96, 132.77, 133.38 (dd, J = 10.97 Hz–3.83 Hz), 135.80, 142.45,

144.67, 150.94, 155.34, 164.0, 164.24, 190.38 (d, J = 3.76 Hz). $[M+H]^+$ calcd for $C_{25}H_{19}F_2N_5OS$: 476.1351; found: 476.1348.

*2.1.14. 2-(4-(4-Methyl-5-(2-(3,4-dihydroxyphenyl)-2-oxo-ethylthio)-4H-1,2,4-triazol-3yl)-phenyl)-5(6)-methyl-1H-benzimidazole **(5i)**.* Yield: 79%. M.p. 249-250°C. FTIR (ATR, cm^{-1}): 3041 (N-H), 1681 (C=O), 854 (1,4-disubstituted benzene C-H). ^{1}H-NMR (300 MHz, DMSO-d_6): δ = 2.43 (3H, s, -CH_3), 3.70 (3H, s, -CH_3), 4.81 (2H, s, -CH_2-), 6.85 (1H, d, J = 8.25 Hz, Dihydroxyphenyl C-H), 7.05 (1H, d, J = 7.95 Hz, Benzimidazole C-H), 7.39–7.46 (4H, m, Dihydroxyphenyl C-H, Benzimidazole C-H), 7.88 (2H, d, J = 8.45 Hz, 1,4-disubstituted benzene C-H), 8.31 (2H, d, J = 8.45 Hz, 1,4-disubstituted benzene C-H), 12.86 (1H, s, Benzimidazole -NH). ^{13}C-NMR (75 MHz, DMSO-d_6): δ (ppm): 21.81, 32.45, 41.07, 111.53, 115.63, 119.09, 119.25, 122.52, 126.26, 127.09, 127.47, 128.31, 129.04, 129.20, 130.44, 131.92, 134.68, 138.29, 145.85, 151.26, 151.91, 155.28, 191.88. $[M+H]^+$ calcd for $C_{25}H_{21}N_5O_3S$: 472.1438; found: 472.1439.

*2.1.15. 2-(4-(4-Ethyl-5-(2-phenyl-2-oxo-ethylthio)-4H-1,2,4-triazol-3yl)-phenyl)-5(6)-methyl-1H-benzimidazole **(5j)**.* Yield: 83%. M.p. 235-236°C. FTIR (ATR, cm^{-1}): 3255 (N-H), 1680 (C=O), 848 (1,4-disubstituted benzene C-H). ^{1}H-NMR (300 MHz, DMSO-d_6): δ = 1.29 (3H, t, J = 7.14, -CH_3), 2.50 (3H, s, -CH_3), 4.15 (2H, q, J = 7.14 Hz, -CH_2), 5.04 (2H, s, -CH_2-), 7.33 (1H, d, J = 8.31 Hz, Benzimidazole C-H), 7.55–7.60 (3H, m, Benzimidazole C-H, Monosubstituted Benzene C-H), 7.69–7.72 (2H, m, Benzimidazole C-H, Monosubstituted Benzene C-H), 7.99 (2H, d, J = 8.24 Hz, 1,4-disubstituted benzene C-H), 8.06 (2H, d, J = 7.44 Hz, Monosubstituted Benzene C-H), 8.39 (2H, d, J = 8.24 Hz, 1,4-disubstituted benzene C-H), 13.24 (1H, s, Benzimidazole -NH). ^{13}C-NMR (75 MHz, DMSO-d6): δ (ppm): 15.49, 21.72, 40.80, 41.21, 114.19, 114.65, 126.88, 127.21, 128.52, 128.91, 129.33, 129.57, 130.75, 132.69, 134.28, 134.52, 135.59, 135.76, 148.69, 151.09, 154.27, 193.63. $[M+H]^+$ calcd for $C_{26}H_{23}N_5OS$: 454.1696; found: 454.1683.

*2.1.16. 2-(4-(4-Ethyl-5-(2-(4-cyanophenyl)-2-oxo-ethylthio)-4H-1,2,4-triazol-3yl)-phenyl)-5(6)-methyl-1H-benzimidazole **(5k)**.* Yield: 80%. M.p. 181-182°C. FTIR (ATR, cm^{-1}): 2987 (N-H), 1676 (C=O), 849 (1,4-disubstituted benzene C-H). ^{1}H-NMR (300 MHz, DMSO-d_6): δ = 1.28 (3H, t, J = 7.20, -CH_3), 2.44 (3H, s, -CH_3), 4.12 (2H, q, J = 7.16 Hz, -CH_2), 5.04 (2H, s, -CH_2-), 7.06 (1H, d, J = 8.28 Hz, Benzimidazole C-H), 7.41 (1H, s, Benzimidazole C-H), 7.52 (1H, d, J = 8.28 Hz, Benzimidazole C-H), 7.82 (2H, d, J = 8.49 Hz, 1,4-disubstituted benzene C-H), 8.07 (2H, d, J = 8.55, 4-cyanophenyl C-H), 8.20 (2H, d, J = 8.55, 4-cyanophenyl C-H), 8.31 (2H, d, J = 8.46 Hz, 1,4-disubstituted benzene C-H), 12.96 (1H, s, Benzimidazole -NH). ^{13}C-NMR (75 MHz, DMSO-d_6): δ (ppm): 15.52, 21.83, 41.13, 116.03, 118.57, 121.93, 123.66, 127.25, 128.42, 129.25, 129.54, 131.74, 132.03, 133.34, 134.03, 137.00, 139.10, 142.84, 150.28, 150.43, 154.86, 193.40. $[M+H]^+$ calcd for $C_{27}H_{22}N_6OS$: 479.1649; found: 479.1640.

*2.1.17. 2-(4-(4-Ethyl-5-(2-(4-fluorophenyl)-2-oxo-ethylthio)-4H-1,2,4-triazol-3yl)-phenyl)-5(6)-methyl-1H-benzimidazole **(5l)**.* Yield: 78%. M.p. 241-242°C. FTIR (ATR, cm^{-1}): 3142 (N-H), 1683 (C=O), 850 (1,4-disubstituted benzene C-H). ^{1}H-NMR (300 MHz, DMSO-d_6): δ = 1.28 (3H, t, J = 7.13, -CH_3), 2.43 (3H, s, -CH_3), 4.12 (2H, q, J = 7.13 Hz, -CH_2), 5.00 (2H, s, -CH_2-), 7.05 (1H, d, J = 8.28 Hz, Benzimidazole C-H), 7.38–7.43 (3H, m, Fluorophenyl C-H, Benzimidazole C-H), 7.56 (1H, d, J = 8.28 Hz, Benzimidazole C-H), 7.82 (2H, d, J = 8.16 Hz, 1,4-disubstituted benzene C-H), 8.12–8.17 (2H, m, Fluorophenyl C-H), 8.33 (2H, d, J = 8.16 Hz, 1,4-disubstituted benzene C-H), 13.05 (1H, s, Benzimidazole -NH). ^{13}C-NMR (75 MHz, DMSO-d_6): δ (ppm): 15.53, 21.80, 40.80, 41.11, 116.37 (d, J = 21.9 Hz), 119.10, 119.22, 123.96, 124.82, 128.37, 129.22, 131.29, 132.02 (d, J = 9.5 Hz), 132.55 (d, J = 3.0 Hz), 135.85, 142.44, 144.65, 150.28, 150.46, 154.83, 165.78 (d, J = 251.1 Hz), 192.41. $[M+H]^+$ calcd for $C_{26}H_{22}FN_5OS$: 472.1602; found: 472.1604.

*2.1.18. 2-(4-(4-Ethyl-5-(2-(4-bromophenyl)-2-oxo-ethylthio)-4H-1,2,4-triazol-3yl)-phenyl)-5(6)-methyl-1H-benzimidazole **(5m)**.* Yield: 81%. M.p. 186-187°C. FTIR (ATR, cm^{-1}): 3122 (N-H), 1680 (C=O), 850 (1,4-disubstituted benzen). ^{1}H-NMR (300 MHz, DMSO-d_6): δ = 1.28 (3H, t, J = 7.16, -CH_3), 2.44 (3H, s, -CH_3), 4.11 (2H, q, J = 7.16 Hz, -CH_2), 4.99 (2H, s, -CH_2-), 7.06 (1H, d, J = 8.18 Hz, Benzimidazole C-H), 7.41 (1H, s, Benzimidazole C-H), 7.52 (1H, d, J = 8.18 Hz, Benzimidazole C-H), 7.78–7.84 (4H, m, 1,4-disubstituted benzene, 4-bromophenyl), 7.98 (2H, d, J = 8.55 Hz, 4-bromophenyl), 8.32 (2H, d, J = 8.37 Hz, 1,4-disubstituted benzene C-H), 12.97 (1H, s, Benzimidazole -NH). ^{13}C-NMR (75 MHz, DMSO-d_6): δ (ppm): 15.53, 21.81, 40.79, 41.06, 112.35, 120.74, 124.48, 127.05, 127.28, 128.39, 128.49, 128.74, 129.26, 130.92, 131.44, 131.89, 132.38, 133.18, 134.81, 140.65, 150.38, 150.41, 154.81, 193.10. $[M+H]^+$ calcd for $C_{26}H_{22}BrN_5OS$: 532.0801; found: 532.0801.

*2.1.19. 2-(4-(4-Ethyl-5-(2-(4-methylphenyl)-2-oxo-ethylthio)-4H-1,2,4-triazol-3yl)-phenyl)-5(6)-methyl-1H-benzimidazole **(5n)**.* Yield: 76%. M.p. 249-250°C. FTIR (ATR, cm^{-1}): 3163 (N-H), 1678 (C=O), 852 (1,4-disubstituted benzen). ^{1}H-NMR (300 MHz, DMSO-d_6): δ = 1.27 (3H, t, J = 7.19, -CH_3), 2.39 (3H, s, -CH_3), 2.43 (3H, s, -CH_3), 4.11 (2H, q, J = 7.17 Hz, -CH_2), 4.98 (2H, s, -CH_2-), 7.06 (1H, d, J = 8.22 Hz, Benzimidazole C-H), 7.37 (2H, d, J = 8.09, 4-methylphenyl C-H), 7.41 (1H, s, Benzimidazole C-H), 7.52 (1H, d, J = 8.22 Hz, Benzimidazole C-H), 7.82 (2H, d, J = 8.37 Hz, 1,4-disubstituted benzene C-H), 7.94 (2H, d, J = 8.09, 4-methylphenyl C-H), 8.32 (2H, d, J = 8.37 Hz, 1,4-disubstituted benzene C-H), 12.71 (1H, s, Benzimidazole -NH). ^{13}C-NMR (75 MHz, DMSO-d_6): δ (ppm): 15.53, 21.68, 21.81, 36.24, 41.22, 114.96, 115.51, 115.96, 124.43, 127.26, 128.50, 129.03, 129.24, 129.84, 130.84, 131.95, 132.24, 133.26, 144.78, 150.42, 150.58, 154.78, 193.21. $[M+H]^+$ calcd for $C_{27}H_{25}N_5OS$: 468.1853; found: 468.1856.

*2.1.20. 2-(4-(4-Ethyl-5-(2-(4-chlorophenyl)-2-oxo-ethylthio)-4H-1,2,4-triazol-3yl)-phenyl)-5(6)-methyl-1H-benzimidazole **(5o)**.* Yield: 79%. M.p. 237-238°C. FTIR (ATR, cm^{-1}): 3045 (N-H), 1695 (C=O), 847 (1,4-disubstituted benzene C-H). ^{1}H-NMR (300 MHz, DMSO-d_6): δ = 1.27 (3H, t, J = 7.10,

$-CH_3$), 2.44 (3H, s, $-CH_3$), 4.11 (2H, br.s, $-CH_2$), 5.00 (2H, s, $-CH_2-$), 7.06 (1H, d, J = 8.01 Hz, Benzimidazole C-H), 7.41 (1H, s, Benzimidazole C-H), 7.51 (1H, d, J = 6.87 Hz, Benzimidazole C-H), 7.65 (2H, d, J = 8.62, 4-chlorophenyl C-H), 7.82 (2H, d, J = 8.35 Hz, 1,4-disubstituted benzene C-H), 8.06 (2H, d, J = 8.62, 4-chlorophenyl C-H), 8.31 (2H, d, J = 8.35 Hz, 1,4-disubstituted benzene C-H), 12.93 (1H, s, Benzimidazole -NH). ^{13}C-NMR (75 MHz, DMSO-d_6): δ (ppm): 15.52, 21.81, 40.80, 41.08, 115.03, 124.41, 124.60, 127.24, 128.43, 129.25, 129.43, 130.85, 131.10, 132.03, 132.27, 134.50, 139.16, 142.37, 150.42, 153.03, 154.83, 192.88. $[M+H]^+$ calcd for $C_{26}H_{22}ClN_5OS$: 488.1306; found: 488.1304.

*2.1.21. 2-(4-(4-Ethyl-5-(2-(2,4-dichlorophenyl)-2-oxo-ethylthio)-4H-1,2,4-triazol-3yl)-phenyl)-5(6)-methyl-1H-benzimidazole (**5p**).* Yield: 75%. M.p. 211-212°C. FTIR (ATR, cm^{-1}): 2920 (N-H), 1670 (C=O), 864 (1,4-disubstituted benzene C-H). ^{1}H-NMR (300 MHz, DMSO-d_6): δ = 1.25 (3H, t, J = 7.19, -CH3), 2.44 (3H, s, $-CH_3$), 4.07 (2H, q, J = 7.10 Hz, $-CH_2$), 4.85 (2H, s, $-CH_2-$), 7.04 (1H, d, J = 8.16 Hz, Benzimidazole C-H), 7.40 (2H, br.s, Benzimidazole C-H), 7.60 (1H, dd, J = 8.37 Hz–1.98 Hz Dichlorophenyl C-H), 7.76 (1H, d, J = 1.92 Hz, Dichlorophenyl C-H), 7.82 (2H, d, J = 8.36 Hz, 1,4-disubstituted benzene C-H), 7.89 (1H, d, J = 8.40 Hz, Dichlorophenyl C-H), 8.32 (2H, d, J = 8.36 Hz, 1,4-disubstituted benzene C-H), 12.91 (1H, s, Benzimidazole -NH). ^{13}C-NMR (75 MHz, DMSO-d_6): δ (ppm): 15.46, 21.81, 43.05, 111.51, 119.19, 120.61, 122.55, 124.00, 127.24, 128.05, 128.32, 129.25, 130.56, 131.98, 132.07, 132.11, 132.80, 135.82, 137.27, 142.48, 150.22, 154.89, 194.94. $[M+H]^+$ calcd for $C_{26}H_{21}Cl_2NOS$: 523.0917; found: 523.0912.

*2.1.22. 2-(4-(4-Ethyl-5-(2-(2,4-difluorophenyl)-2-oxo-ethylthio)-4H-1,2,4-triazol-3yl)-phenyl)-5(6)-methyl-1H-benzimidazole (**5o**).* Yield: 73%. M.p. 239-240°C. FTIR (ATR, cm^{-1}): 3045 (N-H), 1699 (C=O), 847 (1,4-disubstituted benzene C-H). ^{1}H-NMR (300 MHz, DMSO-d_6): δ = 1.28 (3H, t, J = 7.20, $-CH_3$), 2.44 (3H, s, $-CH_3$), 4.11 (2H, q, J = 7.14 Hz, $-CH_2$), 4.90 (2H, s, $-CH_2-$), 7.06 (1H, d, J = 7.20 Hz, Benzimidazole C-H), 7.29 (1H, td, J = 8.44 Hz–2.32 Hz, Difluorophenyl C-H), 7.41 (1H, s, Benzimidazole C-H), 7.41–7.54 (2H, m, Benzimidazole C-H, Difluorophenyl C-H), 7.83 (2H, d, J = 8.42 Hz, 1,4-disubstituted benzene C-H), 7.99–8.07 (1H, m, Difluorophenyl C-H), 8.31 (2H, d, J = 8.42 Hz, 1,4-disubstituted benzene C-H), 12.96 (1H, s, Benzimidazole -NH). ^{13}C-NMR (75 MHz, DMSO-d_6): δ (ppm): 15.50, 21.81, 44.20, 44.30, 105.81 (t, J = 26.8 Hz), 113.04 (dd, J = 21.6 Hz–3.6 Hz), 123.38, 124.40, 127.05, 127.24, 128.42, 128.74 129.26, 132.02, 132.22, 133.37 (dd, J = 11.1 Hz–3.6 Hz), 135.80, 137.99, 142.60, 150.40, 154.84, 163.09 (dd, J = 234.1 Hz–11.7 Hz), 165.86 (dd, J = 255.0 Hz–12.8 Hz), 190.30 (d, J = 4.1 Hz). $[M+H]^+$ calcd for $C_{26}H_{21}F_2N_5OS$: 490.1505; found: 490.1494.

*2.1.23. 2-(4-(4-Ethyl-5-(2-(3,4-dihydroxyphenyl)-2-oxo-ethylthio)-4H-1,2,4-triazol-3yl)-phenyl)-5(6)-methyl-1H-benzimidazole (**5s**).* Yield: 84%. M.p. 252-253°C. FTIR (ATR, cm^{-1}): 3300 (N-H), 1683 (C=O), 852 (1,4-disubstituted benzen). ^{1}H-NMR (300 MHz, DMSO-d_6): δ = 1.27 (3H, t, J = 7.17, $-CH_3$), 2.44 (3H, s, $-CH_3$), 4.11 (2H, q, J = 7.10 Hz, $-CH_2$), 4.88 (2H, s, $-CH_2-$), 6.86 (1H, d, J = 8.25 Hz, Dihydroxyphenyl C-H), 7.05 (1H, d, J = 7.23 Hz, Benzimidazole C-H), 7.35–7.55 (4H, m, Dihydroxyphenyl C-H, Benzimidazole C-H), 7.82 (2H, d, J = 8.40 Hz, 1,4-disubstituted benzene C-H), 8.31 (2H, d, J = 8.40 Hz, 1,4-disubstituted benzene C-H), 9.47 (1H, s, -OH), 10.04 (1H, s, -OH), 12.92 (1H, s, Benzimidazole -NH). ^{13}C-NMR (75 MHz, DMSO-d_6): δ (ppm): 15.53, 21.81, 36.25, 41.07, 111.53, 115.66, 119.10, 119.27, 122.49, 124.03, 124.82, 127.23, 127.54, 128.48, 129.25, 132.04, 132.83, 133.57, 144.63, 145.83, 150.76, 151.85, 154.76, 191.74. $[M+H]^+$ calcd for $C_{26}H_{23}N_5O_3S$: 486.1594; found: 486.1592.

2.2. Antifungal Activity. The antifungal activity of synthesized compounds **5a–5s** against *Candida* species was performed according to EUCAST definitive method EDef 7.1 [29]. Briefly, cultures of four fungi (*C. glabrata* (ATCC 90030), *C. krusei* (ATCC 6258), *C. parapsilosis* (ATCC 22019), and *C. albicans* (ATCC 24433)) were grown in RPMI medium, after an overnight incubation at 37°C. Fungi suspension was adjusted to approximately $0.5–2.5 \times 10^5$ cfu/mL with RPMI medium. The test was carried out for medium at pH = 7 and twofold dilution was carried out in a 96-well plate to obtain the series dilutions with the concentrations ranging from 0.78 to 800 μg/mL. The last well on the microplates, which was containing only the inoculated broth, was kept as control, and the last well with no growth of microorganism was recorded to represent the minimum inhibitory concentration (MIC_{50}) in μg/mL. The 96-well plates were incubated for 24 h, and, at the end of the incubation, resazurin (20 μg/mL) was added to each well to control the growth in the wells. Final plates including microorganism strains were incubated for 2 h. MIC_{50} values were determined using microplate reader at 590 nm excitation and 560 nm emission wavelengths; MIC_{50} readings were performed twice for entire compounds. Ketoconazole and fluconazole were used as controls.

2.3. Quantification of Ergosterol Level. Total intracellular sterols were extracted as described previously [30]. Sabouraud dextrose broth (50 mL, Difco), enclosing 0, 0.78, 1.56, and 3.125 μg/mL of compounds **5i**, **5s**, and positive controls were inoculated with a single *C. albicans* colony from an overnight Sabouraud dextrose agar plate culture. After incubation for 16 h at 35°C, the stationary-phase cells were separated by a centrifugation at 2,700 rpm (Hettich, Rotina 380 R, Germany) for 5 min. Isolated cells were washed with sterile distilled water; alcoholic potassium hydroxide solution (25%, 3 mL) was added to each pellet and vortexed for 1 min. Cell suspensions in sterile borosilicate glass tubes were incubated in a water bath (85°C) for 1 h. After cooling to room temperature, sterols were extracted in the water : chloroform (1 : 3 mL) mixture by a vigorous vortex mixing for 3 min. The chloroform layer was separated and sterol extract (1 μL) was injected to LC-MSMS system (Shimadzu LCMS 8040, Kyoto, Japan). The mass spectrometric analysis conditions were provided as reported in our recent study [31]. Level of ergosterol in negative control samples was considered as 100%. All samples, including different concentrations of compounds (**5i** and **5s**) and reference agents, were analyzed in quadruplicate, and the data were stated as mean ± standard deviation (SD).

2.4. Cytotoxicity Test. NIH/3T3 mouse embryonic fibroblast cell line (ATCC CRL 1658) was used in the cytotoxicity test. The cells were incubated as reported in the supplier's guide. The MTT assay was applied according to our previous studies [32–34]. The IC_{50} values of compounds **5i** and **5s** were calculated from the plots of cell proliferation against concentrations by applying regression analyses on GraphPad Prism Version 5.

2.5. Prediction of ADME Parameters. Physicochemical parameters of synthesized compounds (**5a–5s**) were calculated by using QikProp 4.8. [35].

2.6. Molecular Docking. A structure based in silico procedure was applied to discover the binding modes of most active compounds **5i** and **5s** to 14 alpha-sterol demethylase enzyme active sites. The crystal structure of enzyme (PDB ID: 1EA1) [36], which was crystallized with the reference drug (fluconazole) of antifungal activity assay, was retrieved from the Protein Data Bank server (http://www.pdb.org).

The structure of ligand was built using the Schrödinger Maestro [37] interface and then was submitted to the Protein Preparation Wizard protocol of the Schrödinger Suite 2016 Update 2 [38]. The ligands were prepared by the LigPrep 3.8 [39] to assign the protonation states at pH 7.4 ± 1.0 and the atom types, correctly. Bond orders were assigned and hydrogen atoms were added to the structures. The grid generation was formed using Glide 7.1 [40] program and docking runs were performed with single precision docking mode (SP).

3. Result and Discussion

3.1. Chemistry. In this study, we synthesized benzimidazole-triazole derivatives (**5a–5s**) in five steps. The synthetic path of the target compounds was illustrated in Scheme 1.

The structures of the new compounds (**5a–5s**) were confirmed by FT-IR, ^{1}H NMR, ^{13}C NMR spectroscopy, and mass spectra. The spectroscopic investigations of newly synthesized compounds are accordance with the proposed structure. The IR spectra of the benzimidazole-triazole derivatives (**5a–5s**) exhibited N-H stretching absorption in the region of 3305–2920 cm^{-1} and C=O stretching vibration in the region of 1699–1670 cm^{-1}. The out-of-plane bending of 1,4-disubstituted benzene was assigned at 864–843 cm^{-1}.

^{1}H NMR spectrum showed a broad singlet at 13.32–12.11 ppm due to NH proton of the benzimidazole ring. The aromatic protons belonging to 4-substituted phenyl gave peaks at 8.02–7.82 and 8.31–8.39 as two doublets. Benzimidazole protons were recorded at 7.04–7.67 ppm.

In the ^{13}C NMR spectra carbon of C=O group was assigned at 190.30–194.94 ppm. In the aromatic region, the peaks were seen at estimated areas but the assignments could not be determined, clearly. Chemical shifting of the substituted group and coupling constant of the fluorinated derivatives had coherent results with the theoretical values. HRMS findings were in accordance with the theoretical molecular formula of the compounds (**5a–5s**).

3.2. Antifungal Activity. The *in vitro* antifungal screening for all the synthesized compounds was evaluated against *Candida glabrata* (ATCC 90030), *Candida krusei* (ATCC 6258), *Candida parapsilosis* (ATCC 22019), and *Candida albicans* (ATCC 24433) using twofold serial dilution technique recommended by EUCAST definitive (EDef 7.1) method [29] with the positive control of clinical antifungal drugs ketoconazole and fluconazole. The MIC_{50} value of the compounds and control drugs are summarized in Table 1.

According to antifungal screening, most of the newly synthesized compounds could effectively inhibit the growth of all the tested fungal strains. All of the final compounds exhibited moderate to good antifungal activities with MIC_{50} values ranging from 12.5 to 0.78 μg/mL. Compounds **5i** and **5s**, bearing 3,4-dihydroxyphenyl group, exerted the best activities in inhibiting the growth of all tested fungal strains.

The microbiological results revealed that the antifungal effects of the compounds on *Candida* species did not depend on N-4 position of triazole. The results show that general structure of the synthesized compounds has an impact on antifungal activity and this effect is further enhanced by the 3,4-dihydroxy structure.

3.3. Inhibition of Ergosterol Biosynthesis. Infections caused by eukaryotic organisms like fungus generally present more difficult treatment problems than those of bacterial infections. There are relatively few antifungal drugs that can identify unique targets not shared with human hosts. The fungal cell wall remains an underdeveloped therapeutic target for selective antifungal agents because of its chitin structure, which is absent in human cells [41, 42]. Most therapies are planned to treat fungal infections and target the ergosterol biosynthesis pathway or its end product, ergosterol, a membrane sterol that is unique to fungi. It is the main sterol and thus is essential for growth and normal membrane function of fungal cell. Besides serving as a bioregulator of membrane fluidity, asymmetry, and integrity, it contributes to proper function of membrane-bound enzymes [43].

The present work is a challenge to understand the probable mechanism of antifungal activity of newly synthesized benzimidazole-triazole hybrid compounds **5i** and **5s**. Therefore, we analyzed ergosterol level of *C. albicans* by LC-MSMS studies. Total intracellular sterols were extracted as reported by Breivik and Owades [30]. Ergosterol standard (product number 45480, Sigma-Aldrich, Germany) was used for quantification of ergosterol in both inhibitor-free (negative control) and inhibitor including samples. The most active compounds **5i**, **5s** and reference drugs were used at 0.78 μg/mL, 1.56 μg/mL, and 3.12 μg/mL concentrations. Ergosterol quantity in negative control samples was regarded as 100%. All concentrations were analyzed in quadruplicate, and the results were expressed as mean ± standard deviation (SD) [31] (Figure 2).

Ergosterol quantification studies indicated that compounds **5i**, **5s** and reference agents significantly reduced

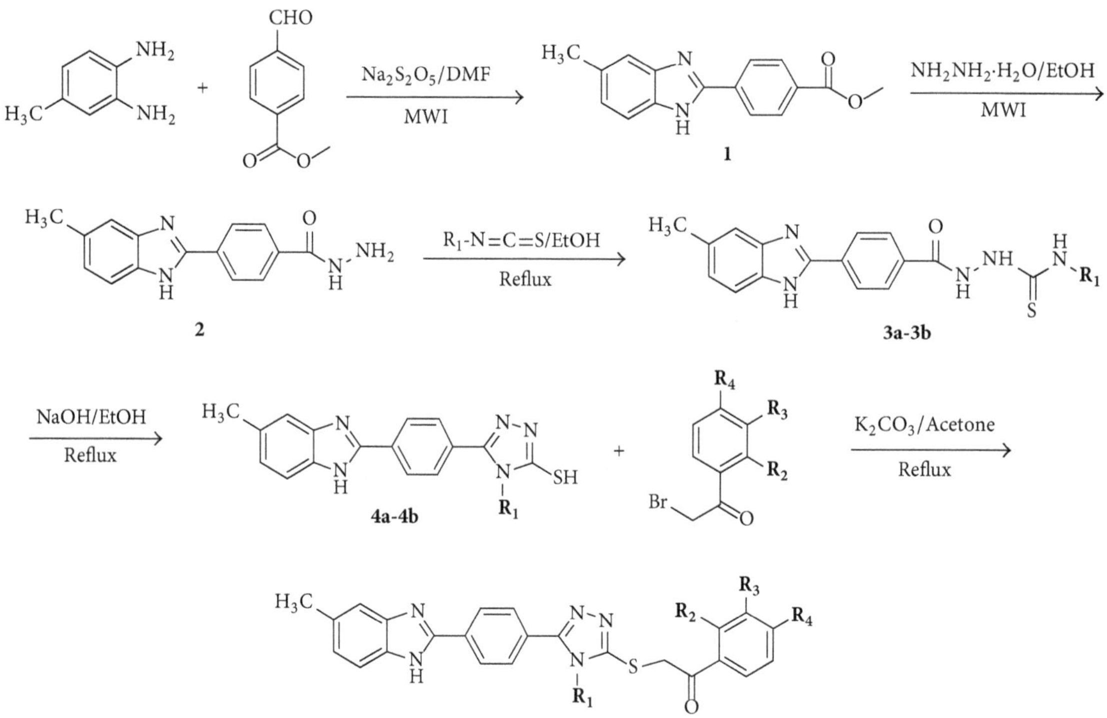

5a–5s

$\mathbf{R_1}$: $-CH_3$, $-C_2H_5$; $\mathbf{R_2}$, $\mathbf{R_3}$, $\mathbf{R_4}$: -H, -Cl, -F, Br, CN, OH, CH_3

Comp.	R_1	R_2	R_3	R_4
5a	$-CH_3$	-H	-H	-H
5b	$-CH_3$	-H	-H	-CN
5c	$- CH_3$	-H	-H	-F
5d	$- CH_3$	-H	-H	-Br
5e	$- CH_3$	-H	-H	$-CH_3$
5f	$- CH_3$	-H	-H	-Cl
5g	$- CH_3$	-F	-H	-F
5h	$- CH_3$	-Cl	-H	-Cl
5i	$- CH_3$	-H	-OH	-OH
5j	$-C_2H_5$	-H	-H	-H
5k	$-C_2H_5$	-H	-H	-CN
5l	$-C_2H_5$	-H	-H	-F
5m	$-C_2H_5$	-H	-H	-Br
5n	$-C_2H_5$	-H	-H	$-CH_3$
5o	$-C_2H_5$	-H	-H	-Cl
5p	$-C_2H_5$	-F	-H	-F
5r	$-C_2H_5$	-Cl	-H	-Cl
5s	$-C_2H_5$	-H	-OH	-OH

Scheme 1: Synthesis way for final compounds.

the level of ergosterol at all tested concentrations. Compound **5i** displayed 61.74%, 85.41%, and 93.71% inhibitions at 0.78 μg/mL, 1.56 μg/mL, and 3.12 μg/mL concentrations, respectively. Inhibitions of 62.05%, 84.48%, and 93.12% were observed for compound **5s** at the same concentrations. Reference drugs ketoconazole and fluconazole indicated similar inhibition potencies to those of compounds **5i** and **5s** in the range of 63.71% to 94.46%. A concentration-dependent

Table 1: MIC_{50} (μg/mL) values of compounds **5a–5s**.

Comp.	*C. albicans*	*C. glabrata*	*C. krusei*	*C. parapsilosis*
5a	12.5	6.25	6.25	12.5
5b	6.25	3.12	6.25	6.25
5c	12.5	6.25	6.25	12.5
5d	6.25	12.5	6.25	6.25
5e	12.5	6.25	12.5	12.5
5f	6.25	3.12	3.12	6.25
5g	3.12	6.25	6.25	6.25
5h	12.5	6.25	12.5	6.25
5i	0.78	1.56	1.56	0.78
5j	12.5	6.25	12.5	12.5
5k	12.5	6.25	12.5	12.5
5l	6.25	12.5	6.25	12.5
5m	3.12	3.12	3.12	6.25
5n	3.12	3.12	1.56	3.12
5o	3.12	3.12	6.25	6.25
5p	12.5	12.5	6.25	6.25
5r	6.25	3.12	3.12	3.12
5s	0.78	1.56	1.56	0.78
Ketoconazole	0.78	1.56	1.56	1.56
Fluconazole	0.78	1.56	1.56	0.78

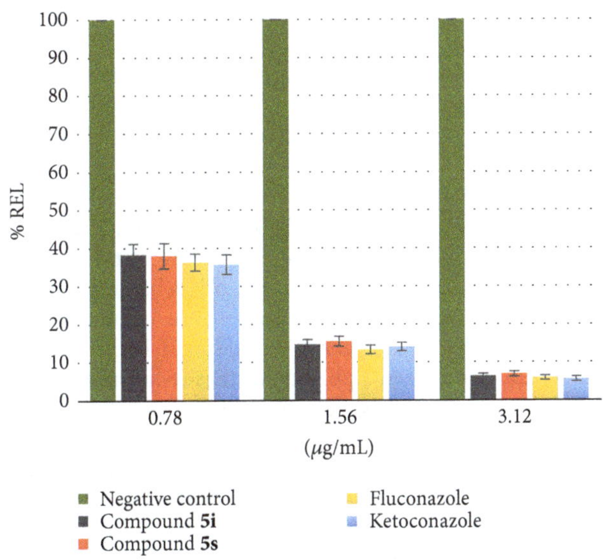

Figure 2: Relative ergosterol level of *C. albicans* after treatment of synthesized compounds and reference agents.

decrease in the ergosterol level was established for all tested agents. Hence, it can be obviously suggested that compounds **5i** and **5s** have a role in the ergosterol biosynthesis pathway.

3.4. Cytotoxicity Test. Toxicity is the main reason for the failure at all stages of the new drug development process. The major part of safety-related attrition occurs at preclinical phases while predicting preclinical safety liabilities earlier in the drug development process. This strategy enables the design and/or selection of improved drug candidates that have more possibilities of becoming commercialized drugs [44]. Therefore, we used the MTT cell viability assay to determine cytotoxicity of the most active compounds **5i** and **5s** against NIH/3T3 mouse embryonic fibroblast cell lines (ATCC CRL1658), which is recommended by ISO (10993-5, 2009) [45]. Cytotoxic concentrations (CC) of test compounds are presented as CC_{50} values in Table 2. Compounds **5i** and **5s** displayed CC_{50} of 86.59 μg/mL and 138.85 μg/mL, respectively. In order to evaluate selectivity profiles of compounds, selectivity indexes (SIs) were expressed as CC_{50} (NIH/3T3)/MIC_{50} (*C. albicans*). Sis were calculated as 111.01 and 178.01 for compounds **5i** and **5s**, respectively. These findings display that the anticandidal activity of the compounds **5i** and **5s** is not related to general toxicity but can be ascribed to their selective action against *Candida* species. Thus, cytotoxicity test findings enhanced the importance of compounds **5i** and **5s** as anticandidal drug candidates.

Table 2: CC_{50} (NIH/3T3), MIC_{50} (*C. albicans*), and SI results of selected compounds.

Comp.	CC_{50} (μg/mL)	MIC_{50} (μg/mL)	SI
5i	86.59	0.78	111.01
5s	138.85	0.78	178.01

3.5. Theoretical Calculation of ADME Parameters. High biological activity and low toxicological effects are not enough for a compound to become a drug candidate. A good pharmacokinetics profile is also very important for the new drug candidates that should be evaluated earlier in the process

TABLE 3: Calculated ADME parameters of compounds **5a–5s**.

Comp.	MW	RB	MV	DHB	AHB	PSA	log *P*	log *S*	*P*Caco	PM	VRT	VRF
5a	439.53	4	1372.27	1	5.5	83.23	5.43	−7.86	754.43	2	1	1
5b	464.54	5	1438.97	1	7	109.02	4.66	−8.79	156.01	2	0	1
5c	457.52	4	1388.38	1	5.5	83.23	5.67	−8.22	754.33	2	1	1
5d	518.43	4	1425.30	1	5.5	83.23	6.00	−8.72	754.38	2	2	1
5e	453.56	4	1431.21	1	5.5	83.23	5.74	−8.42	754.43	3	1	1
5f	473.97	4	1416.39	1	5.5	83.23	5.93	−8.60	754.37	2	1	1
5g	508.42	4	1465.01	1	5.5	80.43	6.39	−9.33	740.09	2	2	1
5h	475.51	4	1410.99	1	5.5	82.91	5.86	−8.58	643.17	2	1	1
5i	471.53	6	1400.25	3	7	128.41	3.79	−6.86	74.95	4	0	1
5j	453.56	5	1412.00	1	5.5	82.77	5.62	−7.82	676.16	2	1	1
5k	478.57	6	1478.70	1	7	108.56	4.85	−8.75	139.83	2	0	1
5l	471.55	5	1428.11	1	5.5	82.77	5.86	−8.18	676.08	2	1	1
5m	532.46	5	1465.03	1	5.5	82.77	6.19	−8.68	676.13	2	2	1
5n	467.59	5	1470.94	1	5.5	82.77	5.93	−8.38	676.16	3	1	1
5o	488.01	5	1456.12	1	5.5	82.77	6.12	−8.56	676.11	2	1	1
5p	522.45	5	1414.92	1	5.5	74.59	6.17	−7.94	1124.54	2	2	1
5r	489.54	5	1449.32	1	5.5	80.90	6.09	−8.56	712.34	2	1	1
5s	485.56	7	1412.81	3	7	124.13	3.93	−6.33	98.80	4	0	1

MW: Molecular weight RB: Number of rotatable bonds MV: Total solvent-accessible volume DHB: Estimated number of hydrogen bond donors AHB: Estimated number of hydrogen bond acceptors PSA: Van der Waals surface area of polar nitrogen and oxygen atoms and carbonyl carbon atoms log *P*: Predicted octanol/water partition coefficient log *S*: Predicted aqueous solubility *P*Caco: Predicted apparent Caco-2 cell permeability log *BB*: Predicted brain/blood partition coefficient PM: Number of likely metabolic reactions VRF: Number of violations of Lipinski's rule of five. The rules are: MW < 500, log $P < 5$, DHB ≤ 5, AHB ≤ 10, Positive PSA value VRT: Number of violations of Jorgensen's rule of three. The three rules are: log $S > -5.7$, *P*Caco > 22 nm/s, PM < 7.

of drug development. Thus, it has become important to examine the ADME profiles of pharmaceuticals as early as possible in the drug discovery process to reduce the amount of wasted time and resources. Moreover, in silico prediction of ADME can be used to identify individual molecules or chemical libraries by evaluating their suitability as potential drug molecules. The determination of the attempted lead optimization before synthesizing the compounds causes less redesign-synthesize-test cycles [46]. Thus, predictions of ADME properties of the all synthesized compounds (**5a–5s**) were performed by QikProp 4.8 software [35]. This software applies Lipinski's rule of five [47] and Jorgensen's rule of three [48], which evaluate the ADME properties of new drug candidates and are essential for the optimization of a biologically active compound. Molecular weight (MW), number of rotatable bonds (RB), molecular volume (MV), number of hydrogen donors (DHB), number of hydrogen acceptors (AHB), polar surface area (PSA), octanol/water partition coefficient (log *P*), aqueous solubility (log *S*), apparent Caco-2 cell permeability (*P*Caco), number of likely primer metabolic reactions (PM), and percent of human oral absorption (% HOA) are presented in Table 3 along with the violations of rules of three (VRT) and five (VRF). According to Lipinski's rule of five and Jorgensen's rule of three, compounds **5i** and **5s** are in accordance with the rule by causing no more than one violation. Thus, it can be suggested that the most active anticandidal compounds **5i** and **5s** may possess a good pharmacokinetic profile, increasing their pharmacological importance.

3.6. Molecular Docking. Docking studies were performed in order to gain more insight into the binding mode of the compounds **5i** and **5s** to lanosterol 14-α-demethylase. Due to the fact that lanosterol 14-α-demethylase is membrane-bound enzyme, its crystallization from the *Candida* species is difficult for X-ray analysis and docking studies. Thus, there is no experimental data or crystal structure of this enzyme in Protein Data Bank server. On the other hand, lanosterol 14-α-demethylase from *Mycobacterium tuberculosis* has high homology to that of lanosterol 14-α-demethylase from *Candida* species. It has been reported that these two enzymes have high degree of similarity between the hydrophobic cavities of the catalytic site [49, 50]. Therefore, docking studies were carried out using X-ray crystal structure of lanosterol 14-α-demethylase from *Mycobacterium tuberculosis* in complex with fluconazole (PDB ID: 1EA1) [36] obtained from Protein Data Bank server (http://www.pdb.org).

As stated in antifungal activity section, compounds **5i** and **5s** were found to be the most active derivatives against *C. krusei*, *C. albicans*, *C. parapsilosis*, and *C. glabrata* with 0.78, 1.56, 1.56, and 0.78 μg/mL MIC values, respectively. Thus, the main purpose of docking studies was to investigate the possible interaction of these compounds with lanosterol 14-α-demethylase enzyme.

The two-dimensional (Figures 3(a) and 4(a)) and three-dimensional (Figures 3(b) and 4(b)) poses of the compounds **5i** and **5s** are presented in Figures 3 and 4. The docking poses on lanosterol 14-α-demethylase reveals that the interactions between the compounds **5i**, **5s** and HEM450 are very important in terms of binding to active site of enzyme. Totally there

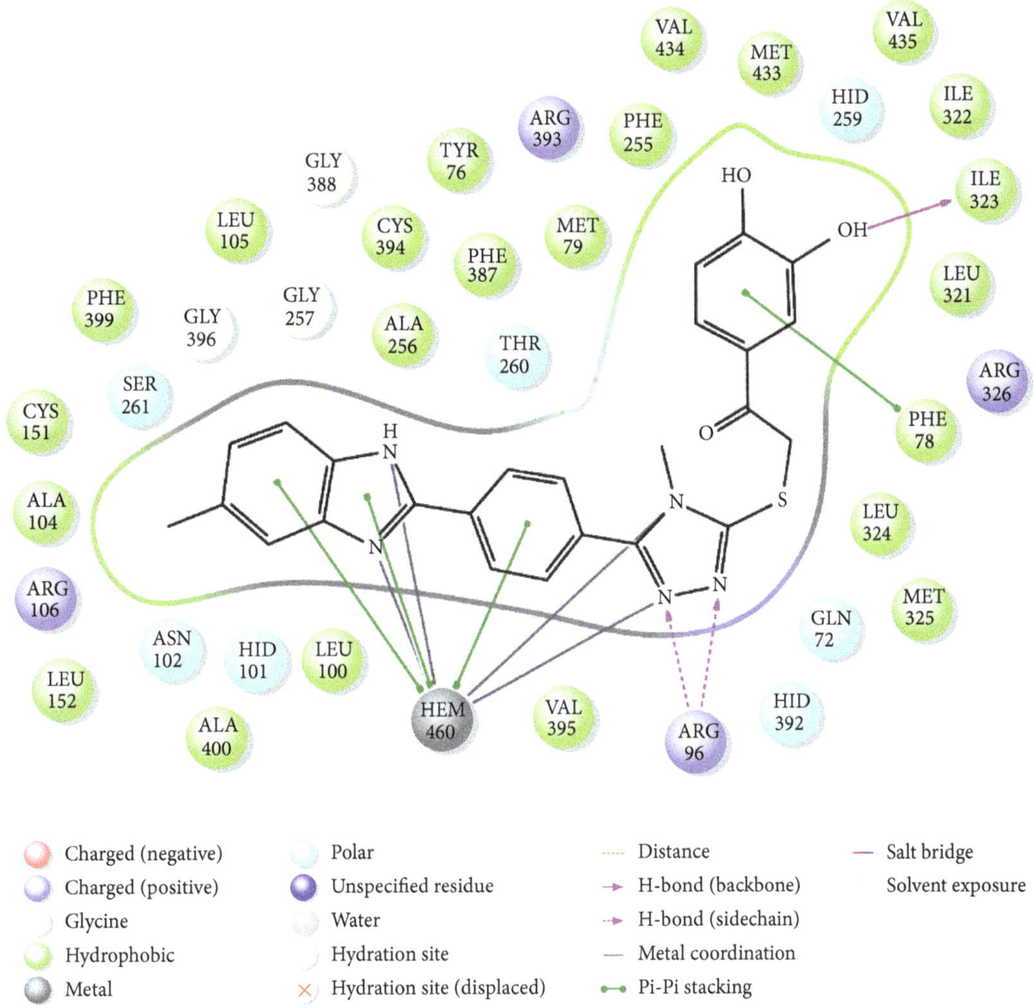

(a)

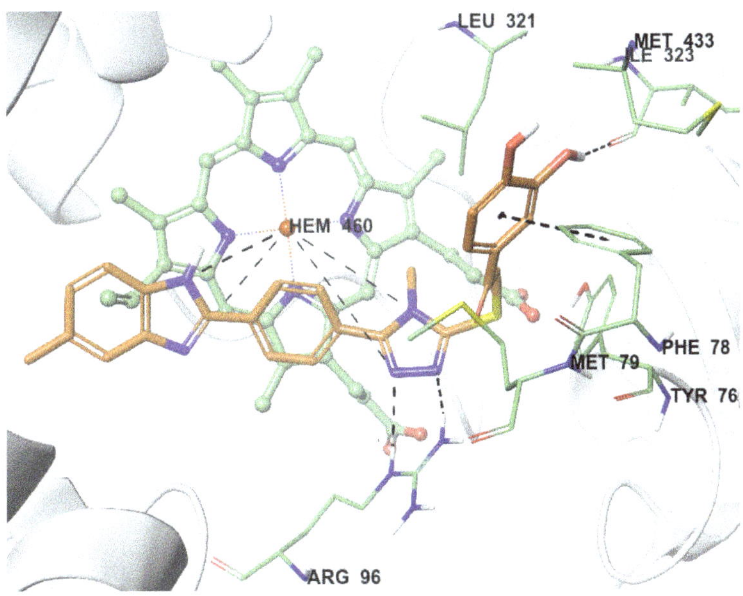

(b)

FIGURE 3: (a) The two-dimensional interacting mode of compound **5i** in the active region of lanosterol 14-α-demethylase. (b) The three-dimensional interacting mode of compound **5i** in the active region of lanosterol 14-α-demethylase. The inhibitor is colored with orange. The HEM molecule is colored green with ball-and-stick model. The binding energy of **5i** is −81.998 kcal/mol.

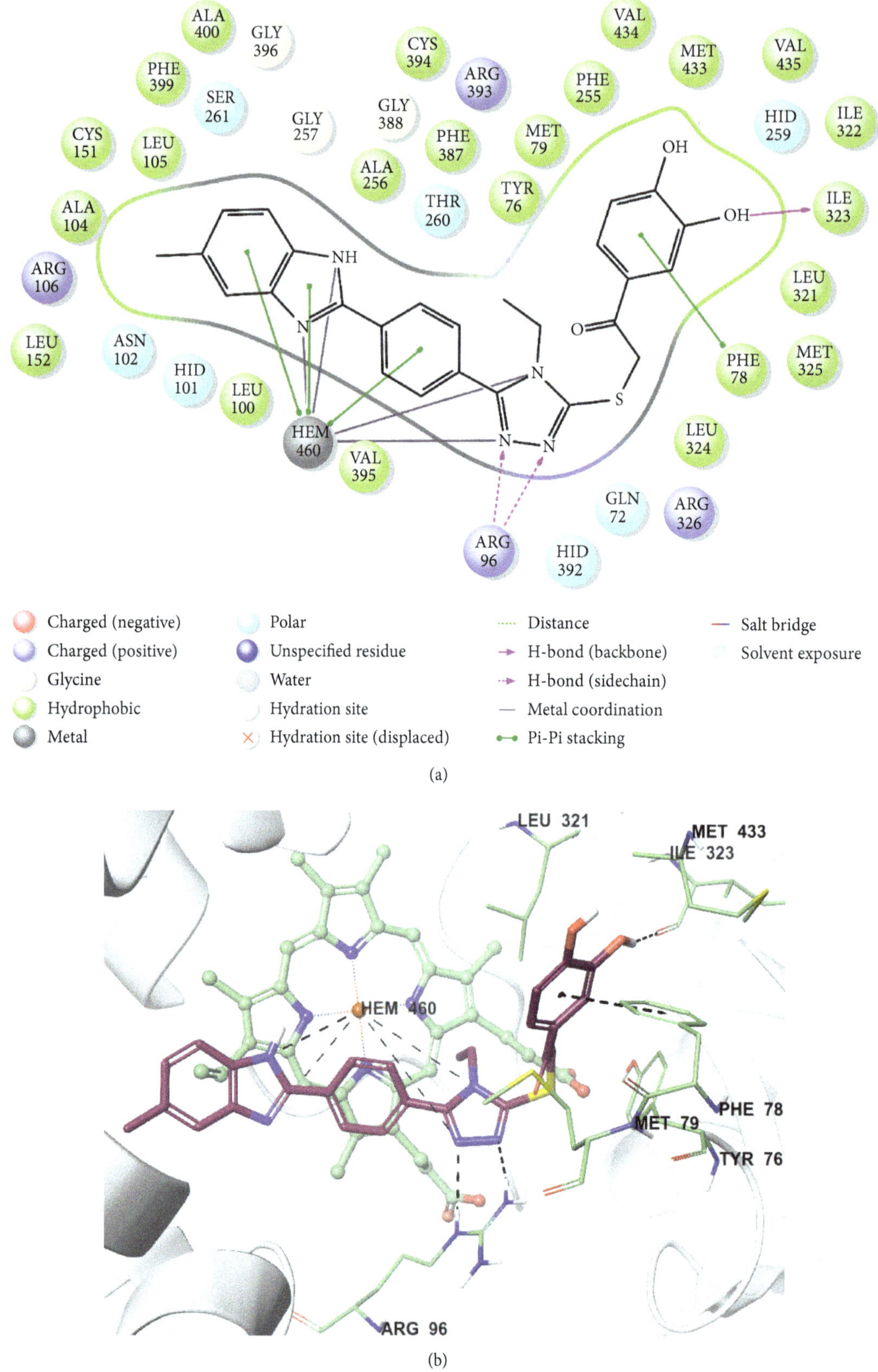

FIGURE 4: (a) The two-dimensional interacting mode of compound **5s** in the active region of lanosterol 14-α-demethylase. (b) The three-dimensional interacting mode of compound **5s** in the active region of lanosterol 14-α-demethylase. The inhibitor is colored with maroon. The HEM molecule is colored green with ball-and-stick model. The binding energy of **5s** is −85.218 kcal/mol.

are seven interactions between Hem450 and compounds. Hem450 creates four metal coordination bonds with the two nitrogen atoms of benzimidazole and two nitrogen atoms triazole (at second and fourth position). Other interactions are seen as π-π interaction between benzimidazole and phenyl ring (near to benzimidazole) in the structure. In addition, triazole provides binding to active site by forming hydrogen bonds. The nitrogen atoms of triazole (at first and second position) establish two different hydrogen bonds with two neighboring nitrogen atoms in the guanidine moiety of Arg96. Also, phenyl ring at the end of the structure is in interaction with phenyl of Phe78 by doing π-π interaction.

Synthesized compounds comprise two different substitutions as methyl or ethyl on N-1 position of triazole ring. Looking at activity results and docking studies, it is observed that the methyl or ethyl substituents at N-4 position of triazole do not cause a meaningful difference on biological activity. Furthermore, docking results explain the reason of stronger anticandidal activity of **5i** and **5s**. In such compounds, there is an additional interaction at C-3 position of phenyl ring differently common chemical structure. The hydroxyl in this position establishes a hydrogen bond with carbonyl of Ile323. Thus, it can be suggested that C-3 of phenyl is very important in terms of binding to enzyme active site and anticandidal activity. As a result, it is considered that this additional interaction could explain more binding capability and stronger activity of compounds **5i** and **5s** than other compounds.

4. Conclusion

In the current work, a series of benzimidazole-triazole derivatives **(5a–5s)** was synthesized and evaluated for their *in vitro* antifungal activities. Results of antifungal study indicated that compounds **5i** and **5s** have a significant anticandidal activity against all the tested microbial strains. Ergosterol quantification studies using *C. albicans* indicated that compounds **5i** and **5s** cause a decrease in the level of ergosterol. Since the selectivity of the compounds is a very important parameter to become a good drug candidate, the *in vitro* toxicity of compounds **5i** and **5s** was evaluated by MTT assay against NIH3T3 cell lines and found as nontoxic at their effective concentrations. ADME prediction and docking studies also enhanced the biological importance of such compounds. As a result it can be declared that synthesized benzimidazole-triazoles are worthwhile to assess in further studies to develop new antifungal agents with higher potency and improved safety profile.

Conflicts of Interest

The authors declare that there are no conflicts of interest regarding the publication of this paper.

Acknowledgments

This study was financially supported by Anadolu University Scientific Projects Fund, Project no. 1606S567.

References

[1] H. Tang, C. Zheng, J. Lv et al., "Synthesis and antifungal activities in vitro of novel pyrazino [2,1-a] isoquinolin derivatives," *Bioorganic & Medicinal Chemistry Letters*, vol. 20, no. 3, pp. 979–982, 2010.

[2] W. Wang, C. Sheng, X. Che et al., "Design, synthesis, and antifungal activity of novel conformationally restricted triazole derivatives," *Archiv der Pharmazie*, vol. 342, no. 12, pp. 732–739, 2009.

[3] H. A. Ghabbour, M. M. Qabeel, W. M. Eldehna, A. Al-Dhfyan, and H. A. Abdel-Aziz, "Design, synthesis, and molecular docking of 1-(1-(4-chlorophenyl)-2-(phenylsulfonyl)ethylidene)-2-phenylhydrazine as potent nonazole anticandidal agent," *Journal of Chemistry*, vol. 2014, Article ID 154357, 2014.

[4] S. Emami, T. Banipoulad, H. Irannejad et al., "Imidazolylchromanones containing alkyl side chain as lanosterol 14α-demethylase inhibitors: Synthesis, antifungal activity and docking study," *Journal of Enzyme Inhibition and Medicinal Chemistry*, vol. 29, no. 2, pp. 263–271, 2014.

[5] R. S. Kankate, P. S. Gide, and D. P. Belsare, "Design, synthesis and antifungal evaluation of novel benzimidazole tertiary amine type of fluconazole analogues," *Arabian Journal of Chemistry*, 2014.

[6] A. Ahmad, A. Khan, N. Manzoor, and L. A. Khan, "Evolution of ergosterol biosynthesis inhibitors as fungicidal against Candida," *Microbial Pathogenesis*, vol. 48, no. 1, pp. 35–41, 2010.

[7] C. Sheng, Z. Miao, H. Ji et al., "Three-dimensional model of lanosterol 14α-demethylase from Cryptococcus neoformans: Active-site characterization and insights into azole binding," *Antimicrobial Agents and Chemotherapy*, vol. 53, no. 8, pp. 3487–3495, 2009.

[8] Q. Zhang, D. Li, P. Wei et al., "Structure-based rational screening of novel hit compounds with structural diversity for cytochrome P450 sterol 14α-demethylase from penicillium digitatum," *Journal of Chemical Information and Modeling*, vol. 50, no. 2, pp. 317–325, 2010.

[9] S. Jacob K, S. Ganguly, P. Kumar, R. Poddar, and A. Kumar, "Homology model, molecular dynamics simulation and novel pyrazole analogs design of Candida albicans CYP450 lanosterol 14 α-demethylase, a target enzyme for antifungal therapy," *Journal of Biomolecular Structure and Dynamics*, vol. 35, no. 7, pp. 1446–1463, 2017.

[10] İ. S. Doğan, S. Saraç, S. Sari et al., "New azole derivatives showing antimicrobial effects and their mechanism of antifungal activity by molecular modeling studies," *European Journal of Medicinal Chemistry*, vol. 130, pp. 124–138, 2017.

[11] K. K. Reddy, S. K. Singh, S. K. Tripathi, C. Selvaraj, and V. Suryanarayanan, "Shape and pharmacophore-based virtual screening to identify potential cytochrome P450 sterol 14α-demethylase inhibitors," *Journal of Receptors and Signal Transduction*, vol. 33, no. 4, pp. 234–243, 2013.

[12] A. Stana, D. C. Vodnar, R. Tamaian et al., "Design, synthesis and antifungal activity evaluation of new thiazolin-4-ones as potential lanosterol 14α-demethylase inhibitors," *International Journal of Molecular Sciences*, vol. 18, no. 1, article no. 177, 2017.

[13] A. G. S. Warrilow, J. G. L. Mullins, C. M. Hull et al., "S279 point mutations in Candida albicans sterol 14-α demethylase (CYP51) reduce in vitro inhibition by fluconazole," *Antimicrobial Agents and Chemotherapy*, vol. 56, no. 4, pp. 2099–2107, 2012.

[14] A. A. Sagatova, M. V. Keniya, R. K. Wilson, M. Sabherwal, J. D. A. Tyndall, and B. C. Monk, "Triazole resistance mediated by

mutations of a conserved active site tyrosine in fungal lanosterol 14α-demethylase," *Scientific Reports*, vol. 6, Article ID 26213, pp. 1–6, 2016.

[15] C. Sheng, W. Zhang, M. Zhang et al., "Homology modeling of lanosterol 14α-demethylase of Candida albicans and Aspergillus fumigatus and insights into the enzyme-substrate interactions," *Journal of Biomolecular Structure and Dynamics*, vol. 22, no. 1, pp. 91–99, 2004.

[16] Y. Jiang, J. Zhang, Y. Cao et al., "Synthesis, in vitro evaluation and molecular docking studies of new triazole derivatives as antifungal agents," *Bioorganic & Medicinal Chemistry Letters*, vol. 21, no. 15, pp. 4471–4475, 2011.

[17] D. J. Sheehan, C. A. Hitchcock, and C. M. Sibley, "Current and emerging azole antifungal agents," *Clinical Microbiology Reviews*, vol. 12, no. 1, pp. 40–79, 1999.

[18] A. A.-M. Abdel-Hafez, "Benzimidazole condensed ring systems: New synthesis and antineoplastic activity of substituted 3,4-dihydro- and 1,2,3,4-tetrahydro-benzo[4,5]imidazo[1, 2-a]pyrimidine derivatives," *Archives of Pharmacal Research*, vol. 30, no. 6, pp. 678–684, 2007.

[19] Z. Ateş-Alagöz, C. Kuş, and T. Çoban, "Synthesis and antioxidant properties of novel benzimidazoles containing substituted indole or 1,1,4,4-tetramethyl-1,2,3,4-tetrahydro-naphthalene fragments," *Journal of Enzyme Inhibition and Medicinal Chemistry*, vol. 20, no. 4, pp. 325–331, 2005.

[20] N. N. Gülerman, H. N. Doan, S. Rollas, C. Johansson, and C. Çelik, "Synthesis and structure elucidation of some new thioether derivatives of 1,2,4-triazoline-3-thiones and their antimicrobial activities," *Il Farmaco*, vol. 56, no. 12, pp. 953–958, 2001.

[21] F. Hernández-Luis, A. Hernández-Campos, R. Castillo et al., "Synthesis and biological activity of 2-(trifluoromethyl)-1H-benzimidazole derivatives against some protozoa and Trichinella spiralis," *European Journal of Medicinal Chemistry*, vol. 45, no. 7, pp. 3135–3141, 2010.

[22] K. S. Jain, V. M. Khedkar, N. Arya, P. V. Rane, P. K. Chaskar, and E. C. Coutinho, "Design, synthesis & evaluation of condensed 2H-4-arylaminopyrimidines as novel antifungal agents," *European Journal of Medicinal Chemistry*, vol. 77, pp. 166–175, 2014.

[23] H.-T. Le, I. B. Lemaire, A.-K. Gilbert et al., "Histogranin-like antinociceptive and anti-inflammatory derivatives of o-phenylenediamine and benzimidazole," *The Journal of Pharmacology and Experimental Therapeutics*, vol. 309, no. 1, pp. 146–155, 2004.

[24] A. T. Mavrova, P. Denkova, Y. A. Tsenov, K. K. Anichina, and D. I. Vutchev, "Synthesis and antitrichinellosis activity of some bis(benzimidazol-2-yl)amines," *Bioorganic & Medicinal Chemistry*, vol. 15, no. 18, pp. 6291–6297, 2007.

[25] J. R. Maxwell, A. C. Wasdahl, D. A. Wolfson, and V. I. Stenberg, "Synthesis of 5-aryl-2H-tetrazoles, 5-aryl-2H-tetrazole-2-acetic acids, and [(4-phenyl-5-aryl-4H-1,2,4-triazol-3-yl)thio]acetic acids as possible superoxide scavengers and antiinflammatory agents," *Journal of Medicinal Chemistry*, vol. 27, no. 12, pp. 1565–1570, 1984.

[26] Y. Özkay, Y. Tunali, H. Karaca, and I. Işikdağ, "Antimicrobial activity and a SAR study of some novel benzimidazole derivatives bearing hydrazone moiety," *European Journal of Medicinal Chemistry*, vol. 45, no. 8, pp. 3293–3298, 2010.

[27] H.-Z. Zhang, G. L. V. Damu, G.-X. Cai, and C.-H. Zhou, "Design, synthesis and antimicrobial evaluation of novel benzimidazole type of Fluconazole analogues and their synergistic effects with Chloromycin, Norfloxacin and Fluconazole," *European Journal of Medicinal Chemistry*, vol. 64, pp. 329–344, 2013.

[28] G. L. V. Damu, Q. Wang, H. Zhang, Y. Zhang, J. Lv, and C. Zhou, "A series of naphthalimide azoles: design, synthesis and bioactive evaluation as potential antimicrobial agents," *Science China Chemistry*, vol. 56, no. 7, pp. 952–969, 2013.

[29] EUCAST, Definitive Document EDef 7.1: Method for the Determination of Broth Dilution MICs of Antifungal Agents for Fermentative Yeasts.

[30] O. N. Breivik and J. L. Owades, "Yeast analysis, spectrophotometric semimicrodetermination of ergosterol in yeast," *Journal of Agricultural and Food Chemistry*, vol. 5, no. 5, pp. 360–363, 1957.

[31] H. K. Gençer, U. A. Çevik, S. Levent et al., "New benzimidazole-1,2,4-triazole hybrid compounds: synthesis, anticandidal activity and cytotoxicity evaluation," *Molecules*, vol. 22, no. 4, article 507, 507 pages, 2017.

[32] N. Ö. Can, D. Osmaniye, S. Levent et al., "Synthesis of new hydrazone derivatives for MAO enzymes inhibitory activity," *Molecules*, vol. 22, no. 8, p. 1381, 2017.

[33] Ü. Demir Özkay, Ö. D. Can, B. N. Sağlık et al., "Design, synthesis, and AChE inhibitory activity of new benzothiazole–piperazines," *Bioorganic & Medicinal Chemistry Letters*, vol. 26, no. 22, pp. 5387–5394, 2016.

[34] B. N. Sağlık, S. Ilgın, and Y. Özkay, "Synthesis of new donepezil analogues and investigation of their effects on cholinesterase enzymes," *European Journal of Medicinal Chemistry*, vol. 124, pp. 1026–1040, 2016.

[35] QikProp Schrödinger, LLC, New York, NY, USA, 2016, version 4.8.

[36] L. M. Podust, T. L. Poulos, and M. R. Waterman, "Crystal structure of cytochrome P450 14α-sterol demethylase (CYP51) from Mycobacterium tuberculosis in complex with azole inhibitors," in *Proceedings of the National Academy of Sciences*, vol. 98, pp. 3068–3073, 2001.

[37] Maestro, version 10.6, Schrödinger, LLC, New York, NY, USA, 2016.

[38] Schrödinger, LLC, New York, NY, USA, 2016.

[39] LigPrep Schrödinger, LLC, New York, NY, USA, 2016, version 3.8.

[40] Glide Schrödinger, LLC, New York, NY, USA, 2016, version 7.1.

[41] M. A. Malik, S. A. Al-Thabaiti, and M. A. Malik, "Synthesis, structure optimization and antifungal screening of novel tetrazole ring bearing acyl-hydrazones," *International Journal of Molecular Sciences*, vol. 13, no. 9, pp. 10880–10898, 2012.

[42] S. Dhingra and R. A. Cramer, "Regulation of sterol biosynthesis in the human fungal pathogen Aspergillus fumigatus: Opportunities for therapeutic development," *Frontiers in Microbiology*, vol. 8, article no. 92, 2017.

[43] A. Lupetti, R. Danesi, M. Campa, M. D. Tacca, and S. Kelly, "Molecular basis of resistance to azole antifungals," *Trends in Molecular Medicine*, vol. 8, no. 2, pp. 76–81, 2002.

[44] J. A. Kramer, J. E. Sagartz, and D. L. Morris, "The application of discovery toxicology and pathology towards the design of safer pharmaceutical lead candidates," *Nature Reviews Drug Discovery*, vol. 6, no. 8, pp. 636–649, 2007.

[45] International Organization for Standardization, *Biological Evaluation of Medical Devices-Part 5: Tests for In Vitro cytotoxicity ISO-10993-5*, 3rd edition, 2009.

[46] D. Butina, M. D. Segall, and K. Frankcombe, "Predicting ADME properties in silico: Methods and models," *Drug Discovery Therapy*, vol. 7, no. 11, pp. S83–S88, 2002.

[47] C. A. Lipinski, F. Lombardo, B. W. Dominy, and P. J. Feeney, "Experimental and computational approaches to estimate solubility and permeability in drug discovery and development settings," *Advanced Drug Delivery Reviews*, vol. 46, no. 1–3, pp. 3–26, 2001.

[48] W. L. Jorgensen and E. M. Duffy, "Prediction of drug solubility from structure," *Advanced Drug Delivery Reviews*, vol. 54, no. 3, pp. 355–366, 2002.

[49] R. González-Chávez, R. Martínez, M. E. Torre-Bouscoulet, M. Gallo, and M. M. González-Chávez, "De novo design of non-coordinating indolones as potential inhibitors for lanosterol 14-α-demethylase (CYP51)," *Chemical & Pharmaceutical Bulletin*, vol. 62, no. 1, pp. 16–24, 2014.

[50] R. Gollapudy, S. Ajmani, and S. A. Kulkarni, "Modeling and interactions of aspergillus fumigatus lanosterol 14-α demethylase 'A' with azole antifungals," *Bioorganic & Medicinal Chemistry*, vol. 12, no. 11, pp. 2937–2950, 2004.

The Oxygen Generation Performance of Hollow-Structured Oxygen Candle for Refuge Space

Weixiang Wang,[1,2] Longzhe Jin,[1,2] Na Gao,[1,2] Jianlin Wang,[1] and Mingyang Liu[1]

[1]*School of Civil and Resource Engineering, University of Science and Technology Beijing, Beijing 100083, China*
[2]*Mine Emergency Technology Research Center, Beijing 100083, China*

Correspondence should be addressed to Na Gao; aggie198308@163.com

Academic Editor: Albert Demonceau

To improve oxygen generation performance, we dissected and analyzed the incompletely reacted oxygen candles and thus proposed the concept of the hollow-structured oxygen candle. We calculated the surface area ratio and designed the mold for hollow-structured oxygen candles at a radius of 0, 5, 9, 12, 15.5, and 20 mm. The structural stability of the oxygen candles was tested by the loading experiment. The oxygen generation rate (OGR) and other properties were explored by combustion experiments. The composition of the oxygen candles and the residual solids after combustion were observed with scanning electron microscope (SEM). The results show that, with the increase of the hollow-structure radius (r), the stability of the hollow-structured oxygen candles gradually weakens, and the oxygen candles cannot be made when r is 20 mm. The hollow structure has a positive influence on the combustion of the oxygen candles, and the extent of combustion reaction can increase to 98.3% at the most, a 28.9% increase compared with that of solid-structured oxygen candles. The analysis of the morphology of cobalt(II,III) oxide and the macroscopic and microscopic morphology of the oxygen candles before and after combustion validates the rationality of the concept of the hollow-structured oxygen candle. The best performance of the oxygen candles is obtained when $r = 9$ mm, where the highest extent of combustion reaction is 98.3%. The total oxygen generated is 48.75 L, and the average rate is 9.2 L/min. The test of the oxygen candles with $r = 9$ mm after combustion by the X-ray diffractometer (XRD) shows that these oxygen candles fully reacted. This study provides a basis for optimizing the performance of oxygen candles and for the development of the oxygen supply system fitted with the oxygen candle in spaceflight, submarine, and underground space.

1. Introduction

Refuge space system can provide the people in distress with a living space, [1–3] which can isolate toxic and harmful gases and supply oxygen, water, food, and other necessary substances. It is installed with the oxygen supply system, carbon dioxide, carbon monoxide, climate control system, etc. [4]. The oxygen supply system is especially more important in the refuge space system [5, 6]. The common oxygen supply methods used in the refuge space include the forced air system, compressed oxygen cylinder, and chemical oxygen supply [7]. Among them, the chemical oxygen supply commonly includes water-electrolyzed, potassium hyperoxide, peroxide, and oxygen candle. Similar to the liquid oxygen density, the oxygen candle is known for lightweight, safety, high oxygen storage capacity, and a long shelf life and thus has been extensively used in the internal life support systems of protective engineering, submarine, civilian aircraft, and fire-fighting [8–11]. Moreover, its oxygen generation performance is not influenced by surrounding temperature and humidity. Hence, it has become the common source of chemical oxygen supply.

Scholars worldwide mainly focused on the oxygen candle formula, catalyst, fuel, burning fluctuation, size of the composition, start-up mode, and molding process which influence the performance of oxygen candles [12–16]. Our team has contributed to the performance of oxygen supply in refuge space for years [17–20]. We concentrated on the oxygen candle formula, catalyst, molding process, and start-up mode by thermogravimetric analysis, orthogonal experiment, and other laboratory experiments. These findings could provide scientific basis for continuous optimization of the oxygen candle. However, few studies focused on the oxygen generation performance of the oxygen candle with different structures. In this paper, according to the dissection and analysis of the incompletely reacted oxygen candle, we put forward the concept of the hollow-structured oxygen candle. Furthermore, we perform laboratory studies to explore the oxygen generation performance of the hollow-structured oxygen candle, investigate the micromorphology of the oxygen candles before and after combustion using SEM, test the chemical composition of solid combustion products by XRD, and describe their influence on the combustion of the oxygen candles. This study can benefit the development of the oxygen candle system in spaceflight, submarine, and underground space.

2. Hollow-Structured Oxygen Candle

2.1. Design of the Hollow-Structured Oxygen Candle. The principle of oxygen generation of the oxygen candle is as follows: the oxygen generation reagent, sodium chlorate, decomposes to generate oxygen at high temperature produced by the combustion of metal fuel Mn powder. Two chemical equations are mainly involved:

$$Mn + O_2 = MnO_2 + \Delta H \quad (1)$$

$$2NaClO_3 \xrightarrow[\Delta H]{\text{catalyzer}} 2NaCl + 3O_2 \quad (2)$$

The previous experiments showed that some oxygen candles made even by the optimal formula [19] cannot be completely combusted, as shown in Figure 1. The simple dissection of these oxygen candles reveals that the top and the periphery are fully combusted, and the unreacted parts are distributed inside the oxygen candles, near the oxygen candle axis. So, the closer the parts to the surface, the fuller they are combusted. The reason is that the substances accumulated in the external reaction hinder the heat transfer to the inside.

Actually, in the reaction, Mn powder is used as a fuel that maintains burning. Therefore, the complete reaction of the oxygen candles mainly depends on whether the Mn powder can fully combust. The combustion of Mn powder is a gas-solid two-phase reaction. In this case, when combustion occurs, the combustion products are continuously generated from the combustion zone, while the fuel enters the combustion zone. Otherwise, the combustion will not sustain. In this process, the departure of the products and the entry of the fuel involve the transfer of materials (gas, heat, etc.) [21, 22]. So, the increase of channels for the material transfer allows complete reaction of Mn powder in the oxygen candle. In addition, the gas-solid two-phase combustion reaction in combustion and heat-transmission science discovered that the specific surface of solid particles, which influences the rate and extent of the reaction, provides a contact surface and reaction area for combustion reaction.

Subsequently, we propose the concept of the hollow-structured oxygen candle [10]. The hollow structure adds the transmission channels of gas and heat, allowing the heat to diffuse as evenly as possible to the unreacted part of the oxygen candle and to preheat the parts to be combusted in no time. Besides, the hollow structure can increase the surface area of the oxygen candle and provide the necessary contact surface and effective reaction area in the combustion reaction. The schematic diagram is shown in Figure 2.

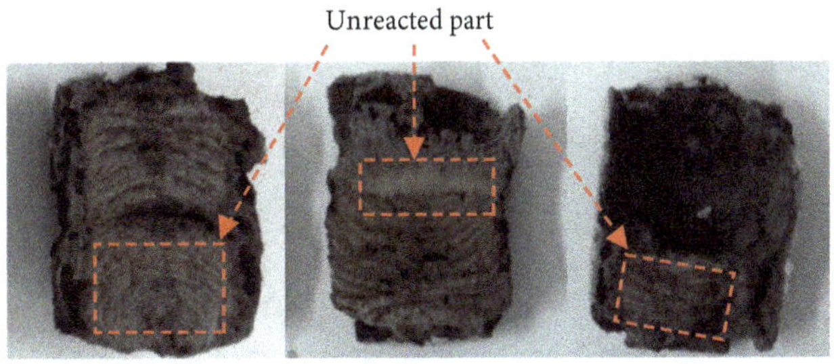

FIGURE 1: Incompletely reacted oxygen candles.

2.2. Determination of the Hollow Structure Size. According to the results of previous studies and the existing pressing process of the oxygen candle [17, 19], we determine that the mass of OCB is 200 g; correspondingly, the height of the pressed OCB is about 40 mm. Under the same stress, loading rate, and holding time, the density of the solid-structured OCB ρ_1 is equal to that of the hollow-structured OCB ρ_2. We control the mass M of the two OCBs to be the same and the maximum radius R of the two OCBs to be the same, at a constant of 30 mm. Thus, the volume of the solid-structured OCBs (V_1) and the hollow-structured OCBs (V_2) is the same. We perform the following calculation based on this premise, and then the surface area ratio of the hollow-structured OCB to the solid-structured OCB is ΔP:

$$\Delta P = \frac{S_2}{S_1} = \frac{2\pi(R^2 - r^2) + 2\pi h_2 (R + r)}{2\pi R^2 + 2\pi R h_1}, \quad (3)$$

where S_1 and S_2 are the surface area of the solid- and hollow-structured OCBs, respectively; r is the radius of the hollow structure; and h_1 and h_2 are the height of the solid- and hollow-structured OCBs. Then, we can know the relationship between ΔP and r, as shown in Figure 3, which shows that with the increase in r, the surface area ratio slowly increases before $r = 20$ mm and then sharply increases. When r is close to 30 mm, the surface area ratio tends to infinity. According to the actual situation of the experiment, we select the hollow-structured OCBs with the following property as research objects: the ΔP of which is equal to 1.0, 1.1, 1.2, 1.3, 1.5, and 2.0 and the theoretical values of r are equal to 0, 5, 9, 12, 15.5, and 20 mm, respectively.

3. Experiment

3.1. Manufacture and Compression Test of Hollow-Structured OCBs. We first weighed the reagent in proportion (Figures 4(a)–4(d)) according to the optimal formula of the oxygen candles and then mixed them using a KE-2L planetary ball mill (with a metal grinding jar and grinding

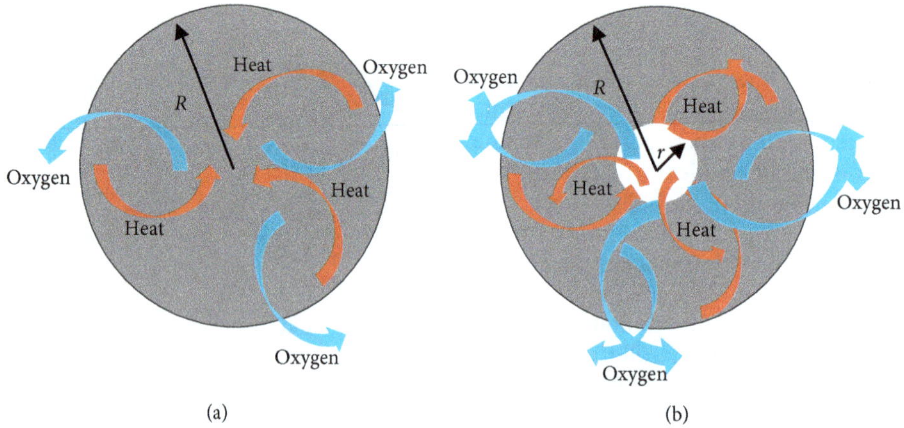

FIGURE 2: Schematic of heat transfer and oxygen release of solid-structured oxygen candles (a) and hollow-structured oxygen candles (b).

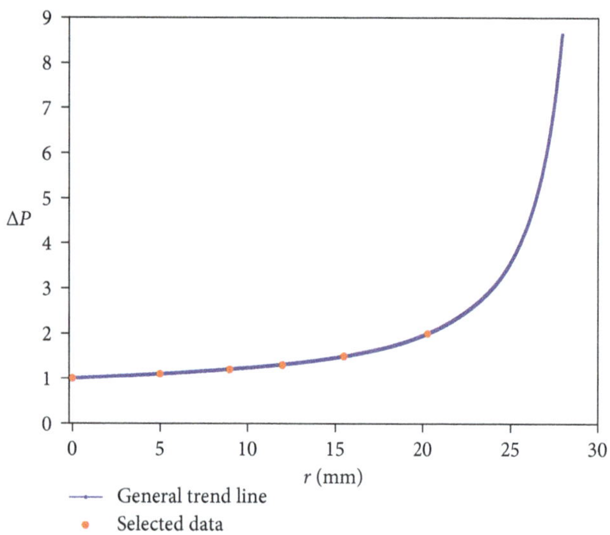

FIGURE 3: Relationship between ΔP and r.

ball) and set the speed at 280 r/min and the time at 5 min. The well-mixed powder (Figure 4(e)) was placed in a vacuum-drying oven to be dried at 60°C for 12 hours, and then the dried powder was poured into the mold (Figure 4(f)) and placed on a press to manufacture the OCBs (Figure 4(g)). In the experiment, the YES-300 press and a self-designed mold were employed. The fast loading rate was 0.5 kN/s, and holding time was 5 minutes. We put the OCBs on the press to test their axial compressive strength and radial compressive strength (Figures 4(h) and 4(i)) and then judged their structural stability.

3.2. Oxygen Generation Experiment of Hollow-Structured Oxygen Candles. The periphery of OCBs was wrapped in an insulating material (ceramic fiber) (Figure 4(j)). The small ignition head was fixed on the OCBs (Figure 4(k)), and 3 g ignition powder was used to cover the small ignition head and the upper surface of the OCBs. Then, the insulating material was used to seal the oxygen candle in order to prevent heat dissipation, maintain the reaction, and ensure that the small ignition head and ignition powder are in close contact when the small ignition head is ignited (Figure 4(k)). Next, we placed the oxygen candle in the reaction vessel and put them together into a 425 L sealed box. We connected the wire of the small ignition head with the igniter and opened the gas detector and temperature logging device. In the end, we closed the door of the sealed box and started the igniter by a remote control.

In this experiment, the oxygen concentration was recorded using the CD-KJ70N monitoring system, which is designed on the basis of the "Coal Mine Safety Regulations" [23] and the actual conditions of underground mines and can collect the environmental parameters such as oxygen concentration, carbon dioxide concentration, carbon monoxide concentration, gas concentration, wind speed, and air pressure. The monitoring system has obtained the coal safety mark and explosion-proof qualification certificate. A wireless, remotely controlled igniter (A01; Liuyang Fireworks Development Co., Ltd., China), composed of a receiver box and remote-control unit, was used to start the ignition. The temperature of the reaction vessel was detected in real time using a temperature-monitoring instrument (Beijing Sailing Technology Co., Ltd., China), which consisted of an STI-AS temperature-inspecting instrument (measurement accuracy: 0.2% FS) and a PC1625 platinum resistance temperature detector (PRTD) (operating temperature: −200 to 600°C; error: ±0.06%). The PRTD was fixed outside the reaction vessel. The experimental data were collected using the computer software and recorded in real time.

3.3. SEM Experiments. We collected the reagent that made up the oxygen candle, some well-mixed oxygen candle powders, and the small particles with fresh section of the OCBs after combustion for the SEM experiment. ZEISS Sigma SEM was employed to research the micromorphology of oxygen candles before and after combustion. Its secondary electron image resolution can reach 1.3 nm (20 kV), the acceleration voltage of the electron source can be adjusted from 100 V to 30 kV, and the magnification range is 12x to 1000,000x.

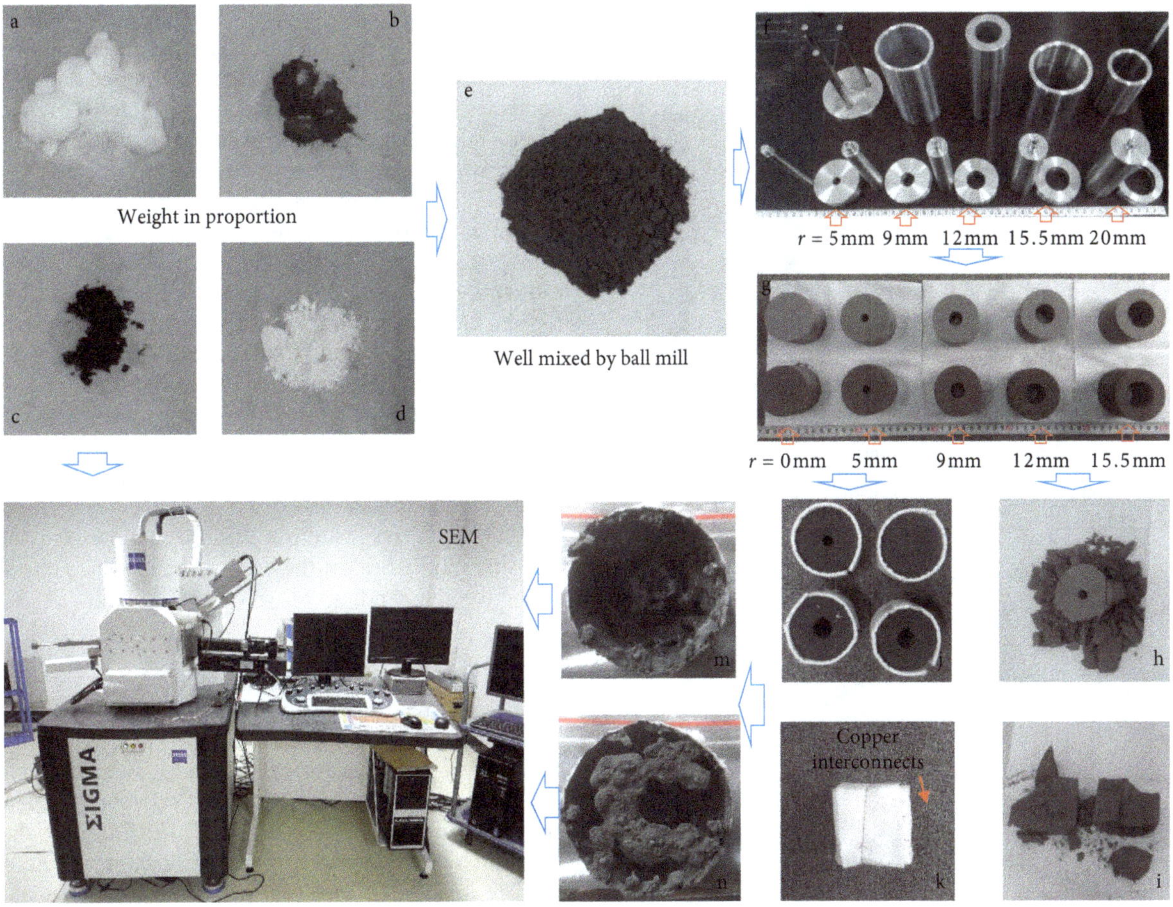

FIGURE 4: The process of experiments: (a) sodium chlorate; (b) Mn powder; (c) cobalt(II,III) oxide; (d) kaolin; (e) well-mixed powder of the oxygen candles; (f) mold manufactured; (g) oxygen candles at different r; (h) axial compressive strength test; (i) radial compressive strength test; (j, k) wrapping of the OCBs; (m, n) the oxygen candles after combustion.

Figure 4 illustrates the flow diagram of the process of the experiments.

4. Analysis of the Results

4.1. The Maximum Load for Ejection. In the manufacture process, we found that the OCBs at r = 0, 5, 9, 12, and 15.5 mm can be made and the OCBs at r = 20 mm failed to be made. After five tries, all OCBs at r = 20 mm cracked, as shown in Figure 5. The reason attributes to the following: (1) at r = 30 mm and r = 20 mm, the effective thickness of the OCBs is only 10 mm. Thus, the structure is unstable. (2) The ejection process is not continuous, which is prone to create stress concentration and physical deformation due to instantaneous release, as shown in the dotted line frame in Figure 5(b). So, we did not study the OCBs at r = 20 mm.

Figure 6 indicates that when $r \leq 9$ mm, the maximum load for ejection slowly increases with the increase of r. When $r \geq 9$ mm, the load for ejection sharply increases as r increases. By comparative analysis, we can find that the larger load ejection required causes the greater damage and difficulties in manufacturing the OCBs. The larger the load for ejection, the bigger the stress concentration and physical deformation due to the instantaneous release. Thus, the large load for ejection is not conducive to the manufacture of OCBs.

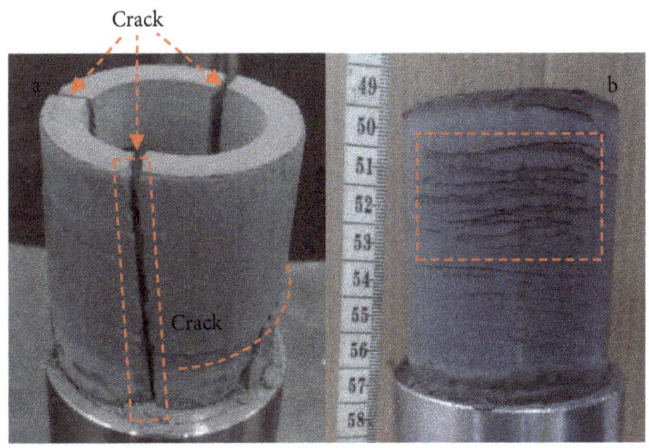

FIGURE 5: Oxygen candle at r = 20 mm.

4.2. The Maximum Compressive Strength. Figure 7 shows that the maximum axial compressive strength of each OCB is greater than the maximum radial compressive strength, and

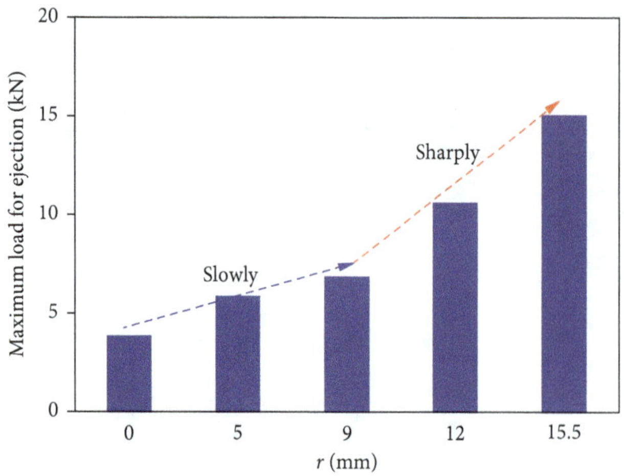

FIGURE 6: Maximum load for ejection on manufacturing hollow-structured OCBs.

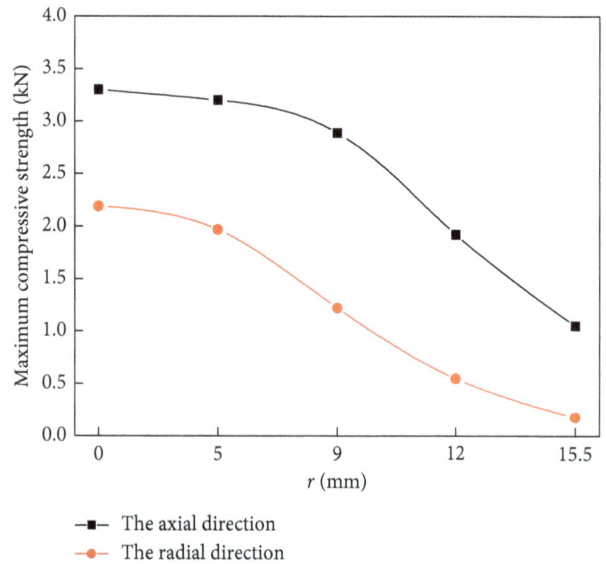

FIGURE 7: Maximum axial compressive strength and maximum radial compressive strength of hollow-structured oxygen candles.

the maximum axial compressive strength and maximum radial compressive strength decrease with the increase in r. At $r = 0$, 5, and 9 mm, the reduction of the maximum axial compressive strength is not obvious. At $r = 12$ and 15.5 mm, the maximum axial compressive strength sharply decreases, and the maximum axial compressive strength of the hollow-structured OCBs at $r = 15.5$ mm is only 31.8% of that at $r = 0$ mm. At $r = 0$ and 5 mm, the reduction of the maximum radial compressive strength is not obvious. At $r = 9$, 12, and 15.5 mm, the reduction of the maximum radial compressive strength is significant, and the maximum radial compressive strength of the hollow-structured OCBs at $r = 15.5$ mm is only 8.2% of that at $r = 0$ mm.

The result of this experiment can be summarized as follows: with the increase in r, the stability of the OCB structure decreases.

4.3. Variation of Oxygen Concentration with Time. Figure 8 illustrates the change of oxygen concentration of each oxygen candle. In the first minute, the oxygen concentration fluctuates slightly, and then it starts to increase dramatically. After a slight decrease, there is an approximate line increase over time, and finally, it reaches a maximum value.

The data in Figure 8 expound that the final concentration of the oxygen candles at $r = 0$, 5, 9, 12, and 15.5 mm is 28.8%, 31.43%, 32.05%, 32.1%, and 32%, respectively, indicating that at $r = 0$ and 5 mm, the reaction extent of the oxygen candles is lower, and at $r = 9$, 12 and 15.5 mm, the extent is higher. In this figure, the time for each curve to reach the maximum value is different, which is 400, 400, 360, 340, and 300 s at $r = 0$, 5, 9, 12, and 15.5 mm, respectively. We can conclude that, with the increase in r, the reaction extent of the oxygen candles increases and the reaction rate accelerates.

The above results show that the hollow structure has a positive influence on the oxygen generation performance of the oxygen candles.

4.4. The Total Oxygen Generation. Table 1 indicates that, with the increase in r, the oxygen generation displays an increasing trend. The difference in the total volume of oxygen generation of the oxygen candles at $r = 9$, 12, and 15.5 mm is very small. The ratio of actual volume to theoretical volume at $r = 9$, 12, and 15.5 mm is close to 100%, and the difference does not exceed 3%, demonstrating that the reaction of the oxygen candles at $r = 9$, 12, and 15.5 mm is more complete, and the highest total oxygen generation is 48.75 L at $r = 9$ mm.

4.5. The Average Rate of Oxygen Generation. The average OGR of the oxygen candles at different r is calculated, as listed in Table 1, which indicates that with the increase in r, the average OGR accordingly increases. So, the hollow structure in the oxygen candles has a positive influence on the oxygen generation; namely, within a certain range, the larger the r, the less the unreacted part of the oxygen candles.

4.6. The Temperature of the Reactor Vessel. Figure 9 exhibits that the temperature of the reactor vessel for each oxygen candle displays a sharp increase at the beginning and then slowly rises. When it reaches the highest value, it begins to slowly decline. In addition, the rate of temperature rise increases as r increases, and the maximum temperature increases accordingly. At $r = 0$ mm, the oxygen candle generates a little heat, leading to a small value of temperature; correspondingly, the maximum temperature is the lowest among the 5 hollow-structured oxygen candles. This indicates that the reaction of this oxygen candle is the least. At $r = 15.5$ mm, the rate of temperature rise is the fastest, and the maximum temperature is the largest, which reveals that the reaction of this oxygen candle is more complete and the rate is more rapid. The temperature difference in the reactor vessel before and after the reaction correspondingly increases with the increase in r, which shows the consistent variation of temperature.

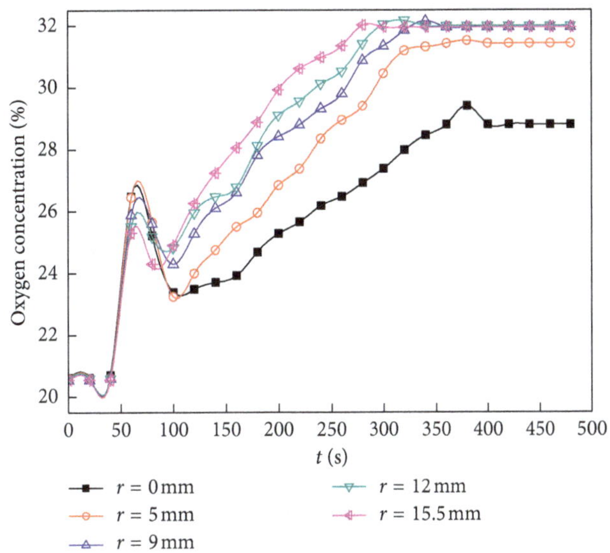

FIGURE 8: Relationship between oxygen concentration and reaction time.

TABLE 1: Oxygen concentration, oxygen generation, and the average OGR.

Radius of the hollow structure (mm)	The initial oxygen concentration (%)	The oxygen concentration at the end of reaction (%)	The reaction time (min)	The actual volume (L)	The theory volume (L)	The ratio of actual volume to theoretical volume (%)	The average OGR ($L\cdot min^{-1}$)
0	20.7	28.8	6	34.43	49.61	69.4	5.74
5	20.62	31.43	5.7	45.94	49.61	92.6	8.06
9	20.58	32.05	5.3	48.75	49.61	98.3	9.2
12	20.6	32.1	5	48.45	49.61	97.7	9.69
15.5	20.62	32	4.3	48.36	49.61	97.5	11.25

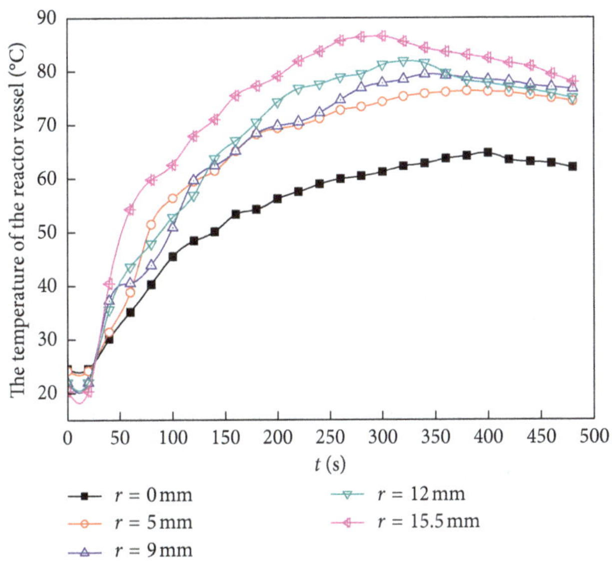

FIGURE 9: Relationship between the temperature and time.

4.7. The Macroscopic and Microscopic Morphology of Oxygen Candles before and after Combustion. Figure 10 shows the macroscopic and microscopic morphology of the oxygen candles before and after combustion. It can be observed from Figures 10(a) and 10(b) that most of the cobalt (II, III) oxide exists in the spheres or ellipsoids at a radius of about 5–10 um, whose morphology is like a flower cluster to be unfolded and has more pores. These pores increase the specific surface area of the cobalt (II, III) oxide and become a pathway for heat and oxygen transport, promoting the combustion reaction. As shown in Figures 10(c) and 10(d), the burned oxygen candles display a porous structure, where some pores penetrate the oxygen candles (red circle portion) and some ones are interconnected (yellow circle portion). Figure 10(e) displays that there are many small holes (red circle portion) with different sizes in the fresh section of the burned oxygen candles, which extend inward. There are also some large pits (yellow circle portion) and small holes in the pits as well. The presence of these small holes and large pits allows heat and oxygen to be transferred and exchanged during the reaction and facilitates the combustion reaction. The above research further validates the rationality of the concept of the hollow-structured oxygen candle proposed in this paper.

5. Determination of the Best Performance of Hollow-Structured Oxygen Candles and Verification by the XRD Experiment

5.1. Determination of the Best Performance. Because the oxygen candle at $r = 0$ mm has low extent of reaction, we will not further discuss it. We focus our discussion on oxygen

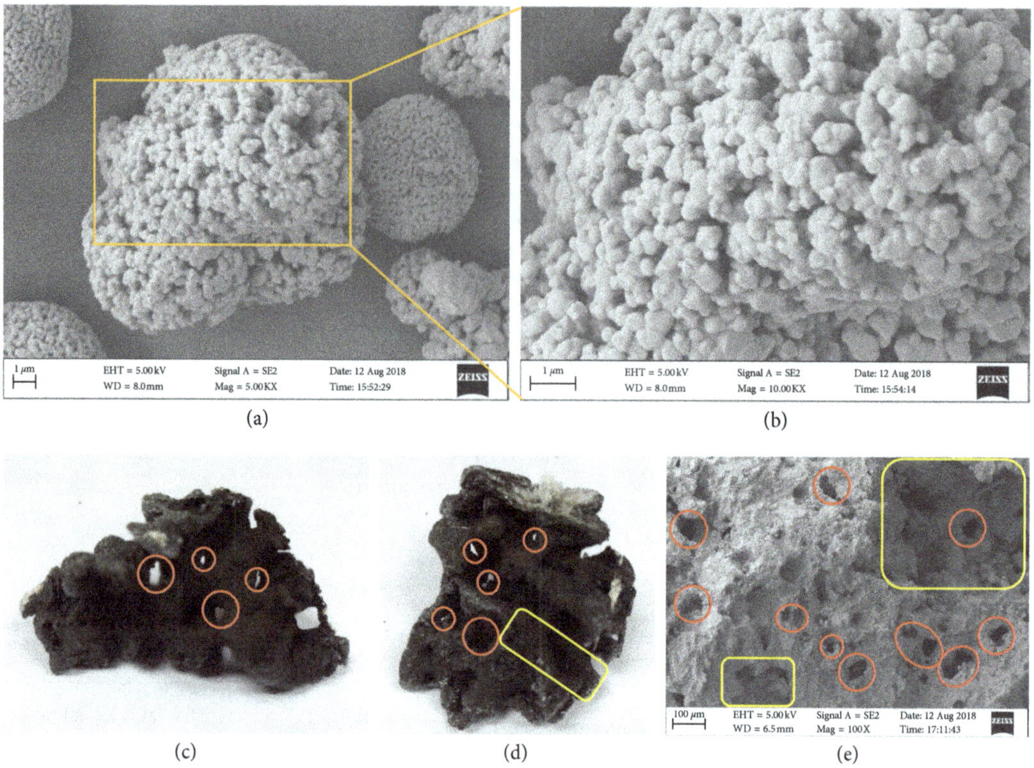

FIGURE 10: The macroscopic and microscopic morphology of the oxygen candles before and after combustion. Microscopic morphology of cobalt(II,III) oxide (a) and its local enlargement (b) by SEM; macroscopic morphology of the oxygen candles after combustion (c, d); microscopic morphology of fresh section of the oxygen candles after combustion (e) by SEM.

candles at $r = 5, 9, 12$, and 15.5 mm, which have high extent of reaction. To determine the best performance of hollow-structured oxygen candles, we analyze the results of the above experiments synthetically. On the basis of four indicators, compressive strength, oxygen generation, average OGR, and the temperature of the reactor vessel, we assign a number to the oxygen candles at different r and identify the weight of the four indicators according to their importance in engineering practices [17–19], as listed in Table 2. The comprehensive score of the oxygen generation performance = $\sum$ score of index × weight. Then, we obtain the comprehensive score of each oxygen candle at different r, as listed in Table 2. So, we determine that the oxygen candle at $r = 9$ mm has the best performance in terms of oxygen generation, and the axial compressive capacity and radial compressive capacity are 2.89 kN and 1.22 kN, respectively. Its oxygen generation is 48.75 L, the average OGR is 9.2 L/min, and the temperature of the reaction vessel is 79.4°C.

5.2. Analysis of the XRD Experiment. We collected the residual solids of the oxygen candles with $r = 9$ mm after combustion and grinded them into powder for XRD analysis. X-ray diffraction analysis was carried out on a Bruker D8 Focus diffractometer, whose light source is radiation from a copper target of an X-ray tube. The tube voltage is 40 kV, the tube electricity is 30 mA, the scanning angle is from 20 to 120°, and the step width (relative to 2θ) is 0.02°, used to test the chemical composition of the oxygen candles after combustion.

It can be seen from Figure 11 that the main compositions of the hollow-structured oxygen candle with $r = 9$ mm after combustion are halite (sodium chloride), hausmannite (manganese tetroxide), manganosite (manganese oxide), and cobalt (II, III) oxide. The results show that all the sodium ions in the combustion products exist in the form of sodium chloride. There are two kinds of combustion products of Mn powder, namely, manganese oxide and manganese tetroxide. The existence of manganese oxide indicates that some Mn powder does not react completely to form a stable compound manganese dioxide. The existence of manganese tetroxide indicates that oxygen produced a violent oxidation reaction of some manganese powder. Because the kaolin is mainly composed of aluminum oxides, silicon oxides, and crystalline water, the relative molecular mass is large, resulting in less aluminum and silicon content, so we did not detect the existence of aluminum oxides and silicon oxides. The above test and analysis results show that the hollow-structured oxygen candle with $r = 9$ mm is completely reacted.

6. Conclusions

We have dissected and analyzed the solid-structured oxygen candles after combustion and proposed the concept of the hollow-structured oxygen candle. According to the laboratory experiment results, we drew the following conclusions:

Table 2: The number, the weight, and the comprehensive score of each indicator.

Index	$r = 5$ mm	$r = 9$ mm	$r = 12$ mm	$r = 15.5$ mm	Weight
The compressive strength	4	3	2	1	0.2
The oxygen generation	1	4	3	2	0.5
The average OGR	4	3	2	1	0.2
The temperature of the reactor vessel	4	3	2	1	0.1
The comprehensive score	2.5	3.5	2.5	1.5	—

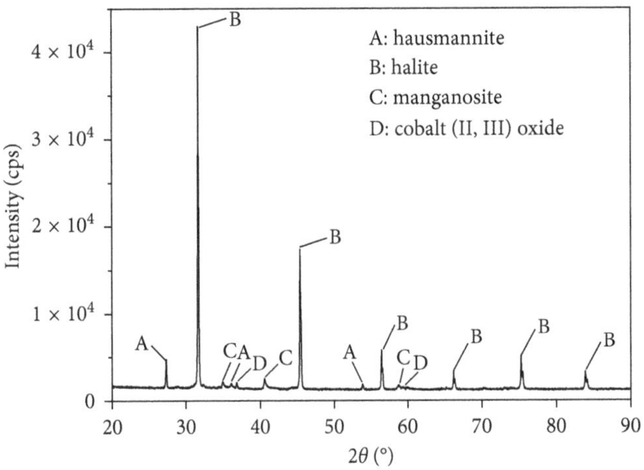

Figure 11: The results of the XRD experiment.

(1) With the increase in r, the compressive strength of the hollow-structured oxygen candle decreases. The lower the compressive strength, the poorer the structural stability, which is adverse for manufacturing oxygen candles. When $r = 15.5$ mm, the axial compressive capacity and radial compressive capacity of the hollow-structured oxygen candle account for 31.8% and 8.2% of those of the solid-structured oxygen candle, respectively. And when $r = 20$ mm, the oxygen candle cannot be formed.

(2) The hollow structure is beneficial in improving the performance of the oxygen candle. In our experiments, the hollow-structured oxygen candle can increase the oxygen generation ratio to 98.3%, an increase of 28.9% compared with that of the solid-structured one.

(3) A comprehensive analysis of the four indicators shows that the oxygen candle at $r = 9$ mm has the best performance of oxygen generation among the six sizes of hollow-structured oxygen candles. Its axial compressive strength and radial compressive strength are 2.89 kN and 1.22 kN, respectively, and the total oxygen generated is 48.75 L. The average OGR is 9.2 L/min, and the highest temperature of its reactor vessel is 79.4°C.

Conflicts of Interest

The authors declare that they have no conflicts of interest.

Acknowledgments

This work was supported by the National Natural Science Foundation of China (Grant No. 51504017), China Postdoctoral Science Foundation (Grant No. 2014T70039), Fundamental Research Funds for the Central Universities (Grant No. FRF-TP-15-043A3), and Fundamental Research Funds of China Academy of Safety Science and Technology (Grant Nos. 2016JBKY07 and 2017JBKY11).

References

[1] L. Z. Jin, Y. Wang, S. Wang, and Z. N. Zhan, "Distribution regularity study of air pressure supply to mine refuge chambers," *Journal of University of Science and Technology Beijing*, vol. 36, no. 8, p. 1007, 2014.

[2] Y. X. Jia, Y. S. Liu, W. H. Liu, and Z. Y. Li, "Study on purification characteristic of CO_2 and CO within closed environment of coal mine refuge chamber," *Separation and Purification Technology*, vol. 130, pp. 65–73, 2014.

[3] J. L. Yang, L. W. Yang, and J. Wei, "Study on open-cycle carbon dioxide refrigerator for movable mine refuge chamber," *Applied Thermal Engineering*, vol. 52, no. 2, pp. 304–312, 2013.

[4] E. I. Trushliakov, "Indoor air comfort for human life support in living compartments of manned submersibles," *SAE Technical Paper Series*, vol. 1, p. 2154, 2006.

[5] M. Bamsey, T. Graham, M. Stasiak et al., "Canadian advanced life support capacities and future directions," *Advances in Space Research*, vol. 44, no. 2, pp. 151–161, 2009.

[6] J. P. Sun, "Research on key technologies of refuge chamber and rescue capsule in underground mine," *Journal of China Coal Society*, vol. 36, no. 5, pp. 713–717, 2011.

[7] N. Gao, L. Z. Jin, L. Wang, and F. You, "Research and application of oxygen supply system in Changcun coal mine refuge haven," *Journal of China Coal Society*, vol. 37, no. 6, pp. 1021–1025, 2012.

[8] W. H. Schechter, R. R. Miller, R. M. Bovard, C. B. Jackson, and J. R. Pappenheimer, "Chlorate candle as a source," *Industrial and Engineering Chemistry*, vol. 42, no. 11, pp. 2348–2353, 1950.

[9] J. Graf, C. Dunlap, J. Haas et al., "Development of a solid chlorate backup oxygen delivery system for the international space station," in *Proceedings of 30th International Conference on Environmental System*, p. 2348, Toulouse, France, July 2000.

[10] E. Shafirovich, A. S. Mukasyan, A. Varma, G. Kshirsagar, Y. Zhang, and J. C. Cannon, "Mechanism of combustion in low-exothermic mixtures of sodium chlorate and metal fuel," *Combustion and Flame*, vol. 128, no. 1-2, pp. 133–144, 2002.

[11] X. Han, "On emergency oxygen-generation by using chlorate candle for passenger cars," *Journal of Safety and Environment*, vol. 5, no. 5, p. 43, 2005.

[12] Y. C. Zhang, G. Kshirsagar, and J. C. Cannon, "Functions of barium peroxide in sodium chlorate chemical oxygen generators," *Industrial and Engineering Chemistry Research*, vol. 32, no. 5, pp. 966–969, 1993.
[13] E. Shafirovich, C. J. Zhou, A. S. Mukasyan et al., "Combustion fluctuations in low-exothermic condensed systems for Emergency Oxygen Generators," *Combustion and Flame*, vol. 135, no. 4, pp. 557–561, 2003.
[14] E. Shafirovich, C. J. Zhou, S. Ekambaram, A. Varma, G. Kshirsagar, and J. E. Ellison, "Catalytic effects of metals on thermal decomposition of sodium chlorate for emergency oxygen generators," *Industrial and Engineering Chemistry Research*, vol. 46, no. 10, pp. 3073–3077, 2007.
[15] A. McCarrick, J. Haas, K. Johnson, and R. Hagar, "US Navy sodium chlorate oxygen candle safety," in *Proceedings of 41st International Conference on Environmental Systems*, p. 21, Portland, OR, USA, July 2011.
[16] X. M. Zhou, X. Hu, S. H. Mao, Y. Zhang, and L. Mao, "Test of oxygen candle as backup oxygen for manned space station," *Space Medicine and Medical Engineering*, vol. 26, no. 5, p. 394, 2013.
[17] L. Z. Jin, S. Wang, S. C. Liu, and Z. Zhang, "Development of a low oxygen generation rate chemical oxygen generator for emergency refuge spaces in underground mines," *Combustion Science and Technology*, vol. 187, no. 8, pp. 1229–1239, 2015.
[18] S. Wang, L. Z. Jin, S. C. Liu, and Z. Zhang, "Study on chemical oxygen source in underground emergency refuge system," in *Proceedings of 2nd International Symposium of Mine Safety Science and Engineering*, p. 571, Taylor & Francis–Balkema, Beijing, China, September 2013.
[19] Z. Zhang, L. Z. Jin, S. C. Liu, S. Wang, N. Gao, and J. Li, "Study on optimization of oxygen candle formula for refuge chamber," *China Safety Science Journal*, vol. 23, no. 9, pp. 129–135, 2013.
[20] J. G. Liu, L. Z. Jin, N. Gao, S. Wang, and H. Zhang, "Catalytic effect of Mn particle size on thermal decomposition of sodium chlorate in oxygen generators," *Chinese Journal of Engineering*, vol. 39, no. 8, p. 1159, 2017.
[21] Y. H. Zhang, Z. A. Huang, and Y. K. Gao, *Combustion and Explosion*, Metallurgical Industry Press, Beijing, China, 2015.
[22] Y. H. Li, *Combustion Theory and Technology*, China Electric Power Press, Beijing, China, 2011.
[23] State Administration of Work Safety, *State Administration of Coal Mine Safety*, Coal Mine Safety Regulations, Beijing, China, 2016.

20

Synthesis and Antitumor Activity of Novel Pyridoxine-Based Bioisosteric Analogs of *trans*-Stilbenes

Mikhail V. Pugachev,[1] Thang T. N. Nguyen,[1] Timur M. Bulatov,[1] Roman S. Pavelyev,[1] Alfia G. Iksanova,[1] Oksana V. Bondar,[1] Konstantin V. Balakin,[1,2] and Yurii G. Shtyrlin[1]

[1]*Kazan (Volga Region) Federal University, Kremlyovskaya 18, Kazan 420008, Russia*
[2]*I.M. Sechenov First Moscow State Medical University, Trubetskaya 8, Bldg 2, Moscow 119991, Russia*

Correspondence should be addressed to Yurii G. Shtyrlin; yurii.shtyrlin@gmail.com

Academic Editor: Ponnurengam Malliappan Sivakumar

A series of *trans*-6-phenylethenyl substituted pyridoxine derivatives, novel bioisosteric analogs of drugs based on *trans*-stilbene scaffold, were synthesized using the Wittig reaction of a bis-triphenylphosphonium pyridoxine derivative with various aromatic aldehydes. Two compounds demonstrated high activity against the estrogen-dependent MCF-7 (breast cancer) cell line with IC_{50} in the range of 1.9–7.9 μM and very good selectivity for other studied normal and tumor cells, including the estrogen receptor negative MDA-MB-231 breast cancer cells. The active compounds possessed an intense blue fluorescence, and this feature allowed us to effectively visualize them in cytoplasm and in nucleus. The obtained results make the described chemotype a promising starting point for the development of new anticancer agents for the therapy of estrogen-dependent malignancies.

1. Introduction

Substituted derivatives of pyridoxine (vitamin B6) carrying alkenyl substituents at various positions of the pyridine ring represent interesting molecular structures with a rich bioactivity potential. In particular, pyridoxine molecules carrying a 5-ethenyl substituent possess anti-inflammatory properties [1], 6-phenylethenyl substituted pyridoxines have been described as purine receptor antagonists [2], and 2-ethenyl derivatives possess antitumor activity [3]. In our group, we have systematically studied the chemistry and biological activity of pyridoxine derivatives (e.g., [4–8]). Recently, we described a number of *cis*- and *trans*-5-alkenyl substituted pyridoxines which demonstrated promising antitumor activity [7]. In a more general context, variously substituted pyridoxine derivatives represent valuable therapeutic and diagnostic agents in actual therapeutic areas (for some examples, see [9–12]).

In this work, we studied a series of *trans*-6-phenylethenyl substituted pyridoxine derivatives for their cytotoxic activity against a panel of normal and cancer cell lines. The underlying idea was to design a novel chemotype of bioisosteric analogs of the well-known estrogen receptor targeting drugs such as diethylstilbestrol, tamoxifen, and raloxifene (Figure 1). All these drugs are bioisosteric mimetics of the natural estrogen receptor (ER) agonists such as estradiol (**I**). Their characteristic feature is the presence in their molecular structures of a *trans*-diarylethylene fragment (structures **II–IV**). These agents also typically contain the hydroxy or alkoxy groups at the remote axial positions of the molecule (e.g., at *para*-positions of phenyl rings). In some cases, for example, in tamoxifen, the hydroxyl groups appear as a result of metabolic transformations in human body. Such molecular features provide electronic and spatial compatibility to the ER active site. The mentioned drugs are clinically important ER modulators, widely used in the therapy of the estrogen-dependent malignancies. For example, diethylstilbestrol is an inexpensive and effective option for prostate cancer treatment before chemotherapy [13]. Persistently increased blood levels of estrogens are associated with an increased risk of breast cancer. Selective estrogen receptor modulators (SERMs), such as tamoxifen and raloxifene, are a class of compounds actively used for breast cancer prevention [14].

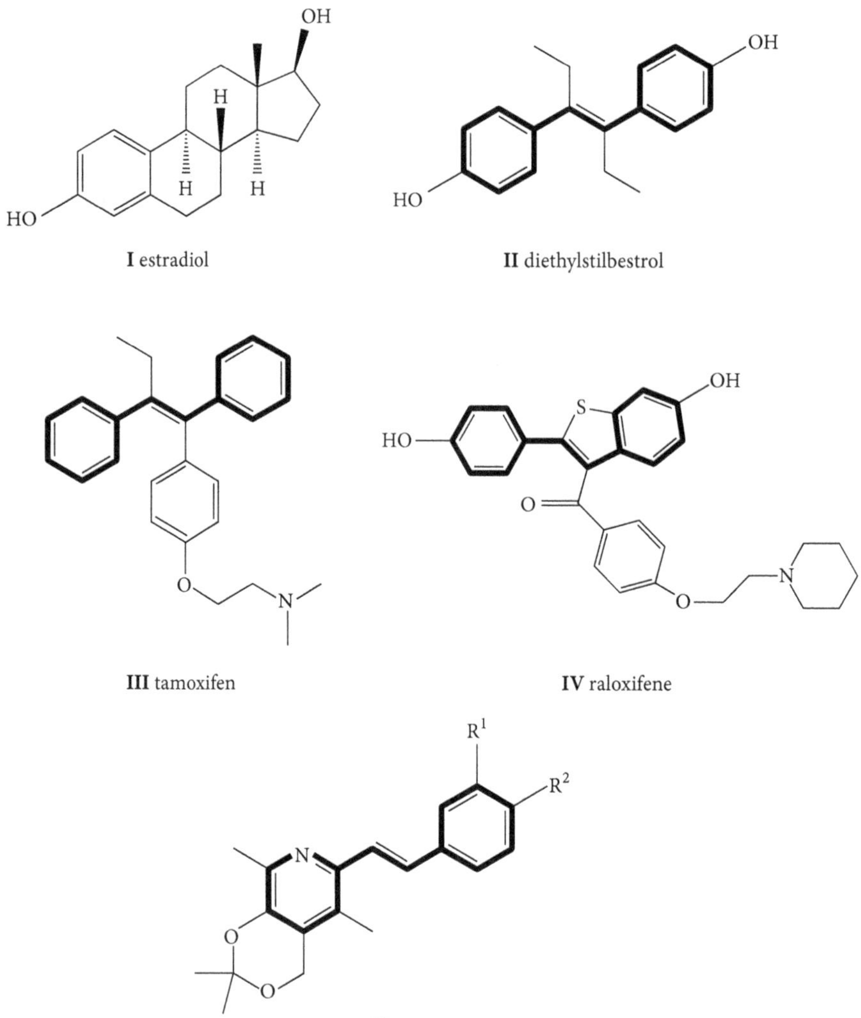

FIGURE 1: The natural ER agonist estradiol (**I**), examples of ER targeting drugs (**II–IV**), and compounds studied in this work (**V**); highlighted is the *trans*-diarylethylene motif characteristic of the bioisosteric analogs of the natural agonist.

Compounds of general formula **V** (Figure 1) studied in this work represent the substituted azaheterocyclic analogs of *trans*-stilbene. The presence in their structures of the characteristic *trans*-diarylethylene fragment makes them novel bioisosteric analogs of the natural and synthetic drugs based on *trans*-stilbene scaffold, and this feature prompted us to investigate their potential as selective estrogen-dependent cytotoxic agents. The synthetic route to compounds **V** has been reported in our recent work [8]. Herein, we report synthesis of one novel compound from this structural series and also describe biological and physicochemical properties of this interesting chemotype. In accordance with the underlying hypothesis, we have found that compounds of general formula **V**, which contain the hydroxy or alkoxy groups at the remote axial positions of the molecule, possess high activity against an estrogen-dependent breast cancer cell line and very good selectivity for other studied normal and tumor cells. The obtained compounds also possess an intense blue fluorescence that makes it possible to visualize them in an intracellular environment. The obtained results suggest promising potential of the active compounds as new anticancer agents potentially useful in the therapy of the estrogen-dependent malignancies such as breast cancer or as novel fluorescent probes for molecular biological applications.

2. Materials and Methods

2.1. Synthetic Procedures. Chromatographic purification of compounds was carried out using column chromatography on Acros silica gel (60–200 mesh). Reaction progress and purity of compounds were monitored by TLC on Sorbfil PTLC-AF-A-UF plates. Melting points of the products were determined using a Stanford Research Systems MPA-100 OptiMelt appliance. ^{1}H and ^{13}C NMR spectra were recorded on a Bruker Avance 400 spectrometer (400.17 and 100.62 MHz). Signals of chloroform (δH 7.26, δC 77.16) were used as references in the ^{1}H and ^{13}C NMR spectra. Coupling constants (J) are reported in Hz. HPLC/MS-experiment was carried out using a Triple TOF 5600, AB Sciex superhigh resolution mass spectrometer from the solution in methanol by turboionic spray (TIS) ionization method with the collision energy with nitrogen molecules of 10 eV. The excitation and emission spectra have been recorded using TECAN M200 Pro plate reader.

Synthesis and analytical characteristics of compounds **4a**, **b**, **5b**, and **6a–c** were reported in our recent papers [7, 8].

(E)-6-(4-Hydroxy-3-methoxystyryl)-2,2,5,8-tetramethyl-4H-[1,3]dioxino[4,5-c]pyridine (4c). Sodium hydride 0.22 g (9.0 mmol) and 4-hydroxy-3-methoxybenzaldehyde 0.46 g (3.00 mmol) were sequentially added to a solution of 1.20 g (1.5 mmol) 5,6-bis(triphenylphosphonio(methyl))-2,2,8-trimethyl-4H-[1,3]dioxino[4,5-c]pyridine dichloride **2** in dichloromethane (30 mL), and the reaction mixture was stirred for 30 hours at reflux. Then, the precipitate was filtered off, and the solvent was removed in vacuo. The dry residue was purified using silica gel column chromatography (eluent ethyl acetate–petroleum ether, 1 : 15). Yield 0.06 g (11%), a yellow crystalline compound, m.p. 99–101°C. R_f 0.5 (ethyl acetate : petroleum ether 1 : 2). ^{1}H NMR (400 MHz, $CDCl_3$) δ, ppm: 1.54 (s, 6H, $C(CH_3)_2$); 2.17 (s, 3H, CH_3); 2.42 (s, 3H, CH_3); 3.91 (s, 3H, CH_3O); 4.76 (s, 2H, CH_2O); 6.89 (ABX, ^3JHH = 8.0 Hz, C_6H_3); 7.04 (ABX, ^4JHH = 1.6 Hz, C_6H_3); 7.10 (ABX, ^3JHH = 8.0 Hz, ^4JHH = 1.6 Hz, C_6H_3); 7.13, 7.52 (AB, ^3JHH = 15.6 Hz, -CH=CH-). ^{13}C NMR (100 MHz, $CDCl_3$) δ, ppm: 12.7 (CH_3); 18.7 (CH_3); 24.8 ($C(CH_3)_2$); 56.0 (CH_3O); 59.8 (CH_2O); 99.3 ($C(CH_3)_2$); 109.4 (CHAr); 114.7 (CHAr); 120.5 (CHAr); 122.0 (-CH=CH-); 123.5 (CAr); 125.4 (CAr); 130.4 (CAr); 131.7 (-CH=CH-); 144.3 (CAr); 144.9 (CAr); 145.1 (CAr); 145.8 (CAr); 146.8 (CAr). MS (HRMS-ESI): found [M+H]+ 342.1700. $C_{20}H_{24}NO_4$. Calculated [M+H] 342.1700.

2.2. In Vitro Cytotoxicity Studies. Breast adenocarcinoma MCF-7 (ATCC® HTB-22™), breast adenocarcinoma MDA-MB-231 (ATCC HTB-26™), glioblastoma SNB-19 (ATCC CRL-2219™), human colon colorectal adenocarcinoma HCT-15 (ATCC CCL-225™), and prostate adenocarcinoma PC-3 (GSM136316); kidney carcinoma A-498 (GSM136294); melanoma M-14 (GSM136320); ovary adenocarcinoma OVCAR-4 (GSM136312); non-small-cell lung carcinoma NCI-H322M (GSM136307); colorectal carcinoma HCT-116 (GSM136288), primary human foreskin fibroblast cells (HSF) isolated from the skin explant according to conventional protocol [15], and human embryonic kidney cells HEK-293 (ATCC CRL_1573™) were cultured in α-MEM (PanEko, Russia) supplemented with 10% fetal bovine serum (PAA, Australia), L-glutamine, and 1% penicillin-streptomycin at 37°C in a 5% humidified CO_2 atmosphere in air. Cells were removed from the culture substrate by treatment with trypsin-EDTA with subsequent inactivation of trypsin by adding α-MEM-containing serum. Cell suspensions were precipitated by centrifugation at 500g and pellet was resuspended in phosphate buffered saline (PBS). Cell viability and density were measured in a Neubauer chamber by using 0.4% solution of trypan blue. Suspensions with the amount of viable cells no less than 90% were used in the experiments.

The effect of synthesized compounds on the proliferation of the cells was measured by MTT-assay (3-(4,5-dimethylthiazol-2-yl)-2,5-diphenyltetrazolium bromide) (Promega, USA). The cells were seeded in a 96-well plate at a concentration of 2000 cells per well and were cultured in supplemented α-MeM (180 μL) according to the standard culturing conditions for 1 day. The cell viability was measured at 72 h after treatment with the test agent (20 μL). Then, the medium was exchanged for the supplemented medium (80 μL). MTT solution (5 mg mL^{-1}) was added to a volume of 20 μL in each well and incubated for 3.5 h. Then the solution was removed and DMSO (100 μL) was added to solubilize the formazan crystals. After 10 min, absorbance was measured at 555 nm (the reference wavelength 650 nm) by a TECAN plate reader (Switzerland). The results were presented as the percent ratio to the control sample treated with PBS. For each compound tested, the IC_{50} values ($P < 0.05$) were generated from the dose-response curves for each cell line.

2.3. Visual Microscopy. Intracellular accumulation and distribution of compound **4b** was assessed by confocal microscopy. Breast cancer cells MCF-7 were seeded into glass slides on a 6-well cultural plate in number of 200000 cells per well. Next day, compound **4b** was added to each well in cultural media from the stock solution in DMSO to reach final concentration of 28 μM and maintained 20 minutes at 37°C. Then the cells were washed by Hanks's solution and fixed for 10 min at 4°C in 4% paraformaldehyde. Then the cover slides with the cells were washed again and transferred to microscope slides with the use of a mounting medium Shandon Consul-Mount (Thermo Scientific). The received slides were analyzed on Zeiss LSM 800 confocal laser scanning microscope with ×100 immersion objective and ×20 air objective. The samples with compound **4b** were excited using 405 nm laser and detected in the range of 410–476 nm.

3. Results and Discussion

The studied compounds were obtained from pyridoxine hydrochloride **1** according to our previously reported approach [8] (Scheme 1). The Wittig reactions of **2** monohydrate with various aromatic aldehydes **3a–c** were performed in CH_2Cl_2 in the presence of NaH (6 equiv.) at reflux for 30 hours. The target 5-methyl-6-alkenyl substituted derivatives **4a–c** with a *trans*-configuration of the alkene fragment were the major products of this reaction. The reaction typically led to a number of additional products such as *cis*-alkenes, 5,6-bis-alkenyl substituted compounds, and some minor hyperconjugated structures. Compounds **4a–c** and a *cis*-alkene compound **5b** could be separated from other products by silica gel column chromatography and were obtained in low to moderate isolated yields (9–30%) as white and yellow crystalline substances. Characteristic features of compounds **4a–c** were the signals of *trans*-alkene protons at 6.50–7.70 ppm (AB-pattern, $^3J_{HH}$ = 15.6 Hz). Synthesis and analytical characteristics of compounds **4a** and **b** and **5b** have been reported in our recent work [8], while compound **4c** is described in this paper for the first time.

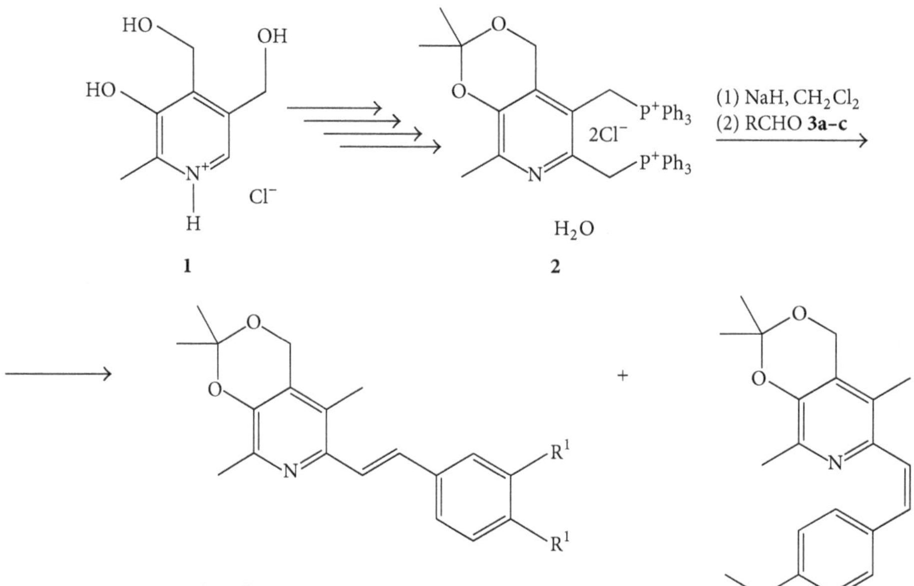

4a, $R^1 = R^2 = H$, 30% [8]
4b, $R^1 = R^2 = OCH_3$, 21% [8]
4c, $R^1 = OCH_3$, $R^2 = OH$, 11%

5b, 9% [8]

SCHEME 1: Synthesis of compounds studied in this work.

TABLE 1: Results of the cytotoxicity screening experiment against two cell lines.

	Cytotoxicity, IC_{50} μM		Selectivity index,
	HEK-293	MCF-7	$IC_{50\ HEK\text{-}293}/IC_{50\ MCF\text{-}7}$
4a	2.7 ± 0.3	81.2 ± 6.5	0.03
4b	22.8 ± 2.4	1.9 ± 0.9	12.0
4c	>293.0	7.9 ± 4.9	37.0
5b	23.1 ± 2.3	32.4 ± 2.6	0.7
6a	36.1 ± 0.1	67.8 ± 1.0	0.5
6b	>306.0	>306.0	1.0
6c	>320.0	>320.0	1.0
Doxorubicin	1.2 ± 0.1	1.2 ± 0.1	1.0

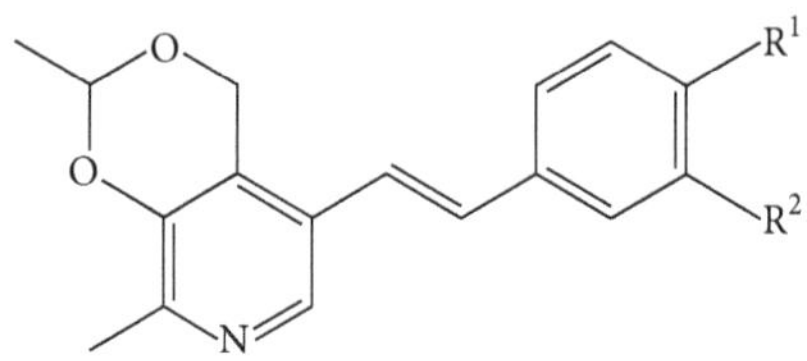

6a, $R^1 = R^2 = H$

6b, $R^1 = R^2 = OCH_3$

6c, $R^1 = OCH_3$, $R^2 = OH$

FIGURE 2: 5-Phenylethenyl substituted pyridoxine derivatives described in our recent work [5] and studied in the cytotoxicity screening experiment.

The obtained compounds were screened for cytotoxicity against two cell lines, the estrogen-dependent MCF-7 (breast cancer) cells and the human embryonic kidney 293 (HEK-293) cells (Table 1). To explore SAR dependencies, we also tested the three structurally related *trans*-5-phenylethenyl substituted pyridoxine derivatives **6a–c** (Figure 2) described in our recent work [7]. The widely used anticancer anthracycline antibiotic doxorubicin was used as a broad-spectrum reference drug. In this screening experiment, compounds **4b** and **4c** demonstrated high cytotoxic activity against the MCF-7 cells with IC_{50} in the range of 1.9–7.9 μM, which is comparable to cytotoxic activity of the reference drug (IC_{50} 1.2 μM). At the same time, the cytotoxic activity of **4b** and **4c** against the normal HEK-293 cells (IC_{50} 22.8 and >293 μM, respectively) was significantly lower than that of doxorubicin (IC_{50} 1.2 μM). As a result, compounds **4b** and **4c** have significantly better selectivity index (SI 12 and 37, respectively) as compared to doxorubicin (SI 1.0).

Of note, the active compounds contain in their structures the hydroxy or methoxy groups in *para*-position of the phenyl ring, which is consistent with the general aforementioned SAR requirements for this class of ER modulators [16].

Compounds **4b** and **4c** were then tested against an extended panel of normal and cancer cell lines (Table 2). In agreement with our underlying theoretical hypothesis, a very good selectivity for the estrogen-dependent MCF-7 cell line was observed for the active compounds as compared to other

TABLE 2: Cytotoxicity of compounds **4b** and **4c** against an extended panel of normal and tumor cells.

	IC_{50}, μM											
	HSF	HEK-293	MCF-7	PC-3	MDA-MB-231	A-498	SNB-19	M-14	OVCAR-4	NCI-H322M	HCT-15	HCT-116
4b	291.0 ± 23.0	22.8 ± 2.4	1.9 ± 0.9	308.0 ± 6.0	18.7 ± 2.3	290.5 ± 12.9	745.7 ± 26.0	340.8 ± 32.3	441.4 ± 0.5	368.1 ± 10.7	234.6 ± 53.9	79.1 ± 0.9
4c	66.5 ± 3.0	>292.9	7.9 ± 4.9	48.6 ± 3.3	n.m.*	43.7 ± 5.1	86.5 ± 2.2	82.4 ± 0.5	89.6 ± 8.5	66.9 ± 1.5	48.1 ± 4.8	80.7 ± 8.0
Doxorubicin	2.2 ± 0.2	1.2 ± 0.1	1.2 ± 0.1	14.1 ± 1.0	0.12 ± 0.01	12.2 ± 0.9	6.1 ± 0.5	3.4 ± 0.2	67.3 ± 3.0	124.0 ± 14.0	2.2 ± 0.3	0.38 ± 0.12

*n.m.: not measured.

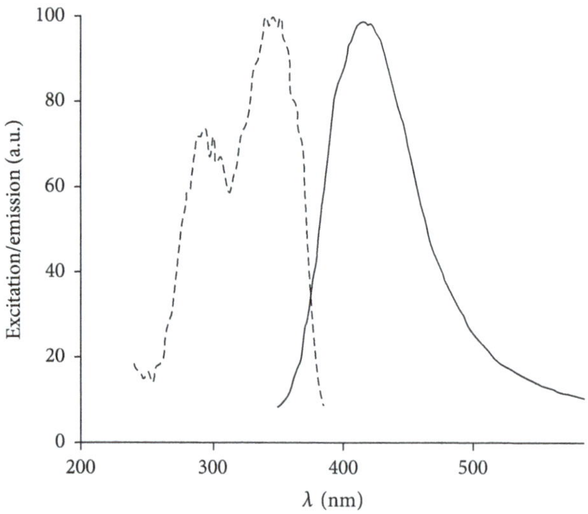

FIGURE 3: The excitation (dashed line) and emission (solid line) spectra of compound **4b** (a.u.: arbitrary units).

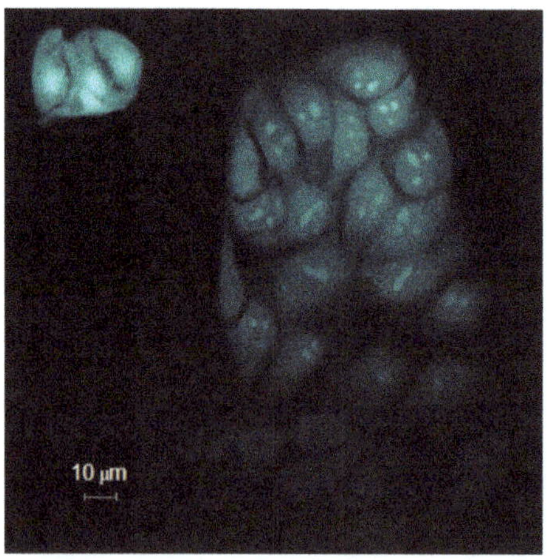

FIGURE 4: Microphotograph of MCF-7 cells incubated with **4b** (28 μM, 20 min), λ_{em} 420 nm.

studied cells including the normal ones such as human fibroblast cells (HSF) and human embryonic kidney 293 (HEK-293) cells. Thus, the MCF-7 versus HSF selectivity indexes were 153 and 8.4 for compounds **4b** and **4c**, respectively (not shown in the table). The leading compound **4b** also demonstrated clear selectivity for MCF-7 as compared to ER negative MDA-MB-231 breast cancer cells with the selectivity index ~10. This important observation suggests a potential role of estrogen receptors as primary targets for the active compounds.

An interesting and practically significant property of the studied compounds is their intense blue fluorescence with the maximum at 416 nm (Figure 3). The fluorescence appears as a result of π-conjugation in the *trans*-diarylethylene scaffold.

This feature allowed us to assess their cellular permeability using the visible light microscopy. Figure 4 demonstrates that one of the leading compounds **4b** effectively penetrates through cellular membrane. It can be effectively visualized in cytoplasm and, moreover, is accumulated in nucleus, the cellular compartment where the ERs are localized.

The reasons for the observed effective penetration of the studied compounds through the cell membrane are probably related to their relatively high lipophilicity (log P in the range of 4.3–4.5) and the presence of a basic aromatic nitrogen atom, which are optimal for effective passive diffusion through the lipid bilayers. On the other hand, the presence of pyridoxine moiety can enhance transmembrane transport of the obtained derivatives via active mechanisms. First of all, many cells have specific pyridoxine transporters in their membranes (e.g., [17]). Pyridoxine is a cofactor for more than 140 essential enzymatic reactions, and pyridoxine-modified molecules are often recognized by many pathogenic cells and microorganisms as endogenous. As a result, they have increased cellular permeability and increased bioactivity. For example, the development of pyridoxine-modified nanoparticles for efficient intracellular delivery of doxorubicin (DOX-B6-SA-NP) was reported [18]. It was demonstrated that the treatment with DOX-B6-SA-NP kept higher doxorubicin accumulation inside the cells than conventional lipid nanoparticles.

4. Conclusion

Herein, we have described a novel pyridoxine-based chemotype of bioisosteric analogs of the natural and synthetic drugs based on *trans*-stilbene scaffold. Despite inherent complications, the synthetic route to these compounds is rather straightforward and can be performed in an industrial scale. The active compounds **4b** and **4c** demonstrate high activity against the estrogen-dependent MCF-7 (breast cancer) cell line with IC_{50} in the range of 1.9–7.9 μM and very good selectivity for other studied tumor and normal cells. The leading compound **4b** also demonstrated clear selectivity for estrogen-dependent breast cancer MCF-7 cells as compared to ER negative MDA-MB-231 breast cancer cells (selectivity index 10), suggesting a potential role of estrogen receptors as targets for the active compounds. The active compounds possessed an intense blue fluorescence with the maximum at 410–420 nm, and this feature allowed us to assess their cellular permeability using the confocal microscopy. Thus, compound **4b** could be effectively visualized in cytoplasm and in nucleus, where the ERs are localized.

The obtained results make the described chemotype a promising starting point for the development of new anticancer agents potentially useful in the therapy of the estrogen-dependent malignancies such as breast cancer or as a novel fluorescent probe for molecular biological applications. The active compounds are currently under active investigation within our group, and their in vitro and in vivo properties will be reported and discussed in a complementary paper.

Conflicts of Interest

The authors declare that there are no conflicts of interest regarding the publication of this manuscript.

Acknowledgments

This work was supported by Program of Competitive Growth of Kazan Federal University.

References

[1] T.-Y. Shen, H. Jones, and C. P. Dorn, *Mercaptoalkylpyridine Disulfides*, U.S. Patent 4053606 A, 1977.

[2] Y.-C. Kim, K.-Y. Jung, and J. H. Cho, *Novel Pyridine Carboxylic Acid Based Compound Used as a p2x1 and p2x3 Receptor Antagonist, a Production Method for the Same and a Composition Comprising the Same*, U.S. Patent 20130040997, 2011.

[3] W. Korytnyk, S. C. Srivastava, N. Angelino, P. G. G. Potti, and B. Paul, "A general method for modifying the 2-methyl group of pyridoxol. Synthesis and biological activity of 2-vinyl- and 2-ethynylpyridoxols and related compounds," *Journal of Medicinal Chemistry*, vol. 16, no. 10, pp. 1096–1101, 1973.

[4] M. V. Pugachev, N. V. Shtyrlin, L. P. Sysoeva et al., "Synthesis and antibacterial activity of novel phosphonium salts on the basis of pyridoxine," *Bioorganic and Medicinal Chemistry*, vol. 21, no. 14, pp. 4388–4395, 2013.

[5] M. V. Pugachev, N. V. Shtyrlin, S. V. Sapozhnikov et al., "Bisphosphonium salts of pyridoxine: the relationship between structure and antibacterial activity," *Bioorganic and Medicinal Chemistry*, vol. 21, no. 23, pp. 7330–7342, 2013.

[6] A. D. Strelnik, A. S. Petukhov, I. V. Zueva et al., "Novel potent pyridoxine-based inhibitors of AChE and BChE, structural analogs of pyridostigmine, with improved in vivo safety profile," *Bioorganic and Medicinal Chemistry Letters*, vol. 26, no. 16, pp. 4092–4094, 2016.

[7] M. V. Pugachev, R. S. Pavelyev, T. N. Nguyen, A. G. Iksanova, O. A. Lodochnikova, and Y. G. Shtyrlin, "Synthesis and antitumor activity of pyridoxine monoalkenyl derivatives," *Russian Chemical Bulletin*, vol. 65, no. 2, pp. 532–536, 2016.

[8] M. V. Pugachev, T. M. Bulatov, T. T. Nguyen et al., "Wittig reactions of a bis-triphenylphosphonium pyridoxine derivative," *Tetrahedron Letters*, vol. 58, no. 8, pp. 766–769, 2017.

[9] B. R. Stranix, J. J. Wu, G. Milot et al., "Pyridoxine hydroxamic acids as novel HIV-integrase inhibitors," *Bioorganic and Medicinal Chemistry Letters*, vol. 26, no. 4, pp. 1233–1236, 2016.

[10] J.-X. Yu, W. Cui, V. A. Bourke, and R. P. Mason, "6-trifluoromethylpyridoxine: novel 19F NMR pH indicator for in vivo detection," *Journal of Medicinal Chemistry*, vol. 55, no. 15, pp. 6814–6821, 2012.

[11] W. Zhang, J. Yao, V. Pham et al., "Pyridoxine as a template for the design of antiplatelet agents," *Bioorganic and Medicinal Chemistry Letters*, vol. 14, no. 18, pp. 4747–4750, 2004.

[12] V. Pham, W. Zhang, V. Chen et al., "Design and synthesis of novel pyridoxine 5′-phosphonates as potential antiischemic agents," *Journal of Medicinal Chemistry*, vol. 46, no. 17, pp. 3680–3687, 2003.

[13] R. Turo, M. Smolski, R. Esler et al., "Diethylstilboestrol for the treatment of prostate cancer: Past, present and future," *Scandinavian Journal of Urology*, vol. 48, no. 1, pp. 4–14, 2014.

[14] F. Li, J. Dou, L. Wei, S. Li, and J. Liu, "The selective estrogen receptor modulators in breast cancer prevention," *Cancer Chemotherapy and Pharmacology*, vol. 77, no. 5, pp. 895–903, 2016.

[15] L. F. G. Rittie, *Isolation and Culture of Skin Fibroblasts*, Humana Press, Totowa, NJ, USA, 2010.

[16] A. Jameera Begam, S. Jubie, and M. Nanjan, "Estrogen receptor agonists/antagonists in breast cancer therapy: a critical review," *Bioorganic Chemistry*, vol. 71, pp. 257–274, 2017.

[17] N. Szydlowski, L. Bürkle, L. Pourcel, M. Moulin, J. Stolz, and T. B. Fitzpatrick, "Recycling of pyridoxine (vitamin B6) by PUP1 in Arabidopsis," *Plant Journal*, vol. 75, no. 1, pp. 40–52, 2013.

[18] S. Sharma, A. Verma, J. Singh et al., "Vitamin B6 tethered endosomal PH responsive lipid nanoparticles for triggered intracellular release of doxorubicin," *ACS Applied Materials and Interfaces*, vol. 8, no. 44, pp. 30407–30421, 2016.

Synthesis and Anticancer Activities of 4-[(Halophenyl)diazenyl]phenol and 4-[(Halophenyl)diazenyl]phenyl Aspirinate Derivatives against Nasopharyngeal Cancer Cell Lines

Boon Kui Ho,[1] Zainab Ngaini,[1] Paul Matthew Neilsen,[2,3] Siaw San Hwang,[2] Reagan Entigu Linton,[2] Ee Ling Kong,[2] and Boon Kiat Lee[2]

[1] *Department of Chemistry, Faculty of Resource Science and Technology, Universiti Malaysia Sarawak, 94300 Kota Samarahan, Sarawak, Malaysia*
[2] *Cancer Research and Advanced Therapeutic Group (CREATE), Faculty of Engineering, Computing and Science, Swinburne University of Technology Sarawak Campus, Jalan Simpang Tiga, 93350 Kuching, Sarawak, Malaysia*
[3] *School of Health, Medical and Applied Sciences, Central Queensland University, North Rockhampton, QLD 4702, Australia*

Correspondence should be addressed to Zainab Ngaini; nzainab@unimas.my

Academic Editor: Liviu Mitu

Aspirin and azo derivatives have been widely studied and have drawn considerable attention due to diverse biological activities. In this study, a series of 4-[(halophenyl)diazenyl]phenyl aspirinate derivatives were synthesized from the reaction of aspirin with 4-[(halophenyl)diazenyl]phenol via esterification, in the presence of DCC/DMAP in DCM with overall yield of 45–54%. 4-[(Halophenyl)diazenyl]phenol was prepared prior to esterification from coupling reaction of aniline derivatives and phenol in basic solution. All compounds were characterized using elemental analysis, FTIR, and ^{1}H and ^{13}C NMR spectroscopies. All compounds were screened for their anticancer activities against nasopharyngeal cancer (NPC) HK-1 cell lines and the viability of cultured cells was determined by MTS [3-(4,5-dimethylthiazol-2-yl)-5-(3-carboxylmethoxylphenyl)-2-(4-sulfophenyl)-2H-tetrazolium]-based colorimetric assay. 4-[(E)-(Fluorophenyl)diazenyl]phenol showed the highest anticancer activity against NPC HK-1 cell lines compared to other synthesized compounds. 4-[(Halophenyl)diazenyl]phenyl aspirinate showed low cytotoxicity against NPC HK-1 cell lines compared to 4-[(halophenyl)diazenyl]phenol but better anticancer activity than aspirin alone.

1. Introduction

Cancer has become a serious clinical problem worldwide [1]. It is because cancer cells have the ability to grow rapidly and develop into malignancy tumor [2]. Nasopharyngeal cancer is the most common malignant tumor and occurs in the nasopharynx, the upper part of the throat near to the neck [3]. Nasopharyngeal cancer (NPC) is commonly developed from infection with Epstein-Barr virus, foods, and family history [4]. NPC, which also known as HK-1, was found to have high level of epidermal growth factor receptor (EGFR) protein [5]. Hence, by targeting the EGFR protein, it might become a new therapeutic strategy to inhibit the HK-1 cells [6]. The usage of drugs could stop the growth of the malignant cells in one's body by inhibiting its EGFR pathway of the cancer cells itself [7].

The development of new drug for cancer treatment has gained much interest in recent years. The modification of pharmacologically active compounds may lead to a new finding of novel drugs with diverse pharmacological activities [8]. Chemical modification of aspirin, for instance, has been reported to reduce gastrointestinal toxicity and exhibited diverse biological activities such as anticancer [9], anti-inflammatory [10], and antibacterial [11, 12] activities. Aspirin bearing active functional groups such as diethyl phosphate and nitro groups has been reported to have anticancer activity against pancreatic and colon cancers [13, 14].

Azo is another active compound reported for its pharmaceutical importance as antidiabetic [15], antineoplastic [16], antibacterial [17], and anticancer [18] agent. Azo molecules are known to be involved in the inhibition of DNA, RNA, carcinogenesis, and protein synthesis [19]. The presence of -N=N- in the molecular structure of azo is responsible for the interaction with the active site of protein [20]. The presence of halogen substituents in azo molecule also played an important role in the inhibition of receptor enzyme [21].

In this study, a series of azo derivatives (**1a–d**) and aspirin azo derivatives (**2a–d**) have been synthesized and demonstrated for potential anticancer agent against nasopharyngeal cancer cell lines. The viability of the cultured cells was determined by MTS-based colorimetric assay and discussed.

2. Materials and Methods

2.1. Materials. Chemical agents used in the present study included dimethyl sulfoxide (DMSO, Sigma-Aldrich), penicillin-streptomycin (GIBCO), fetal bovine serum (FBS, Sigma-Aldrich), and RPMI 1640 (GIBCO).

2.2. Characterizations. Melting point was measured by Stuart SMP3 melting point apparatus and the elemental analysis was determined by flash EA1112 analyzer. IR spectra (v/cm^{-1}) were recorded on a Perkin Elmer GX spectrophotometer able to show infrared spectra of functional groups, potassium bromide pellet (KBr) solid material. ^1H NMR spectra were recorded on a JEOL.ECA 500 spectrometer at 500 MHz and ^{13}C NMR at 125 MHz. Tetramethylsilance (TMS) was used as a reference for both spectra while $CDCl_3$ and DMSO-d_6 were used as solvents.

*2.3. General Method for the Synthesis of Halogenated Azo Compounds **1a–d**.* A mixture of 4-fluoroaniline (0.56 g, 5 mmol) and HCl (2 M, 6 mL) was added slowly to $NaNO_2$ in water (1 M, 2 mL) at 0–5°C. Phenol (0.47 g, 5 mmol) was dissolved in NaOH (2 M, 10 mL) and cooled to 0–5°C. The diazo salt was added slowly to phenol solution at 0–5°C and stirred for 40 min. HCl (2 M, 10 mL) was added to form precipitate and filtered and washed with cold water (2 × 20 mL). The crude product was recrystallized from hot ethanol to give (**1a**) (0.56 g, 52%) as purple solid; m.p. 158-159°C. (Found: C, 66.58; H, 4.12; N, 12.80% $C_{12}H_9N_2OF$ Requires C 66.67; H, 4.17; N, 12.96%); R_f 0.54 (1 : 4 EA/hexane) IR: v_{max} (thin film/cm^{-1}) 3135 (OH), 1591 (C=C aromatic), 1471 (N=N), 1235 (C-N), 1102 (C-O), 839 (C-F) ^1H NMR (500 MHz, DMSO-d_6) ppm 6.94 (2H, d, J = 6.85 Hz, Ar-$H_{3'}$) 7.37 (2H, d, J = 8.78 Hz, Ar-H_3) 7.78–7.88 (4H, m, Ar-$H_{2,2'}$) 10.32 (1H, s, OH) ^{13}C NMR (125 MHz, $CDCl_3$) 116.00 (Ar-C_3), 116.21 (Ar-$C_{3'}$), 124.58 (Ar-$C_{2'}$), 124.65 (Ar-C_2), 125.13 (Ar-$C_{1'}$), 147.02 (Ar-C_1), 149.32 (Ar-$C_{4'}$), 158.63 (Ar-C_4). The others, **1b–d,** were synthesized by repeating the same procedure as **1a**.

4-[(E)-(Chlorophenyl)diazenyl]phenol (**1b**) (0.62 g, 53%) as yellow solid; m.p. 154-155°C. (Found: C, 61.68; H, 3.86; N, 11.98% $C_{12}H_9N_2OCl$ Requires C, 61.94; H, 3.87; N, 12.04%); R_f 0.57 (1 : 4 EA/hexane) IR: v_{max} (thin film/cm^{-1}) 3186 (OH), 1587 (C=C aromatic), 1474 (N=N), 1221 (C-N), 1080 (C-O), 836 (C-Cl). ^1H NMR (500 MHz, $CDCl_3$) ppm 5.46 (1H, s, OH) 6.96 (2H, d, J = 8 Hz, Ar-$H_{3'}$) 7.46 (2H, d, J = 8 Hz, Ar-$H_{2'}$) 7.82 (2H, d, J = 8 Hz, Ar-H_2) 7.86 (2H, d, J = 8 Hz, Ar-H_3) ^{13}C NMR (125 MHz, $CDCl_3$) 116.0 (Ar-$C_{3'}$), 123.98 (Ar-$C_{2'}$), 125.25 (Ar-C_2), 129.42 (Ar-C_3), 136.46 (Ar-C_4), 147.17 (Ar-$C_{1'}$), 151.42 (Ar-C_1), 158.76 (Ar-$C_{4'}$).

4-[(E)-(Bromophenyl)diazenyl]phenol (**1c**) (0.89 g, 64%) as red solid; m.p. 165-166°C. (Found: C, 51.20; H, 3.16; N, 10.05% $C_{12}H_9N_2OBr$ Requires C, 51.99; H, 3.25; N, 10.11%); R_f 0.56 (1 : 4 EA/hexane) IR: v_{max} (thin film/cm^{-1}) 3076 (OH), 1588 (C=C aromatic), 1473 (N=N), 1225 (C-N), 1101 (C-O), 833 (C-Br). ^1H NMR (500 MHz, $CDCl_3$) ppm 5.54 (1H, s, OH) 6.94 (2H, d, J = 8.6 Hz, Ar-$H_{3'}$) 7.62 (2H, d, J = 8.6 Hz Ar-$H_{2'}$) 7.74 (2H, d, J = 8.6 Hz, Ar-H_2) 7.86 (2H, d, J = 8.6 Hz, Ar-H_3). ^{13}C NMR (125 MHz, $CDCl_3$) ppm 116.02 (Ar-$C_{3'}$), 124.23 (Ar-$C_{2'}$), 124.79 (Ar-C_4), 125.28 (Ar-C_2), 132.39 (Ar-C_3), 147.10 (Ar-$C_{1'}$), 151.48 (Ar-C_1), 158.76 (Ar-$C_{4'}$).

4-[(E)-(Iodophenyl)diazenyl]phenol (**1d**) (0.68 g, 42%) as purple solid; m.p. 168–170°C. (Found: C, 44.06; H, 2.76; N, 8.44% $C_{12}H_9N_2OI$ Requires C, 44.44; H, 2.78; N, 8.64%); R_f 0.54 (1 : 4 EA/hexane) IR: v_{max} (thin film/cm^{-1}) 3172 (OH), 1591 (C=C aromatic), 1471 (N=N), 1242 (C-N), 1074 (C-O), 837 (C-I) ^1H NMR (500 MHz, DMSO-d_6) ppm 6.93 (2H, d, J = 8.6 Hz, CH, Ar-$H_{3'}$) 7.58 (2H, d, J = 8 Hz, CH, Ar-H_2) 7.79 (2H, d, J = 8.6 Hz, CH, Ar-$H_{2'}$) 7.92 (2H, d, J = 8 Hz, CH, Ar-H_3) 10.41 (1H, s, OH) ^{13}C NMR (125 MHz, DMSO-d_6) ppm 97.45 (Ar-C_4), 116.07 (Ar-$C_{3'}$), 124.06 (Ar-$C_{2'}$), 125.11 (Ar-C_2), 138.29 (Ar-C_3), 145.15 (Ar-$C_{1'}$), 151.46 (Ar-C_1), 161.32 (Ar-$C_{4'}$).

*2.4. General Method for the Synthesis of Aspirin-Halogenated Azo Derivatives **2a–d**.* Azo (**1a**) (0.65 g, 3 mmol) was dissolved in dry DCM (20 mL) and added to a round bottom flask containing acetylsalicylic acid (0.54 g, 3 mmol) in dry DCM (20 mL) at 0–10°C. DCC (0.62 g, 3 mmol) and DMAP (0.37 g, 3 mmol) were added to the mixture and stirred for 4 h at 0–10°C. The mixture was filtered and evaporated. The crude was recrystallized from hot ethanol to give (**2a**) (0.58 g, 51%) as pale yellow solid; m.p. 147-148°C. (Found: C, 66.65; H, 3.96; N, 7.22% $C_{21}H_{15}N_2O_4F$ Requires C, 66.67; H, 3.97; N, 7.41%); R_f 0.62 (1 : 4 EA/hexane) IR: v_{max} (thin film/cm^{-1}) 1755, 1745 (C=O), 1590 (C=C aromatic), 1482 (N=N), 1222 (C-N), 1184 (C-O), 841 (C-F) ^1H NMR (500 MHz, DMSO-d_6) ppm 2.27 (3H, s, -CH_3) 7.35 (1H, d, J = 7.80 Hz, Ar-$H_{5''}$) 7.44 (2H, d, J = 8.60 Hz, Ar-H_3) 7.47 (2H, d, J = 8.60 Hz, Ar-$H_{3'}$) 7.51 (1H, t, J = 7.55 Hz, Ar-$H_{4''}$) 7.79 (1H, t, J = 7.8 Hz, Ar-$H_{3''}$) 7.98–8.01 (4H, m, Ar-$H_{2,2'}$) 8.20 (1H, d, J = 7.80 Hz, Ar-$H_{2''}$) ^{13}C NMR (125 MHz, DMSO-d_6) 20.73 (-CH_3), 116.43 (Ar-C_3), 116.61 (Ar-$C_{1''}$), 121.98 (Ar-$C_{3'}$), 122.98 (Ar-$C_{2'}$), 123.98 (Ar-$C_{5''}$), 124.23 (Ar-C_2), 125.03 (Ar-$C_{3''}$), 126.57 (Ar-$C_{2''}$), 131.87 (Ar-$C_{4''}$), 135.33 (Ar-$C_{1'}$), 148.57 (Ar-C_1), 149.70 (Ar-$C_{6''}$), 150.49 (Ar-$C_{4'}$), 152.43 (Ar-C_4), 162.41 (C_a=O), 169.24 (C_b=O). The others, **2b–d,** were synthesized following the same procedure as **2a** with yields given below.

4-[(E)-(Chlorophenyl)diazenyl]phenyl aspirinate (**2b**) (0.54 g, 46%) as orange solid; m.p. 161-162°C. (Found: C, 63.58; H, 3.73 N, 7.44% $C_{21}H_{15}N_2O_4Cl$ Requires C, 63.88; H, 3.80; N, 7.10%); R_f 0.64 (1:4 EA/hexane) IR: ν_{max} (thin film/cm^{-1}) 1766, 1740 (C=O), 1588 (C=C aromatic), 1479 (N=N), 1246 (C-N), 1182 (C-O), 833 (C-Cl) ^{1}H NMR (500 MHz, DMSO-d$_6$) ppm 2.27 (3H, s, -CH$_3$) 7.35 (1H, d, J = 8.6 Hz, Ar-H$_{5''}$) 7.49 (2H, d, J = 9.2 Hz, Ar-H$_{3'}$) 7.53 (1H, t, J = 7.5 Hz, Ar-H$_{4''}$) 7.68 (2H, d, J = 8.6 Hz, Ar-H$_3$) 7.81 (1H, t, J = 7.7 Hz, Ar-H$_{3''}$) 7.93 (2H, d, J = 9.2 Hz, Ar-H$_2$) 8.01 (2H, d, J = 8.6 Hz, Ar-H$_{2'}$) 8.20 (1H, d, J = 8 Hz, Ar-H$_{2''}$) ^{13}C NMR (125 MHz, DMSO-d$_6$) 20.74 (-CH$_3$), 121.96 (Ar-C$_{1''}$), 123.04 (Ar-C$_{3'}$), 124.14 (Ar-C$_{2'}$), 124.23 (Ar-C$_2$), 124.31 (Ar-C$_{5''}$), 126.57 (Ar-C$_{3''}$), 128.78 (Ar-C$_3$), 129.67 (Ar-C$_{2''}$), 131.88 (Ar-C$_{4''}$), 134.99 (Ar-C$_4$), 136.09 (Ar-C$_{1'}$), 149.71 (Ar-C$_1$), 150.50 (Ar-C$_{6''}$), 152.50 (Ar-C$_{4'}$), 162.38 (C$_a$=O), 169.35 (C$_b$=O).

4-[(E)-(Bromophenyl)diazenyl]phenyl aspirinate (**2c**) (0.59 g, 45%) as yellow solid; m.p. 149–151°C. (Found: C, 57.31; H, 3.39; N, 6.02% $C_{21}H_{15}N_2O_4Br$ Requires C, 57.40; H, 3.42; N, 6.38%); R_f 0.62 (1:4 EA/hexane) IR: ν_{max} (thin film/cm^{-1}) 1763, 1734 (C=O), 1581 (C=C aromatic), 1477 (N=N), 1250 (C-N), 1179 (C-O), 829 (C-Br) ^{1}H NMR (500 MHz, DMSO-d$_6$) ppm 2.27 (3H, s, -CH$_3$) 7.34 (1H, d, J = 8.6 Hz, Ar-H$_{5''}$) 7.48 (2H, d, J = 9.15 Hz, Ar-H$_{3'}$) 7.51 (1H, t, J = 8.3 Hz, Ar-H$_{4''}$) 7.79–7.87 (5H, m, Ar-H$_{2,3,3''}$) 8.01 (2H, d, J = 6.9 Hz, Ar-H$_{2'}$) 8.20 (1H, d, J = 9.15 Hz, Ar-H$_{2''}$) ^{13}C NMR (125 MHz, DMSO-d$_6$) 20.78 (-CH$_3$), 121.91 (Ar-C$_{1''}$), 123.09 (Ar-C$_{3'}$), 124.20 (Ar-C$_{2'}$), 124.26 (Ar-C$_{5''}$), 124.55 (Ar-C$_2$), 125.14 (Ar-C$_4$), 126.62 (Ar-C$_{3''}$), 131.92 (Ar-C$_{2''}$), 132.66 (Ar-C$_3$), 135.34 (Ar-C$_{4''}$), 149.67 (Ar-C$_{1'}$), 150.12 (Ar-C$_{6''}$), 150.63 (Ar-C$_{4'}$), 152.31 (Ar-C$_1$), 162.38 (C$_a$=O), 169.31 (C$_b$=O).

3-[(E)-(Iodophenyl)diazenyl]phenyl aspirinate (**2d**) (0.79 g, 54%) as orange solid; m.p. 180-181°C. (Found: C, 51.63; H, 3.06; N, 5.43% $C_{21}H_{15}N_2O_4I$ Requires C, 51.85; H, 3.09; N, 5.76%); R_f 0.63 (1:4 EA/hexane) IR: ν_{max} (thin film/cm^{-1}) 1764, 1732 (C=O), 1578 (C=C aromatic), 1475 (N=N), 1250 (C-N), 1176 (C-O), 826 (C-I) ^{1}H NMR (500 MHz, DMSO-d$_6$) 2.27 (3H, s, -CH$_3$) ppm 7.35 (1H, d, J = 8.0 Hz, Ar-H$_{5''}$) 7.48 (2H, d, J = 8.60 Hz, Ar-H$_{3'}$) 7.51 (1H, t, J = 7.45 Hz, Ar-H$_{4''}$) 7.69 (2H, d, J = 8.0 Hz, Ar-H$_2$) 7.79 (1H, t, J = 8.0 Hz, Ar-H$_{3''}$). 7.99–8.03 (4H, m, Ar-H$_{3,2'}$) 8.20 (1H, d, J = 8.0 Hz, Ar-H$_{2''}$) ^{13}C NMR (125 MHz, DMSO-d$_6$) 20.78 (-CH$_3$), 99.40 (Ar-C$_4$), 121.96 (Ar-C$_{1''}$), 123.08 (Ar-C$_{3'}$), 124.18 (Ar-C$_{2'}$), 124.48 (Ar-C$_{5''}$), 126.62 (Ar-C$_2$), 131.92 (Ar-C$_{3''}$), 135.33 (Ar-C$_{2''}$), 137.56 (Ar-C$_{4''}$), 138.54 (Ar-C$_3$), 149.60 (Ar-C$_{1'}$), 150.43 (Ar-C$_{6''}$), 151.11 (Ar-C$_{4'}$), 152.72 (Ar-C$_1$), 162.21 (C$_a$=O), 169.17 (C$_b$=O).

2.5. Cell Culture. The cell viability of nasopharyngeal cancer cell lines HK-1 was determined by MTS-based colorimetric assay. The cells were kept at 37°C in a 5% CO_2 in RPMI 1640 (GIBCO) supplemented with 10% fetal bovine serum (FBS; Sigma-Aldrich). The NPC cells were assayed with a volume of 100 μL to each well plate at a concentration of 5000 cells/well. The drug dissolved in DMSO and a compound dilution 96-well plate was prepared at different concentrations of which 100 μL/well of the diluted mixtures was then added. A volume of 0.5 μL (0.1% DMSO) was added with cells as a control. The final drug concentrations were ranging from 0.39 to 50 μM at 200 μL. The cells were incubated for 72 h at 37°C. After incubation, the media were discarded from each well and added with 10 μL MTS solution and 50 μL fresh media. The plates were further incubated for 30 minutes at 37°C. The absorbance was measured by a multiwell plate reader at 490 nm. This experiment was performed in triplicate [22].

2.6. Statistical Analyses. The quantitative data are expressed as mean ± standard deviation. The cell viability against log IC$_{50}$ values of all compounds was determined using Graphpad prism 6 program.

3. Results and Discussion

The synthesis of 4-[(halophenyl)diazenyl]phenyl aspirinate (**2a–d**) was carried out via esterification of aspirin and 4-[(halophenyl)diazenyl]phenol (**1a–d**) in the presence of DCC/DMAP in DCM. 4-[(Halophenyl)diazenyl]phenol (**1a–d**) was earlier synthesized from the reaction of aniline derivatives and phenol via direct coupling reaction. The synthesis of **2a–d** is depicted in Scheme 1. Characteristic spectroscopic data of the synthesised compounds are provided in the supplementary data in Supplementary Material available online at https://doi.org/10.1155/2017/6760413.

All compounds were characterized using CHN, FTIR, and ^{1}H and ^{13}C NMR spectroscopy. FTIR spectra of **1a–d** and **2a–d** showed the absorption peaks at 1474–1471 cm^{-1} and 1591–1587 cm^{-1} attributed to ν(-N=N-) and ν(C=C), respectively. The formation of **2a–d** was confirmed by the disappearance of ν(OH) peak at 3186–3076 cm^{-1} and the appearance of new absorption peaks at 1765–1745 cm^{-1} corresponded to ν(C=O) ester. The absorption frequencies at 841–826 cm^{-1} in both FTIR spectra of **1a–d** and **2a–d** corresponded to *para*-halogenated substitution [23].

^{1}H NMR spectra of **1a–d** and **2a–d** showed the presence of aromatic protons at 8.20–6.93 ppm, while CH$_3$ in **2a–d** was observed at 2.27 ppm. ^{13}C NMR spectra of **1a–d** and **2a–d** showed resonance of aromatic carbons at 152.7–99.4 ppm. The presence of C=O and CH$_3$ in **2a–d** was observed at 169.4–162.4 ppm and 20.8–20.7 ppm, respectively.

3.1. Anticancer Activity. Compounds **1a–d** and **2a–d** were screened for anticancer activity against NPC HK-1 cell lines. The anticancer activities of **1a–d** and **2a–d** on NPC HK-1 cell lines were studied by MTS based on colorimetric assay. The cells were seeded onto 96-well culture plates and incubated with different concentrations of compounds (up to 50 μM) for 72 h. The relative influence of **1a–d** and **2a–d** was determined via viability of the NPC cell line against log [concentration, μM] as shown in Figure 1. The IC$_{50}$ of **1a–d** and **2a–d** is summarized in Table 1. Among four halogen substituents of **1a–d**, **1a** substituted with fluorine showed good anticancer activity against HK-1 cell lines with IC$_{50}$ of 19.1 ± 0.9 μM. In comparison to IC$_{50}$ of positive control (standard drug), 5-fluorouracil, and cisplatin with 12.5 ± 1.4 μM and 8.9 ± 1.9 μM, respectively, IC$_{50}$ of **1a**

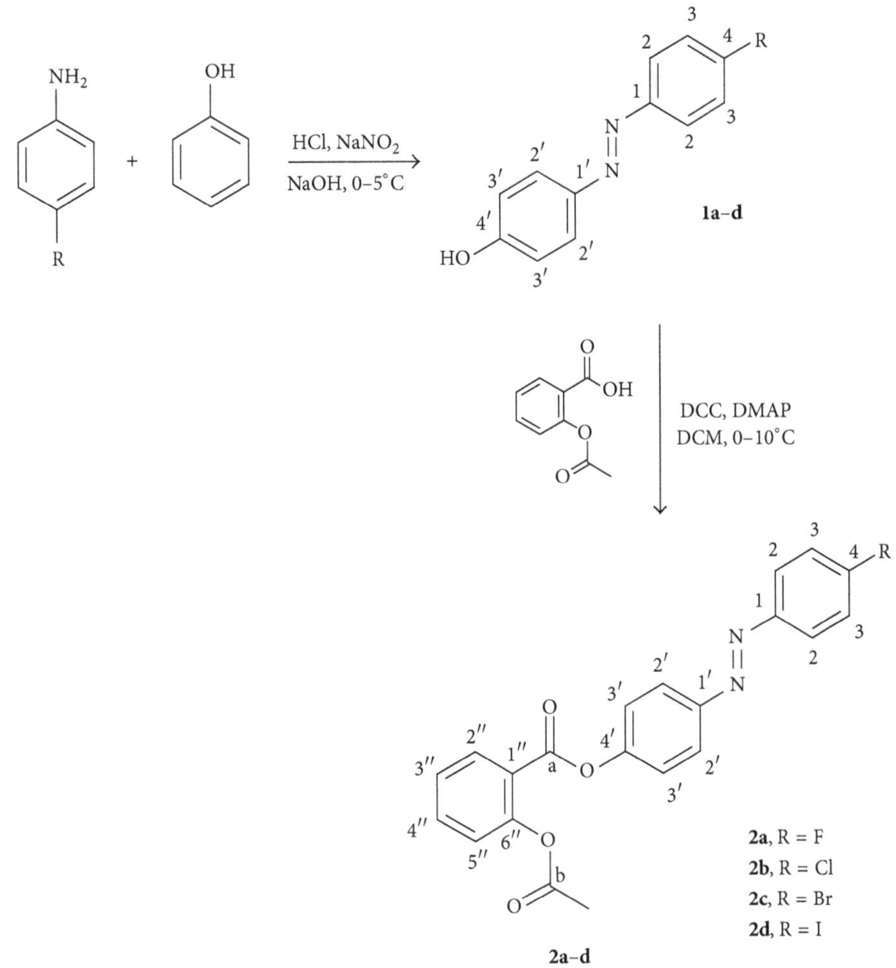

SCHEME 1: Synthesis of **1a–d** and **2a–d**.

was comparable to 5-fluorouracil. Compound **1a–d** bearing halogen substituents has good lipophilic property and thus improved the binding properties towards the active site of cells enzyme [21, 24]. Among all halogens, fluorine is the most selectively bound halogen to specific enzyme and remains covalently and irreversibly bound to the active site of enzyme, hence providing good anticancer activity [25]. In addition, the presence of -N=N- group in the molecular structure has also contributed to anticancer activity by interaction with the active site of protein via hydrogen bonding resulting in the inhibition growth of cells [20]. The presence of phenyl ring is also important to provide more lipophilic character [26]. Compound with high lipophilicity has potential to inhibit proteins for cell growth [27].

Anticancer activity of **2a–d** against NPC HK-1 cell lines only showed **2a** (38.7 ± 0.8 μM) and **2d** (44.9 ± 5.4 μM) with average anticancer activity compared to the aspirin alone. The presence of azo moiety in the molecular structure has contributed to the anticancer activity [20]. The weak inhibition on NPC HK-1 cell lines on **2c-2d** was due to higher molecular weight with larger size of bromine and iodine atoms resulting in bulky molecules [28] and gave low cytotoxicity towards cancer cells. High molecular weight

TABLE 1: IC_{50} of **1a–d** and **2a–d**, aspirin, 5-fluorouracil, and cisplatin against NPC HK-1 cells lines.

Compound	IC_{50} (μM)
Aspirin	>50
cisplatin	8.9 ± 3.2
5-Fluorouracil	12.5 ± 2.5
1a	19.1 ± 1.5
1b	36.7 ± 8.2
1c	21.9 ± 7.3
1d	20.8 ± 6.6
2a	38.7 ± 1.5
2b	44.9 ± 9.4
2c	>50
2d	>50

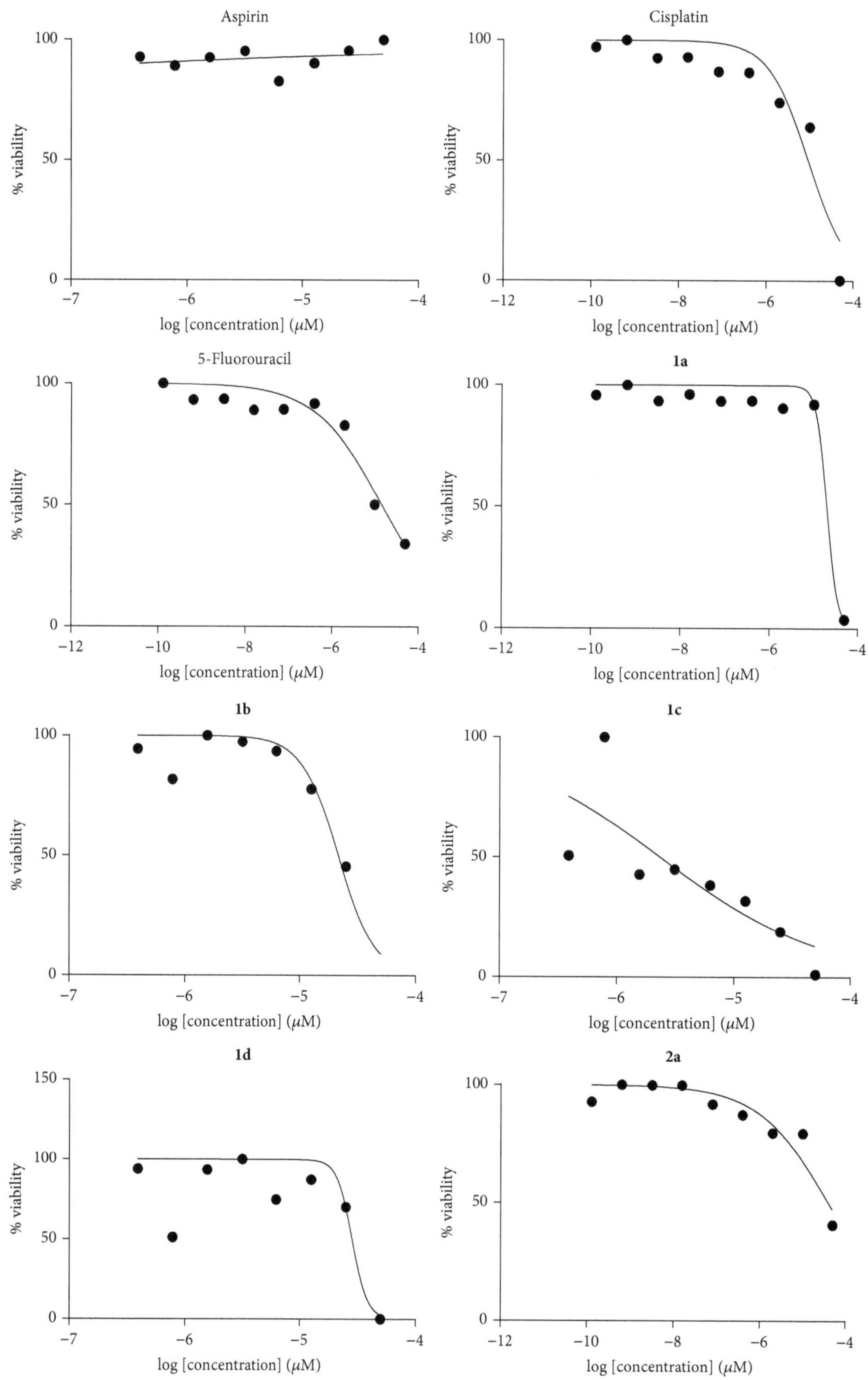

FIGURE 1: Continued.

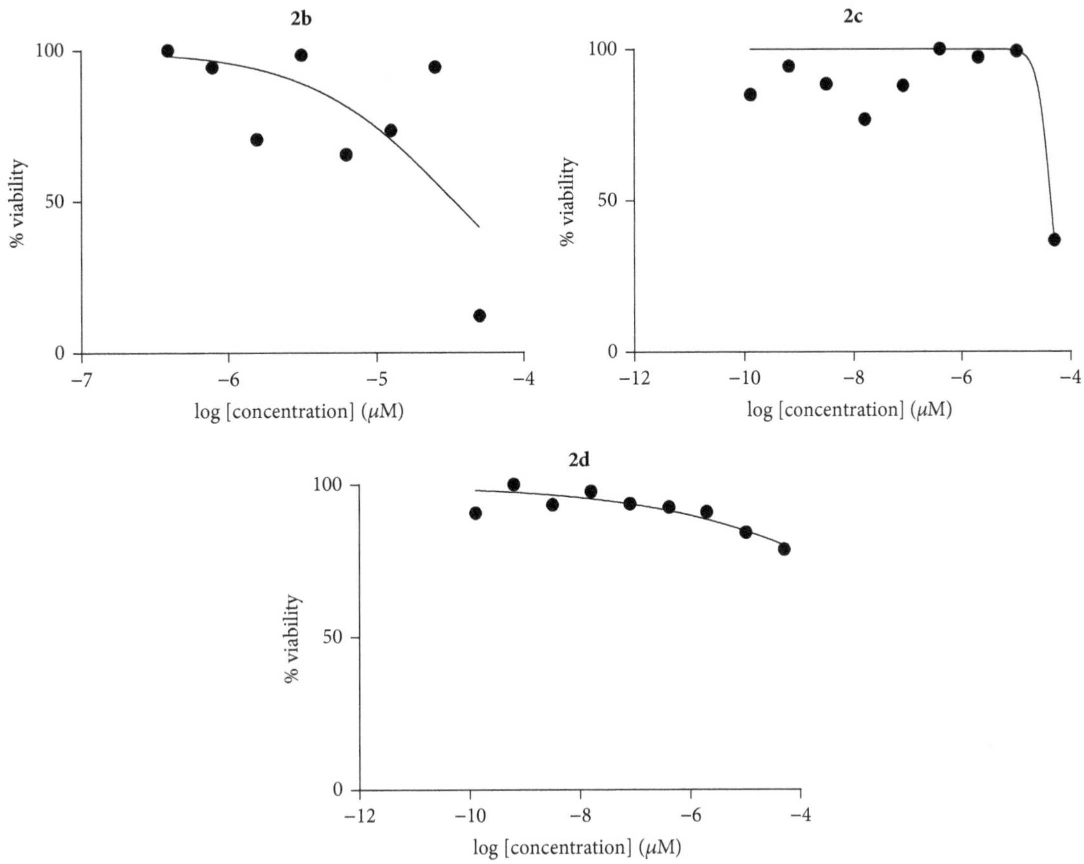

FIGURE 1: Cell viability of NPC cells, HK1 treated with **1a–d** and **2a–d**, aspirin, and standards (cisplatin and 5-fluorouracil).

and bulky molecules contributed to steric hindrance, thus preventing **2a–d** from reaching the active site of enzyme [29].

4. Conclusion

In conclusion, a series of **1a–d** and **2a–d** were synthesized and screened for their anticancer activities based on MTS assay. Compound **1a–d** demonstrated good anticancer activities compared to **2a–d**, where **1a** exhibited the highest cytotoxicity against NPC HK-1 cells lines, comparable to the standard drug (5-fluorouracil). These results indicated that **1a–d** could be used as potential lead compounds for anticancer drug development. Compounds **2a–d** provide low cytotoxicity due to the occurrence of steric hindrance which prevent them from reaching the receptor of enzymes.

Conflicts of Interest

The authors declare that they have no conflicts of interest.

Acknowledgments

The authors wish to acknowledge the University Malaysia Sarawak and Ministry of Higher Education for financial support of this project through Grants FRGS/ST02(01)/1115/2014(01) and FRGS/ST01(01)/1298/2015(15).

References

[1] "World Health Organization. Health at a glance: Asia/Pacific 2014: Measuring progress towards universal health coverage, Paris, OECD Publishing, 2014".

[2] G. M. Cooper, *The cell: A molecular approach*, Sinauer Associates, Sunderland, UK, 2nd edition, 2000.

[3] B.-J. Feng, M. Jalbout, W. B. Ayoub et al., "Dietary risk factors for nasopharyngeal carcinoma in Maghrebian countries," *International Journal of Cancer*, vol. 121, no. 7, pp. 1550–1555, 2007.

[4] E. T. Chang and H. Adami, "The enigmatic epidemiology of nasopharyngeal carcinoma," *Cancer Epidemiology Biomarkers and Prevention*, vol. 15, no. 10, pp. 1765–1777, 2006.

[5] F. L. Sung, T. C. W. Poon, and E. P. Hui, "Antitumor effect and enhancement of cytotoxic drug activity by cetuximab in nasopharyngeal carcinoma cells," *In Vivo*, vol. 19, no. 1, pp. 237–246, 2005.

[6] P. M. Harari, "Epidermal growth factor receptor inhibition strategies in oncology," *Endocrine-Related Cancer*, vol. 11, no. 4, pp. 689–708, 2004.

[7] M. Gleeson, G. G. Browning, M. J. Burton et al., *Scott-Browns otorhinolaryngology, head and neck surgery*, Hodder Arnold, London, UK, 7th edition, 2008.

[8] Y. C. Mayur, G. J. Peters, V. V. S. Rajendra Prasad, C. Lemos, and N. K. Sathish, "Design of new drug molecules to be used in reversing multidrug resistance in cancer cells," *Current Cancer Drug Targets*, vol. 9, no. 3, pp. 298–306, 2009.

[9] A. I. Francisco, M. D. Vargas, T. P. Fragoso et al., "Theoretical studies of the tautomerism in 3-(2-R-phenylhydrazono)-naphthalene-1,2,4-triones: Synthesis of copper(II) complexes and studies of antibacterial and antitumor activities," *Journal of the Brazilian Chemical Society*, vol. 21, no. 7, pp. 1293–1302, 2010.

[10] L. Huang, G. G. MacKenzie, N. Ouyang et al., "The novel phospho-non-steroidal anti-inflammatory drugs, OXT-328, MDC-22 and MDC-917, inhibit adjuvant-induced arthritis in rats," *British Journal of Pharmacology*, vol. 162, no. 7, pp. 1521–1533, 2011.

[11] N. A. Nordin, T. W. Chai, B. L. Tan et al., "Novel synthetic monothiourea aspirin derivatives bearing alkylated amines as potential antimicrobial agents," *Journal of Chemistry*, vol. 2017, pp. 1–7, 2017.

[12] Z. Ngaini, M. A. Mohd Arif, H. Hussain, E. S. Mei, D. Tang, and D. H. A. Kamaluddin, "Synthesis and antibacterial activity of acetoxybenzoyl thioureas with aryl and amino acid side Chains," *Phosphorus, Sulfur, and Silicon and the Related Elements*, vol. 187, no. 1, pp. 1–7, 2012.

[13] K. Kashfi and B. Rigas, "The mechanism of action of nitric oxide-donating aspirin," *Biochemical and Biophysical Research Communications*, vol. 35, no. 8, pp. 1096–1101, 2007.

[14] A. Cemel and S. Ahmet, "Rapid and simultaneous determination of acetylsalicylic acid, paracetamol and their degradation and toxic impurity product by HPLC in pharmaceutical dosage forms," *Turkish Journals Medical Science*, vol. 38, pp. 167–173, 2008.

[15] H. G. Garg and C. Praksh, "Preparation of 4-arylazo-3,5-disubstituted-(2H)-1,2,6-thiadiazine 1,1-dioxides," *Journal Medicinal Chemistry*, vol. 15, no. 4, pp. 435–439, 1972.

[16] R. G. Child, R. G. Wilkinson, and A. Tomcu-Fucik, "Effect of substrate orientation of the adhesion of polymer joints," *Journal of Chemical Society*, vol. 87, pp. 6031–6038, 1977.

[17] A. Khalid, M. Arshad, and D. E. Crowley, "Accelerated decolorization of structurally different azo dyes by newly isolated bacterial strains," *Applied Microbiology and Biotechnology*, vol. 78, no. 2, pp. 361–369, 2008.

[18] T. A. Farghaly and Z. A. Abdallah, "Synthesis, azo-hydrazone tautomerism and antitumor screening of N-(3-ethoxycarbonyl-4,5,6,7-tetrahydro-benzo[b]thien-2-yl)-2-arylhydrazono-3- oxobutanamide derivatives," *Archieve for Organic Chemistry*, vol. 17, no. 17, pp. 295–305, 2008.

[19] C. Park, J.-S. Lim, Y. Lee et al., "Optimization and morphology for decolorization of reactive black 5 by Funalia trogii," *Enzyme Microbial Technology*, vol. 40, no. 7, pp. 1758–1764, 2007.

[20] N. Raman, L. Mitu, A. Sakthivel, and M. S. S. Pandi, "Studies on DNA cleavage and antimicrobial screening of transition metal complexes of 4-aminoantipyrine derivatives of N_2O_2," *Journal of the Iranian Chemical Society*, vol. 6, no. 4, pp. 738–748, 2009.

[21] M. Z. Hernandes, S. M. T. Cavalcanti, D. R. M. Moreira, W. F. De Azevedo Jr., and A. C. L. Leite, "Halogen atoms in the modern medicinal chemistry: hints for the drug design," *Current Drug Targets*, vol. 11, no. 3, pp. 303–314, 2010.

[22] G. K. Y. Chan, T. L. Kleinheinz, D. Peterson, and J. G. Moffat, "A simple high-content cell cycle assay reveals frequent discrepancies between cell number and ATP and MTS proliferation assays," *PLoS ONE*, vol. 8, no. 5, Article ID e63583, 2013.

[23] T. W. G. Solomons and C. B. Fryhle, *Organic Chemistry*, John Wiley Sons, 8th edition, 2003.

[24] N. R. Modi, R. J. Shah, M. J. Patel, M. Suthar, B. F. Chauhan, and L. J. Patel, "Design, synthesis, and QSAR study of novel 2-(2, 3-dioxo-2, 3-dihydro-1H-indol-1-yl)-N- phenyl acetamide derivatives as cytotoxic agents," *Medicinal Chemistry Research*, vol. 20, pp. 615–625, 2011.

[25] J. Zhang, P. L. Yang, and N. S. Gray, "Targeting cancer with small molecule kinase inhibitors," *Nature Review Cancer*, vol. 9, no. 1, pp. 28–39, 2009.

[26] A. J. Hoey, C. M. Jackson, G. G. Pegg, and M. N. Sillence, "Characteristics of cyanopindolol analogues active at the beta(3)-adrenoceptor in rat ileum," *British Journal of Pharmacology*, vol. 119, no. 3, pp. 564–568, 1996.

[27] J. C. Lopez-Romero, H. Gonzalez-Rios, A. Borges, and M. Simoes, "Antibacterial effects and mode of action of selected essential oils components against *Escherichia coli* and *Staphylococcus aureus*," *Evidence-Based Complementary and Alternative Medicine*, vol. 2015, Article ID 795435, 9 pages, 2015.

[28] M. Hum, B. E. McLaughlin, G. Roman, J. Z. Vlahakis, W. A. Szarek, and K. Nakatsu, "The effects of azole-based heme oxygenase inhibitors on rat cytochromes P450 2E1 and 3A1/2 and human cytochromes P450 3A4 and 2D6," *Journal of Pharmacology and Experimental Therapeutics*, vol. 334, no. 3, pp. 981–987, 2010.

[29] C. H. Yun, K. E. Mengwasser, A. V. Toms, M. S. Woo, and G. Hetal, "The T790M mutation in EGFR kinase causes drug resistance by increasing the affinity for ATP," in *Proceedings of the National Academy of Sciences*, vol. 105, pp. 2070–2075, 2008.

22

Synthesis and Growth Stimulant Properties of 2-Acetyl-3,7-dimethyl-5H-thiazolo[3,2-a]pyrimidin-5-one Derivatives

Vergush A. Pivazyan,[1] Emma A. Ghazaryan,[2] Roza S. Shainova,[1,2] Rafael A. Tamazyan,[3] Armen G. Ayvazyan,[3] and Aleksandr P. Yengoyan[1,2]

[1]*National Agrarian University of Armenia, 74 Teryan Str., 0009 Yerevan, Armenia*
[2]*Armenian-Russian (Slavonic) State University, 123 H. Emin Str., 0051 Yerevan, Armenia*
[3]*STC OPC Molecule Structure Research Center of NAS RA, 26 Azatutyan Av., 0014 Yerevan, Armenia*

Correspondence should be addressed to Aleksandr P. Yengoyan; ayengoyan@mail.ru

Academic Editor: Xinyong Liu

A convenient, accessible, and high yield method for preparing of 6-methyl-2-thioxo-2,3-dihydropyrimidin-4(1H)-one (**1**) by treatment of acetoacetic acid ethyl ester with thiourea in sodium methylate was developed. The alkylation of the latter with 3-chloropentane-2,4-dione and further regioselective cyclization of intermediate compound (**2**) in high yield afforded 2-acetyl-3,7-dimethyl-5H-thiazolo[3,2-a]pyrimidin-5-one (**3**). The halogenation and some transformations of synthesized thiazolo[3,2-a]pyrimidine (**3**) due to its ketone group were carried out to obtain the corresponding carboxamide, carbothioamide, sulfonohydrazide, and oxime and its alkylated derivatives (**5**). At preliminary biological studies the synthesized compounds have shown growth stimulant properties. The activity of four of them was higher than 70%, compared with heteroauxin.

1. Introduction

Thiazole and pyrimidine are classes of heterocycles that are of considerable interest because of the wide range of their biological properties. Among them there are many pharmaceutically important molecules and chemical means of plants protection. Thiazolopyrimidines, which consist of these two active pharmacophore heterocycles, can also be of interest as potential bioactive molecules and have acquired a growing importance in the field of medicinal chemistry. Due to the great potential of both of the moiety, different scientists synthesized a large number of thiazolopyrimidines derivatives in particular thiazolo[3,2-a]pyrimidines with different substituents to evaluate their various pharmacological activities. The survey of literature reveals that these fused biheterocyclic compounds show pharmacological properties like anti-inflammatory [1–3], bactericidal [4–9], analgesic and antiparkinsonian [10], antifungal [6, 7, 9, 11–14], antioxidant [6, 9, 15], anticancer [16–20], antimicrobial [9, 21–26], hypoglycemic [27], psychopharmacological [28], antituberculosis [29], antidepressant [30], and antiviral [31, 32]. Some derivatives are offered as acetylcholinesterase inhibitors [3], antipsychotic agents [33], calcium antagonists [34], phosphate inhibitors [35, 36], and 5-HT2A receptor antagonists [37].

However, to our surprise, there are virtually no data in the literature about pesticidal and plant growth regulatory properties of thiazolo[3,2-a]pyrimidines. Only a few studies of pesticidal [38, 39], herbicidal [40], and growth regulatory activities [41] of these systems derivatives may be noted.

Prompted by the potential importance of this condensed biheterocyclic system the present investigation was undertaken to synthesize a series of novel thiazolo[3,2-a]pyrimidines derivatives to obtain the new chemical means of plants protection and growth regulators [42].

2. Results and Discussion

2.1. Chemistry. As precursors for the synthesis of thiazolo[3,2-a]pyrimidines the substituted pyrimidines and

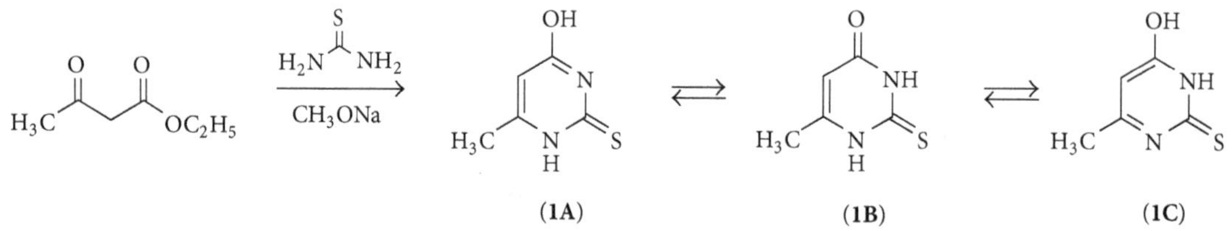

SCHEME 1: Synthesis of 6-methyl-2-thioxo-2,3-dihydropyrimidin-4(1H)-one.

aminothiazoles are used. In most of the papers the synthesis of initial pyrimidine-2-thione derivatives is carried out *via* Biginelli multicomponent condensation reaction of substituted aldehydes with 1,3-dicarbonyl compounds and thiourea [3, 6, 8, 9, 13, 20, 23]. In some cases, pyrimidines were obtained in two steps [20]. In these syntheses the yield of pyrimidine derivatives was 50–85%. Their further condensation with substituted acyl chlorides, halocarboxylic acids esters, or dicarbonyl compounds affords thiazolo[3,2-a]pyrimidine biheterocyclic systems. In order to accelerate the synthesis of targeted thiazolo[3,2-a]pyrimidines and increase their yield, some authors used MW irradiation [9, 14, 43] or catalysts [44, 45].

The cyclization of pyrimidine derivatives may yield several isomers, due to substitution position in pyrimidine cycle in the intermediate product. To achieve regioselectivity, the cyclization reaction was carried out with bromomalononitrile [9] or *via* S-propargyl derivatives [44, 45].

To obtain thiazolo[3,2-a]pyrimidines some authors use 2-aminothiazoles as precursors. In these cases targeted compounds were obtained by cyclization of 2-aminothiazole derivatives with bis(methylthio)methylene malononitrile [15], benzylidene malononitrile [9], 2-acetylbutyrolactone [37], and formamide and formic acid [41]. In [14, 43] 2-aminothiazole was used in Biginelli reaction instead of thiourea. A series of thiazolo[3,2-a]pyrimidinones were synthesized in a two-step procedure, using Eaton's reagent to effect the cyclization of 2-aminothiazoles [46].

In order to develop a convenient, accessible, and high yield method for preparing of 6-methyl-2-thioxo-2,3-dihydropyrimidin-4(1H)-one, in present investigation acetoacetic acid ethyl ester was treated with thiourea in sodium methylate (Scheme 1). Obtained compound can exist in different tautomeric forms (**1A–1C**). In ^{13}C NMR spectra the signal of double C=S bond at 175.83 ppm and in ^{1}H NMR spectra one broaden average absorption at 11.98 ppm are observed. These data indicate that 6-methyl-2-thioxo-2,3-dihydropyrimidin-4(1H)-one has structure (**1B**).

To obtain targeted thiazolo[3,2-a]pyrimidine, compound **1** interacted with 3-chloro-pentane-2,4-dione (Scheme 2). At this process the substitution may occur at sulfur or one of cyclic nitrogen atoms, due to position of hydrogen atoms in pyrimidine moieties. In ^{13}C NMR spectrum of compound **2** the signal of C=S carbon atom has disappeared and in ^{1}H NMR spectrum besides the signal of enol hydrogen atom (17.30 ppm, integral intensity equal to 0.9) also the second absorption for cyclic OH or NH group is observed (12.20 ppm). These data indicated that the substitution

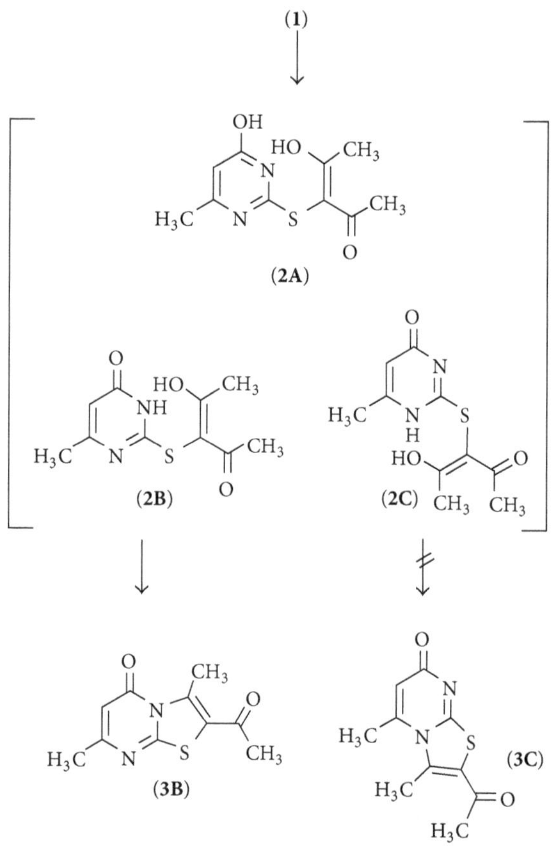

SCHEME 2: Synthesis of 2-acetyl-3,7-dimethyl-5H-thiazolo[3,2-a]pyrimidin-5-one.

occurred at sulfur atom and depending on tautomerism of heterocyclic ring and pentadione moiety the obtained compound may be described by one of the structures (**2A**, **2B**) or (**2C**). The heating of the latter in toluene at 125–130°C in the presence of toluene-4-sulfonic acid afforded compound (**3**). The intramolecular dehydration and cyclization can occur from tautomers (**2B**) or (**2C**) and as a result 2-acetyl-3,7-dimethyl-5H-thiazolo[3,2-a]pyrimidin-5-one (**3B**) or 2-acetyl-3,5-dimethyl-thiazolo[3,2-a]pyrimidin-7-one (**3C**) can be obtained.

Since the further transformations have been carried out on the basis of this compound (**3**), its structure was strictly proved by X-ray diffraction analysis, whose data unambiguously agree with the structure (**3B**) (Figure 1).

To introduce some pharmacophore groups into molecule structure, some transformations due to the ketone

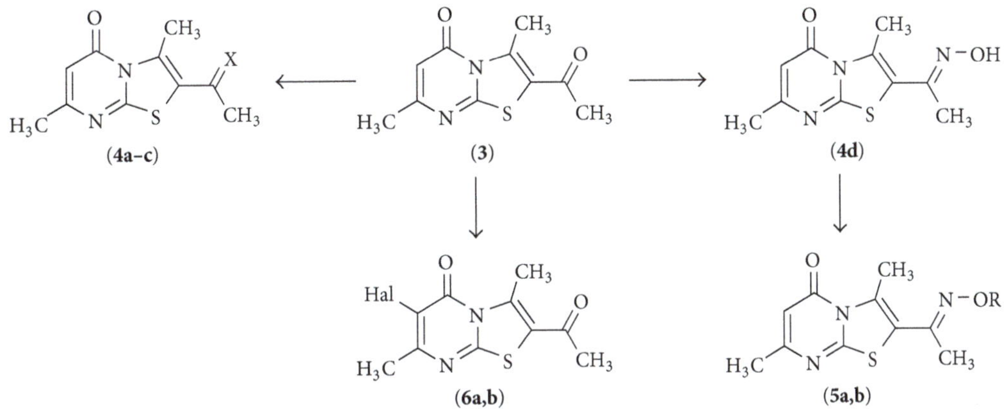

Scheme 3: Transformations of 2-acetyl-3,7-dimethyl-5H-thiazolo[3,2-a]pyrimidin-5-one.

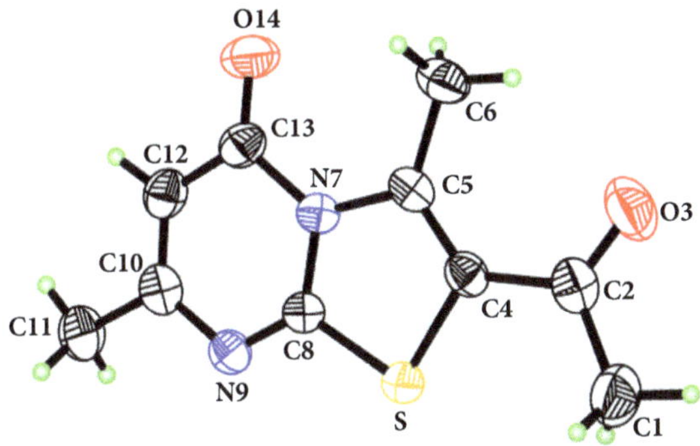

Figure 1: The structure of 2-acetyl-3,7-dimethyl-5H-thiazolo[3,2-a]pyrimidin-5-one (**3**).

group were carried out. By interaction of ketone (**3**) with semicarbazide, thiosemicarbazide, and 4-methylbenzenesulfonohydrazide the corresponding 2-(1-(3,7-dimethyl-5-oxo-5H-thiazolo[3,2-a]pyrimidin-2-yl)ethylidene)hydrazine-1-carboxamide (**4a**), 2-(1-(3,7-dimethyl-5-oxo-5H-thiazolo[3,2-a]pyrimidin-2-yl)ethylidene)hydrazine-1-carbothioamide (**4b**), and N′-(1-(3,7-dimethyl-5-oxo-5H-thiazolo[3,2-a]pyrimidin-2-yl)ethylidene)-4-methylbenzene-sulfonohydrazide (**4c**) were obtained (Scheme 3).

In accordance with literature data some heterocyclic substituted oximes and their ethers show fungicidal [47–49], herbicidal [50, 51], insecticidal [52], and growth stimulant [49] activities. Therefore, by reaction of ketone (**3**) with hydroxylamine hydrochloride, the corresponding 2-(1-(hydroxyimino)ethyl)-3,7-dimethyl-5H-thiazolo[3,2-a]pyrimidin-5-one (**4d**) was synthesized. The latter with alkyl halides formed 3,7-dimethyl-2-(1-(propoxyimino)ethyl)-5H-thiazolo[3,2-a]pyrimidin-5-one (**5a**) and 3,7-dimethyl-2-(1-((2-phenoxyethoxy)imino)ethyl)-5H-thiazolo[3,2-a]pyrimidin-5-one (**5b**).

We also carried out an electrophilic substitution in the pyrimidine ring to synthesize the halogenated derivatives (**6a** and **6b**) of ketone (**3**) (Scheme 3).

To synthesize the condensed system with a saturated thiazole cycle the reaction of starting 6-methyl-2-thioxo-2,3-dihydropyrimidin-4(1H)-one with 1,2-dichloroethane was carried out. As a result 2-((2-chloroethyl)thio)-6-methylpyrimidin-4(3H)-one or 2-((2-chloroethyl) thio)-6-methylpyrimidin-4(1H)-one could be formed (Scheme 4).

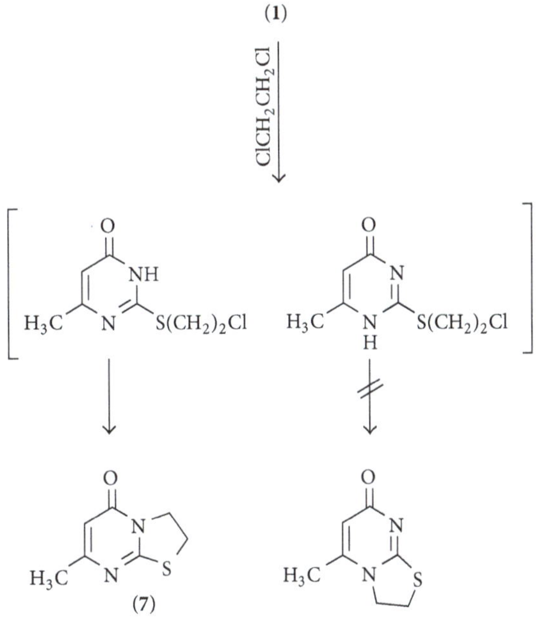

Scheme 4: Synthesis of 7-methyl-2,3-dihydro-5H-thiazolo[3,2-a]pyrimidin-5-one.

In ^{13}C NMR spectrum of the synthesized compound the signal of carbon atom of C=O bond is retained, and C=S bond has disappeared. In ^{1}H NMR spectrum the values of chemical shift of methylene groups' signals are conformed to S-substitution. At the same time, in accordance with elemental analysis and mass spectrum data (M + 1 = 169), there are no chlorine atoms in the molecule of resulting product. Therefore, even at room temperature, the intermediate S-substituted derivative is subjected to heterocyclization, and 7-methyl-2,3-dihydro-5H-thiazolo[3,2-a]pyrimidin-5-one (**7**) or 5-methyl-2,3-dihydro-7H-thiazolo[3,2-a]pyrimidin-7-one may be formed. The data of ^{1}H and ^{13}C NMR spectra of the reaction product are similar to those of compound (**3**), which conform to structure (**7**).

2.2. Biological Evaluation. At preliminary screening the herbicidal, fungicidal, and growth regulatory activities of novel synthesized compounds were studied. All preparations did not possess noticeable herbicidal or antifungal properties, but they showed the pronounced growth stimulant activity.

For the growth regulatory properties evaluation the action of aqueous emulsions (25 mg/L and 50 mg/L) of synthesized compounds on the germination, growth, and survivability of seeds and seedlings of dicotyledonous bean (*Phaseolus vulgaris* L.) was studied and compared with that of heteroauxin (IAA).

Two series of bean seeds were incubated for 24 hours in an appropriate medium in the dark at 25°C. Then the seeds were transplanted into soil and watered daily. The experimental data calculations were produced in 20–25 days. The number of the plants roots of each series, their length and weight in the moist and dry forms, and their average values were calculated. The results were compared with similar data of plants placed in IAA solutions, and the activities of preparations in comparison with IAA (in %) were determined (Table 1).

Practically all targeted substances (**3–6**) have shown growth stimulant properties. The most active compounds (more than 70% compared to heteroauxin) were (**3**), (**4b**), (**4d**), and (**6b**).

TABLE 1: Growth stimulant activity of synthesized compounds **3–6**.

Number	Growth stimulant activity compared with IAA (%)
IAA	100[a]
	100[b]
3	79[a]
	49[b]
4a	64[a]
	10[b]
4b	75[a]
	13[b]
4c	39[a]
	66[b]
4d	72[a]
	57[b]
5a	54[a]
	66[b]
5b	55[a]
	42[b]
6a	43[a]
	48[b]
6b	66[a]
	74[b]

[a]50 mg/L.
[b]25 mg/L.

3. Experimental

The ^{1}H NMR and ^{13}C NMR spectra were recorded on Varian Mercury-300 MHz (Varian, USA) spectrometer, in the mixture of solvents DMSO-d_6 + CCl_4 (1:3), using tetramethylsilane as internal standard and IR spectra on Avatar 330FT-IR (Thermo Nicolet, USA) spectrometer, using attenuated total reflectance (ATP) method. Mass spectra were recorded on LC-MS system (Agilent, USA). The X-ray analysis was done at room temperature on automated Enraf-Nonius CAD-4 diffractometer. The reaction progress and purity of the obtained substances were checked by using TLC method on UV-254 plates ("Silufol," Czech Republic) with acetone/hexane mixture (2:1) as eluent. All melting points were determined in open capillaries and are uncorrected.

3.1. Synthesis of 6-Methyl-2-thioxo-2,3-dihydropyrimidin-4(1H)-one (1). To methanol (10 mL) was added sodium (30 mmol) and then thiourea (10 mmol) and acetoacetic acid ethyl ester (10 mmol). The mixture was boiled for 7 h, methanol was evaporated, and the precipitate was dissolved in water, neutralized with acetic acid, and filtered off.

Yield (1.26 g, 90%); light yellow crystals; m.p. 275–277°C. ^{1}HNMR: δ = 2.09 (s, 3H, 6-CH_3); 5.52 (s, 1H, CH-pyrim.); 11.98 [b.s, 2H, $(NH)_2$]. ^{13}C NMR: δ = 17.88; 103.38 151.91; 160.32; 175.83. Anal. Calcd for $C_5H_6N_2OS$: C, 42.24; H, 4.25; N, 19.70; S, 22.55; found: C, 42.24; H, 4.25; N, 19.70; S, 22.55.

3.2. Synthesis of 3-((4-Hydroxy-6-methylpyrimidin-2-yl)thio)pentane-2,4-dione (2). The mixture of 6-methyl-2-thioxo-2,3-dihydropyrimidin-4(1H)-one (**1**) (10 mmol) and KOH (10 mmol) in dimethylformamide (DMFA) (15 mL) was stirred at room temperature for 3 h; then at cooling 3-chloro-pentane-2,4-dione (10 mmol) was added. The reaction mixture was allowed to stand for 24 h at 25°C; then DMFA was evaporated at low pressure and the residue was processed with water and filtered off.

Yield (1.68 g, 70%); white crystal; m.p. 188–190°C. ^{1}HNMR: δ = 2.18 (s, 3H, 6-CH_3); 2.22 [s, 6H, $(CH_3)_2$]; 5.92 (s, 1H, CH-pyrim.); 12.20 (b.s., 1H, OH-pyrim.); 17.30 (s, 0.9H, OH-enol). Anal. Calcd for $C_{10}H_{12}N_2O_3S$: C, 49.99; H, 5.03; N, 11.66; S, 13.34; found: C, 49.81; H, 4.91; N, 11.22; S, 13.00.

3.3. Synthesis of 2-Acetyl-3,7-dimethyl-5H-thiazolo[3,2-a]pyrimidin-5-one (3). A solution of compound **2** (10 mmol) and catalytic amounts of toluene-4-sulfonic acid in toluene (10 mL) was heated at 125–130°C for 4 h. After toluene evaporation water (20 mL) was added and compound **3** was filtered off from the mixture.

Yield (1.65 g, 75%); light yellow crystals; m.p. 126–128°C. IR (KBr): ν = 1695 and 1705 (C=O). ^{1}HNMR: δ = 2.25 (s, 3H, 7-CH_3); 2.57 (s, 3H, $COCH_3$); 3.08 (s, 3H, 3-CH_3); 5.96 (s, 1H, CH-pyrim.). ^{13}C NMR: δ = 16.06; 22.85; 30.30; 105.22; 120.99; 141.69; 160.16; 160.94; 162.60; 189.40. Anal. Calcd for $C_{10}H_{10}N_2O_2S$: C, 54.04; H, 4.54; N, 12.60; S, 14.42; found: C, 53.90; H, 4.39; N, 12.29; S, 14.08.

X-Ray Diffraction Analysis. Crystal data: $C_{10}H_{10}N_2O_2S$, monoclinic, space group $P21/n$, a = 14.106(3) Å, b = 4.0190(8) Å, c = 18.062(4) Å, β = 103.32(3)°, V = 996.4(4) Å^3, Z = 4,

$T = 293(2)$ K, $dx = 1.482\ \mathrm{g\,cm^{-3}}$, $\mu(\mathrm{MoK_\alpha}) = 0.304\ \mathrm{mm^{-1}}$, 3010 data were collected up to $\theta_{max} = 30°$ ($R_{int} = 0.025$, $R_\sigma = 0.0678$). Final R-indices for 1591 reflections with $I > 2\sigma(I)$ and 139 refined parameters are $R_1 = 0.0562$, $wR_2 = 0.1080$ ($R_1 = 0.1283$, $wR_2 = 0.1328$ for all 2902 data). Crystallographic data for compound (**3**) have been deposited with the Cambridge Crystallographic Data Centre, deposition number CCDC 991572.

3.4. Synthesis of 2-(1-(3,7-Dimethyl-5-oxo-5H-thiazolo[3,2-a]pyrimidin-2-yl)ethylidene)hydrazine-1-carboxamide (***4a***). The mixture of compound **3** (10 mmol) and semicarbazide hydrochloride (10 mmol) in water (10 mL) was heated at 100°C for 4 h.

Yield (2.6 g, 89%); yellow crystals; m.p. 310–312°C. ^{1}HNMR: δ = 2.25 (s, 3H, 7-CH_3); 2.41 (s, 3H, C=NCH_3); 2.85 (s, 3H, 3-CH_3); 5.92 (s, 1H, CH-pyrim.); 7.25 and 8.20 (b.s, 2H, NH_2); 10.64 (s, 1H, NH). Anal. Calcd for $C_{11}H_{13}N_5O_2S$: C, 47.30; H, 4.69; N, 25.07; S, 11.48; found: C, 47.20; H, 4.55; N, 24.79; S, 11.08.

General Synthesis of Compounds (***4b,c***). To a suspension of thiosemicarbazide or toluene-4-sulfohydrazide (10 mmol) in water (10 mL), 36% HCl (12 mmol) and compound **3** (10 mmol) were added. The mixture was heated at 100°C for 4 h, cooled, and filtered off. The precipitate was purified by boiling in ethanol (50%).

3.4.1. 2-(1-(3,7-Dimethyl-5-oxo-5H-thiazolo[3,2-a]pyrimidin-2-yl)ethylidene)hydrazine-1-carbothioamide (***4b***). Yield (2.45 g, 84%); yellow crystals; m.p. 248–250°C. ^{1}HNMR: δ = 2.24 (s, 3H, 7-CH_3); 2.39 (s, 3H, C=NCH_3); 2.88 (s, 3H, 3-CH_3); 5.90 (s, 1H, CH-pyrim.); 7.19 and 8.27 (b.s, 2H, NH_2); 10.55 (s, 1H, NH). ^{13}C NMR: δ = 16.48; 17.43; 22.85; 105.06; 121.77; 133.73; 140.41; 159.96; 161.66; 161.86; 178.99. Anal. Calcd for $C_{11}H_{13}N_5OS_2$: C, 44.73; H, 4.44; N, 23.71; S, 21.71; found: C, 44.55; H, 4.29; N, 23.40; S, 21.39.

3.4.2. N′-(1-(3,7-Dimethyl-5-oxo-5H-thiazolo[3,2-a]pyrimidin-2-yl)ethylidene)-4-methylbenzenesulfonohydrazide (***4c***). Yield (3.25 g, 90%); white crystals; m.p. 208–210°C. ^{1}HNMR: δ = 2.21 and 2.22 (s, 3H, 3H, CH_3-C_6H_5 and 7-CH_3); 2.44 (s, 3H, C=NCH_3); 2.75 (s, 3H, 3-CH_3); 5.86 (s, 1H, CH-pyrim.); 7.32–7.80 (m, 4H, C_6H_4); 10.74 (s, 1H, NH). ^{13}C NMR: δ = 16.14; 17.17; 20.94; 22.79; 104.92; 121.34; 127.52; 128.82; 133.48; 135.88; 142.64; 145.08; 159.95; 161.42; 161.79. Anal. Calcd for $C_{17}H_{18}N_4O_3S_2$: C, 52.29; H, 4.65; N, 14.35; S, 16.42; found: C, 52.11; H, 4.53; N, 14.02; S, 16.11.

3.5. Synthesis of 2-(1-(Hydroxyimino)ethyl)-3,7-dimethyl-5H-thiazolo[3,2-a]pyrimidin-5-one (***4d***). To a solution of hydroxyl amine (11 mmol) in water (10 mL), at cooling a solution of NaOH (11 mmol) in water (5 mL) was added dropwise and then compound **3** (10 mmol) and ethanol (10 mL). The mixture was allowed to stand for 24 h and then heated at 50–60°C for 2 h, cooled, processed with water, and filtered off.

Yield (1.65 g, 71%); white crystals; m.p. 214–216°C. ^{1}HNMR: δ = 2.23 (s, 6H, C=N$CH_3$and 7-CH_3); 2.86 (s, 3H, 3-CH_3); 5.89 (s, 1H, CH-pyrim.); 11.66 (s, 1H, OH). ^{13}C NMR: δ = 14.13; 16.17; 22.77; 104.93; 119.48; 132.27; 146.48; 159.93; 161.59; 161.67. Anal. Calcd for $C_{10}H_{11}N_3O_2S$: C, 50.62; H, 4.67; N, 17.71; S, 13.51; found: C, 50.49; H, 4.51; N, 17.39; S, 13.18.

3.6. General Synthesis of Compounds (***5a,b***). In DMFA (20 mL) consecutively KOH (10 mmol) and compound **4d** (10 mmol) were diluted. The solution was stirred at 20°C for 3 h till salt formation. After cooling propyl bromide or (2-bromoethoxy)benzene (11 mmol) was added. The mixture was allowed to stand for 24 h and then heated at 70–80°C for 2 h. DMFA was evaporated at low pressure; the residue was processed with water and filtered off. Compounds (**5a,b**) were purified by recrystallization from hexane.

3.6.1. 3,7-Dimethyl-2-(1-(propoxyimino)ethyl)-5H-thiazolo[3,2-a]pyrimidin-5-one (***5a***). Yield (2.36 g, 85%); white crystals; m.p. 70–72°C. ^{1}HNMR: δ = 0.98 (t, J = 7.4, 3H, CH_3-Pr); 1.72 (m, 2H, CH_2), 2.24 (s, 6H, C=NCH_3 and 7-CH_3); 2.83 (s, 3H, 3-CH_3); 4.11 (t, J = 6.6, OCH_2); 5.90 (s, 1H, CH-pyrim.). ^{13}C NMR: δ = 9.85; 14.86; 16.18; 21.77; 22.76; 75.50; 104.97; 118.02; 133.26; 147.11; 159.82; 161.53; 161.77. Anal. Calcd for $C_{13}H_{17}N_3O_2S$: C, 55.89; H, 6.13; N, 15.04; S, 11.48; found: C, 55.90; H, 6.01; N, 14.69; S, 11.11.

3.6.2. 3,7-Dimethyl-2-(1-((2-phenoxyethoxy)imino)ethyl)-5H-thiazolo[3,2-a]pyrimidin-5-one (***5b***). Yield (2.4 g, 67%); white crystals; m.p. 125–127°C. IR (KBr): ν = 1679 (C=O). ^{1}HNMR: δ = 2.24 and 2.26 (s, 3H, 3H, C=NCH_3 and 7-CH_3); 2.83 (s, 3H, 3-CH_3); 4.23 and 4.49 (t,t, 4H, OCH_2CH_2O); 5.91 (s, 1H, CH-pyrim.); 6.82–7.28 (m, 5H, C_6H_5). ^{13}C NMR: δ = 15.14; 16.23; 22.76; 65.43; 72.54; 105.03; 114.02; 120.18; 128.76; 133.50; 148.06; 158.01; 159.82; 1161.55; 161.83. Anal. Calcd for $C_{18}H_{19}N_3O_3S$: C, 60.49; H, 5.36; N, 11.76; S, 8.97; found: C, 60.52; H, 5.22; N, 11.49; S, 8.56.

3.7. General Synthesis of Compounds (***6a,b***). The mixture of compound **4d** (10 mmol), chlorosuccinimide, or bromosuccinimide (12 mmol) in acetic acid (20 mL) was heated at 100–110°C for 3 h. After acid evaporation the residue was processed with water, neutralized with $NaHCO_3$ till pH 8 and filtered off. Compounds **6a,b** were purified by recrystallization from hexane-benzene mixture (1 : 1).

3.7.1. 2-Acetyl-6-chloro-3,7-dimethyl-5H-thiazolo[3,2-a]pyrimidin-5-one (***6a***). Yield (2.0 g, 80%); light brown crystals; m.p. 140–142°C. ^{1}HNMR: δ = 2.43 (s, 3H, 7-CH_3); 2.58 (s, 3H, COCH_3); 3.10 (s, 3H, 3-CH_3). Anal. Calcd for $C_{10}H_9ClN_2O_2S$: C, 46.79; H, 3.53; N, 10.91; S, 12.49; found: C, 46.58; H, 3.41; N, 10.57; S, 12.06.

3.7.2. 2-Acetyl-6-bromo-3,7-dimethyl-5H-thiazolo[3,2-a]pyrimidin-5-one (***6b***). Yield (2.52 g, 84%); light brown crystals; m.p. 90–92°C. ^{1}HNMR: δ = 2.48 (s, 3H, 7-CH_3); 2.57 (s, 3H, COCH_3); 3.09 (s, 3H, 3-CH_3). Anal. Calcd for

$C_{10}H_9BrN_2O_2S$: C, 39.88; H, 3.01; N, 9.30; S, 10.65; found: C, 39.69; H, 2.89; N, 8.91; S, 10.21.

3.8. Synthesis of 7-Methyl-2,3-dihydro-5H-thiazolo[3,2-a]pyrimidin-5-one (7). A suspension of compound **3** (10 mmol) and KOH (20 mmol) in DMFA (10 mL) was stirred for 3 h till salt formation; then at cooling (0°C) 1,2-dichloroethane (10 mmol) was added dropwise and allowed to stand for 24 h at room temperature. The mixture was filtered off, DMFA was evaporated at low pressure, and the residue was processed with $CHCl_3$ and then with ether.

Yield (1.2 g, 70%); white crystals; m.p. 118–120°C. IR (KBr): ν = 1660 (C=O). ^{1}HNMR: δ = 2.15 (s, 3H, 7-CH_3); 3.49 (t, J = 7.7, 2H, SCH_2); 4.34 (t, J = 7.7, 2H, NCH_2); 5.83 (s, 1H, CH-pyrim.). ^{13}C NMR: δ = 22.76; 25.49; 48.02; 106.68; 159.46; 163.22; 163.73. MS: 169 (M + 1). Anal. Calcd for $C_7H_8N_2OS$: C, 49.98; H, 4.79; N, 16.65; S, 19.06; found: C, 49.81; H, 4.65; N, 16.33; S, 18.80.

4. Conclusions

6-Methyl-2-thioxo-2,3-dihydropyrimidin-4(1H)-one (**1**) can be obtained with high yield (90%) on treatment of acetoacetic acid ethyl ester with thiourea in sodium methylate. When the latter reacts with 3-chloro-pentane-2,4-dione, the substitution occurs at sulfur atom (**2**) and the further regioselective cyclization with good yields afforded 2-acetyl-3,7-dimethyl-5H-thiazolo[3,2-a]pyrimidin-5-one. This three-step synthesis is convenient and efficient and can be used for the syntheses of different biologically active thiazolo[3,2-a]pyrimidine derivatives.

The preliminary biological assay indicates that investigated biheterocyclic system derivatives show growth stimulant activity and can be of interest to obtain new growth regulators.

Competing Interests

The authors declare no conflict of interests.

References

[1] B. Tozkoparan, M. Ertan, P. Kelicen, and R. Demirdamar, "Synthesis and anti-inflammatory activities of some thiazolo[3,2-a]pyrimidine derivatives," *Il Farmaco*, vol. 54, no. 9, pp. 588–593, 1999.

[2] O. Alam, S. A. Khan, N. Siddiqui, and W. Ahsan, "Synthesis and pharmacological evaluation of newer thiazolo [3,2-a] pyrimidines for anti-inflammatory and antinociceptive activity," *Medicinal Chemistry Research*, vol. 19, no. 9, pp. 1245–1258, 2010.

[3] H. Zhi, L.-M. Chen, L.-L. Zhang et al., "Design, synthesis, and biological evaluation of 5H-thiazolo[3,2-a] pyrimidine derivatives as a new type of acetylcholinesterase inhibitors," *Arkivoc*, no. 13, pp. 266–277, 2008.

[4] J. M. Parmar and A. R. Parikh, "Synthesis of thiazolo[3,2-a]pyrimidine derivatives as a biologically active heterocycles," *Heterocyclic Communications*, vol. 4, no. 5, pp. 463–466, 1998.

[5] H. H. Sayed, A. H. Shamroukh, and A. E. Rashad, "Synthesis and biological evaluation of some pyrimidine, pyrimido[2,1-b][1,3]thiazine and thiazolo[3,2-a]pyrimidine derivatives," *Acta Pharmaceutica*, vol. 56, no. 2, pp. 231–244, 2006.

[6] S. Maddila, G. L. V. Damu, E. O. Oseghe, O. A. Abafe, C. Venakata Rao, and P. Lavanya, "Synthesis and biological studies of novel biphenyl-3,5-dihydro-2H-thiazolopyrimidines derivatives," *Journal of the Korean Chemical Society*, vol. 56, no. 3, pp. 334–340, 2012.

[7] T. D. Oza, A. Parikh, and V. Gudaparthi, "Synthesis and biological screening of some novel 3-amino-4-arylidene-5-pyrazolones and thiazolo[3,2-α]pyrimidines," *Inernational Journal of Pharmaceutical Sciences*, vol. 3, no. 4, pp. 2775–2785, 2012.

[8] M. A. Salama and S. A. El-Essa, "Synthesis and reactions of some new 2,3-dihydro-5H-5,7-diarylthiazolo-[3,2-a]pyrimidine-3-one derivatives and their antibacterial and fungicidal activity," *Indian Journal of Chemistry Section*, vol. 42, no. 1, pp. 173–179, 2003.

[9] M. S. K. Youssef, R. A. Ahmed, M. S. Abbady, S. A. Abdel-Mohsen, and A. A. Omar, "Reactions of 4-(2-aminothiazole-4-yl)-3-methyl-5-oxo-1-phenyl-2-pyrazoline. Synthesis of thiazolo[3,2-a]pyrimidine and imidazo[2,1-b]thiazole derivatives," *Monatshefte für Chemie*, vol. 139, no. 5, pp. 553–559, 2008.

[10] A.-E.-G. Amr, S. S. Maigali, and M. M. Abdulla, "Microwave assisted synthesis of some new thiazolopyrimidine, thiazolodipyrimidine and thiazolopyrimidine derivatives with potential antioxidant and antimicrobial activity," *Monatshefte für Chemie*, vol. 139, pp. 1409–1415, 2008.

[11] M. J. Solanki and M. M. Chavda, "Synthesis and antimicrobial activity of some novel derivatives of thiazolo[2,3-b]dihydropyrimidine containing 4-pyrazolyl moiety," *Rasayan Journal of Chemistry*, vol. 4, no. 3, pp. 605–608, 2011.

[12] Q. Chen, X.-L. Zhu, L.-L. Jiang, Z.-M. Liu, and G.-F. Yang, "Synthesis, antifungal activity and CoMFA analysis of novel 1,2,4-triazolo[1,5-a]pyrimidine derivatives," *European Journal of Medicinal Chemistry*, vol. 43, no. 3, pp. 595–603, 2008.

[13] V. Patel, "Synthesis, characterization and antifungal activity of substituted ethyl 5,7- dimethyl-3-oxo-2,3-dihydro-5H-[1,3]-thiazolo[3,2-a]pyrimidie-6-carboxylate derivatives," *Der Pharmacia Sinica*, vol. 4, no. 5, pp. 72–78, 2013.

[14] M. Bansal, R. Kaur, and B. Kaur, "Synthesis and antifungal evaluation of thiazolo[3,2-α]pyrimidine," *Journal of Applicable Chemistry*, vol. 2, no. 3, pp. 391–397, 2013.

[15] S. P. Vartale, D. B. Kadam, N. K. Halikar, and M. M. Pund, "An efficient method for synthesis of novel iminothiazolopyrimidines and plausible antioxidant potential," *International Journal of Drug Development and Research*, vol. 5, no. 1, pp. 128–134, 2013.

[16] E. E. Flefel, M. A. Salama, M. El-Shahat, M. A. El-Hashash, and A. F. El-Farargy, "A novel synthesis of some new pyrimidine and thiazolopyrimidine derivatives for anticancer evaluation," *Phosphorus, Sulfur, and Silicon and the Related Elements*, vol. 182, no. 8, pp. 1739–1756, 2007.

[17] A. E.-F. G. Hammam, M. A. Sharaf, and N. A. Abd El-Hafez, "Synthesis and anti-cancer activity of pyridine and thiazolopyrimidine derivatives using 1-ethylpiperidone as a synthon," *Indian Journal of Chemistry B*, vol. 40, no. 3, pp. 213–221, 2001.

[18] M. Said, K. Abouzid, A. Mouneer, A. Ahmedy, and A. M. Osman, "Synthesis and biological evaluation of new thiazolopyrimidines," *Archives of Pharmacal Research*, vol. 27, no. 5, pp. 471–477, 2004.

[19] A. M. Mohamed, A.-G. E. Amr, M. A. Alsharari, H. R. M. Al-Qalawi, M. O. Germoush, and R. A. Al-Salahi, "Anticancer activities of some new synthesized thiazolo[3,2-a]pyrido[4,3-d]pyrimidine derivatives," *American Journal of Biochemistry and Biotechnology*, vol. 7, no. 2, pp. 43–54, 2011.

[20] T. P. Selvam, V. Karthick, P. V. Kumar, and M. A. Ali, "Synthesis and structure-activity relationship study of 2-(substituted benzylidene)-7-(4-fluorophenyl)-5-(furan-2-yl)-2*H*-thiazolo[3,2-*a*]pyrimidin-3(7*H*)-one derivatives as anticancer agents," *Drug Discoveries & Therapeutics*, vol. 6, no. 4, pp. 198–204, 2012.

[21] A. Nagaraj and C. Sanjeeva Reddy, "Synthesis and biological study of novel bis-chalcones, bis-thiazines and bis-pyrimidines," *Journal of the Iranian Chemical Society*, vol. 5, no. 2, pp. 262–267, 2008.

[22] D. B. A. Kumar, G. K. Prakash, M. N. Kumaraswamy, and B. P. Nandeshwarappa, "Microwave assisted facile synthesis of amino pyrimidines bearing benzofuran and investigation of their antimicrobial activity," *Indian Journal of Chemistry Section B*, vol. 45, no. 7, pp. 1699–1703, 2006.

[23] M. B. Buddh, A. H. Bapodra, and K. D. Ladva, "Synthesis and biological evaluation of thiazolo[3, 2-α] pyrimidine derivatives as a new type of potential antimicrobial agents," *Rasayan Journal of Chemistry*, vol. 4, no. 4, pp. 824–828, 2011.

[24] A. E. Rashad, A. H. Shamroukh, R. E. Abdel-Megeid, and W. A. El-Sayed, "Synthesis, reactions, and antimicrobial evaluation of some polycondensed thienopyrimidine derivatives," *Synthetic Communications*, vol. 40, no. 8, pp. 1149–1160, 2010.

[25] T. I. El-Emary and S. A. Abdel-Mohsen, "Synthesis and antimicrobial activity of some new 1,3-diphenylpyrazoles bearing pyrimidine, pyrimidinethione, thiazolopyrimidine, triazolopyrimidine, thio- and alkylthiotriazolopyrimidinone moieties at the 4-position," *Phosphorus, Sulfur and Silicon and the Related Elements*, vol. 181, no. 11, pp. 2459–2474, 2006.

[26] M. Bansal, R. Kaur, and B. Kaur, "Synthesis and antimicrobial activities of thiazolo[3, 2-a]pyrimidine derivatives," *Asian Journal of Research in Chemistry*, vol. 4, no. 1, pp. 116–119, 2011.

[27] H. W. Lee, B. Y. Kim, J. B. Ahn et al., "Molecular design, synthesis, and hypoglycemic and hypolipidemic activities of novel pyrimidine derivatives having thiazolidinedione," *European Journal of Medicinal Chemistry*, vol. 40, no. 9, pp. 862–874, 2005.

[28] M. Van Laar, E. Volkerts, and M. Verbaten, "Subchronic effects of the GABA-agonist lorazepam and the 5-$HT_{2A/2C}$ antagonist ritanserin on driving performance, slow wave sleep and daytime sleepiness in healthy volunteers," *Psychopharmacology*, vol. 154, no. 2, pp. 189–197, 2001.

[29] K. S. Nimavat and K. H. Popat, "Synthesis, anticancer, antitubercular and antimicrobial activities of 1-substituted, 3-aryl-5-(3′-bromophenyl)pyrazoline," *Indian Journal of Heterocyclic Chemistry*, vol. 16, p. 333, 2007.

[30] W. J. Houlihan, R. Road, and E. Robert, "Substituted thiazolo[3, 2-a]pyrimidines," US Pat. 3,635,173, 1972.

[31] S. F. Mohamed, E. M. Flefel, A. E.-G. E. Amr, and D. N. Abd El-Shafy, "Anti-HSV-1 activity and mechanism of action of some new synthesized substituted pyrimidine, thiopyrimidine and thiazolopyrimidine derivatives," *European Journal of Medicinal Chemistry*, vol. 45, no. 4, pp. 1494–1501, 2010.

[32] K. Danel, E. B. Pedersen, and C. Nielsen, "Synthesis and anti-HIV-1 activity of novel 2,3-dihydro-7*H*-thiazolo[3,2-*a*]pyrimidin-7-ones," *Journal of Medicinal Chemistry*, vol. 41, no. 2, pp. 191–198, 1998.

[33] E. Jeanneau-Nicolle, M. Benoit-Guyod, A. Namil, and G. Leclerc, "New thiazolo[3,2-*a*]pyrimidine derivatives, synthesis and structure-activity relationships," *European Journal of Medicinal Chemistry*, vol. 27, no. 2, pp. 115–120, 1992.

[34] A. Balkan, S. Uma, M. Ertan, and W. Wiegrebe, "Thiazolo[3,2-a]pyrimidine derivatives as calcium antagonists," *Pharmazie*, vol. 47, no. 9, pp. 687–688, 1992.

[35] R. Duval, S. Kolb, E. Braud, D. Genest, and C. Garbay, "Rapid discovery of triazolobenzylidene-thiazolopyrimidines (TBTP) as CDC25 phosphatase inhibitors by parallel click chemistry and in situ screening," *Journal of Combinatorial Chemistry*, vol. 11, no. 6, pp. 947–950, 2009.

[36] S. Kolb, O. Mondésert, M.-L. Goddard et al., "Development of novel thiazolopyrimidines as CDC25B phosphatase inhibitors," *ChemMedChem*, vol. 4, no. 4, pp. 633–648, 2009.

[37] F. M. Awadallah, "Synthesis, pharmacophore modeling, and biological evaluation of novel 5*H*-thiazolo[3,2-a]pyrimidin-5-one derivatives as 5-HT_{2A} receptor antagonists," *Scientia Pharmaceutica*, vol. 76, no. 3, pp. 415–438, 2008.

[38] S. Tetsuo, T. Mikio, H. Hidetoshi, H. Daijiro, and I. Akira, "2-Benzylthieno(2,3-d)pyrimidin-4(3H)-one derivative," Japanese Kokai Tokyo Koho JP, 62,132,884, 1987, (Chemical Abstracts, vol. 107, 198350h, 1987).

[39] W. Linder and W. Brandes, "Pesticidal thiazolopyrimidine derivatives," US Pat. 367,820, 1991.

[40] C. J. Shishoo and K. S. Jain, "Synthesis of some novel azido/tetrazolothienopyrimidines and their reduction to 2,4-diaminothieno[2,3-d]pyrimidines," *Journal of Heterocyclic Chemistry*, vol. 29, no. 4, pp. 883–893, 1992.

[41] F. Azam, I. A. Alkskas, and M. A. Ahmed, "Synthesis of some urea and thiourea derivatives of 3-phenyl/ethyl-2-thioxo-2,3-dihydrothiazolo[4,5-d]pyrimidine and their antagonistic effects on haloperidol-induced catalepsy and oxidative stress in mice," *European Journal of Medicinal Chemistry*, vol. 44, no. 10, pp. 3889–3897, 2009.

[42] P. K. Chakaravorty, W. J. Grelnlee, K. Dooseap, N. B. Mantlo, and A. A. Patchett, "New pyrimidinone derivatives are angiotensin II antagonists," PCT-The International Patent System WO 92.20.687.156, (Chemical Abstracts 1993, 118, 213104d) 1992.

[43] B. Zhao, L.-L. Jiang, Z. Liu et al., "A microwave assisted synthesis of highly substituted 7-methyl-5H-thiazolo[3,2-a]pyrimidine-6-carboxylate derivatives via one-pot reaction of aminothiazole, aldehyde and ethyl acetoacetate," *Heterocycles*, vol. 87, no. 10, pp. 2093–2102, 2013.

[44] M. Bakherad and F. Gholipoor, "Regioselective syntheses of 3-benzyl-substituted 7H-thiazolo[3,2-*a*]pyrimidine-7-ones through palladium-catalyzed heteroannulation of acetylenic compounds," *Organic Chemistry International*, vol. 2012, Article ID 956584, 5 pages, 2012.

[45] M. M. Heravi, A. N. Asar, and M. Bakavoli, "Selective cyclization of 2-propargylthio-6-methyl-pyrimidin-4(1H)-one to thiazolo[3,2-a]pyrimidin-7-one," *Journal of Sciences, Islamic Republic of Iran*, vol. 9, no. 1, pp. 27–29, 1998.

[46] N. M. Ahmad and K. Jones, "An efficient synthesis of thiazolo[3,2-a]pyrimidinones," *Tetrahedron Letters*, vol. 51, no. 25, pp. 3263–3265, 2010.

[47] G. Massolini, M. Kitsos, C. Gandini, and G. Caccialanza, "Fungicidal activity of a series of phenyl pyridyl ketoximes and their O–acetyl derivatives," *Pesticide Science*, vol. 26, no. 2, pp. 209–214, 1989.

[48] I. H. Aspinall and P. A. Worthington, "β–Methoxyacrylates; synthesis of new types of strobilurin fungicides with extended

side chains," *Pest Management Science*, vol. 55, no. 2, pp. 197–198, 1999.

[49] K. A. Eliazyan, A. M. Knyazyan, V. A. Pivazyan, E. A. Ghazaryan, S. V. Harutyunyan, and A. P. Yengoyan, "Synthesis of 3,4-dialkyl-5-(1-oxo-ethyl)-3H-thiazole-2-thiones derivatives and their pesticidal activity," *Journal of Heterocyclic Chemistry*, vol. 50, no. 5, pp. 1083–1088, 2013.

[50] Suk Jin Koo, S.-C. Ahn, Jae Suk Lim et al., "Biological activity of the new herbicide LGC-40863 benzophenone O-[2,6-bis[(4,6-dimethoxy-2-pyramdinyl)oxy]benzoyl]oxime," *Pesticide Science*, vol. 51, no. 2, pp. 109–114, 1997.

[51] P. J. De Fraine and J. M. Clough, "A new series of broad spectrum of fungicides," *Pesticide Science*, vol. 44, no. 1, pp. 77–79, 1995.

[52] A. Liu, X. Ou, M. Huang et al., "Synthesis and insecticidal activities of novel oxime ether pyrethroids," *Pest Management Science*, vol. 61, no. 2, pp. 166–170, 2005.

23

Synthesis of Peripherally Tetrasubstituted Phthalocyanines and Their Applications in Schottky Barrier Diodes

Semih Gorduk,[1] Ozge Koyun,[1] Oguzhan Avciata,[2] Ahmet Altindal,[3] and Ulvi Avciata[1,4]

[1] *Faculty of Arts and Science, Department of Chemistry, Yildiz Technical University, Esenler, 34210 Istanbul, Turkey*
[2] *Faculty of Chemical and Metallurgical Engineering, Department of Metallurgical and Materials Engineering, Yildiz Technical University, Esenler, 34210 Istanbul, Turkey*
[3] *Faculty of Arts and Science, Department of Physics, Yildiz Technical University, Esenler, 34210 Istanbul, Turkey*
[4] *Department of Occupational Health and Safety, Esenyurt University, Esenyurt, 34510 Istanbul, Turkey*

Correspondence should be addressed to Semih Gorduk; semih_grdk@hotmail.com and Ulvi Avciata; uavciata@gmail.com

Academic Editor: Maria F. Carvalho

New metal-free and metallophthalocyanine compounds (Zn, Co, Ni, and Cu) were synthesized using 2-hydroxymethyl-1,4-benzodioxan and 4-nitrophthalonitrile compounds. All newly synthesized compounds were characterized by elemental analysis, FT-IR, UV-Vis, ^{1}H-NMR, MALDI-TOF MS, and GC-MS techniques. The applications of synthesized compounds in Schottky barrier diodes were investigated. Ag/Pc/p–Si structures were fabricated and charge transport mechanism in these devices was investigated using dc technique. It was observed from the analysis of the experimental results that the charge transport can be described by Ohmic conduction at low values of the reverse bias. On the other hand, the voltage dependence of the measured current for high values of the applied reverse bias indicated that space charge limited conduction is the dominant mechanism responsible for dc conduction. From the observed voltage dependence of the current density under forward bias conditions, it has been concluded that the charge transport is dominated by Poole-Frenkel emission.

1. Introduction

Phthalocyanines (Pcs) have drawn special interest due to their particular unique physicochemical properties and exhibit a variety of superb properties, such as architectural flexibility, diverse coordination properties, increased stability, improved spectroscopic characteristics, and both semi- and photoconductive characteristics [1, 2]. These properties make them of considerable interest in diverse technological and scientific areas such as gas sensors [3], optoelectronic devices [4], static induction transistors [5], Langmuir–Blodgett films [6], electrophotographic applications [7], optical data storage [8], solar cells [9], organic field-effect transistors (OFETs) [10], and nonlinear optics [11]. In the fields of electronics and optoelectronics, Pcs are very promising candidates for future technology. The application of Pc compounds for electronics and optoelectronics devices is related to the dielectric relaxation process and electrical conductivity in the Pc compounds and these are frequently the deciding factors about the compatibility of the material for a specific device application [12, 13].

As main building blocks of various electronic and optoelectronic devices such as microwave diodes, solar cells, and photodetectors [14, 15], Schottky barrier diodes (SBDs) play a crucial role in the further development of the electronic devices. High barrier height and low leakage current are the most important performance parameters for a SBD. Although important advances have been made toward high performance SBD, a lot of improvement is still needed for practical applications. There are currently a vast number of experimental studies on the realization and the manipulation of the Schottky barrier height and the leakage current using an organic interlayer. In this respect, Pc compounds offer several advantages over inorganic counterparts such as strontium titanate ($SrTiO_3$) [16] and hafnium dioxide (HfO_2) [17] because of their low cost, suitability for synthetic modification, and ease of device fabrication. Therefore, the understanding of the physical behavior of these compounds under dc condition is essential in order to decide suitability and improving the quality and performance of SBD based electronic devices.

Herein, our study presents the synthesis and characterization of new peripherally tetrasubstituted phthalocyanines. After their widespread use as active layer in many different applications, 2(3),9(10),16(17),23(24)-tetrakis-{(2,3-dihydrobenzo[b][1,4]dioxin-2-yl)methoxy} phthalocyanines are proposed as an interlayer in SBDs. The effects of the central metal ion in the Pc layer on the leakage current in p-Si based SBDs have also been investigated.

2. Experimental Section

2.1. Materials and Equipment. *N,N′*-Dimethylformamide (DMF), dimethyl sulfoxide (DMSO), chloroform ($CHCl_3$), diethyl ether, methanol (MeOH), tetrahydrofuran (THF), ethanol (EtOH), dichloromethane (DCM), n-hexane, n-pentanol, and acetone were purchased from MERCK. 4-Nitrophthalonitrile, 2-hydroxymethyl-1,4-benzodioxan, 1,8-diazabicyclo[5.4.0]undec-7-ene (DBU), potassium carbonate (K_2CO_3), zinc(II) acetate, cobalt(II) acetate, nickel(II) acetate, copper(II) acetate were purchased from Aldrich. All chemicals were of reagent grade. All solvents were dried and stored over molecular sieves. The progress of the reactions was monitored by thin layer chromatography (TLC). Melting points of the substances were determined using an Electrothermal Gallenkamp device. The IR spectra were recorded using a Perkin Elmer spectrophotometer with ATR sampling accessory. UV-Vis spectra were recorded on a Agilent 8453 UV/Vis spectrophotometer. Elemental analyses were carried out by a LECO CHNS 932 instrument. A Varian Unity Inova 500 MHz spectrophotometer was used for ^1H NMR and ^{13}C NMR spectra. Mass spectra of the synthesized substances were acquired by using a Bruker Microflex LT MALDI-TOF MS. GC-MS spectrum was obtained on an Agilent Technologies 6890N GC-System-5973 IMSD.

2.2. Synthesis of Compounds

*2.2.1. 4-((2,3-Dihydrobenzo[b][1,4]dioxin-2-yl)methoxy)phthalonitrile (**1**).* 2-Hydroxymethyl-1,4-benzodioxan (0.52 g, 2.89 mmol) was dissolved in DMF (30 ml) under nitrogen and 4-nitrophthalonitrile (0.5 g, 2.89 mmol) was added. After stirring for 20 min at room temperature, finely ground anhydrous K_2CO_3 (1.96 g, 14.45 mmol) was added to this mixture in portions over 2 h with efficient stirring. The reaction mixture was stirred under a nitrogen atmosphere at 50°C for a total of 24 h. The reaction mixture was cooled to room temperature and was then poured into 200 ml ice-water, and the precipitate that formed was filtered off, washed with water, n-hexane, and diethyl ether, and then dried. The creamy crude product was recrystallized from MeOH. Finally the pure powder was dried in a vacuum. This compound was soluble in MeOH, EtOH, acetone, DCM, and $CHCl_3$. Yield: 0.76 g (90%). Mp: 145-146°C. *Anal.* Calc. for $C_{17}H_{12}N_2O_3$: C, 69.86; H, 4.14; N, 9.58. Found: C, 69.55; H, 4.02; N, 9.30%. IR (ATR, cm^{-1}): 3048 (Ar-CH), 2935–2889 (Aliphatic-CH, CH_2), 2228 (C≡N), 1604–1493 (C=C), 1275–1253 (C-O-C), 1091, 848, 760. ^1H NMR (Acetone-d_6), (δ: ppm): 7.88–7.86, 7.60–7.59 (d, 2H, Ar-H), 7.43–7.41 (dd, 1H, Ar-H), 6.75–6.71 (m, 4H, Ar-H), 4.56–4.52 (s, 1H, CH), 4.42–4.09 (m, 4H, CH_2). ^{13}C NMR ($CDCl_3$), (δ: ppm): 161.35, 142.88, 142.32, 135.37, 122.21, 122.02, 119.79, 119.46, 117.66, 117.48, 115.49 (C≡N), 115.08 (C≡N), 108.25, 70.90, 67.30, 64.64. MS (GC–MS) *m/z*, Calc.: 292.08, Found: 292 $[M]^+$.

*2.2.2. General Synthesis Procedure for Phthalocyanine Derivatives (**2–6**).* The mixture of phthalonitrile compound (**1**) (0.2 g. 0.69 mmol), n-pentanol (4 mL), 1,8-diazabicyclo[4.5.0]undec-7-ene (DBU) (5 drops), no metal salt for compound (**2**), and equivalent amounts of anhydrous $Zn(CH_3COO)_2$ for compound (**3**), $Co(CH_3COO)_2$ for compound (**4**), $Ni(CH_3COO)_2$ for compound (**5**), and $Cu(CH_3COO)_2$ for compound (**6**) were heated to 160°C and stirred for 24 h at this temperature under N_2 atmosphere. Then, after cooling to room temperature, the reaction mixture was precipitated by the addition of n-hexane and filtered off. After washing with hot MeOH and hot EtOH, the product was purified with column chromatography by using silica gel and THF/$CHCI_3$ solvent system. The all phthalocyanines (**2–6**) were soluble in THF, $CHCI_3$, DMF and DMSO.

*(1) 2(3),9(10),16(17),23(24)-Tetrakis-{(2,3-dihydrobenzo[b][1,4]dioxin-2-yl)methoxy} Phthalocyanine (**2**).* Solvent system for column chromatography was THF : $CHCI_3$ (100 : 3). Yield: 65 mg (32%), Mp: >200°C, Anal. calc. for $C_{68}H_{50}N_8O_{12}$: C, 69.24; H, 4.30; N, 9.57; Found: C, 69.01; H, 4.05; N, 9.43%. IR (ATR, cm^{-1}): 3289 (single bond NH), 3046 (Ar-CH), 2929–2871 (Aliphatic-CH, CH_2), 1610–1491 (C=C), 1264–1235 (C-O-C), 1095, 837, 742. ^1H-NMR ($CDCI_3$), (δ: ppm): 7.67–7.58 (bm, 6H, Ar-H), 7.40–7.15 (bm, 6H, Ar-H), 6.98–6.82 (bm, 16H, Ar-H), 4.95 (bs, 4H, Aliphatic-CH), 4.60–4.20 (bm, 16H, Aliphatic-CH_2). UV-vis (THF, 1×10^{-5} M): λ_{max}/nm (log ε): 701 (4.99), 664 (5.01), 639 (4.82), 607 (4.68), 332 (5.07). MS (MALDI-TOF), (*m/z*): Calc.: 1171.17, Found: 1171.31 $[M]^+$.

*(2) 2(3),9(10),16(17),23(24)-Tetrakis-{(2,3-dihydrobenzo[b][1,4]dioxin-2-yl)methoxy} Phthalocyaninato Zinc(II) (**3**).* Solvent system for column chromatography was THF : $CHCI_3$ (100 : 3). Yield: 96 mg (45%), Mp: >200°C, Anal. calc. for $C_{68}H_{48}N_8O_{12}Zn$: C, 66.16; H, 3.92; N, 9.08; Found: C, 65.97; H, 3.55; N, 8.90%. IR (ATR, cm^{-1}): 3045 (Ar-CH), 2923–2876 (Aliphatic-CH, CH_2), 1606–1489 (C=C), 1265–1232 (C-O-C), 1091, 843, 742. ^1H-NMR ($CDCI_3$), (δ: ppm): 7.74–7.63 (bm, 6H, Ar-H), 7.40–7.15 (bm, 6H, Ar-H), 6.98–6.44 (bm, 16H, Ar-H), 4.94 (bs, 4H, Aliphatic-CH), 4.54–4.17 (bm, 16H, Aliphatic-CH_2). UV-vis (THF, 1×10^{-5} M): λ_{max}/nm (log ε): 675 (5.00), 609 (4.31), 351 (4.68). MS (MALDI-TOF), (*m/z*): Calc.: 1234.54, Found: 1234.51 $[M]^+$.

(3) 2(3),9(10),16(17),23(24)-Tetrakis-{(2,3-dihydrobenzo[b][1,4]dioxin-2-yl)methoxy} Phthalocyaninato Cobalt(II) (4). Solvent system for column chromatography was THF : $CHCI_3$ (100 : 2). Yield: 85 mg (40%), Mp: >200°C, Anal. calc. for

$C_{68}H_{48}CoN_8O_{12}$: C, 66.50; H, 3.94; N, 9.12; Found: C, 66.05; H, 3.55; N, 8.85%. IR (ATR, cm^{-1}): 3064 (Ar-CH), 2954–2870 (Aliphatic-CH, CH_2), 1609–1492 (C=C), 1265–1232 (C-O-C), 1093, 838, 743. UV-vis (THF, 1×10^{-5} M): λ_{max}/nm ($\log \varepsilon$): 663 (5.01), 603 (4.54), 329 (5.08). MS (MALDI-TOF), (m/z): Calc.: 1227.27, Found: 1227.15 $[M]^+$.

*(4) 2(3),9(10),16(17),23(24)-Tetrakis-{(2,3-dihydrobenzo[b][1, 4]dioxin-2-yl)methoxy} Phthalocyaninato Nickel(II) (**5**).* Solvent system for column chromatography was THF : $CHCl_3$ (100 : 2.5). Yield: 74 mg (35%), Mp: >200°C, Anal. calc. for $C_{68}H_{48}N_8NiO_{12}$: C, 66.52; H, 3.94; N, 9.13; Found: C, 66.10; H, 3.58; N, 8.85%. IR (ATR, cm^{-1}): 3044 (Ar-CH), 2927–2871 (Aliphatic-CH, CH_2), 1608–1491 (C=C), 1264–1237 (C-O-C), 1092, 840, 742. ^{1}H-NMR ($CDCl_3$), (δ: ppm): 7.74–7.66 (bm, 6H, Ar-H), 7.39–7.19 (bm, 6H, Ar-H), 6.98–6.44 (bm, 16H, Ar-H), 4.95 (bs, 4H, Aliphatic-CH), 4.57–4.21 (bm, 16H, Aliphatic-CH_2). UV-vis (THF, 1×10^{-5} M): λ_{max}/nm ($\log \varepsilon$): 662 (5.10), 612 (4.67), 327 (5.08). MS (MALDI-TOF), (m/z): Calc.: 1227.85, Found: 1227.79 $[M]^+$.

*(5) 2(3),9(10),16(17),23(24)-Tetrakis-{(2,3-dihydrobenzo[b][1, 4]dioxin-2-yl)methoxy} Phthalocyaninato Copper(II) (**6**).* Solvent system for column chromatography was THF : $CHCl_3$ (100 : 2.5). Yield: 77 mg (37%), Mp: >200°C, Anal. calc. for $C_{68}H_{48}CuN_8O_{12}$: C, 66.26; H, 3.92; N, 9.09; Found: C, 65.90; H, 3.56; N, 8.87%. IR (ATR, cm^{-1}): 3044 (Ar-CH), 2929–2870 (Aliphatic-CH, CH_2), 1606–1491 (C=C), 1264–1235 (C-O-C), 1091, 841, 743. UV-vis (THF, 1×10^{-5} M): λ_{max}/nm ($\log \varepsilon$): 675 (5.01), 611 (4.57), 332 (4.82). MS (MALDI-TOF), (m/z): Calc.: 1232.70, Found: 1232.34 $[M]^+$.

2.3. Fabrication and Characterization of SBDs. The p-type Si wafers with (100) orientation and resistivity in the 8–10 Ω cm range were used as substrate. After cleaning the Si wafers using standard RCA cleaning procedure, Al metal was thermally evaporated under 4×10^{-5} mbar pressure onto the Si wafers and followed by a heat treatment at 500°C for 3 min in N_2 atmosphere in order to form the Ohmic contacts. After the removing of native oxide layer on the polished surface of the Si wafers by HF/H_2O (1 : 8) solution, thin films of the Pc compounds were formed onto the polished surface of Si substrate by spin coating technique. The thickness of the films was determined by ellipsometric technique. After the spin coating of the interface layer, 1 mm in diameter and 2000 Å thick Ag Schottky contacts were deposited on the the Pc layer by thermal evaporation technique. In this way, Ag/Pc film/p-Si Schottky barrier diodes were fabricated. Current-voltage (I-V) characteristics of the devices were measured under vacuum (<10^{-3} mbar) using an electrometer (Keithley, Model 617).

3. Results and Discussion

3.1. Synthesis and Characterization of Compounds. The general synthetic pathway for the synthesis of 4-((2,3-dihydrobenzo[b][1,4]dioxin-2-yl)methoxy)phthalonitrile (**1**) and its metal-free (**2**), zinc(II) (**3**), cobalt(II) (**4**), nickel(II) (**5**), and copper(II) (**6**) phthalocyanine complexes was given in Scheme 1. Initial compound **1** was synthesized by the nucleophilic aromatic substitution reaction of compound 2-hydroxymethyl-1,4-benzodioxan with 4-nitrophthalonitrile, in DMF at 50°C for 24 h. New phthalocyanines (**2–6**) were synthesized using corresponding anhydrous metal salts and/or without metal salt in 3 mL n-pentanol for 24 h at 160°C under N_2 atmosphere in the presence of DBU. Structures of synthesized compounds were verified by elemental analysis, UV-vis, FT-IR, ^{1}H NMR, ^{13}C NMR, and mass spectroscopy techniques. According to the IR spectrum of **1**, a new vibration that appeared at 2228 cm^{-1} proved the formation of **1**. The FT-IR spectrum of **1** showed aromatic bond -CH peaks at around 3013–3048 cm^{-1} and aromatic -C=C peaks at around 1604–1592 cm^{-1}. The stretching peaks at 2935–2880 cm^{-1} confirmed the presence of -CH and -CH_2 groups. ^{1}H NMR spectrum of **1** (Figure S1) exhibited aromatic protons, integrating for a total of 7 protons, at 7.88–7.86, 7.60–7.59, 7.43–7.41, and 6.75–6.71 ppm. In addition, CH and CH_2 protons of **1** were observed at 4.56–4.52 and 4.42–4.09 ppm which integrated for 5 protons. ^{13}C NMR of **1** (Figure S2) is another evidence for proposed structure; the new bands observed at 115.49 and 115.08 ppm could be defined as carbons of nitrile groups. The molecular ion peak for **1** was observed at 292 m/z in its GC-MS spectrum which verified the formation of this phthalonitrile.

FT-IR spectra of the phthalocyanine compounds are very similar to each other. The proposed target structures of all new phthalocyanines were affirmed in the FT-IR spectra by the disappearance of the -C≡N vibration at 2228 cm^{-1} for phthalonitrile (**1**). For metal-free phthalocyanine (**2**), inner core -NH vibration was observed at 3289 cm^{-1}, and this peak was the essential difference between the IR spectral data of metallophthalocyanines and metal-free ones. In the FT-IR spectra of phthalocyanines (**2–6**), stretching vibrations of aromatic CH groups around 3064–3044 cm^{-1}, aliphatic-CH, CH_2 groups around 2954–2870 cm^{-1}, and aromatic -C=C groups around 1610–1489 cm^{-1} appeared at expected frequencies. In the ^{1}H NMR spectra of **2**, **3**, and **5** phthalocyanines (Figures S3, S4, and S5), the bands belonging to the phthalocyanine and substitute 2-hydroxymethyl-1,4-benzodioxan aromatic protons were observed between 7.67 and 6.82 ppm for compound **2**, between 7.74 and 6.44 ppm for compound **3**, and between 7.74 and 6.44 ppm for compound **5** as broad multiplet peaks. In addition, the aliphatic-CH protons for **2**, **3**, and **5** were assigned at around 4.95, 4.94, and 4.95 ppm, respectively, as broad single peaks. The aliphatic CH_2 protons for **2**, **3**, and **5** were also assigned at around 4.60–4.20, 4.54–4.17, and 4.57–4.21 ppm, respectively, as broad multiplet peaks. For **2**, the typical shielding of inner core protons could not be observed due to the strong aggregation between Pc molecules [18]. The ^{1}H NMR spectrum of compounds **4** and **6** could not be determined because of the presence of paramagnetic copper and cobalt ions [19]. The MALDI-TOF mass spectra of the phthalocyanines (**2–6**) show the presence of characteristic peaks at m/z = 1171.31 $[M]^+$ for **2**, 1234.51 $[M]^+$ for **3**, 1227.15 $[M]^+$ for **4**, 1227.79 $[M]^+$ for **5**, and 1232.34 $[M]^+$ for **6** confirming the proposed structures (Figure 1).

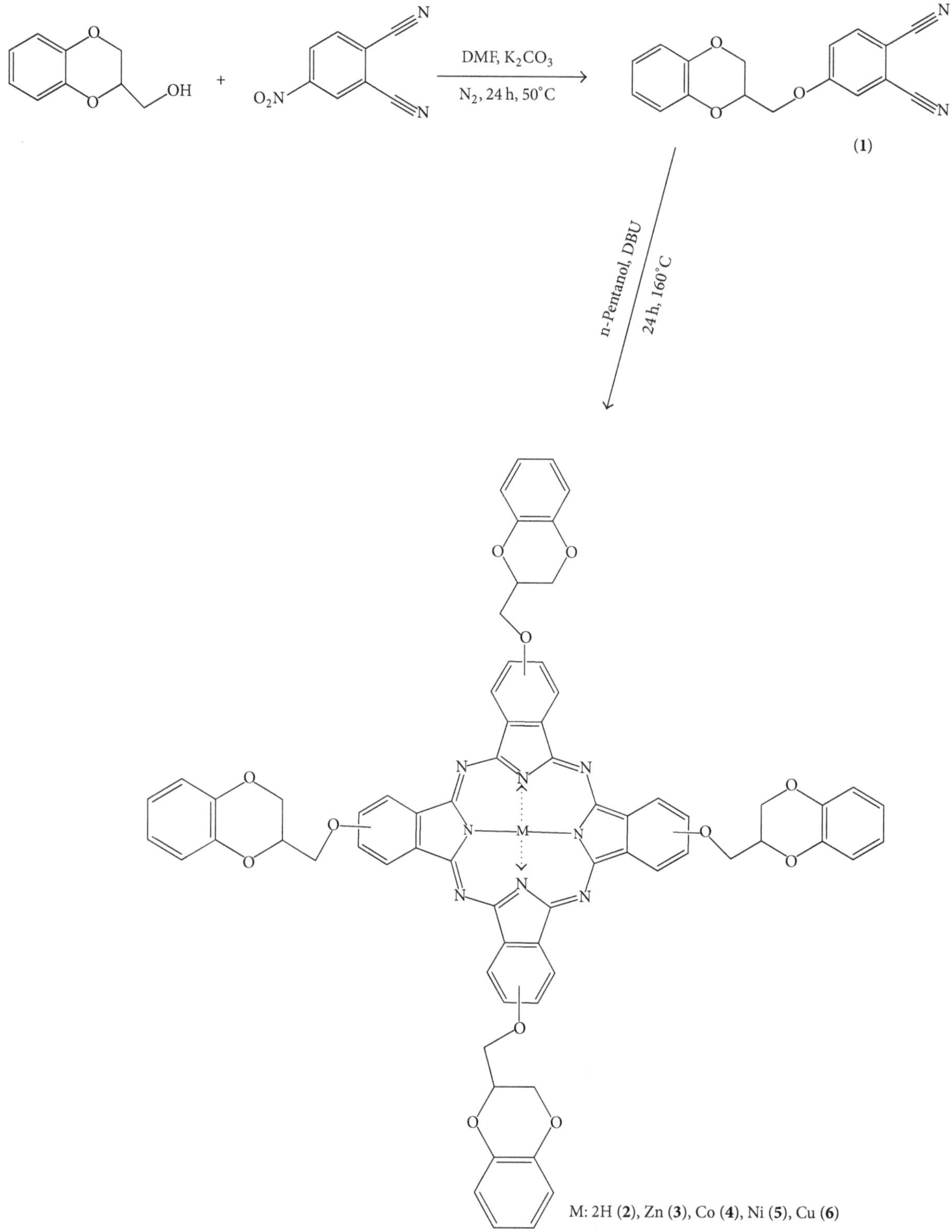

SCHEME 1: Schematic representation of synthesized compounds (**1–6**).

UV-Vis spectroscopy is thought to be the basic method to verify the formation of phthalocyanines. The phthalocyanine compounds show two main electronic transitions named as Q band (600–700 nm in the visible region), assigned to the π-π^* transition from the highest occupied molecular orbital (HOMO) to the lowest unoccupied molecular orbital (LUMO) of the Pc ring, and B band (300–350 nm in the UV region), resulting from the deeper π-π^* transitions [20–22]. The UV spectra of the metallophthalocyanines (**3–6**) display intense Q band at 675 nm for **3**, 663 nm for **4**, 671 nm for **5**,

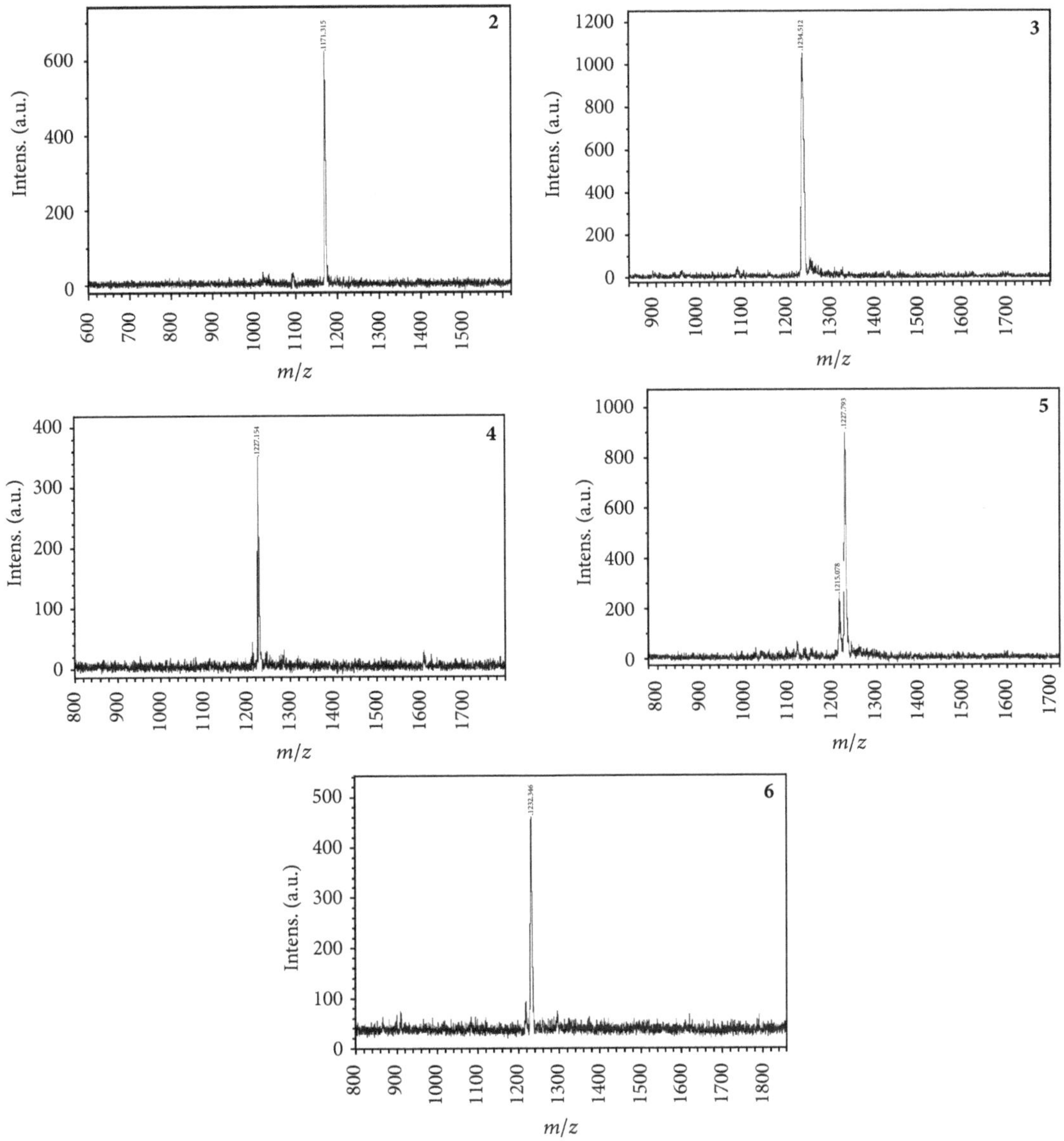

FIGURE 1: The MALDI-TOF MS spectra of the phthalocyanines (**2**–**6**).

and 675 nm for **6** in THF. The shoulders of phthalocyanines **3**–**6** were observed at 609, 603, 617, and 611 nm, respectively. For the Q band of phthalocyanines, the longer wavelength absorptions are owing to the monomeric species and shorter wavelengths (shoulders) are due to the aggregated species [2]. The B bands were observed at 351 nm for **3**, 329 nm for **4**, 327 nm for **5**, and 341 nm for **6**. For the metal-free phthalocyanine (**2**), a split Q band was observed at 701 and 664 nm, while B band was observed at 332 nm (Figure 2).

3.2. Characterization of SBDs. A device without Pc interface layer was also fabricated in order to verify the efficacy of the insertion of Pc layer on the charge transfer characteristic of the devices. For comparison, the room temperature current density (J) voltage characteristics of the devices are presented in Figure 3. As is clear from Figure 3, all the devices exhibit rectification behavior with different rectification ratio between 7 for Ag/p-Si and 310 for Ag/**2**/P-Si structure at ±5. This results prove the good rectification performance for the Pc layer inserted devices. Leakage current is an important factor influencing the performance of an SBD. It was found that the value of the leakage current density for the Pc layer inserted devices varies between 2.29 $\times 10^{-7}$ A/cm^2 and 6.56 $\times 10^{-7}$ A/cm^2, while the observed leakage current density was 1.38 $\times 10^{-4}$ A/cm^2 for Ag/p-Si structure. It should be mentioned here that the observed level of leakage current density is significantly low when compared with the reported values with similar structure such as Au/PDI/p-Si [23] and Al/DNA/p-Si [24] Schottky devices. This findings clearly show that the main performance parameters of a SBD such as rectification ratio and leakage current can be improved by modifying p-type silicon surface with 2(3),9(10),16(17),23(24)-tetrakis-{(2,3-dihydrobenzo[b][1,4]dioxin-2-yl)methoxy}phthalocyanines.

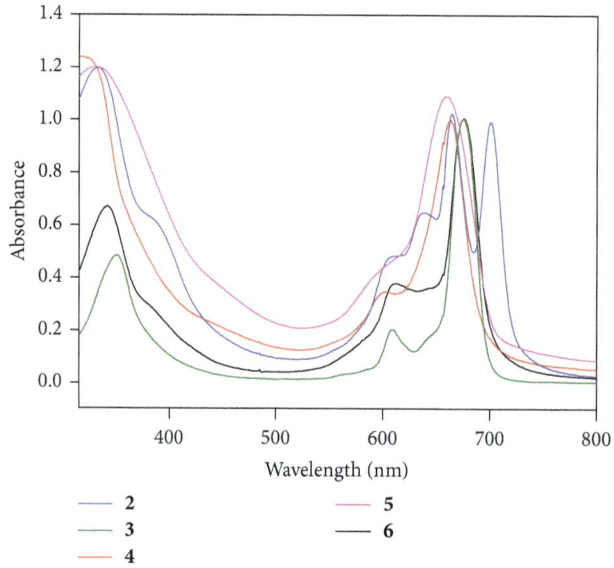

FIGURE 2: UV-Vis spectra of phthalocyanines (**2–6**).

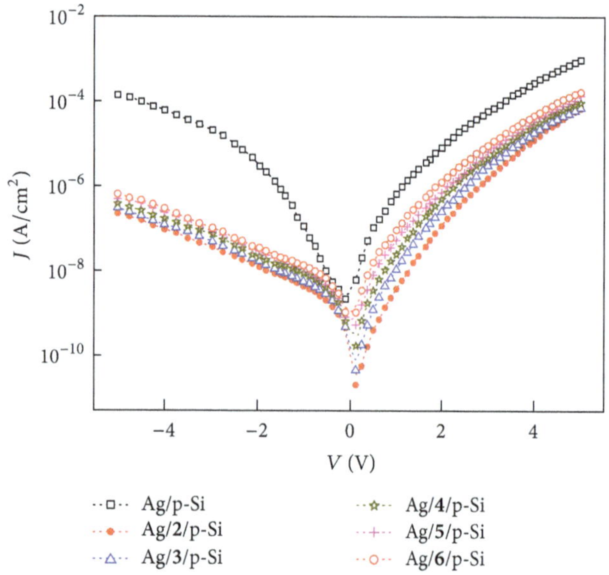

FIGURE 3: Room temperature J-V characteristics of the devices.

Various models, such as Fowler-Nordheim tunneling, Poole-Frenkel emission, and Ohmic and space charge limited (SCL) current, have widely been used to understand the nature of the charge transport through an organic thin film [25–27]. To investigate the dc conduction mechanism in fabricated devices, the obtained J-V data were replotted on double logarithmic scale for forward and reverse bias conditions in Figure 4.

It will be clear from the analysis of the data presented in Figures 4(a) and 4(b) that there are two regions with different slopes in double logarithmic J-V plots for both forward and reverse bias conditions, which is consistent with earlier reports on Metal/Phthalocyanine/Metal (MIM) structures [28, 29]. A relatively small slope in J-V curves at low voltage regions is clear for both directions. Special attention must be given to interpreting the presented data in Figure 4, because different conduction mechanism can give rise to this type of J-V characteristics. A close analysis of the slopes of the curves shown in Figure 4 indicates that the slopes of the curves vary between 0.92 and 1.16 for reverse bias conditions, which is indicative of the Ohmic conduction in this voltage region. On the other hand, it was observed that the slopes of the J-V curves for forward bias vary between 2.15 and 2.68, which cannot be considered in the framework of Ohmic conduction. Therefore, in order to identify the conduction mechanism taking place in Ag/Pc/p-Si structures, the J-V curves of the fabricated devices were analyzed separately for forward and reverse bias directions. As mentioned before, the reverse bias J-V characteristic of the devices at lower voltage range can be explained by the Ohmic behavior of the device. However, a power law dependence in the form of $J\alpha V^s$ with $s > 2$ reveals the presence of space charge limited conductivity (SCLC) controlled by exponentially distributed trapping levels. This type of J-V characteristic has been observed in some other Pc films such as copper phthalocyanine thin film [30]. It should be mentioned here that the reverse bias voltage at which the exponent s deviates from unity is nearly the same for all Pc modified devices. In order to investigate whether our experimental data can be explained using Poole-Frenkel emission, we have plotted, in Figure 5, $\ln(J/V)$ versus $V^{1/2}$ curves for all samples.

In the case of Poole-Frenkel emission, the relationship between the current density and the applied voltage can be expressed as given in [15]

$$J = A\frac{V}{d}\exp\left(\frac{q\varphi_{\mathrm{PF}} - \beta_{\mathrm{PF}}\sqrt{V/d}}{akT}\right), \quad (1)$$

where A is a fitting parameter, d is the thickness of the film, $q\varphi_{\mathrm{PF}}$ is the required energy for an electron to escape from the

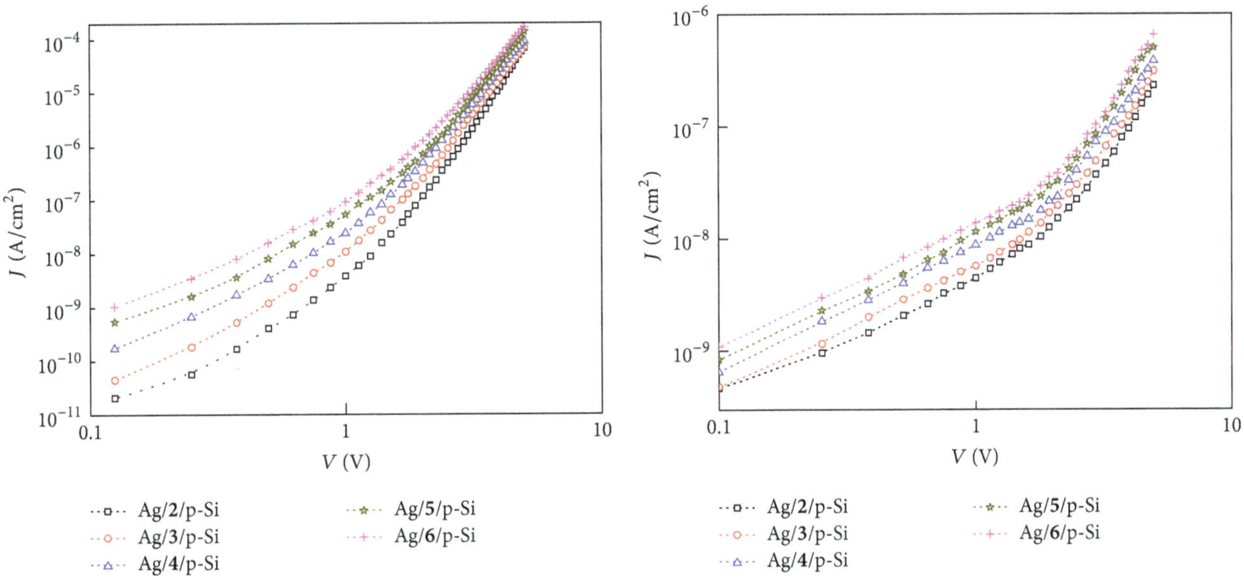

(a) (b)

FIGURE 4: J-V plots for forward (a) and reverse bias (b) conditions.

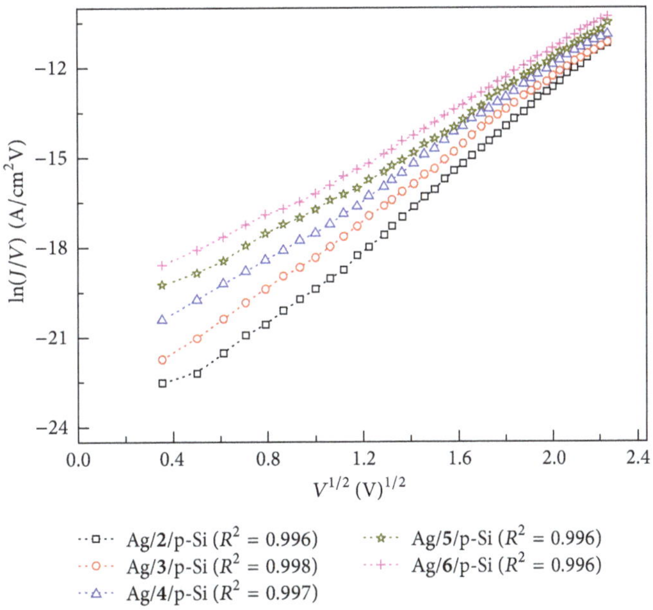

FIGURE 5: Poole-Frenkel plots for the device investigated under forward bias condition.

trap, a is the slope parameter, and $\beta_{PF}\sqrt{V}$ is barrier lowering. According to (1), the plot of $\ln(J/V)$ versus $\sqrt{V}$ should be linear if the charge transport takes place through Poole-Frenkel emission. Good correlation coefficients (R^2) obtained from $\ln(J/V)$ versus $\sqrt{V}$ plots reveal that the mechanism responsible for conduction in Ag/Pc/p-Si structures under forward bias conditions can be described by the Poole-Frenkel emission.

4. Conclusion

In this work, we have defined the synthesis and characterization of peripherally tetrasubstituted metal-free, zinc, cobalt, nickel, and copper phthalocyanine derivatives. These new five compounds were characterized by various techniques. In order to get more quantitative information about charge transport mechanism and the usability of these compounds to modify the main SBD parameters such as rectification ratio and the leakage current, SBDs with the structure of Ag/Pc/p-Si were fabricated and characterized. Our experimental results showed that charge transport in these compounds take place via different mechanism in reverse and forward bias conditions. Results from this preliminary analysis indicated that the compounds can be considered, among other candidates, as a potential passivation layer for the new SBD diodes with high rectification ratio and low leakage current.

Conflicts of Interest

The authors declare that they have no conflicts of interest.

Acknowledgments

This study was supported by the Research Fund of Yildiz Technical University (Project no. 2016-01-02-DOP01).

References

[1] N. B. McKeown, *Phthalocyanine Materials: Synthesis, Structure and Function*, Cambridge University Press, Cambridge, UK, 1998.

[2] C. C. Leznoff, A. B. P. Lever, P. Stuzhin, O. Khelevina, and B. Berezin, *Phthalocyanines: Properties and Applications*, VCH Publishers, New York, NY, USA, 1996.

[3] M. Bouvet, P. Gaudillat, and J.-M. Suisse, "Phthalocyanine-based hybrid materials for chemosensing," *Journal of Porphyrins and Phthalocyanines*, vol. 17, no. 10, pp. 913–919, 2013.

[4] C. G. Claessens, U. Hahn, and T. Torres, "Phthalocyanines: From outstanding electronic properties to emerging applications," *The Chemical Record*, vol. 8, no. 2, pp. 75–97, 2008.

[5] M. Kraus, S. Richler, A. Opitz et al., "High-mobility copper-phthalocyanine field-effect transistors with tetratetracontane passivation layer and organic metal contacts," *Journal of Applied Physics*, vol. 107, no. 9, Article ID 094503, 2010.

[6] T. Basova, A. Hassan, M. Durmuş, A. G. Gürek, and V. Ahsen, "Liquid crystalline metal phthalocyanines: Structural organization on the substrate surface," *Coordination Chemistry Reviews*, vol. 310, pp. 131–153, 2016.

[7] D. S. Weiss and M. Abkowitz, "Advances in organic photoconductor technology," *Chemical Reviews*, vol. 110, no. 1, pp. 479–526, 2010.

[8] R. Ao, L. Kümmerl, and D. Haarer, "Present limits of data storage using dye molecules in solid matrices," *Advanced Materials*, vol. 7, no. 5, pp. 495–499, 1995.

[9] M.-E. Ragoussi, M. Ince, and T. Torres, "Recent advances in phthalocyanine-based sensitizers for dye-sensitized solar cells," *European Journal of Organic Chemistry*, no. 29, pp. 6475–6489, 2013.

[10] Y. Zhang, X. Cai, Y. Bian, and J. Jiang, "Organic semiconductors of phthalocyanine compounds for field effect transistors (FETs)," in *Functional Phthalocyanine Molecular Materials*, J. Jiang, Ed., vol. 135 of *Structure and Bonding*, pp. 275–321, Springer, Berlin, Germany, 2010.

[11] D. Dini, M. J. F. Calvete, and M. Hanack, "Nonlinear optical materials for the smart filtering of optical radiation," *Chemical Reviews*, vol. 116, no. 22, pp. 13043–13233, 2016.

[12] A. Altindal, Ş. Abdurrahmanoğlu, M. Bulut, and Ö. Bekaroğlu, "Charge transport mechanism in bis(double-decker lutetium(III) phthalocyanine) (Lu2Pc4) thin film," *Synthetic Metals*, vol. 150, no. 2, pp. 181–187, 2005.

[13] T. S. Shafai and T. D. Anthopoulos, "Junction properties of nickel phthalocyanine thin film devices utilising indium injecting electrodes," *Thin Solid Films*, vol. 398-399, pp. 361–367, 2001.

[14] N. N. Halder, P. Biswas, S. Kundu, and P. Banerji, "Au/p-Si Schottky junction solar cell: Effect of barrier height modification by InP quantum dots," *Solar Energy Materials & Solar Cells*, vol. 132, pp. 230–236, 2015.

[15] E. Rhoderick and R. Williams, *Metal-Semiconductor Contacts*, Clarendon, Oxford, UK, 2nd edition, 1988.

[16] S. Yadav and S. Ghosh, "Amorphous strontium titanate film as gate dielectric for higher performance and low voltage operation of transparent and flexible organic field effect transistor," *ACS Applied Materials & Interfaces*, vol. 8, no. 16, pp. 10436–10442, 2016.

[17] J. Tardy, M. Erouel, A. L. Deman et al., "Organic thin film transistors with HfO2 high-k gate dielectric grown by anodic oxidation or deposited by sol-gel," *Microelectronics Reliability*, vol. 47, no. 2-3, pp. 372–377, 2007.

[18] Ü. Demirbaş, M. Pişkin, B. Barut, R. Bayrak, M. Durmuş, and H. Kantekin, "Metal-free, zinc(II) and lead(II) phthalocyanines functioning with 3-(2H-benzo[d][1,2,3]triazol-2-yl)-4-hydroxyphenethyl methacrylate groups: Synthesis and investigation of photophysical and photochemical properties," *Synthetic Metals*, vol. 220, pp. 276–285, 2016.

[19] Ü. Demirbaş, D. Akyüz, B. Barut, R. Bayrak, A. Koca, and H. Kantekin, "Electrochemical and spectroelectrochemical properties of thiadiazole substituted metallo-phthalocyanines," *Spectrochimica Acta Part A: Molecular and Biomolecular Spectroscopy*, vol. 153, pp. 71–78, 2016.

[20] G. Gümrükçü, M. Ü. Özgür, A. Altindal, A. R. Özkaya, B. Salih, and Ö. Bekaroğlu, "Synthesis and electrochemical, electrical and gas sensing properties of novel mononuclear metal-free, Zn(II), Ni(II), Co(II), Cu(II), Lu(III) and double-decker Lu(III) phthalocyanines substituted with 2-(2H-1,2,3-benzotriazol- 2-yl)-4-(1,1,3,3-tetramethylbutyl) phenoxy," *Synthetic Metals*, vol. 161, no. 1-2, pp. 112–123, 2011.

[21] Ö. Koyun, S. Gördük, B. Keskin, A. Çetinkaya, A. Koca, and U. Avciata, "Microwave-assisted synthesis, electrochemistry and spectroelectrochemistry of phthalocyanines bearing tetra terminal-alkynyl functionalities and click approach," *Polyhedron*, vol. 113, pp. 35–49, 2016.

[22] G. K. Karaoğlan, G. Gümrükçü, S. Gördük, N. Can, and A. Gül, "Novel homoleptic, dimeric zinc(II) phthalocyanines as gate dielectric for OFET device," *Synthetic Metals*, vol. 230, pp. 7–11, 2017.

[23] Ö. F. Yüksel, N. Tuğluoğlu, H. Şafak, and M. Kuş, "The modification of Schottky barrier height of Au/p-Si Schottky devices by perylene-diimide," *Journal of Applied Physics*, vol. 113, no. 4, Article ID 044507, 2013.

[24] H. M. J. Al-Ta'ii, Y. M. Amin, and V. Periasamy, "Calculation of the electronic parameters of an Al/DNA/p-Si schottky barrier diode influenced by alpha radiation," *Sensors*, vol. 15, no. 3, pp. 4810–4822, 2015.

[25] B. Keskin, A. Altindal, U. Avciata, and A. Gül, "A.C. and D.C. Conduction processes in octakis[(4-tert-butylbenzylthio)-porphyrazinato]Cu(II) thin films with gold electrodes," *Bulletin of Materials Science*, vol. 37, no. 3, pp. 461–468, 2014.

[26] B. Keskin, S. Arda Öztürkcan, and A. Altindal, "Ultrasound-assisted rapid one-pot synthesis, characterization and electrical properties of a β-aminoketone with a ferrocenyl moiety," *Polyhedron*, vol. 69, pp. 135–140, 2014.

[27] F.-C. Chiu, "A review on conduction mechanisms in dielectric films," *Advances in Materials Science and Engineering*, vol. 2014, Article ID 578168, 18 pages, 2014.

[28] T. G. Abdel Malik and R. M. Abdel-Latif, "Ohmic and space-charge limited conduction in cobalt phthalocyanine thin films," *Thin Solid Films*, vol. 305, no. 1-2, pp. 336–340, 1997.

[29] A. C. Varghese and C. S. Menon, "Electrical properties of nickel phthalocyanine thin films using gold and lead electrodes," *Journal of Materials Science: Materials in Electronics*, vol. 17, no. 2, pp. 149–153, 2006.

[30] A. K. Hassan and R. D. Gould, "The electrical properties of copper phthalocyanine thin films using indium electrodes," *Journal of Physics D: Applied Physics*, vol. 22, no. 8, pp. 1162–1168, 1989.

24

Study of the Molecular Properties of Mono- and Binuclear Metal *s*-Indacenyl Complexes with Ir, Rh, and Re: A Theoretical Approach

Ximena Zarate,[1] Eduardo Schott,[2] Emilio Bunel,[3] Juan M. Manríquez,[2] and Ivonne Chávez[2]

[1]*Instituto de Ciencias Químicas Aplicadas, Facultad de Ingeniería, Universidad Autónoma de Chile, Av. Pedro de Valdivia 425, Santiago, Chile*
[2]*Departamento de Química Inorgánica, Facultad de Química, Pontificia Universidad Católica de Chile, Casilla 306, Santiago, Chile*
[3]*Argonne National Laboratory, Argonne, IL 60439, USA*

Correspondence should be addressed to Ximena Zarate; jazminac@gmail.com and Ivonne Chávez; ichavez@uc.cl

Academic Editor: Henryk Kozlowski

Density functional theory (DFT) calculations were performed on a new family of mono- and bimetallic complexes, containing 4,8-([10]paracyclophane)-1,5-dihydro-*s*-indacene as the bridging ligand between the two metallic centers and different ancillary ligands. The *s*-indacene was blocked by substitution of the central benzene ring with the [10]paracyclophane to obtain the *syn*-conformations. This would force the metallic centers to be close together. It is proposed, due to the calculated molecular and electronic properties such as the reactivity indexes, the delocalized nature of the *s*-indacenyl ligand, and the electron-rich metals, that these systems could be reactive in a catalytic reaction. The results indicate that the systems with Rh and Re holding ancillary ligands such as bipy and CO show the best properties to be active in a chemical reaction. In this sense, by the assessed geometrical and electronic properties, when compared with a previously reported system, these complexes could be candidates for the reduction of CO_2 to oxalate.

1. Introduction

s-Indacene and *as*-indacene belong to a class of conjugated hydrocarbons characterized by having $4n\pi$ electrons ($n = 2, 3$). Therefore, according to Hückel's Rule, they should be antiaromatic [1]. Nevertheless, they have been for decades of significant interest for organic chemists as they have focused on them upon their diverse and fascinating reactivity. Due to the synthetic challenges that these kinds of compounds present and the questions that have arisen over their aromaticity and structure, the potential of these systems in different applications, such as semiconducting devices and catalysis, has been slowly explored.

Even when the unsubstituted molecules are unstable, their dianions and several substituted derivatives stabilize both sterically and electronically, resulting in thermally stable structures at room temperature [2].

Moreover, *s*-indacene may connect two or more organometallic centers, which may have different oxidation states, and allow interaction between them. Binuclear metal complexes bearing an *as*-indacenyl or *s*-indacenyl ligand were first reported in the literature in 1988 [3, 4]. A large variety of transition metals homobinuclear complexes have been described including Fe, Co, Ni, Ru, Rh, Re, and Mn. Also, heterobinuclear complexes with combinations of metals such as Rh-Ru, Fe-Ru, Fe-Rh, Ru-Rh, F-Co, and Cr-Ni have been reported [5–8]. The most striking feature of this family of binuclear complexes is the ability to delocalize charge among metal centers upon oxidation. In this sense, we believe that bimetallic complexes are useful compounds that might induce catalytic reactions.

On the other hand, literature reports on the selective conversion of CO_2 to oxalate $(C_2O_4)^{2-}$ point to the occurrence of a two-metal-centers mechanism. Bi- or polymetallic

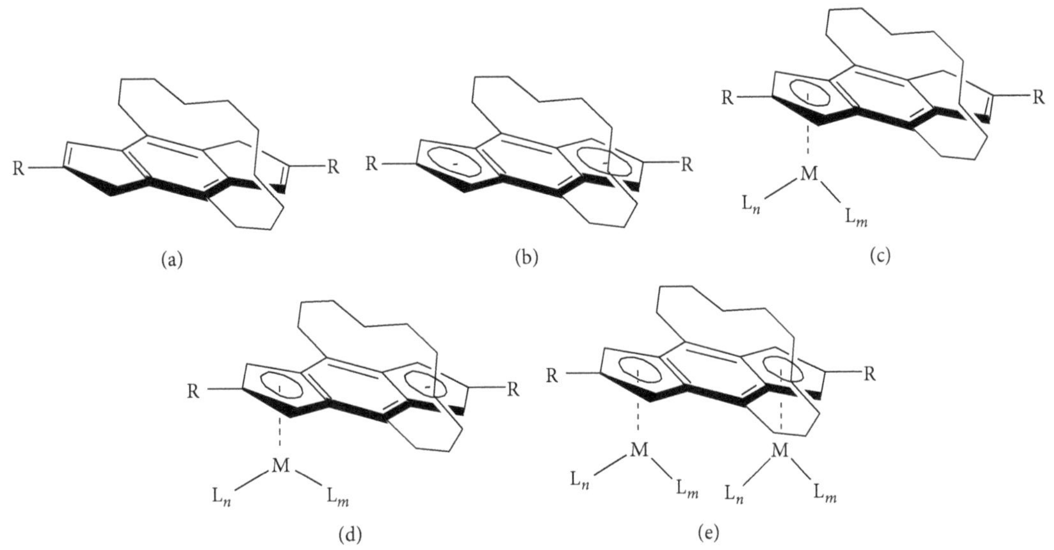

FIGURE 1: Structures of the studied systems. (a) Neutral ligand, (b) dianion ligand, and (c) neutral monometallic, (d) monoanion monometallic, and (e) bimetallic systems. R = H, CH_3, and CH_2CH_3; M = Ir, Rh, and Re; L_m and L_n = bipy, CO, with m and n = 0, 1, 2, 3.

compounds have been studied showing a wide variation of results [9–12]. The most important conclusion of these reports is that, for systems where the coordination of CO_2 is not reversible, the ability to transform CO_2 into other compounds depends on the extent of the charge transfer from the transition metal to CO_2. If the metal transfers an electron, the formation of C-C bonds is possible and the rigidity of the bimetallic system seems to play an essential role in the formation of this bond [13, 14]. In this sense, interestingly, one of the most innovative works in the area of CO_2 reduction reported by Angamuthu et al. in 2010 consists in the synthesis of a tetranuclear complex of Cu(I) with the ligand N-(2-mercaptopropyl)-N,N-bis(2-pyridylmethyl)amine, which spontaneously and selectively captures CO_2 from the air [15]. Cu(I) gets oxidized to Cu(II) and a bridge of oxalate is formed, which links two binuclear complexes, forming a tetramer. Adding a lithium salt to the solution allows the quantitative precipitation of lithium oxalate. Finally, the Cu(I) complex can be recovered by electrochemical reduction. The crystallographic study of the Cu complex showed that the Cu-Cu distance in the complex was 5.43 Å. This distance is similar to the Rh-Rh distance observed in a *syn* binuclear complex synthesized by our group [(2,6-diethyl-4,8-dimethyl-*s*-indacenediide) {Rh (COD)}$_2$] [16] of about 5.5 Å.

The large majority of the examples of binuclear complexes with *s*-indacene reported in the literature show the two metals centers in an *anticonfiguration*. This is due mostly to the steric bulk imposed by the ancillary ligands. In this work, we propose that it is possible to avoid the *anticonfiguration* by blocking one of the faces of the *s*-indacene introducing a group that can effectively obstruct the access of the metals to both faces forcing them to adopt the *syn* configuration. This could be easily achieved by constructing a cyclophane structure over the plane of the rings as shown in Figure 1.

Therefore, in order to have the right template structure for coupling reactions in those compounds and to favor the reaction, the right Metal-Metal (M-M) distance, electronic configuration, and lability of the ligand coordinated to the metal center (ancillary ligands) are needed. Specifically, binuclear metal frameworks could play an important role as templates for these types of reactions, such as selective reductive coupling of two molecules of CO_2 to oxalate.

Inspired by the work of Bouwman et al. and the versatile indacene derivatives, we decided to propose templates with synthetical feasibility, consisting in a family of binuclear metal complexes.

The aim of this work is to evaluate the effect of different transition metals (Ir, Rh, and Re), different ancillary ligands (bipy and CO), and -R groups (CH_3 and CH_2CH_3) (Figure 1) over the electronic properties, assessing the Frontier Molecular Orbitals (FMOs) location, the geometrical parameters highlighting the M-M distance, the global reactivity indexes, the atomic metal charges, and the Fukui functions. This preliminary study could lead experimentalist to have a hint about which metallic *s*-indacenyl complexes, ancillary ligands, and peripheral substituents might show suitable properties for a variety of chemical applications.

2. Computational Details

Quantum chemical calculations were performed in order to further investigate the electronic structure of a family of compounds in the ground state. Density functional theory (DFT) as implemented in Gaussian 09 [17] software was used to perform the calculations of the studied systems. The Becke three-parameter hybrid functional combined with Lee-Yang-Parr correlation B3LYP [18, 19] was employed. The relativistic effects of the heavy metals were incorporated using the pseudopotential LANL2DZ [20–22] and the Gaussian basis sets 6-31G(d) were used for the nonmetallic atoms [23]. With the aim of getting the minima points over the potential energy surface of the proposed compounds, the molecular structures

TABLE 1: Selected geometrical parameters of the ground state conformers. Bond lengths are given in Angstrom (Å) (see Scheme 1).

(a)

Par. Geo.	M = Ir, L_n = bipy			M = Rh, L_n = bipy			M = Re, $L_{n,m}$ = bipy, CO		
	H	CH_3	CH_2CH_3	H	CH_3	CH_2CH_3	H	CH_3	CH_2CH_3
M1-C1	2.22	2.22	2.21	2.23	2.22	2.22	2.25	2.24	2.24
M1-C2	2.15	2.16	2.16	2.20	2.20	2.20	2.22	2.24	2.24
M1-C3	2.22	2.21	2.22	2.23	2.22	2.22	2.32	2.31	2.31
M1-C9	2.61	2.58	2.59	2.57	2.56	2.56	2.52	2.51	2.51
M1-C10	2.62	2.59	2.60	2.57	2.56	2.57	2.53	2.52	2.52
M2-C5	2.22	2.22	2.22	2.23	2.23	2.23	2.25	2.25	2.25
M2-C6	2.16	2.16	2.16	2.20	2.21	2.21	2.22	2.24	2.24
M2-C7	2.22	2.22	2.21	2.23	2.22	2.22	2.32	2.31	2.31
M2-C11	2.60	2.60	2.61	2.57	2.56	2.55	2.52	2.51	2.51
M2-C12	2.61	2.60	2.61	2.57	2.56	2.56	2.52	2.52	2.52
M1-M2	5.58	5.53	5.59	5.45	5.43	5.41	5.40	5.43	5.44

(b)

Par. Geo.	M = Ir, $L_{n,m}$ = CO			M = Rh, $L_{n,m}$ = CO			M = Re, L = $(CO)_3$		
	H	CH_3	CH_2CH_3	H	CH_3	CH_2CH_3	H	CH_3	CH_2CH_3
M1-C1	2.29	2.28	2.28	2.28	2.27	2.27	2.33	2.33	2.32
M1-C2	2.25	2.26	2.26	2.27	2.29	2.28	2.32	2.34	2.34
M1-C3	2.29	2.28	2.28	2.28	2.27	2.27	2.34	2.32	2.33
M1-C9	2.65	2.64	2.64	2.60	2.58	2.59	2.50	2.50	2.49
M1-C10	2.65	2.65	2.65	2.60	2.59	2.59	2.52	2.50	2.50
M2-C5	2.29	2.28	2.28	2.28	2.27	2.27	2.32	2.33	2.32
M2-C6	2.25	2.26	2.26	2.27	2.28	2.28	2.32	2.34	2.34
M2-C7	2.29	2.28	2.28	2.28	2.27	2.27	2.34	2.32	2.33
M2-C11	2.66	2.65	2.65	2.59	2.59	2.59	2.49	2.50	2.49
M2-C12	2.66	2.65	2.65	2.60	2.59	2.59	2.52	2.49	2.51
M1-M2	5.58	5.60	5.61	5.41	5.40	5.44	5.08	5.05	5.05

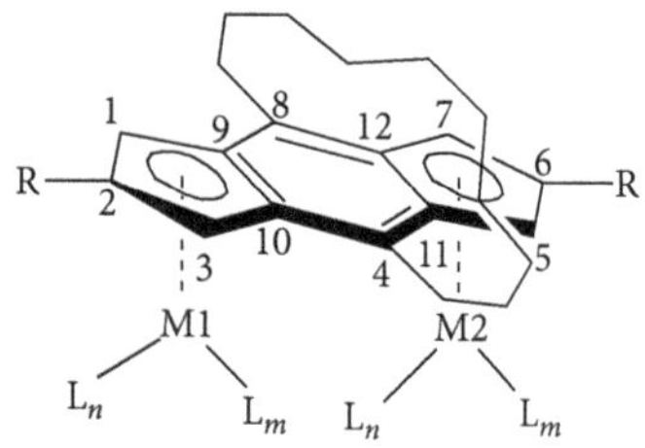

SCHEME 1

were fully optimized without symmetry constriction and the frequency calculations were performed at the same level of theory to confirm that all the optimized systems are stationary minima points. Implicit solvation effects were incorporated using the polarized continuum model (PCM, ε = 7.426) for tetrahydrofuran [24]. The local reactivity indexes and the *Fukui* functions computed at the ground state geometry are also assessed and analyzed.

3. Results and Discussion

The optimized ground state minima of the series of compounds displayed in Figure 1, ligand (s-Ic) in the neutral and dianion states, and the mononuclear and binuclear coordination systems were obtained at the B3LYP/6-31G(d) + PCM (ε = 7.426) level of theory. Selected geometrical parameters of the corresponding minima are listed in Table 1.

The ligand s-Ic shows a planar structure and one plane face is blocked by substitution of the central benzene ring with a saturated alkyl chain formed by 10 -CH_2- units. This substitution is a proposal for experimentalist to synthesize the compounds adopting exclusively the *syn* conformation after metallic coordination.

Geometrical parameters such as C-C and C-H of the s-Ic coordinating different metals (not shown in Table 1) do not change significantly when all the studied compounds are compared as some values show differences in the third decimal digit or in less proportion in the second decimal digit. This indicates that the different metals and ancillary ligands do not lead to significant changes in the structural geometry of the s-Ic.

Regarding the M-M distances, as observed in Table 1, the binuclear transition metals frameworks show the appropriate length between the two metal centers to act as a template for the selective reduction of two molecules of CO_2 to oxalate. Specifically, the results show that the systems with Rh and

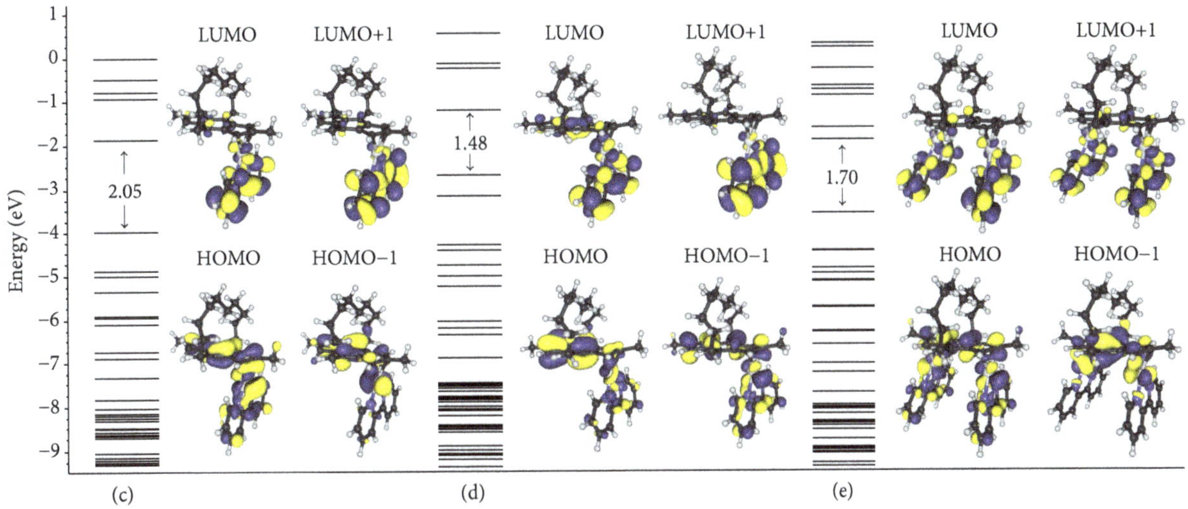

FIGURE 2: Compounds with Rh: (c) neutral monometallic, (d) monometallic monoanion, and (e) bimetallic systems with L = bipy and R = CH_3.

Re holding ancillary ligands such as bipy and CO, containing $-CH_3$ as -R substituent, present the most accurate M-M distance around 5.43 Å (Table 1). This distance corresponds to the Cu-Cu distance of the complex reported by Bouwman et al. that catalyzed the transformation of CO_2 to oxalate.

The ancillary ligands contribute steric effects towards the distances of the η bonds between the metal and the *s*-Ic, while as bigger the ancillary ligand is, the shorter the distances among the metal and the carbon atoms are (Table 1). In this sense, it is also worth mentioning that while the -R substituents in the *s*-Ic are changed, some differences over the M-M distances were observed; however, no clear trend is found.

These facts lead us to conclude that the M-M distance can be tuned by adjusting different ancillary ligands and substituents over the *s*-Ic ligand.

In terms of the electronic structure, as consequence of the formation of the *s*-Ic dianion species with different R groups, the HOMO-LUMO gaps get smaller in comparison with the gap of the neutral species (Figures S1-S2 in Supplementary Material available online at https://doi.org/10.1155/2017/9101720). For the isolated ligand, it is also observed that the FMOs structures present contribution of the *s*-Ic ligand, except for the LUMO+1 of the dianionic compounds which is localized over the alkyl chain binding across the six-edge ring of *s*-Ic (Figures S1-S2). The change of the R groups does not show transcendental effect over the FMOs.

In cases of the metallic complexes, the localization of the FMOs does not change considerably when the comparison is made between the monometallic and the bimetallic species (see complexes with Rh as example in Figure 2), as they are formed by contribution of the orbitals extended over the *s*-Ic and over the transition metals.

The energy levels diagrams display that the FMOs of all the mononuclear and binuclear complexes are more stable than the free substituted *s*-Ic ligand. It is worth noticing that all studied systems with ligands CO show more stable occupied orbitals than the system with bipy.

It was also observed that the different studied peripheral R substituents of the *s*-Ic ring (-H, $-CH_3$, and $-CH_2CH_3$) show no contribution to the FMOs.

In general, all the complexes that have a bimetallic structure exhibit FMOs localized mainly over the ancillary ligands and the metal centers, although a slight variation of the FMO isosurfaces is observed specifically in the LUMO. This LUMO, in cases of voluminous ancillary ligands, is composed mainly for the ancillary ligands orbitals and the contribution of the metals decreases (Figures S3–S20 in the Supplementary Material).

In these systems, the mononuclear and the binuclear, the metals would have a dual role. The metal would be receiving electrons, therefore stabilizing the reduced fragments for the formation of the C-C bond between two CO_2 molecules. This is supported by the fact that the HOMO is localized over the fragment containing the metal and the ancillary ligands (Figure 2 and Figures S3–S20 in Supplementary Material). The localization of the FMOs over these regions of the molecules suggests their high reactivity towards any catalytic activity.

To assess the atomic charges in the metals of the studied family of compounds, a Mülliken analysis was carried out and the values are reported in Table 2. The formation of the complexes involves a charge transfer from the *s*-Ic and ancillary ligands to the metals as the atomic charges get less positive with respect to the oxidation state of the metals (+1 for all metals in all cases). Significant charge transfer is observed for compounds containing Rh and CO as ancillary ligands, where it is known that the back-bonding effect of the CO ligand is present. In this sense, it is possible to state that in these systems the electron pair of each CO ligand donates more charge than it receives, which is evidenced in the slight negative charge over the metals. This behavior is followed by Re compounds containing CO which show atomic charges around 0.16 a.u. These values are significantly lower than the values obtained with Ir as metal center.

On the other hand, in cases of monometallic complexes, the charge transfer to the metals is bigger for the mononuclear

TABLE 2: Atomic charges in a.u. for the metal centers of the monometallic (anion and neutral) and bimetallic systems.

(a)

R	M = Ir, L_n = bipy				M = Rh, L_n = bipy				M = Re, $L_{n,m}$ = bipy, CO			
	Mono		Bim		Mono		Bim		Mono		Bim	
	M1 ani	M1 neu	M1	M2	M1 ani	M1 neu	M1	M2	M1 ani	M1 neu	M1	M2
H	0.351	0.300	0.297	0.300	0.156	0.157	0.152	0.157	0.289	0.312	0.315	0.323
CH_3	0.351	0.299	0.290	0.295	0.162	0.159	0.148	0.152	0.284	0.308	0.304	0.312
CH_2CH_3	0.349	0.298	0.287	0.295	0.151	0.048	0.142	0.147	0.278	0.301	0.300	0.309

(b)

R	M = Ir, $L_{n,m}$ = CO				M = Rh, $L_{n,m}$ = CO				M = Re, $L_{n,m}$ = CO			
	Mono		Bim		Mono		Bim		Mono		Bim	
	M1 ani	M1 neu	M1	M2	M1 ani	M1 neu	M1	M2	M1 ani	M1 neu	M1	M2
H	0.142	0.200	0.221	0.225	−0.202	−0.085	−0.065	−0.062	0.090	0.134	0.165	0.171
CH_3	0.148	0.198	0.216	0.220	−0.201	−0.089	−0.072	−0.067	0.089	0.132	0.160	0.166
CH_2CH_3	0.149	0.197	0.214	0.219	−0.203	−0.093	−0.077	−0.071	0.086	0.130	0.158	0.164

TABLE 3: Estimated energies for the HOMO and LUMO, chemical potential (μ), chemical hardness (η), and electrophilicity (ω) calculated for the optimized ground state minima of the metallic complexes.

Compound		HOMO	LUMO	HL GAP	μ	η	ω
M = Rh, L_n = bipy	M1 ani	−2.68	−1.20	1.48	−1.94	0.74	2.55
	M1 neu	−3.97	−1.82	2.16	−2.90	1.08	3.89
	Bim	−3.61	−1.90	1.71	−2.76	0.85	4.46
M = Rh, $L_{n,m}$ = CO	M1 ani	−3.15	−0.64	2.52	−1.89	1.26	1.42
	M1 neu	−5.13	−1.44	3.69	−3.28	1.84	2.92
	Bim	−5.18	−1.77	3.41	−3.47	1.70	3.54
M = Re, $L_{n,m}$ = bipy, CO	M1 ani	−2.76	−1.43	1.33	−2.10	0.66	3.31
	M1 neu	−4.27	−1.95	2.32	−3.11	1.16	4.17
	Bim	−4.11	−1.87	2.24	−2.99	1.12	3.99
M = Re, $L_{n,m}$ = CO	M1 ani	−3.08	−0.64	2.44	−1.86	1.22	1.42
	M1 neu	−5.53	−1.38	4.15	−3.46	2.08	2.88
	Bim	−5.62	−2.39	3.23	−4.01	1.62	4.97

anionic systems than for the neutral mononuclear ones, except the anionic mononuclear system containing Ir and bipy as ancillary ligands.

Taking into account these results, it is observed that Rh and Re bimetallic compounds might act as templates to carry out catalytic applications such as CO_2 reduction reactions. These metals when forming part of these types of systems get rich in electron charge, which is in concordance with the fact that the occupied FMOs are localized over the metals. These facts lead us to consider these systems as better reducing agents than the compounds with Ir.

In order to rationalize the chemical reactivity, a set of global reactivity descriptors were computed in the framework of the conceptual density functional theory. The estimated chemical potential (μ), chemical hardness (η), and electrophilicity (ω) have been analyzed [25–29]. This set of values is reported in Table 3. Moreover, the HOMO and LUMO energies and the HOMO-LUMO energy gaps are reported. For this analysis, we restricted the discussion to the complexes that are considered the best templates for catalytic reactions as stated above, that is, systems with Rh and Re, CO and bipy as ancillary ligands, and -CH_3 as peripheral substituent.

Parr defined μ in (1) as the infinitesimal change of energy of the system with respect to the electron number N at a constant external potential of the nuclei ($\nu(\vec{r})$), that is, the potential created by the nuclei. μ is related with its electronegativity as $-\chi$ and is associated with the feasibility of a system to exchange electron density with the environment at the ground state [25–29].

η is calculated using the second-order derivative of the chemical potential (see (2)) and can be interpreted as the resistance of a molecule to change the electron density in presence of other species.

$$\mu = \left(\frac{\partial E}{\partial N}\right)_{\nu(\vec{r})} = -\chi, \tag{1}$$

$$\eta = \frac{1}{2}\left(\frac{\partial^2 E}{\partial N^2}\right)_{\nu(\vec{r})} = \frac{1}{2}\left(\frac{\partial \mu}{\partial N}\right)_{\nu(\vec{r})}. \tag{2}$$

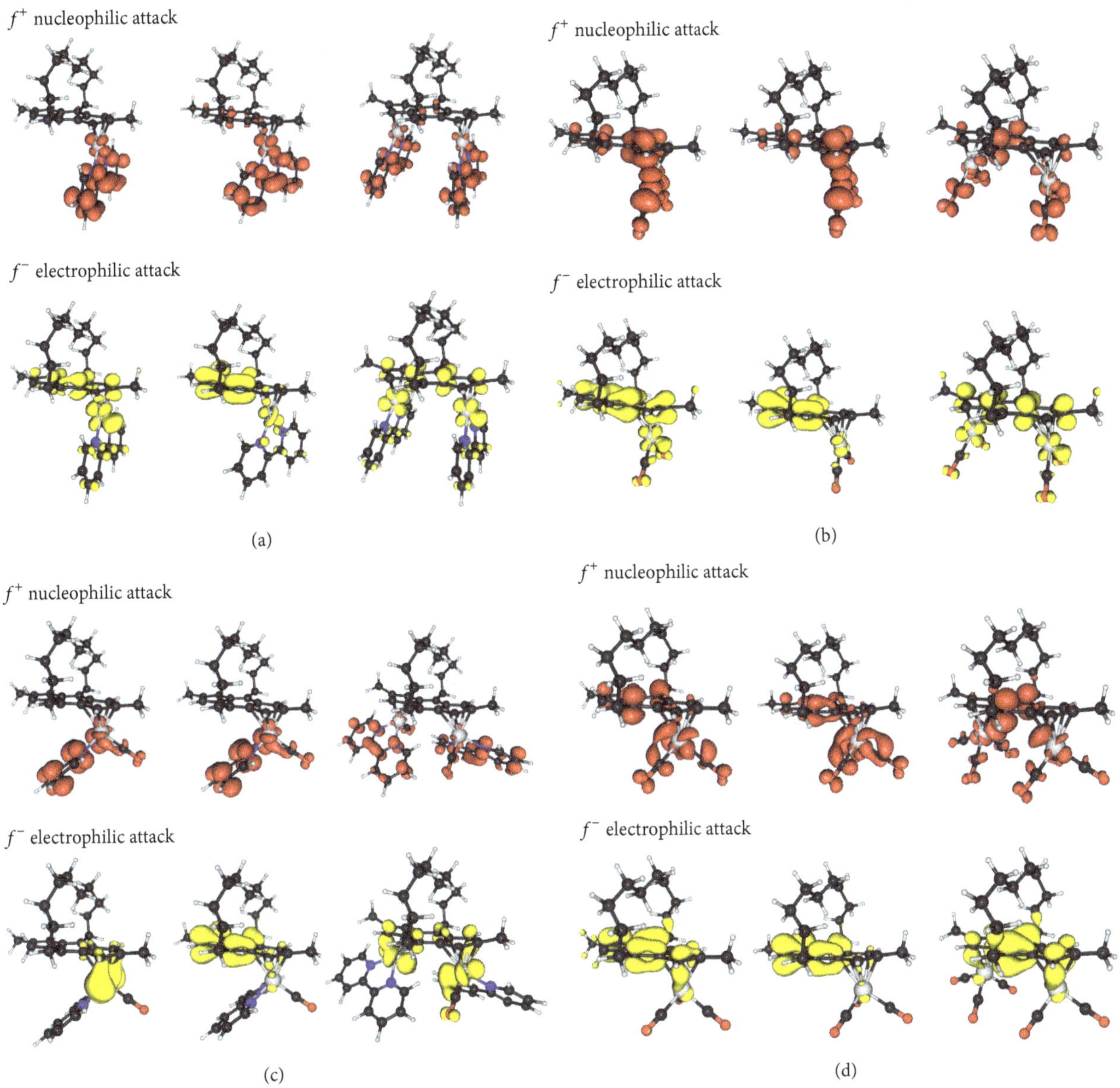

FIGURE 3: Condensed *Fukui* functions for nucleophilic (f^+) and electrophilic (f^-) attacks for compounds with metals (M) and ancillary ligands as follows: (a) Rh and bipy, (b) Rh and CO, (c) Re and bipy, CO, and (d) Re and CO. The neutral monometallic, anionic monometallic, and bimetallic complexes are shown.

In numerical applications, these reactivity indexes are calculated following approximations using Koopmans' theory and finite differences leading to (3). Here, I and A correspond to the ionization energy and electron affinity, while $E(\pi_H)$ and $E(\pi_L)$ correspond to the orbital energies of the HOMO and LUMO, respectively [25–28]:

$$\mu \approx -\frac{1}{2}(I + A) \approx \frac{1}{2}(E(\pi_L) + E(\pi_H)),$$
$$\eta \approx \frac{1}{2}(I - A) \approx \frac{1}{2}(E(\pi_L) - E(\pi_H)). \quad (3)$$

The ω index as defined in (4) measures the tendency of a molecule to receive electronic charge from a donor species. Therefore, ω is considered as a sort of "electrophilicity power" [27–29]:

$$\omega = \frac{\mu^2}{2\eta}. \quad (4)$$

When bimetallic complexes are compared (see Table 3), the values of μ of systems with Rh and bipy (-2.76 eV) and complex with Re and bipy and CO (-2.99 eV) are higher than values presented by systems with Rh, Re, and CO as ancillary ligands (-3.47 eV and -4.01 eV, resp.). This suggests bigger reactivity of the first two named complexes towards electron donation. Hence, these complexes tend to react as source of electrons that enhance their possible behavior as catalytic templates for reduction reactions, such as the reduction of $2CO_2$ to $C_2O_4^{2-}$.

It is worth mentioning that μ of the monometallic neutral species adopts values similar to μ of the bimetallic complexes, which indicates that these systems can also act as strong electron donors. Finally, the anionic mononuclear species display the highest μ contemplated in the range of -1.86 to -2.10 eV.

The η decreases in the order mononuclear neutral > bimetallic > anionic mononuclear are reasonable with respect to the impact of the global reactivity/polarization indexes. The obtained trend indicates that the last two named systems are more reactive. In case of ω, as expected, the anionic mononuclear systems show the lowest ω values and increase for the neutral and bimetallic compounds.

In order to clarify the dependency of the reactivity with the molecular structure, we also analyzed the condensed *Fukui* functions [25, 30, 31]. In Figure 3, we included the wavefunctions for nucleophilic (f^+) and electrophilic (f^-) attacks. Considering that molecules may react with sites that show high or low electron density to carry out a reaction, the f^- plots depict that the metallic centers are able to follow a reaction pathway to carry out a reduction, as they have high density of electrons. These results are consistent with the known reactivity of ferrocene (and many other organometallic complexes) with any reducing species, where the reaction involves the formation of ferrocenium. Therefore, the reactivity of the herein studied molecules will depend on the interactions occurring in the metal of the complex with other species. Although these results are not a definitive evidence on how the mechanism of the chemical reactions might be, they are useful to understand the chemical behavior of the family of complexes.

4. Conclusions

A theoretical study of a new family of organometallic complexes was carried out. As it is observed, the results indicate that the bimetallic systems with Rh and Re holding ancillary ligands such as bipy and CO show the best properties to be active in a chemical reaction. This is supported by the fact that the most electron-rich portion of the molecule corresponds to the metallic centers and the s-Ic ligand. This is shown by the localization of the FMOs, the Mülliken charges, and the Fukui function plots. Furthermore, these complexes show the highest reactivity indexes. Specifically, the results display that the bimetallic complexes containing Rh and bipy and Re with bipy and CO would tend to be strong electron donors in a chemical reaction. This behavior is also observed for their monometallic neutral species.

Finally, in addition to all previously discussed results, it has been stated that an appropriate M-M distance around 5.43 Å is an important parameter for a bimetallic compound to be able to perform the reductive coupling of two molecules of CO_2 to oxalate. All the studied complexes show an M-M distance around this value, which would suggest that these complexes might show activity in this type of reactions.

This study will continue employing more ligands and metals, including the formation of oxalate, and then experiments will be carried out.

Conflicts of Interest

The authors declare that they have no conflicts of interest.

Acknowledgments

The authors acknowledge the financial support from the following projects: Fondecyt (Grants 1141138 and 1161416), CONICYT, Operating Expenses (Grant no. 21110092), and Milenio (RC120001 and REDES150042).

References

[1] K. Hafner, "Struktur und aromatischer charakter nichtbenzoider cyclisch konjugierter systeme," *Angewandte Chemie*, vol. 75, pp. 1041–1044, 1963.

[2] K. Hafner, "Masaichi saito; simetría," *Pure and Applied Chemistry*, vol. 2, pp. 950–969, 2010.

[3] E. E. Bunel, L. Valle, N. L. Jones et al., "Bis((pentamethylcyclopentadienyl)metal)pentalenes. A new class of highly delocalized, fused metallocenes," *Journal of the American Chemical Society*, vol. 110, no. 19, pp. 6596–6598, 1988.

[4] J. M. Manriquez, M. D. Ward, W. M. Reiff et al., "Structural and physical properties of delocalized mixed-valent [Cp*M(pentalene)M′Cp*]n+ and [Cp*M(indacene)M′Cp*]n+ (M, M′ = Fe, Co, Ni; n = 0, 1, 2) complexes," *Journal of the American Chemical Society*, vol. 117, no. 23, pp. 6182–6192, 1995.

[5] E. Esponda, C. Adams, F. Burgos et al., "New Rh derivatives of s-indacene active in dehydrogenative silylation of styrene," *Journal of Organometallic Chemistry*, vol. 691, no. 13, pp. 3011–3017, 2006.

[6] C. Morales-Verdejo, I. Martínez-Díaz, C. Adams et al., "New mono and bimetallic iron complexes derived from partially methylated s-indacene. Evidence of a trinuclear iron s-indacene complex," *Polyhedron*, vol. 69, pp. 15–24, 2014.

[7] C. Adams, C. Morales-Verdejo, V. Morales et al., "Heterobinuclear s-indacene rhodium complexes: synthesis and characterization," *European Journal of Inorganic Chemistry*, no. 6, pp. 784–791, 2009.

[8] C. Morales-Verdejo, I. Martinez, D. Mac-Leod Carey et al., "Synthesis and structure of some heterobimetallic complexes having a polyalkyl-s-indacenyl spacer," *Inorganica Chimica Acta*, vol. 394, pp. 752–756, 2013.

[9] C. Song, "Global challenges and strategies for control, conversion and utilization of CO2 for sustainable development involving energy, catalysis, adsorption and chemical processing," *Catalysis Today*, vol. 115, no. 1–4, pp. 2–32, 2006.

[10] J. Notni, S. Schenk, H. Görls, H. Breitzke, and E. Anders, "Formation of a unique zinc carbamate by CO2 fixation: implications for the reactivity of tetra-azamacrocycle ligated Zn(II) complexes," *Inorganic Chemistry*, vol. 47, no. 4, pp. 1382–1390, 2008.

[11] B. Verdejo, J. Aguilar, E. García-España et al., "CO2 fixation by Cu2+ and Zn2+ complexes of a terpyridinophane aza receptor. Crystal structures of Cu2+ complexes, pH-metric, spectroscopic, and electrochemical studies," *Inorganic Chemistry*, vol. 45, no. 9, pp. 3803–3815, 2006.

[12] A. Gennaro, A. A. Isse, J.-M. Savéant, M.-G. Severin, and E. Vianello, "Homogeneous electron transfer catalysis of the electrochemical reduction of carbon dioxide. Do aromatic

anion radicals react in an outer-sphere manner?" *Journal of the American Chemical Society*, vol. 118, no. 30, pp. 7190–7196, 1996.

[13] T. Fujihara and Y. Tsuji, "Transition metal-catalyzed fixation of Carbon dioxide via carbon-carbon bond formation," *Journal of the Japan Petroleum Institute*, vol. 59, no. 3, pp. 84–92, 2016.

[14] E. García-España, P. Gaviña, J. Latorre, C. Soriano, and B. Verdejo, "CO2 fixation by copper(II) complexes of a ter-pyridinophane aza receptor," *Journal of the American Chemical Society*, vol. 126, no. 16, pp. 5082-5083, 2004.

[15] R. Angamuthu, P. Byers, M. Lutz, A. L. Spek, and E. Bouwman, "Electrocatalytic CO_2 conversion to oxalate by a copper complex," *Science*, vol. 327, no. 5963, pp. 313–315, 2010.

[16] G. A. Van Albada, I. Mutikainen, O. Roubeau, U. Turpeinen, and J. Reedijk, "Ferromagnetic trinuclear carbonato-bridged and tetranuclear hydroxo-bridged Cu(II) compounds with 4,4′-dimethyl-2,2′-bipyridine as ligand. X-ray structure, spectroscopy and magnetism," *Inorganica Chimica Acta*, vol. 331, no. 1, pp. 208–215, 2002.

[17] M. J. Frisch, G. W. Trucks, H. B. Schlegel et al., *Gaussian 09, Revision E.01*, Gaussian Inc., Wallingford, Conn, USA, 2009.

[18] A. D. Becke, "Density-functional thermochemistry. III. The role of exact exchange," *The Journal of Chemical Physics*, vol. 98, no. 7, pp. 5648–5652, 1993.

[19] P. J. Stephens, F. J. Devlin, C. F. Chabalowski, and M. J. Frisch, "Ab Initio calculation of vibrational absorption and circular dichroism spectra using density functional force fields," *Journal of Physical Chemistry*, vol. 98, no. 45, pp. 11623–11627, 1994.

[20] W. R. Wadt and P. J. Hay, "*Ab initio* effective core potentials for molecular calculations. Potentials for main group elements Na to Bi," *The Journal of Chemical Physics*, vol. 82, no. 1, pp. 284–298, 1985.

[21] P. J. Hay and W. R. Wadt, "*Ab initio* effective core potentials for molecular calculations. Potentials for K to Au including the outermost core orbitale," *The Journal of Chemical Physics*, vol. 82, no. 1, pp. 299–310, 1985.

[22] P. J. Hay and W. R. Wadt, "*Ab initio* effective core potentials for molecular calculations. Potentials for the transition metal atoms Sc to Hg," *The Journal of Chemical Physics*, vol. 82, no. 1, pp. 270–283, 1985.

[23] M. J. Frisch, J. A. Pople, and J. S. Binkley, "Self-consistent molecular orbital methods 25. Supplementary functions for Gaussian basis sets," *The Journal of Chemical Physics*, vol. 80, no. 7, pp. 3265–3269, 1984.

[24] M. Cossi, N. Rega, G. Scalmani, and V. Barone, "Energies, structures, and electronic properties of molecules in solution with the C-PCM solvation model," *Journal of Computational Chemistry*, vol. 24, no. 6, pp. 669–681, 2003.

[25] L. R. Domingo, M. Ríos-Gutiérrez, and P. Pérez, "Applications of the conceptual density functional theory indices to organic chemistry reactivity," *Molecules*, vol. 21, no. 6, article 748, 2016.

[26] R. G. Pearson, "The electronic chemical potential and chemical hardness," *Journal of Molecular Structure: THEOCHEM*, vol. 255, pp. 261–270, 1992.

[27] R. G. Parr, L. V. Szentpály, and S. Liu, "Electrophilicity index," *Journal of the American Chemical Society*, vol. 121, no. 9, pp. 1922–1924, 1999.

[28] X. Zarate, E. Schott, T. Gomez, and R. Arratia-Pérez, "Theoretical study of sensitizer candidates for dye-sensitized solar cells: peripheral substituted dizinc pyrazinoporphyrazine-phthalocyanine complexes," *The Journal of Physical Chemistry A*, vol. 117, pp. 430–438, 2013.

[29] P. K. Chattaraj and S. Giri, "Electrophilicity index within a conceptual DFT framework," *Annual Reports on the Progress of Chemistry—Section C*, vol. 105, pp. 13–39, 2009.

[30] C. E. Diaz-Uribe, W. Vallejo, W. Castellar et al., "Novel (E)-1-(pyrrole-2-yl)-3-(aryl)-2-(propen-1-one) derivatives as efficient singlet oxygen quenchers: Kinetics and quantum chemical calculations," *RSC Advances*, vol. 5, no. 88, pp. 71565–71572, 2015.

[31] L. Arrue, T. Barra, M. B. Camarada, X. Zarate, and E. Schott, "Electrochemical and theoretical characterization of the electro-oxidation of dimethoxycurcumin," *Chemical Physics Letters*, vol. 677, pp. 35–40, 2017.

Permissions

All chapters in this book were first published in JC, by Hindawi Publishing Corporation; hereby published with permission under the Creative Commons Attribution License or equivalent. Every chapter published in this book has been scrutinized by our experts. Their significance has been extensively debated. The topics covered herein carry significant findings which will fuel the growth of the discipline. They may even be implemented as practical applications or may be referred to as a beginning point for another development.

The contributors of this book come from diverse backgrounds, making this book a truly international effort. This book will bring forth new frontiers with its revolutionizing research information and detailed analysis of the nascent developments around the world.

We would like to thank all the contributing authors for lending their expertise to make the book truly unique. They have played a crucial role in the development of this book. Without their invaluable contributions this book wouldn't have been possible. They have made vital efforts to compile up to date information on the varied aspects of this subject to make this book a valuable addition to the collection of many professionals and students.

This book was conceptualized with the vision of imparting up-to-date information and advanced data in this field. To ensure the same, a matchless editorial board was set up. Every individual on the board went through rigorous rounds of assessment to prove their worth. After which they invested a large part of their time researching and compiling the most relevant data for our readers.

The editorial board has been involved in producing this book since its inception. They have spent rigorous hours researching and exploring the diverse topics which have resulted in the successful publishing of this book. They have passed on their knowledge of decades through this book. To expedite this challenging task, the publisher supported the team at every step. A small team of assistant editors was also appointed to further simplify the editing procedure and attain best results for the readers.

Apart from the editorial board, the designing team has also invested a significant amount of their time in understanding the subject and creating the most relevant covers. They scrutinized every image to scout for the most suitable representation of the subject and create an appropriate cover for the book.

The publishing team has been an ardent support to the editorial, designing and production team. Their endless efforts to recruit the best for this project, has resulted in the accomplishment of this book. They are a veteran in the field of academics and their pool of knowledge is as vast as their experience in printing. Their expertise and guidance has proved useful at every step. Their uncompromising quality standards have made this book an exceptional effort. Their encouragement from time to time has been an inspiration for everyone.

The publisher and the editorial board hope that this book will prove to be a valuable piece of knowledge for researchers, students, practitioners and scholars across the globe.

List of Contributors

Aurelio Ortiz and Estibaliz Sansinenea
Facultad de Ciencias Químicas, Benemèrita Universidad Autònoma de Puebla, C.P. 72570, Puebla, PUE, Mexico

Jingkai Zhang and Meiling Li
State Key Laboratory of Food Nutrition and Safety, Tianjin University of Science & Technology, Tianjin 300457, China

Zhenjing Li, Yibin Xue and Cheng Luo
State Key Laboratory of Food Nutrition and Safety, Tianjin University of Science & Technology, Tianjin 300457, China
Key Laboratory of Food Nutrition and Safety, Ministry of Education, School of Food Engineering and Biotechnology, Tianjin University of Science and Technology, Tianjin 300457, China

Xiaohong Zhang
State Key Laboratory of Food Nutrition and Safety, Tianjin University of Science & Technology, Tianjin 300457, China
China CFAPA Testing Technology Company Limited of Dalian, Dalian 116000, China

Seol Lee, YongJae Cho and Kee Hag Lee
Department of Chemistry and Nanoscale Sciences and Technology Institute, Wonkwang University, Iksan, Jeonbuk 570-749, Republic of Korea

Yuxiang Ma, Longkai Shi, Yulan Liu and Qiyu Lu
College of Food Science and Technology, Henan University of Technology, Zhengzhou 450001, China

Naglaa F. H. Mahmoud, Eman A. El-Bordany and Galal A. Elsayed
Chemistry Department, Faculty of Science, Ain Shams University, Abbassia, Cairo 11566, Egypt

Joseph C. Sloop
School of Science and Technology, Georgia Gwinnett College, 1000 University Center Lane, Lawrenceville, GA 30043, USA

Monica Iliş and Viorel Círcu
Department of Inorganic Chemistry, University of Bucharest, 23 Dumbrava Rosie St., Sector 2, Bucharest 020464, Romania

Zhen Ma and Huisheng Zhuang
School of Environmental Science and Engineering, Shanghai Jiao Tong University, Shanghai 200240, China

Aymard Didier Fouegue Tamafo, Julius Numbonui Ghogomu and Nyiang Kennet Nkungli
Department of Chemistry, Faculty of Science, University of Dschang, Dschang, Cameroon

Désiré Bikélé Mama
Department of Chemistry, Faculty of Science, University of Douala, Douala, Cameroon

Elie Younang
Department of Inorganic Chemistry, Faculty of Science, University of Yaoundé I, Yaoundé, Cameroon

Cuixia Li, Xiaoping Qin, Liangchuan Li, Fei Yan, Jinzhong Wang and Yixin Zhang
Drilling and Production Technology Research Institute, Petro China Jidong Oilfield Company, Tangshan 063000, China

Da Xue, Yuna Li, Jisheng Liu, Jianwei Xue, Fuxiang Li and Zhiping Lv
College of Chemistry and Chemical Engineering, Taiyuan University of Technology, Taiyuan 030024, China

Chuan Gao
Faculty of Arts and Social Sciences, The University of Sydney, Sydney, NSW, Australia

Xiaobing Liu
School of Chemistry and Chemical Engineering, Jinggangshan University, Ji'an 343009, China

Bin Liu
School of Basic Medicine and Pharmacy, Jinggangshan University, Ji'an 343009, China

Zhiling Liang, Houhe Liu, Dandan Song, Yun Zhang, Hong Huang and Guodong Ye
Guangdong Provincial Key Laboratory of Molecular Target & Clinical Pharmacology, School of Pharmaceutical Sciences, Guangzhou Medical University and The Fifth Affiliated Hospital of Guangzhou Medical University, Guangzhou 510700, China

Nianjun Su
Department of Reproductive Health and Infertility, Guangdong Women and Children Hospital, Guangzhou 510010, China

Jianqi Zheng
The First Clinical School of Guangzhou Medical University, Guangzhou 510275, China

Cheng Zhong
College of Chemistry and Molecular Sciences, Wuhan University, Wuhan 430072, China

Liuzhuang Xing, YongHai Hui, Jinghui Yang, Xuejian Xing, Yadong Hou, Yang Wu and Kui Fan
Key Laboratory of Oil & Gas Fine Chemicals, Ministry of Education & Xinjiang Uyghur Autonomous Region, College of Chemistry and Chemical Engineering, Xinjiang University, Urumqi 830046, China

Wei Wang
Department of Chemistry and Centre for Pharmacy, University of Bergen, 5020 Bergen, Norway

Zubi Sadiq, Erum Akbar Hussain, Narjis Naz, Kausar Yasmeen and Fizzah Iqbal
Department of Chemistry, Lahore College forWomen University, Lahore 54000, Pakistan

Ambreen Ghani
Department of Chemistry, University of Education, D.G. Khan Campus, Dera Ghazi Khan, Pakistan

Mohamed E. Assal, Mufsir Kuniyil, Mohammed Rafi Shaik, Mujeeb Khan, Abdulrahman Al-Warthan, Mohammed Rafiq H. Siddiqui and Syed Farooq Adil
Department of Chemistry, College of Science, King Saud University, Riyadh 11451, Saudi Arabia

Eloisa Román-Maldonado, Miguel A. Sanchez-Carmona, Nelly González-Rivas and Erick Cuevas-Yañez
Centro Conjunto de Investigaciòn en Química Sustentable UAEM-UNAM, Universidad Autònoma del Estado de Mèxico, Carretera Toluca-Atlacomulco Km 14.5, 50200 Toluca, MEX, Mexico

Horacio Reyes
Divisiòn de Ingeniería Química, Instituto de Estudios Superiores de Jocotitlàn, Carretera Toluca-Atlacomulco Km 44.8, Ejido de San Juan y San Agustin, Jocotitlàn, MEX, Mexico

Nafiz Öncü Can
Department of Analytical Chemistry, Faculty of Pharmacy, Anadolu University, 26470 Eskişehir, Turkey

Zafer Asım Kaplancıklı
Department of Pharmaceutical Chemistry, Faculty of Pharmacy, Anadolu University, 26470 Eskişehir, Turkey

Ulviye Acar Çevik, Begüm Nurpelin Sağlık, Serkan Levent and Yusuf Özkay
Department of Pharmaceutical Chemistry, Faculty of Pharmacy, Anadolu University, 26470 Eskişehir, Turkey
Doping and Narcotic Compounds Analysis Laboratory, Faculty of Pharmacy, Anadolu University, 26470 Eskişehir, Turkey

Büşra Korkut
Department of Pharmaceutical Toxicology, Faculty of Pharmacy, Anadolu University, 26470 Eskişehir, Turkey

Ali Savaş Koparal
Department of Environmental Engineering, Faculty of Engineering, Anadolu University, Eskişehir, Turkey

Jianlin Wang and Mingyang Liu
School of Civil and Resource Engineering, University of Science and Technology Beijing, Beijing 100083, China

Weixiang Wang, Longzhe Jin and Na Gao
School of Civil and Resource Engineering, University of Science and Technology Beijing, Beijing 100083, China
Mine Emergency Technology Research Center, Beijing 100083, China

Mikhail V. Pugachev, Thang T. N. Nguyen, Timur M. Bulatov, Roman S. Pavelyev, Alfia G. Iksanova, Oksana V. Bondar and Yurii G. Shtyrlin
Kazan (Volga Region) Federal University, Kremlyovskaya 18, Kazan 420008, Russia

Konstantin V. Balakin
Kazan (Volga Region) Federal University, Kremlyovskaya 18, Kazan 420008, Russia
I.M. Sechenov First Moscow State Medical University, Trubetskaya 8, Bldg 2, Moscow 119991, Russia

Boon Kui Ho and Zainab Ngaini
Department of Chemistry, Faculty of Resource Science and Technology, Universiti Malaysia Sarawak, 94300 Kota Samarahan, Sarawak, Malaysia

Siaw San Hwang, Reagan Entigu Linton, Ee Ling Kong and Boon Kiat Lee
Cancer Research and Advanced Therapeutic Group (CREATE), Faculty of Engineering, Computing and Science, Swinburne University of Technology Sarawak Campus, Jalan Simpang Tiga, 93350 Kuching, Sarawak, Malaysia

Paul Matthew Neilsen
Cancer Research and Advanced Therapeutic Group (CREATE), Faculty of Engineering, Computing and Science, Swinburne University of Technology Sarawak Campus, Jalan Simpang Tiga, 93350 Kuching, Sarawak, Malaysia
School of Health, Medical and Applied Sciences, Central Queensland University, North Rockhampton, QLD 4702, Australia

Vergush A. Pivazyan
National Agrarian University of Armenia, 74 Teryan Str., 0009 Yerevan, Armenia

Roza S. Shainova and Aleksandr P. Yengoyan
National Agrarian University of Armenia, 74 Teryan Str., 0009 Yerevan, Armenia
Armenian-Russian (Slavonic) State University, 123 H. Emin Str., 0051 Yerevan, Armenia

Emma A. Ghazaryan
Armenian-Russian (Slavonic) State University, 123 H. Emin Str., 0051 Yerevan, Armenia

Rafael A. Tamazyan and Armen G. Ayvazyan
STC OPC Molecule Structure Research Center of NAS RA, 26 Azatutyan Av., 0014 Yerevan, Armenia

Semih Gorduk and Ozge Koyun
Faculty of Arts and Science, Department of Chemistry, Yildiz Technical University, Esenler, 34210 Istanbul, Turkey

Oguzhan Avciata
Faculty of Chemical and Metallurgical Engineering, Department of Metallurgical and Materials Engineering, Yildiz Technical University, Esenler, 34210 Istanbul, Turkey

Ahmet Altindal
Faculty of Arts and Science, Department of Physics, Yildiz Technical University, Esenler, 34210 Istanbul, Turkey

Ulvi Avciata
Department of Occupational Health and Safety, Esenyurt University, Esenyurt, 34510 Istanbul, Turkey

Ximena Zarate
Instituto de Ciencias Quìmicas Aplicadas, Facultad de Ingenierìa, Universidad Autònoma de Chile, Av. Pedro de Valdivia 425, Santiago, Chile

Eduardo Schott, Juan M. Manríquez and Ivonne Chávez
Departamento de Quìmica Inorg´anica, Facultad de Quìmica, Pontificia Universidad Catòlica de Chile, Casilla 306, Santiago, Chile

Emilio Bunel
Argonne National Laboratory, Argonne, IL 60439, USA

Index

S

T

V

X

www.ingramcontent.com/pod-product-compliance
Lightning Source LLC
LaVergne TN
LVHW060524210926
848485LV00005B/1139